I0027511

Svetlin G. Georgiev, Khaled Zennir
Functional Analysis with Applications

Also of Interest

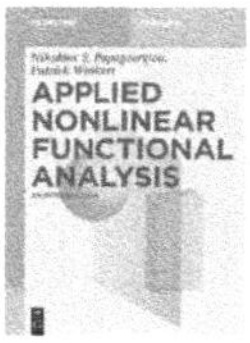

Applied Nonlinear Functional Analysis. An Introduction
Papageorgiou, Nikolaos S. / Winkert, Patrick, 2018
ISBN 978-3-11-051622-7, e-ISBN (PDF) 978-3-11-053298-2,
e-ISBN (EPUB) 978-3-11-053183-1

Elementary Functional Analysis
Markin, Marat V., 2018
ISBN 978-3-11-061391-9, e-ISBN (PDF) 978-3-11-061403-9,
e-ISBN (EPUB) 978-3-11-061409-1

Functional Analysis. A Terse Introduction
Chacón, Gerardo / Rafeiro, Humberto / Vallejo, Juan Camilo, 2016
ISBN 978-3-11-044191-8, e-ISBN (PDF) 978-3-11-044192-5,
e-ISBN (EPUB) 978-3-11-043364-7

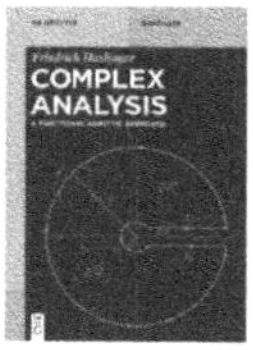

Complex Analysis. A Functional Analytic Approach
Haslinger, Friedrich, 2017
ISBN 978-3-11-041723-4, e-ISBN (PDF) 978-3-11-041724-1,
e-ISBN (EPUB) 978-3-11-042615-1

Real Analysis. Measure and Integration
Markin, Marat V., 2019
ISBN 978-3-11-060097-1, e-ISBN (PDF) 978-3-11-060099-5,
e-ISBN (EPUB) 978-3-11-059882-7

Svetlin G. Georgiev, Khaled Zennir

Functional Analysis with Applications

DE GRUYTER

Mathematics Subject Classification 2010
46-00, 46-01, 46B25, 46E15, 46E20, 46E30

Authors
Prof. Dr. Svetlin G. Georgiev
Kliment Ohridski University of Sofia
Department of Differential Equations
Faculty of Mathematics and Informatics
1126 Sofia
Bulgaria
svetlingeorgiev1@gmail.com

Dr. Khaled Zennir
Qassim University
Department of Mathematics
Buraydah Al-Qassim 51452
Saudi Arabia
khaledzennir4@yahoo.com

ISBN 978-3-11-065769-2
e-ISBN (PDF) 978-3-11-065772-2
e-ISBN (EPUB) 978-3-11-065804-0

Library of Congress Control Number: 2019937688

Bibliographic information published by the Deutsche Nationalbibliothek
The Deutsche Nationalbibliothek lists this publication in the Deutsche Nationalbibliografie; detailed bibliographic data are available on the Internet at http://dnb.dnb.de.

Cover image: -strizh- / iStock / Getty Images
Typesetting: VTeX UAB, Lithuania
Printing and binding: CPI books GmbH, Leck

www.degruyter.com

Preface

Functional analysis means analysis on function spaces. This is a field of mathematics that developed in the first half of the 20th century thanks in particular to the work of M. Frechet, S. Banach, and D. Hilbert. Examples of the efficiency of functional analysis has been the introduction of Sobolev's spaces (1935) and L. Schwartz's invention of the theory of distributions (1945–1950). These spaces have made great progress in solving the problems of partial differential equations and provide the main tools still used today in this field of both theoretical and numerical studies.

Classical analysis focuses on finite dimensional spaces on $\mathbb{R}$ or $\mathbb{C}$. This is suitable, for example, for solving linear differential equations. In order to solve more complicated equations, like nonlinear differential equations, integral equations, and partial differential equations, the solutions have to be sought a priori in vector spaces of an infinite number of dimensions. The computation of explicit solutions is often out of reach and one tries to describe the structure of these solutions by their belonging to spaces adapted to the problem posed. The study of stability naturally leads to considering spaces with topologies defined by norms, semi-norms or distances. From a purely mathematical point of view, functional analysis can also be seen as an extension to infinite dimensions of Euclidean geometry in finite dimensions. The transition from finite dimension to infinite dimension is not always easy because we lose a part of the geometric intuition. Whereas on a finite dimensional vector space there is only one "reasonable" topology, on a space of infinite dimension we must often consider several topologies simultaneously.

The main goal, in realizing this textbook, is to present a useful tool to junior researchers and beginning graduate students of engineering and science courses in order to acquire elementary knowledge and solid tools that are fundamental to the understanding of mathematics and the particular disciplines (geometry, probabilities, partial differential equations) within physics, in mechanics, or in the applications of mathematics to the analysis of large systems. It contains ten chapters, and each chapter consists of results with their detailed proofs, numerous examples, and exercises with solutions. Each chapter concludes with a section featuring advanced practical problems with solutions followed by a section on notes and references, explaining its context within existing literature. We will present here in a detailed way the contents of each of them.

Chapter 1 is entirely devoted to the presentation of definitions and results necessary for proceeding in this work. We first recall a few basic results on the linear, metric, normed and Banach spaces and its properties. These are used in particular to introduce the various concepts of weak solutions to PDEs. We will see regularly links and relationships between function analysis and applications on PDEs. Chapter 2 is titled *Lebesgue integration*. It is devoted to the study of measure and integration, Lebesgue measurable functions and general measure spaces, where there are many proved re-

https://doi.org/10.1515/9783110657722-201

sults. The purpose of Chapter 3 is to present results according to the L^p spaces, which contains, definitions, separability, duality and general L^p spaces with its norms. The results, presented in Chapter 4, concern linear operators, inverse operators in normed linear spaces and their properties. Chapter 5 is titled *Linear functionals*; here we introduce and treat the linear functionals in their general form and related the adjoint operators. Chapter 6 is reserved for topological studies; it is followed by Chapter 7 titled *Self-adjoint operators in Hilbert spaces*. The method of the small parameter will be the main subject of Chapter 8 and the parameter continuation method will be the subject of Chapter 9.

So we realize that the fixed-point theorems are essential in the applications of the function analysis. They are the basic mathematical tools in showing the existence of solutions in various kinds of equations. Fixed-point theory is at the heart of nonlinear analysis and provides the necessary tools to study existence theorems in many different nonlinear problems. The aim of Chapter 10 is the study of some fixed-point theorems. We start with the simplest and best known of them: Banach's fixed-point theorem for contraction maps. Then we address the Brinciari fixed-point theorem, which is a generalization of this theorem. We will then see more powerful and somewhat deeper theorems. We can thus study successively the theorem of the fixed point of Brouwer (valid in finite dimension) and then the theorem of the fixed point of Schauder (which is the generalization in infinite dimension). Unlike Banach's theorem, the proofs of the latter two results are not constructive, which explains why they require somewhat more sophisticated tools. Many different proofs of these results exist and one may be interested in one or more of them. We finish this chapter by giving applications in many problems.

This is the first volume of a series of at least two volumes; the remainder of the series will be prepared later.

Svetlin G. Georgiev
Khaled Zennir

Contents

1 Vector, metric, normed and Banach spaces

1.1 Vector spaces

With **F** we will denote the field of real numbers **R** or the field of complex numbers **C**.

Definition 1.1. A vector space (also called a linear space) over the field **F** is a set **E** with two operations:

1. *Addition*: Takes any two elements x and y and assigns for them a third element of **E**, which is completely written as $x + y$ and which is called the sum of these two elements,
2. *Scalar multiplication*: Takes any scalar a of the field **F** and an element x of **E** and gives another element $a \cdot x$ (shortly ax) of **E**.

These two operations satisfy the following axioms:

(L1) (*Associativity of addition*)

$$x + (y + z) = (x + y) + z$$

for any elements $x, y, z \in \mathbf{E}$.

(L2) (*Commutativity of addition*)

$$x + y = y + x$$

for any elements $x, y \in \mathbf{E}$.

(L3) (*Identity element of addition*) There exists an element 0 of **E**, called the zero element, such that

$$x + 0 = x$$

for any $x \in \mathbf{E}$.

(L4) (*Inverse elements of addition*) For any $x \in \mathbf{E}$ there exists an element $-x \in \mathbf{E}$, called the additive inverse of x, such that

$$x + (-x) = 0.$$

(L5) (*Compatibility of the scalar multiplication with field multiplication*)

$$(ab)x = a(bx)$$

for any $a, b \in \mathbf{F}$ and for any $x \in \mathbf{E}$.

(L6) (*Distributivity of the scalar multiplication with respect to the addition of elements of* **E**)

$$a(x + y) = ax + ay$$

for any $a \in \mathbf{F}$ and for any $x, y \in \mathbf{E}$.

https://doi.org/10.1515/9783110657722-001

(L7) (*Distributivity of scalar multiplication with respect to field addition*)

$$(a+b)x = ax + bx,$$

for any $a, b \in \mathbf{F}$ and for any $x \in \mathbf{E}$.

(L8) $1 \cdot x = x$, for any $x \in \mathbf{E}$.

The elements of **E** are commonly called vectors. The elements of **F** are commonly called scalars. When the scalar field is the field of the real numbers **R**, then the vector space **E** is called a real vector space. When the scalar field is the field of the complex numbers **C**, we say that the vector space **E** is a complex vector space.

Below we will write $x - y$ instead of $x + (-y)$ for any $x, y \in \mathbf{E}$.

Example 1.1 (The space $\mathbf{m}$ of bounded number sequences). With $\mathbf{m}$ we will denote the set of bounded number sequences $x = \{\xi_l\}_{l\in\mathbf{N}}$ implying that for every $x \in \mathbf{m}$ there exists a positive constant K_x such that $|\xi_l| \le K_x$ for any $l \in \mathbf{N}$. For $x = \{\xi_l\}_{l\in\mathbf{N}} \in \mathbf{m}$ and $y = \{\eta_l\}_{l\in\mathbf{N}} \in \mathbf{m}$, and $a \in \mathbf{F}$, we define addition as follows:

$$x + y = \{\xi_l + \eta_l\}_{l\in\mathbf{N}}, \tag{1.1}$$

and scalar multiplication by

$$ax = \{a\xi_l\}_{l\in\mathbf{N}}. \tag{1.2}$$

These operations are well defined. In fact, let $x = \{\xi_l\}_{l\in\mathbf{N}}$, $y = \{\eta_l\}_{l\in\mathbf{N}} \in \mathbf{m}$ and $a \in \mathbf{F}$ be arbitrarily chosen and fixed. Then there exist constants K_x and K_y such that

$$|\xi_l| \le K_x \quad \text{and} \quad |\eta_l| \le K_y \quad \text{for} \quad \text{any} \quad l \in \mathbf{N}.$$

Hence,

$$|\xi_l + \eta_l| \le |\xi_l| + |\eta_l| \le K_x + K_y \quad \text{for} \quad \text{any} \quad l \in \mathbf{N},$$

i. e., $x + y \in \mathbf{m}$. Also, we have

$$|a\xi_l| = |a||\xi_l| \le |a|K_x,$$

i. e., $ax \in \mathbf{m}$. Consequently the operations (1.1) and (1.2) are well defined. We will prove that $\mathbf{m}$ is a vector space over **F**. Suppose that $a, b \in \mathbf{F}$ and $x = \{\xi_l\}_{l\in\mathbf{N}}$, $y = \{\eta_l\}_{l\in\mathbf{N}}$, $z = \{\zeta_l\}_{l\in\mathbf{N}} \in \mathbf{m}$ be arbitrarily chosen and fixed.

1. Because

$$\xi_l + (\eta_l + \zeta_l) = (\xi_l + \eta_l) + \zeta_l \quad \text{for} \quad \text{any} \quad l \in \mathbf{N},$$

we have

$$x + (y + z) = (x + y) + z.$$

2. Since

$$\xi_l + \eta_l = \eta_l + \xi_l \quad \text{for} \quad \text{any} \quad l \in \mathbf{N},$$

we have

$$x + y = y + x.$$

3. Let $0 = \{0\}_{l\in\mathbf{N}}$. Then $0 \in \mathbf{m}$ and

$$x + 0 = \{\xi_l\}_{l\in\mathbf{N}} + \{0\}_{l\in\mathbf{N}} = \{\xi_l + 0\}_{l\in\mathbf{N}} = \{\xi_l\}_{l\in\mathbf{N}} = x.$$

4. We define $-x = \{-\xi_l\}_{l\in\mathbf{N}}$. Then

$$x + (-x) = \{\xi_l + (-\xi_l)\}_{l\in\mathbf{N}} = \{0\}_{l\in\mathbf{N}} = 0.$$

5. We have

$$(ab)\xi_l = a(b\xi_l) \quad \text{for} \quad \text{any} \quad l \in \mathbf{N}.$$

Then

$$(ab)x = \{(ab)\xi_l\}_{l\in\mathbf{N}} = \{a(b\xi_l)\}_{l\in\mathbf{N}} = a\{b\xi_l\}_{l\in\mathbf{N}} = a(b\{\xi_l\}_{l\in\mathbf{N}}) = a(bx).$$

6. We have

$$\begin{aligned} a(x+y) &= a\{\xi_l + \eta_l\}_{l\in\mathbf{N}} = \{a(\xi_l + \eta_l)\}_{l\in\mathbf{N}} = \{a\xi_l + a\eta_l\}_{l\in\mathbf{N}} \\ &= \{a\xi_l\}_{l\in\mathbf{N}} + \{a\eta_l\}_{l\in\mathbf{N}} = a\{\xi_l\}_{l\in\mathbf{N}} + a\{\eta_l\}_{l\in\mathbf{N}} = ax + ay. \end{aligned}$$

7. We have

$$\begin{aligned} (a+b)x &= (a+b)\{\xi_l\}_{l\in\mathbf{N}} = \{(a+b)\xi_l\}_{l\in\mathbf{N}} = \{a\xi_l + b\xi_l\}_{l\in\mathbf{N}} \\ &= \{a\xi_l\}_{l\in\mathbf{N}} + \{b\xi_l\}_{l\in\mathbf{N}} = a\{\xi_l\}_{l\in\mathbf{N}} + b\{\xi_l\}_{l\in\mathbf{N}} = ax + bx. \end{aligned}$$

8. We have

$$1 \cdot x = \{1 \cdot \xi_l\}_{l\in\mathbf{N}} = \{\xi_l\}_{l\in\mathbf{N}} = x.$$

Example 1.2 (The space **c** of convergent number sequences). By **c** we will denote the set of all convergent number sequences. For $x = \{\xi_l\}_{l\in\mathbf{N}}$, $y = \{\eta_l\}_{l\in\mathbf{N}} \in \mathbf{c}$ and $a \in \mathbf{F}$ we define the operations addition

$$x + y = \{\xi_l + \eta_l\}_{l\in\mathbf{N}} \tag{1.3}$$

and scalar multiplication

$$ax = \{a\xi_l\}_{l\in\mathbf{N}}. \tag{1.4}$$

These operations are well defined. In fact, let $x = \{\xi_l\}_{l\in\mathbf{N}} \in \mathbf{c}$ and $y = \{\eta_l\}_{l\in\mathbf{N}} \in \mathbf{c}$ be arbitrarily chosen. Then there exist $\xi, \eta \in \mathbf{F}$ such that

$$\lim_{l\to\infty} \xi_l = \xi \quad \text{and} \quad \lim_{l\to\infty} \eta_l = \eta.$$

Hence,

$$\xi + \eta = \lim_{l\to\infty} (\xi_l + \eta_l),$$

whereupon we conclude that $x + y \in \mathbf{c}$. Also,

$$a\xi = \lim_{l\to\infty} (a\xi_l), \quad a \in \mathbf{F}.$$

Therefore $ax \in \mathbf{c}$ for $a \in \mathbf{F}$. Consequently the operations (1.3) and (1.4) are well defined.

Now we will prove that $\mathbf{c}$ is a vector space. Take $x = \{\xi_l\}_{l\in\mathbf{N}} \in \mathbf{c}$, $y = \{\eta_l\}_{l\in\mathbf{N}} \in \mathbf{c}$ and $z = \{\zeta_l\}_{l\in\mathbf{N}} \in \mathbf{c}$ arbitrarily.

1. Since

$$\xi_l + (\eta_l + \zeta_l) = (\xi_l + \eta_l) + \zeta_l, \quad l \in \mathbf{N},$$

we conclude that

$$x + (y + z) = (x + y) + z.$$

2. Since

$$\xi_l + \eta_l = \eta_l + \xi_l, \quad l \in \mathbf{N},$$

we get

$$x + y = y + x.$$

3. Let $0 = \{0\}_{l\in\mathbf{N}}$. Then 0 is the zero element in $\mathbf{c}$. In fact, we have

$$x + 0 = \{\xi_l + 0\}_{l\in\mathbf{N}} = \{\xi_l\}_{l\in\mathbf{N}} = x.$$

4. We define $-x = \{-\xi_l\}_{l\in\mathbf{N}}$. Note that $x \in \mathbf{c}$ and

$$x + (-x) = \{\xi_l\}_{l\in\mathbf{N}} + \{-\xi_l\}_{l\in\mathbf{N}} = \{\xi_l + (-\xi_l)\}_{l\in\mathbf{N}} = \{0\}_{l\in\mathbf{N}}.$$

5. Since for any $a, b \in \mathbf{F}$ we have

$$(ab)\xi_l = a(b\xi_l), \quad l \in \mathbf{N},$$

we conclude that

$$(ab)x = a(bx).$$

6. For any $a \in \mathbf{F}$ we have

$$\begin{aligned}a(x+y) &= a\{\xi_l + \eta_l\}_{l\in\mathbf{N}} = \{a\xi_l + a\eta_l\}_{l\in\mathbf{N}} = \{a\xi_l\}_{l\in\mathbf{N}} + \{a\eta_l\}_{l\in\mathbf{N}} \\ &= a\{\xi_l\}_{l\in\mathbf{N}} + a\{\eta_l\}_{l\in\mathbf{N}} = ax + ay.\end{aligned}$$

7. For any $a, b \in \mathbf{F}$ we have

$$\begin{aligned}(a+b)x &= (a+b)\{\xi_l\}_{l\in\mathbf{N}} = \{(a+b)\xi_l\}_{l\in\mathbf{N}} = \{a\xi_l + b\xi_l\}_{l\in\mathbf{N}} \\ &= \{a\xi_l\}_{l\in\mathbf{N}} + \{b\xi_l\}_{l\in\mathbf{N}} = a\{\xi_l\}_{l\in\mathbf{N}} + b\{\xi_l\}_{l\in\mathbf{N}} = ax + bx.\end{aligned}$$

8. We have

$$1 \cdot x = \{1 \cdot \xi_l\}_{l\in\mathbf{N}} = \{\xi_l\}_{l\in\mathbf{N}} = x.$$

Example 1.3 (The space $\mathbf{M}[0,1]$ of bounded real functions). Consider the set $\mathbf{M}[0,1]$ of all bounded functions defined on $[0,1]$. Let $f, g \in \mathbf{M}[0,1]$ and $a \in \mathbf{F}$ be arbitrarily chosen. We define addition in $\mathbf{M}[0,1]$ as follows:

$$(f+g)(x) = f(x) + g(x) \quad \text{for} \quad \text{any} \quad x \in [0,1].$$

We define scalar multiplication in $\mathbf{M}[0,1]$ as follows:

$$(af)(x) = a(f(x)) \quad \text{for} \quad \text{any} \quad x \in [0,1].$$

Firstly, we will prove that the operations addition and scalar multiplication are well defined in $\mathbf{M}[0,1]$. In fact, let $f, g \in \mathbf{M}[0,1]$ and $a \in \mathbf{F}$ be arbitrarily chosen. Then there exist positive constants l_1 and l_2 such that

$$|f(x)| \le l_1 \quad \text{and} \quad |g(x)| \le l_2 \quad \text{for} \quad \text{any} \quad x \in [0,1].$$

Hence,

$$\begin{aligned}|af(x)| &\le |a| l_1, \\ |ag(x)| &\le |a| l_2, \\ |f(x) + g(x)| &\le |f(x)| + |g(x)| \le l_1 + l_2\end{aligned}$$

for any $x \in [0,1]$.

Now we will prove that $\mathbf{M}[0,1]$ is a vector space. Let $f, g, h \in \mathbf{M}[0,1]$ be arbitrarily chosen.

1. Let $x \in [0,1]$ be arbitrarily chosen and fixed. Then $f(x), g(x), h(x) \in \mathbf{F}$. Hence,

$$f(x) + (g(x) + h(x)) = (f(x) + g(x)) + h(x). \tag{1.5}$$

Because $x \in [0,1]$ was arbitrarily chosen, we conclude that (1.5) holds for any $x \in [0,1]$.

2. Let $x \in [0,1]$ be arbitrarily chosen and fixed. Then $f(x), g(x) \in \mathbf{F}$. Hence,

$$f(x) + g(x) = g(x) + f(x). \tag{1.6}$$

   Because $x \in [0,1]$ was arbitrarily chosen, we conclude that (1.6) holds for any $x \in [0,1]$.
3. Define

$$0(x) = 0$$

   for any $x \in [0,1]$. Let $x \in [0,1]$ be arbitrarily chosen and fixed. Then $f(x) \in \mathbf{F}$ and

$$0 + f(x) = f(x). \tag{1.7}$$

   Because $x \in [0,1]$ was arbitrarily chosen, we see that (1.7) holds for every $x \in \mathbf{F}$.
4. Define

$$(-f)(x) = -(f(x))$$

   for any $x \in [0,1]$. Let $x \in [0,1]$ be arbitrarily chosen and fixed. Then $f(x) \in \mathbf{F}$ and $-f(x) \in \mathbf{F}$. Hence,

$$f(x) + (-f(x)) = 0. \tag{1.8}$$

   Because $x \in [0,1]$ was arbitrarily chosen, we see that (1.8) holds for any $x \in [0,1]$.
5. Let $x \in [0,1]$ and $a, b \in \mathbf{F}$ be arbitrarily chosen and fixed. Let also, $a, b \in \mathbf{F}$. Then $f(x) \in \mathbf{F}$ and

$$(ab)f(x) = a(bf(x)). \tag{1.9}$$

   Because $x \in [0,1]$ was arbitrarily chosen, we conclude that (1.9) holds for every $x \in [0,1]$.
6. Let $x \in [0,1]$ and $a \in \mathbf{F}$ be arbitrarily chosen and fixed. Then $f(x), g(x) \in \mathbf{F}$ and

$$a(f(x) + g(x)) = af(x) + ag(x). \tag{1.10}$$

   Because $x \in [0,1]$ was arbitrarily chosen, we see that (1.10) holds for any $x \in [0,1]$.
7. Let $x \in [0,1]$ be arbitrarily chosen and fixed. Let also, $a, b \in \mathbf{F}$. Then $f(x) \in \mathbf{F}$ and

$$(a+b)f(x) = af(x) + bf(x). \tag{1.11}$$

   Because $x \in [0,1]$ was arbitrarily chosen, we conclude that (1.11) holds for every $x \in [0,1]$.

8. Let $x \in [0,1]$ be arbitrarily chosen and fixed. Then $f(x) \in \mathbf{F}$ and

$$1 \cdot f(x) = f(x). \tag{1.12}$$

Because $x \in [0,1]$ was arbitrarily chosen, we conclude that (1.12) holds for any $x \in [0,1]$.

Exercise 1.1. Let $\mathbf{E}_n$ be the set of all n-tuple of elements of $\mathbf{F}$. In $\mathbf{E}_n$ we introduce the operations addition

$$x + y = (\xi_1 + \eta_1, \ldots, \xi_n + \eta_n),$$

for $x = (\xi_1, \ldots, \xi_n)$, $y = (\eta_1, \ldots, \eta_n) \in \mathbf{E}_n$, and scalar multiplication,

$$ax = (a\xi_1, \ldots, a\xi_n) \quad \text{for} \quad a \in \mathbf{F} \quad \text{and} \quad x = (\xi_1, \ldots, \xi_n) \in \mathbf{E}_n.$$

Prove that $\mathbf{E}_n$ is a vector space. The space $\mathbf{E}_n$ will be called the n-dimensional Euclidean space.

Exercise 1.2. With $\mathbf{C}[0,1]$ we will denote the set of all continuous functions defined on $[0,1]$. Prove that $\mathbf{C}[0,1]$ is a vector space.

Exercise 1.3. With $\mathbf{s}$ we will denote the set of all sequences of elements of $\mathbf{F}$. Prove that $\mathbf{s}$ is a vector space.

Below by $\mathbf{E}$ we will denote a vector space.

Corollary 1.1. *The zero element is unique.*

Proof. Suppose that there are two zero elements 0_1 and 0_2 in $\mathbf{E}$. Then

$$0_1 + 0_2 = 0_2 \quad \text{and} \quad 0_2 + 0_1 = 0_1.$$

Since $\mathbf{E}$ is a vector space we have

$$0_1 + 0_2 = 0_2 + 0_1.$$

Therefore

$$0_1 = 0_2,$$

which completes the proof. □

Corollary 1.2. *Let $x \in \mathbf{E}$ be arbitrarily chosen. If $y \in \mathbf{E}$ is such that $x + y = 0$, then $y = -x$.*

Proof. We have

$$x + (y + (-x)) = (x + y) + (-x) = 0 + (-x) = -x$$

and

$$x + (y + (-x)) = x + ((-x) + y) = (x + (-x)) + y = 0 + y = y.$$

Therefore $y = -x$, which completes the proof. □

Corollary 1.3. *For every $x \in \mathbf{E}$ we have*

$$0 \cdot x = 0.$$

Proof. We have

$$x = 1 \cdot x = (1 + 0) \cdot x = 1 \cdot x + 0 \cdot x = x + 0 \cdot x.$$

Hence,

$$\begin{aligned} x + (-x) &= (x + 0 \cdot x) + (-x) = x + (0 \cdot x + (-x)) = x + ((-x) + 0 \cdot x) \\ &= (x + (-x)) + 0 \cdot x = 0 + 0 \cdot x = 0 \cdot x. \end{aligned}$$

Because $x + (-x) = 0$, we conclude $0 = 0 \cdot x$. This completes the proof. □

Corollary 1.4. *For every $x \in \mathbf{E}$ we have*

$$(-1) \cdot x = -x.$$

Proof. Using Corollary 1.3, we have

$$(-1)x + x = (-1) \cdot x + 1 \cdot x = (-1 + 1) \cdot x = 0 \cdot x = 0.$$

Hence, by Corollary 1.2, we obtain $(-1) \cdot x = -x$. This completes the proof. □

Corollary 1.5. *If $x \in \mathbf{E}$, $x \neq 0$, and $ax = bx$, $a, b \in \mathbf{F}$, then $a = b$.*

Proof. From $ax = bx$ and $bx - bx = 0$, we get

$$0 = ax - bx = (a - b)x.$$

Assume that $a \neq b$. Then

$$x = \frac{1}{a - b}((a - b)x) = \frac{1}{a - b}0 = 0,$$

which is a contradiction. This completes the proof. □

Corollary 1.6. *If $a \in \mathbf{F}$, $a \neq 0$, and $ax = ay$, $x, y \in \mathbf{E}$, then $x = y$.*

Proof. We have

$$ax - ay = ay - ay = 0,$$

whereupon

$$a(x - y) = 0.$$

Hence, using $a \neq 0$, we obtain

$$x - y = \frac{1}{a}(a(x - y)) = \frac{1}{a}0 = 0.$$

Therefore $x = y$, which completes the proof. □

Definition 1.2. The elements $x_1, \dots, x_n$ of the vector space $\mathbf{E}$ are said to be linearly independent, if any relation

$$a_1 x_1 + \cdots + a_n x_n = 0,$$

implies that $a_1 = \cdots = a_n = 0$.

Definition 1.3. The elements $x_1, \dots, x_n$ of the vector space $\mathbf{E}$ are said to be linearly dependent, if the relation

$$a_1 x_1 + \cdots + a_n x_n = 0,$$

is possible with at least one of the coefficients $a_1, \dots, a_n$ is not zero.

Example 1.4. Consider $\mathbf{E}_3$ and

$$x_1 = (1, 1, 1), \quad x_2 = (2, 2, 2), \quad x_3 = (-3, -3, -3).$$

Then

$$x_1 + x_2 + x_3 = 0.$$

Here $a_1 = a_2 = a_3 = 1$. Therefore the elements x_1, x_2 and x_3 are linearly dependent in $\mathbf{E}_3$.

Example 1.5. Consider the set $E[a, b]$ of all real-valued functions defined on $[a, b]$ and its elements

$$1, \quad x, \quad \dots, \quad x^{p-1}, \quad p \in \mathbf{N}.$$

Assume that there are constants $b_0, \dots, b_{p-1} \in \mathbf{F}$ such that

$$b_0 + b_1 x + \cdots + b_{p-1} x^{p-1} = 0 \quad \text{in} \quad [a, b], \tag{1.13}$$

and $(b_0, \dots, b_{p-1}) \neq (0, \dots, 0)$. Hence, equation (1.13) could hold for at most $p-1$ values of x, whereas it must hold for all $x \in [a, b]$. Therefore the elements $1, \dots, x^{p-1}$ of $\mathbf{E}[a, b]$ are linearly independent.

Example 1.6. Consider the set $E[a,b]$ and its elements

$$0 \quad \text{and} \quad \sin x.$$

Then

$$1 \cdot 0 + 0 \cdot \sin x = 0 \quad \text{for} \quad \text{all} \quad x \in [a,b].$$

Here $a_1 = 1$ and $a_2 = 0$. Therefore 0 and $\sin x$ are linearly dependent.

Exercise 1.4. Consider the set $E[a,b]$. Prove that e^x and e^{2x} are linearly independent.

Definition 1.4. The vector space $\mathbf{E}$ will be called m-dimensional if in it there are m linearly independent elements and every $m+1$ its elements are linearly dependent. The dimension of the vector space $\mathbf{E}$ will be denoted by $\dim(\mathbf{E})$.

Example 1.7. Consider $\mathbf{E}_m$. Let

$$\begin{aligned} x_1 &= (1,0,\dots,0),\\ x_2 &= (0,1,\dots,0),\\ &\vdots\\ x_m &= (0,0,\dots,1). \end{aligned}$$

Assume that there are $a_i \in \mathbf{F}$, $i \in \{1,\dots,m\}$, such that $\sum_{i=1}^{m} a_i x_i = 0$. Then

$$(a_1,\dots,a_m) = (0,\dots,0),$$

whereupon $a_i = 0$, $i \in \{1,\dots,m\}$. Therefore $x_1,\dots,x_m$ are linearly independent. Let now

$$\begin{aligned} y_1 &= (y_1^1,\dots,y_m^1),\\ &\vdots\\ y_{m+1} &= (y_1^{m+1},\dots,y_m^{m+1}) \end{aligned}$$

be arbitrarily chosen elements of $\mathbf{E}_m$. Since

$$\operatorname{rank}\begin{pmatrix} y_1^1 & \cdots & y_1^{m+1}\\ \vdots & \vdots & \vdots\\ y_m^1 & \cdots & y_m^{m+1} \end{pmatrix} \le m,$$

we see that $y_1,\dots,y_{m+1}$ are linearly dependent. Therefore $\mathbf{E}_m$ is m-dimensional.

Example 1.8. Let $n \in \mathbf{N}$. With $\mathbf{P}^n$ we will denote the set of all polynomials of degree n with coefficients which are elements of the field $\mathbf{F}$. Consider $1, t, \ldots, t^n$. Assume that there are $a_i \in \mathbf{F}$, $i \in \{1, \ldots, n+1\}$, such that

$$\sum_{i=1}^{n+1} a_i t^{i-1} = 0 \quad \text{for} \quad \text{all} \quad t \in \mathbf{R}. \tag{1.14}$$

Because (1.14) has at most n solutions with respect to t, we conclude that (1.14) holds if $a_i = 0$, $i \in \{1, \ldots, n+1\}$. Therefore $1, t, \ldots, t^n$ are linearly independent. Let now $p_i \in \mathbf{P}^n$, $i \in \{1, \ldots, n+2\}$, be arbitrarily chosen and

$$p_i(t) = \sum_{j=1}^{n+1} \alpha_j^i t^{j-1}, \quad \alpha_j^i \in \mathbf{F},$$

$i \in \{1, \ldots, n+2\}$, $j \in \{1, \ldots, n+1\}$. Assume that there are $b_i \in \mathbf{F}$, $i \in \{1, \ldots, n+2\}$, such that

$$\sum_{i=1}^{n+2} b_i p_i(t) = 0 \quad \text{for} \quad \text{all} \quad t \in \mathbf{R}.$$

Hence,

$$\sum_{i=1}^{n+2} b_i \alpha_j^i = 0, \quad j \in \{1, \ldots, n+1\}. \tag{1.15}$$

Since

$$\operatorname{rank}\begin{pmatrix} \alpha_1^1 & \ldots & \alpha_{n+1}^1 \\ \vdots & \vdots & \vdots \\ \alpha_1^{n+2} & \ldots & \alpha_{n+1}^{n+2} \end{pmatrix} \le n+1,$$

we conclude that (1.15) holds if $b_i \neq 0$ for some $i \in \{1, \ldots, n+2\}$. Therefore every $n+2$ elements of $\mathbf{P}^n$ are linearly dependent. Hence, $\dim(\mathbf{P}^n) = n+1$.

Exercise 1.5. Let $m \in \mathbf{N}$. With $\mathbf{M}_m$ we will denote the set of all $m \times m$ matrices whose entries are elements of $\mathbf{F}$. Prove that $\mathbf{M}_m$ is a vector space and $\dim(\mathbf{M}_m) = m^2$.

Definition 1.5. Let $\mathbf{E}$ be a vector space with $\dim(\mathbf{E}) = m$. Then every system of m linearly independent elements of $\mathbf{E}$ will be called a basis of $\mathbf{E}$.

Theorem 1.1. *Let* $\mathbf{E}$ *be a vector space with* $\dim(\mathbf{E}) = m$ *and* $\{e_1, \ldots, e_m\}$ *be its basis. Then every* $x \in \mathbf{E}$ *can be represented in a unique way in the form*

$$x = \alpha_1 e_1 + \cdots + \alpha_m e_m, \quad \alpha_i \in \mathbf{F}, \quad i \in \{1, \ldots, m\}. \tag{1.16}$$

Proof. Since $e_1, \dots, e_m$, x are linearly dependent, there are $\tilde{\alpha}_i \in \mathbf{F}$, $i \in \{1, \dots, m+1\}$, such that

$$\tilde{\alpha}_1 e_1 + \cdots + \tilde{\alpha}_m e_m + \tilde{\alpha}_{m+1} x = 0, \quad (\tilde{\alpha}_1, \dots, \tilde{\alpha}_{m+1}) \neq (0, \dots, 0).$$

Assume that $\tilde{\alpha}_{m+1} = 0$. Then

$$\tilde{\alpha}_1 e_1 + \cdots + \tilde{\alpha}_m e_m = 0, \quad (\tilde{\alpha}_1, \dots, \tilde{\alpha}_m) \neq (0, \dots, 0),$$

which is a contradiction because $\{e_1, \dots, e_m\}$ is a basis in $\mathbf{E}$. Therefore $\tilde{\alpha}_{m+1} \neq 0$. Hence,

$$x = -\frac{\tilde{\alpha}_1}{\tilde{\alpha}_{m+1}} e_1 - \cdots - \frac{\tilde{\alpha}_m}{\tilde{\alpha}_{m+1}} e_m.$$

We set $\alpha_i = -\frac{\tilde{\alpha}_i}{\tilde{\alpha}_{m+1}}$, $i \in \{1, \dots, m\}$, and we get the representation (1.16). Now we assume that

$$x = \beta_1 e_1 + \cdots + \beta_m e_m, \quad \beta_i \in \mathbf{F}, \quad i \in \{1, \dots, m\}.$$

Then, using (1.16), we obtain

$$0 = (\alpha_1 - \beta_1) e_1 + \cdots + (\alpha_m - \beta_m) e_m.$$

Since $\{e_1, \dots, e_m\}$ is a basis in $\mathbf{E}$, from the last equality we conclude that $\alpha_i = \beta_i$, $i \in \{1, \dots, m\}$. This completes the proof. □

Definition 1.6. A vector space $\mathbf{E}$ is infinite-dimensional if for every natural n there is a system of n linearly independent elements in $\mathbf{E}$.

Example 1.9. Consider $\mathcal{C}([a, b])$. Because for every $n \in \mathbf{N}$ the elements

$$1, \quad t, \quad \dots, \quad t^n$$

is linearly independent, we conclude that $\mathcal{C}([a, b])$ is an infinite-dimensional vector space.

Exercise 1.6. Let $k \in \mathbf{N}$ be arbitrarily chosen. Prove that $\mathcal{C}^k([a, b])$ is an infinite-dimensional vector space.

Definition 1.7. Let $\mathbf{E}$ be a vector space. Suppose that $\mathbf{V} \subseteq \mathbf{E}$. If $\mathbf{V}$ is a vector space itself with the same vector space operations as $\mathbf{E}$ has, then $\mathbf{V}$ is called a linear subspace of $\mathbf{E}$.

Theorem 1.2. *Let $\mathbf{E}$ be a m-dimensional vector space and $\mathbf{V}$ be its linear subspace. Then* $\dim(\mathbf{V}) \leq m$.

Proof. Suppose that $\dim(\mathbf{V}) > m$. Then there are $\dim(\mathbf{V})$ linearly independent elements in $\mathbf{V}$. Since $\mathbf{V}$ is a linear subspace of $\mathbf{E}$, we conclude that in $\mathbf{E}$ there are $\dim(\mathbf{V})$ linearly independent elements, which is a contradiction. This completes the proof. □

Theorem 1.3. *Let* $\mathbf{E}$ *be a vector space and* $\mathbf{V}$ *be a subset of* $\mathbf{E}$. *Then* $\mathbf{V}$ *is a nonempty linear subspace if and only if* $\mathbf{V}$ *satisfies the following properties.*

1. $0 \in \mathbf{V}$,
2. *if* $x, y \in \mathbf{V}$, *then* $x + y \in \mathbf{V}$,
3. *if* $x \in \mathbf{V}$ *and* $a \in \mathbf{F}$, *then* $ax \in \mathbf{V}$.

Proof.

1. Let $\mathbf{V}$ satisfies (1), (2) and (3). By (1) it follows that $\mathbf{V}$ is nonempty. The properties (2) and (3) ensure closure of $\mathbf{V}$ under addition and scalar multiplication. Since the elements of $\mathbf{V}$ are necessarily elements of $\mathbf{E}$, the axioms (L1), (L2), (L3), (L5), (L6), (L7) and (L8) are satisfied. By the closure of $\mathbf{V}$ under scalar multiplication and Corollary 1.4, it follows that $-x = (-1) \cdot x \in \mathbf{V}$ and $x + (-x) = 0$ for every $x \in \mathbf{V}$, i. e., the axiom (L4) is satisfied.
2. Let $\mathbf{V}$ be a nonempty linear subspace of $\mathbf{E}$. Then $\mathbf{V}$ is itself a vector space under the operations induced by $\mathbf{E}$. So, the properties (2) and (3) are satisfied. By property (3) and Corollary 1.4, we have $-x \in \mathbf{V}$ for every $x \in \mathbf{V}$. Hence, it follows that $\mathbf{V}$ is closed under subtraction as well. Because $\mathbf{V}$ is nonempty, there is an element $x \in \mathbf{V}$. For this we have $x - x = 0 \in \mathbf{V}$, i. e., the property (1) is satisfied. This completes the proof. □

Theorem 1.4. *Let* $\mathbf{E}$ *be a vector space and* $\mathbf{V}$ *be its nonempty subset. Then* $\mathbf{V}$ *is a linear subspace of* $\mathbf{E}$ *if and only if every linear combination of finitely many elements of* $\mathbf{V}$ *also belongs to* $\mathbf{V}$.

Proof.

1. Let $\mathbf{V}$ be a subspace. Then the properties (1), (2) and (3) of Theorem 1.3 are satisfied. Let $x_1, \ldots, x_k \in \mathbf{V}$ and $a_1, \ldots, a_k \in \mathbf{F}$ be arbitrarily chosen. Then, by the property (3) of Theorem 1.3, it follows that

$$a_l x_l \in \mathbf{V} \quad \text{for} \quad \text{any} \quad l \in \{1, \ldots, k\}.$$

Hence, by the property (2) of Theorem 1.3, it follows that

$$a_1 x_1 + a_2 x_2 \in \mathbf{V}.$$

Again, by the property (2) of Theorem 1.3, it follows that

$$a_1 x_1 + a_2 x_2 + a_3 x_3 \in \mathbf{V}.$$

And so on,

$$a_1 x_1 + \cdots + a_k x_k \in \mathbf{V}.$$

2. Let **V** is closed under linear combination of finitely many its elements. Then, if $x, y \in \mathbf{V}$ and $a \in \mathbf{F}$ be arbitrarily chosen, we get

$$x + y \in \mathbf{V} \quad \text{and} \quad ax \in \mathbf{V},$$

i. e., the properties (2) and (3) of Theorem 1.3 are satisfied. Now, using the same arguments as in the second part of the proof of Theorem 1.3, we conclude that the property (1) of Theorem 1.3 is satisfied. This completes the proof. □

Definition 1.8. Let **E** be a vector space and $x_1, \ldots, x_k \in \mathbf{E}$. The set

$$\operatorname{Span}\{x_1, \ldots, x_k\} = \{a_1 x_1 + \cdots + a_k x_k : a_l \in \mathbf{F}, l \in \{1, \ldots, k\}\}$$

will be called the span of the elements $x_1, \ldots, x_k$.

Theorem 1.5. *Let* **E** *be a vector space and* $x_1, \ldots, x_k$ *be its elements. Then* $\operatorname{Span}\{x_1, \ldots, x_k\}$ *is a linear subspace of* **E**.

Proof. Note that $0 \in \operatorname{Span}\{x_1, \ldots, x_k\}$ because

$$0 = 0x_1 + \cdots + 0x_k \in \operatorname{Span}\{x_1, \ldots, x_k\}.$$

Let

$$x = a_1 x_1 + \cdots + a_k x_k, \quad y = b_1 x_1 + \cdots + b_k x_k, \quad a_l, b_l \in \mathbf{F}, \quad l \in \{1, \ldots, k\}.$$

Then, for any $a, b \in \mathbf{F}$, we have $aa_l, bb_l \in \mathbf{F}, l \in \{1, \ldots, k\}$, and

$$\begin{aligned} ax + by &= aa_1 x_1 + \cdots + aa_k x_k + bb_1 x_1 + \cdots + bb_k x_k \\ &= (aa_1 + bb_1) x_1 + \cdots + (aa_k + bb_k) x_k \in \operatorname{Span}\{x_1, \ldots, x_k\}. \end{aligned}$$

Also, $ax \in \operatorname{Span}\{x_1, \ldots, x_k\}$. Hence, by Theorem 1.3, it follows that $\operatorname{Span}\{x_1, \ldots, x_k\}$ is a linear subspace. This completes the proof. □

Theorem 1.6. *The intersection of any collection of linear subspaces of a vector space* **E** *is a linear subspace of* **E**.

Proof. Let $\{\mathbf{A}_n\}_{n \in \mathcal{A}}$ be a collection of linear subspaces of the vector space **E**. Here $\mathcal{A}$ is an index set. We set

$$\mathbf{B} = \bigcap_{n \in \mathcal{A}} \mathbf{A}_n.$$

Since $0 \in \mathbf{A}_n$ for any $n \in \mathcal{A}$, we see that $0 \in \mathbf{B}$. Let $x, y \in \mathbf{B}$ and $a, b \in \mathbf{F}$ be arbitrarily chosen. Then $x, y \in \mathbf{A}_n$ for any $n \in \mathcal{A}$. Because $\mathbf{A}_n$, $n \in \mathcal{A}$, are linear subspaces of **E**, we conclude that $ax + by, ax \in \mathbf{A}_n$ for any $n \in \mathcal{A}$. Therefore $ax + by, ax \in \mathbf{B}$. Hence, by Theorem 1.3, we see that **B** is a linear subspace of **E**. This completes the proof. □

Definition 1.9. Suppose that $\mathbf{E}$ is a vector space and $\mathbf{S}$ is its subset. There exist linear subspaces $\mathbf{A}_n$ containing the set $\mathbf{S}$. For example, the space $\mathbf{E}$ is a such subspace. By $\mathbf{B}$ we will denote the intersection of any linear subspaces containing the set $\mathbf{S}$. Then $\mathbf{B}$ contains $\mathbf{S}$ and it is the smallest such linear subspace. The subspace $\mathbf{B}$ is called the linear subspace spanned by $\mathbf{S}$ or the span of $\mathbf{S}$.

Theorem 1.7. *If* $\mathbf{S}$ *is any set of elements in a vector space* $\mathbf{E}$ *and if* $\mathbf{B}$ *is the linear subspace spanned by* $\mathbf{S}$*, then* $\mathbf{B}$ *is the same as the set of all linear combinations of elements of* $\mathbf{S}$.

Proof. Let $\mathbf{C}$ be the set of all linear combinations of elements of $\mathbf{S}$. Note that $\mathbf{C}$ is a linear subspace containing $\mathbf{S}$. Therefore

$$\mathbf{B} \subseteq \mathbf{C}. \tag{1.17}$$

On the other hand, since $\mathbf{B}$ is spanned by $\mathbf{S}$ it must contain all linear combinations of elements of $\mathbf{S}$. Consequently

$$\mathbf{C} \subseteq \mathbf{B}.$$

From the last inclusion and from (1.17), we conclude that $\mathbf{B} = \mathbf{C}$. This completes the proof. □

Theorem 1.8. *Let* $\mathbf{A}$ *and* $\mathbf{B}$ *be linear subspaces of a vector space* $\mathbf{E}$ *and* $\mathbf{C}$ *be the linear subspace spanned by* $\mathbf{A}$ *and* $\mathbf{B}$ *together. Then* $\mathbf{C}$ *is the same as the set of all elements of the form* $x + y$ *with* $x \in \mathbf{A}$ *and* $y \in \mathbf{B}$.

Proof. Let $\mathbf{D}$ be the set of all elements of the form $x + y$, $x \in \mathbf{A}$ and $y \in \mathbf{B}$. Then

$$\mathbf{C} \subseteq \mathbf{D}. \tag{1.18}$$

Since $\mathbf{C}$ is spanned by $\mathbf{A}$ and $\mathbf{B}$ together it must contain all elements of the form $x + y$, $x \in \mathbf{A}$ and $y \in \mathbf{B}$. Therefore

$$D \subseteq \mathbf{C}.$$

From the last inclusion and from (1.18), we conclude that $\mathbf{C} = \mathbf{D}$. This completes the proof. □

Prompted by Theorem 1.8, we shall use the notation $\mathbf{A} + \mathbf{B}$ for the linear subspace $\mathbf{C}$ spanned by the linear subspaces $\mathbf{A}$ and $\mathbf{B}$ together.

Definition 1.10. We will say that a linear subspace $\mathbf{A}$ of a vector space $\mathbf{E}$ is a complement of a linear subspace $\mathbf{B}$ of $\mathbf{E}$ if

$$\mathbf{A} \cap \mathbf{B} = \{0\} \quad \text{and} \quad \mathbf{A} + \mathbf{B} = \mathbf{E}.$$

Example 1.10. Let **A** and **B** be finite-dimensional linear subspaces of the same dimension. Suppose that $\mathbf{A} \subseteq \mathbf{B}$. We will prove that $\mathbf{A} = \mathbf{B}$. In fact, let $\dim(\mathbf{A}) = \dim(\mathbf{B}) = m$. Then there are m linearly independent elements of **A**, $\{x_1, \dots, x_m\}$. Since $\mathbf{A} \subseteq \mathbf{B}$, we see that $\{x_1, \dots, x_m\}$ are linearly independent elements of **B**. Hence, using $\dim(\mathbf{B}) = m$, we conclude that every element of **B** is a linear combination of $\{x_1, \dots, x_m\}$. Therefore $\mathbf{B} \subseteq \mathbf{A}$, which completes the proof.

Exercise 1.7. Let x_1, x_2 and x_3 be elements of the vector space **E** for which $x_1+x_2+x_3 = 0$. Prove that x_1 and x_2 span the same linear subspace as x_2 and x_3.

Exercise 1.8. A polynomial y is called even if $y(-x) = y(x)$ identically in x and odd if $y(-x) = -y(x)$ identically in x.
1. Prove that the sets $\mathbf{P}_{\text{even}}$ and $\mathbf{P}_{\text{odd}}$ of even and odd polynomials, respectively, are linear subspaces of the space **P** of all polynomials with coefficients in **F**.
2. Prove that $\mathbf{P}_{\text{even}}$ and $\mathbf{P}_{\text{odd}}$ are each other's complements.

Definition 1.11. Two vector spaces **U** and **V**, over the same field **F**, are called isomorphic if there is a one-to-one correspondence between the elements x of **U** and the elements y of **V**, say $y = T(x)$, such that

$$T(a_1x_1 + a_2x_2) = a_1T(x_1) + a_2T(x_2)$$

for any $a_1, a_2 \in \mathbf{F}$ and any $x_1, x_2 \in \mathbf{U}$. In other words, **U** and **V** are isomorphic if there is an isomorphism between them, where an isomorphism is a one-to-one correspondence that preserves all linear relations.

Theorem 1.9. *Let* **U** *and* **V** *be vector spaces that are isomorphic and* T *is an isomorphism between them,* $T(x) = y$, $x \in \mathbf{U}$, $y \in \mathbf{V}$. *Then* $T(0) = 0$.

Proof. For $x \in \mathbf{U}$ we have

$$T(0) = T(x - x) = T(x) - T(x) = 0,$$

which completes the proof. □

Theorem 1.10. *Let* **U** *and* **V** *be vector spaces that are isomorphic and* T *is an isomorphism between them,* $T(x) = y$, $x \in \mathbf{U}$, $y \in \mathbf{V}$.
1. *If* $\{x_1, \dots, x_n\}$ *are linearly dependent, then* $\{T(x_1), \dots, T(x_n)\}$ *are linearly dependent.*
2. *If* $\{x_1, \dots, x_n\}$ *are linearly independent, then* $\{T(x_1), \dots, T(x_n)\}$ *are linearly independent.*

Proof.
1. Since $\{x_1, \dots, x_n\}$ are linearly dependent elements of **U**, there are constants $\alpha_1, \dots, \alpha_n \in \mathbf{F}$ such that

$$\alpha_1x_1 + \cdots + \alpha_nx_n = 0, \quad (\alpha_1, \dots, \alpha_n) \neq (0, \dots, 0).$$

Hence,

$$0 = T(\alpha_1 x_1 + \cdots + \alpha_n x_n) = \alpha_1 T(x_1) + \cdots + \alpha_n T(x_n).$$

Therefore $\{T(x_1), \ldots, T(x_n)\}$ are linearly dependent elements of V.

2. Let $\{x_1, \ldots, x_n\}$ be linearly independent elements of $\mathbf{U}$. Assume that $\{T(x_1), \ldots, T(x_n)\}$ are linearly dependent elements of $\mathbf{V}$. Then there are $\alpha_1, \ldots, \alpha_n \in \mathbf{F}$, $(\alpha_1, \ldots, \alpha_n) \neq (0, \ldots, 0)$, such that

$$\alpha_1 T(x_1) + \cdots + \alpha_n T(x_n) = 0.$$

Hence,

$$T(\alpha_1 x_1 + \cdots + \alpha_n x_n) = 0.$$

Therefore, using Theorem 1.9, we get

$$\alpha_1 x_1 + \cdots + \alpha_n x_n = 0,$$

which is a contradiction. This completes the proof. □

Corollary 1.7. *Let* $\mathbf{U}$ *and* $\mathbf{V}$ *be finite-dimensional vector spaces that are isomorphic. Then* $\dim(\mathbf{U}) = \dim(\mathbf{V})$.

Theorem 1.11. *Every n-dimensional vector space* $\mathbf{U}$ *over the field* $\mathbf{F}$ *is isomorphic to* $\mathbf{E}_n$.

Proof. Let $\{x_1, \ldots, x_n\}$ be any basis in $\mathbf{U}$. Then every $x \in \mathbf{U}$ can be represented in the form

$$x = \xi_1 x_1 + \cdots + \xi_n x_n,$$

where $\xi_i \in \mathbf{F}$, $i \in \{1, \ldots, n\}$, are uniquely determined by x. We consider the one-to-one correspondence

$$x \mapsto (\xi_1, \ldots, \xi_n)$$

between $\mathbf{U}$ and $\mathbf{E}_n$. If

$$y = \eta_1 x_1 + \cdots + \eta_n x_n, \quad \eta_l \in \mathbf{F}, \quad l \in \{1, \ldots, n\},$$

then for any $a, b \in \mathbf{F}$ we have

$$\begin{aligned} ax + by &= a(\xi_1 x_1 + \cdots + \xi_n x_n) + b(\eta_1 x_1 + \cdots + \eta_n x_n) \\ &= (a\xi_1 + b\eta_1)x_1 + \cdots + (a\xi_n + b\eta_n)x_n, \end{aligned}$$

this establishes the desired isomorphism. □

Theorem 1.12. *Let* **U** *and* **V** *be n-dimensional vector spaces. Then they are isomorphic.*

Proof. By Theorem 1.11 it follows that there exist isomorphisms T_1 and T_2 between **U** and $\mathbf{E}_n$ and $\mathbf{E}_n$ and **V**, respectively. Then $T_2 \circ T_1$ is an isomorphism between **U** and **V**, which completes the proof. □

Definition 1.12. A linear functional on a vector space **E** is a scalar-valued function y defined for every element $x \in \mathbf{E}$ with the property

$$y(a_1x_1 + a_2x_2) = a_1y(x_1) + a_2y(x_2)$$

for any $a_1, a_2 \in \mathbf{F}$ and any $x_1, x_2 \in \mathbf{E}$. The space of all linear functionals on **E** will be denoted by $\mathbf{E}'$. We define the zero in $\mathbf{E}'$ as follows: $y(x) = 0$ for any $x \in \mathbf{E}$. If y_1 and y_2 are two linear functionals on **E** and if $a_1, a_2 \in \mathbf{F}$, then $a_1y_1 + a_2y_2$ is a linear functional on **E**. With these concepts (zero, addition, scalar multiplication), the set $\mathbf{E}'$ forms a vector space, the dual space of **E**.

Example 1.11. Consider $\mathbf{E}_n$. Let $\alpha_1, \ldots, \alpha_n, \beta \in \mathbf{F}$, $\beta \neq 0$, be fixed. For any $x \in \mathbf{E}_n$, $x = (\xi_1, \ldots, \xi_n)$, we define the scalar-valued function

$$y(x) = \alpha_1\xi_1 + \cdots + \alpha_n\xi_n + \beta.$$

If $x_1 = (\xi_1^1, \ldots, \xi_n^1)$, $x_2 = (\xi_1^2, \ldots, \xi_n^2) \in \mathbf{E}_n$ and $a_1, a_2 \in \mathbf{F}$ be arbitrarily chosen, then

$$\begin{aligned}
a_1x_1 &= (a_1\xi_1^1, \ldots, a_1\xi_n^1),\\
a_2x_2 &= (a_2\xi_1^2, \ldots, a_2\xi_n^2),\\
a_1x_1 + a_2x_2 &= (a_1\xi_1^1 + a_2\xi_1^2, \ldots, a_1\xi_n^1 + a_2\xi_n^2),\\
a_1y(x_1) &= a_1\alpha_1\xi_1^1 + \cdots + a_1\alpha_n\xi_n^1 + a_1\beta,\\
a_2y(x_2) &= a_2\alpha_1\xi_1^2 + \cdots + a_2\alpha_n\xi_n^2 + a_2\beta,\\
y(a_1x_1 + a_2x_2) &= \alpha_1(a_1\xi_1^1 + a_2\xi_1^2) + \cdots + \alpha_n(a_1\xi_n^1 + a_2\xi_n^2) + \beta.
\end{aligned}$$

Since $a_1\beta + a_2\beta \neq \beta$ unless $a_1 + a_2 = 1$, we conclude that y is not a linear functional on $\mathbf{E}_n$.

Example 1.12. Consider **C** over **R**. For every $x = a + ib \in \mathbf{C}$, $a, b \in \mathbf{R}$, we define the scalar-valued function y as follows:

$$y(x) = a.$$

Let

$$x_1 = a_1 + ib_1, \quad x_2 = a_2 + ib_2, \quad a_1, a_2, b_1, b_2 \in \mathbf{R}, \quad \alpha_1, \alpha_2 \in \mathbf{R}$$

be arbitrarily chosen. Then

$$\alpha_1y(x_1) = \alpha_1a_1, \quad \alpha_2y(x_2) = \alpha_2a_2,$$

$$\alpha_1 x_1 + \alpha_2 x_2 = (\alpha_1 a_1 + \alpha_2 a_2) + i(\alpha_1 b_2 + \alpha_2 b_2),$$
$$y(\alpha_1 x_1 + \alpha_2 x_2) = \alpha_1 a_1 + \alpha_2 a_2 = \alpha_1 y(x_1) + \alpha_2 y(x_2).$$

Consequently y is a linear functional on $\mathbf{C}$.

Exercise 1.9. Consider $\mathbf{C}$. For $x = a + ib \in \mathbf{C}$, $a, b \in \mathbf{R}$, we define the scalar-valued function y as follows:

$$y(x) = \sqrt{a^2 + b^2}.$$

Check if y is a linear functional on $\mathbf{C}$.

Answer. No.

Definition 1.13. Suppose that $\mathbf{E}$ is a vector space and $\mathbf{E}'$ is its dual space. We make correspond $[x, y]$, $x \in \mathbf{E}$, $y \in \mathbf{E}'$, to be the value of y at x. In terms of the symbol $[x, y]$ the defining property of a linear functional is

$$[a_1 x_1 + a_2 x_2, y] = a_1 [x_1, y] + a_2 [x_2, y]$$

and the definition of the linear operations for linear functionals is

$$[x, a_1 y_1 + a_2 y_2] = a_1 [x, y_1] + a_2 [x, y_2].$$

The two relations together are expressed by saying that $[x, y]$ is a bilinear functional of the elements $x \in \mathbf{E}$ and $y \in \mathbf{E}'$.

Theorem 1.13. *Let $\mathbf{E}$ be a n-dimensional vector space and $\{x_1, \ldots, x_n\}$ is its basis. Let also, $\{a_1, \ldots, a_n\}$ be a set of n scalars. Then there exists a unique linear functional y such that $[x_i, y] = a_i$, $i \in \{1, \ldots, n\}$.*

Proof. For any $x \in \mathbf{E}$ and any linear functional $y \in \mathbf{E}'$ we have

$$x = \xi_1 x_1 + \cdots + \xi_n x_n \quad \text{and} \quad [x, y] = \xi_1 [x_1, y] + \cdots + \xi_n [x_n, y], \quad \xi_l \in \mathbf{F}, \quad l \in \{1, \ldots, n\}.$$

We define a linear functional $\tilde{y}$ such that $[x_i, \tilde{y}] = a_i$, $i \in \{1, \ldots, n\}$. Then the value of $[x, \tilde{y}]$ is determined by

$$[x, \tilde{y}] = a_1 \xi_1 + \cdots + a_n \xi_n.$$

Suppose that there are two linear functionals $\tilde{y}$ and $\tilde{\tilde{y}}$ such that

$$[x_i, \tilde{y}] = [x_i, \tilde{\tilde{y}}] = a_i, \quad i \in \{1, \ldots, n\}.$$

Then $\tilde{y} - \tilde{\tilde{y}}$ is a linear functional and

$$[x, \tilde{y} - \tilde{\tilde{y}}] = 0.$$

Therefore $\tilde{y} = \tilde{\tilde{y}}$. This completes the proof. □

Theorem 1.14. *Let $\mathbf{E}$ be a vector space with a basis $\{x_1, \dots, x_n\}$. Then there exists a uniquely determined basis in $\mathbf{E}'$, $\{y_1, \dots, y_n\}$, such that $[x_i, y_j] = \delta_{ij}$, $i, j \in \{1, \dots, n\}$. Here $\delta_{ij} = 1$ if $i = j$ and $\delta_{ij} = 0$ if $i \neq j$, $i, j \in \{1, \dots, n\}$.*

Proof. By Theorem 1.13, it follows that there are uniquely determined functionals $\{y_1, \dots, y_n\}$ such that $[x_i, y_j] = \delta_{ij}$, $i, j \in \{1, \dots, n\}$. We will prove that $\{y_1, \dots, y_n\}$ is a basis in $\mathbf{E}'$. Suppose that

$$a_1 y_1 + \cdots + a_n y_n = 0$$

for some $a_i \in \mathbf{F}$, $i \in \{1, \dots, n\}$. Then for any $x \in \mathbf{E}$ we have

$$[x, a_1 y_1 + \cdots + a_n y_n] = a_1 [x, y_1] + \cdots + a_n [x, y_n] = 0.$$

In particular, if $x = x_i$, $i \in \{1, \dots, n\}$, we get

$$[x_i, a_1 y_1 + \cdots + a_n y_n] = \sum_{j=1}^{n} a_j [x_i, y_j] = \sum_{j=1}^{n} a_j \delta_{ij} = a_i, \quad i \in \{1, \dots, n\}.$$

Therefore $a_i = 0$, $i \in \{1, \dots, n\}$, and hence, $\{y_1, \dots, y_n\}$ are linearly independent. Let $y \in \mathbf{E}'$ be arbitrarily chosen. We set $[x_i, y] = \alpha_i$, $i \in \{1, \dots, n\}$, and let $x = \sum_{i=1}^{n} \xi_i x_i$. Then

$$[x, y] = \left[\sum_{i=1}^{n} \xi_i x_i, y \right] = \sum_{i=1}^{n} \xi_i [x_i, y] = \sum_{i=1}^{n} \alpha_i \xi_i.$$

On the other hand,

$$[x, y_j] = \left[\sum_{i=1}^{n} \xi_i x_i, y_j \right] = \sum_{i=1}^{n} \xi_i [x_i, y_j] = \sum_{i=1}^{n} \xi_i \delta_{ij} = \xi_j, \quad j \in \{1, \dots, n\}.$$

Therefore

$$[x, y] = \sum_{i=1}^{n} \alpha_i [x, y_i] = \left[x, \sum_{i=1}^{n} \alpha_i y_i \right].$$

Consequently $y = \sum_{i=1}^{n} \alpha_i y_i$ and the proof of the theorem is complete. □

Theorem 1.15. *Let $\mathbf{E}$ be a n-dimensional vector space. Then, for any $x \in \mathbf{E}$, $x \neq 0$, there corresponds an $y \in \mathbf{E}'$ such that $[x, y] \neq 0$.*

Proof. Let $\{x_1, \dots, x_n\}$ be a basis in $\mathbf{E}$. Then, using Theorem 1.14, there exists a uniquely determined basis in $\mathbf{E}'$, $\{y_1, \dots, y_n\}$. If $x = \sum_{i=1}^{n} \xi_i x_i \in \mathbf{E}$, then $[x, y_j] = \xi_j$, $j \in \{1, \dots, n\}$. Hence, if $[x, y] = 0$ for any $y \in \mathbf{E}'$, then $\xi_j = 0$, $j \in \{1, \dots, n\}$, and hence $x = 0$, which completes the proof. □

Definition 1.14. If **U** and **V** be vector spaces over the same field, their direct sum is the vector space **W**, denoted by $\mathbf{U} \oplus \mathbf{V}$, whose elements are all ordered pairs $\langle x, z \rangle$ with $x \in \mathbf{U}$ and $z \in \mathbf{V}$, with linear operations defined by

$$a_1 \langle x_1, z_1 \rangle + a_2 \langle x_2, z_2 \rangle = \langle a_1 x_1 + a_2 x_2, a_1 z_1 + a_2 z_2 \rangle.$$

Consider $\mathbf{W} = \mathbf{U} \oplus \mathbf{V}$. The set of all elements of the form $\langle u, 0 \rangle$ is a linear subspace of **W**. The correspondence $\langle u, 0 \rangle \longmapsto u$ shows that this linear subspace is isomorphic to **U**. It is convenient to identify u and $\langle u, 0 \rangle$, to speak of **U** as a linear subspace of **W**. Similarly, the elements $v \in \mathbf{V}$ may be identified with the elements of the form $\langle 0, v \rangle$ in **W**. In the case in which **U** and **V** have no non-zero elements in common, we could have defined the direct sum of **U** and **V** as the set consisting of all $u \in \mathbf{U}$ and all $v \in \mathbf{V}$, and all those pairs $\langle u, v \rangle$ for which $u \neq 0$ and $v \neq 0$.

Theorem 1.16. *If* **U** *and* **V** *are linear subspaces of the vector space* **E**, *then the following three conditions are equivalent:*
1. $\mathbf{E} = \mathbf{U} \oplus \mathbf{V}$,
2. $\mathbf{U} \cap \mathbf{V} = \{0\}$ *and* $\mathbf{U} + \mathbf{V} = \mathbf{E}$,
3. *Every element* $z \in \mathbf{W}$ *may be written in the form* $z = u + v$, *with* $u \in \mathbf{U}$ *and* $v \in \mathbf{V}$, *in one and only one way.*

Proof.
$1 \Longrightarrow 2$. We assume that $\mathbf{E} = \mathbf{U} \oplus \mathbf{V}$. Suppose that $z = \langle u, v \rangle \in \mathbf{U}$, $z = \langle u, v \rangle \in \mathbf{V}$. Then $u = v = 0$ and $z = 0$. Therefore $\mathbf{U} \cap \mathbf{V} = \{0\}$. Since the representation

$$z = \langle u, 0 \rangle + \langle 0, v \rangle$$

is valid for every z, it follows that $\mathbf{U} + \mathbf{V} = \mathbf{E}$.

$2 \Longrightarrow 3$. We assume 2. Then for every element $z \in \mathbf{E}$ we have the representation

$$z = u + v.$$

Assume that $z = u_1 + v_1$ for $u_1 \in \mathbf{U}$ and $v_1 \in \mathbf{V}$. Then

$$u + v = u_1 + v_1,$$

whereupon

$$u - u_1 = v_1 - v.$$

Since **U** and **V** are vector spaces, we have $u - u_1 \in \mathbf{U}$ and $v_1 - v \in \mathbf{V}$. Hence,

$$u - u_1 = 0 = v_1 - v \quad \text{or} \quad u = u_1, \quad v = v_1.$$

3 $\Longrightarrow$ 1. We form the direct sum $\mathbf{U} \oplus \mathbf{V}$ and then identify $\langle u, 0\rangle$ and $\langle 0, v\rangle$ with u and v, respectively. We committed to identifying the sum

$$\langle u, v\rangle = \langle u, 0\rangle + \langle 0, v\rangle$$

with what we have assuming to be the general element $z = u + v$ of $\mathbf{E}$. From the hypothesis that the representation of z in the form $u + v$ is unique, we conclude that the correspondence between $\langle u, 0\rangle$ and u, also between $\langle 0, v\rangle$ and v, is one-to-one. This completes the proof. □

Theorem 1.17. *Let* $\mathbf{U}$ *and* $\mathbf{V}$ *be vector spaces with* $\dim(\mathbf{U}) = m$ *and* $\dim(\mathbf{V}) = n$. *Then* $\dim(\mathbf{U} \oplus \mathbf{V}) = m + n$.

Proof. Let $\{x_1, \ldots, x_m\}$ be a basis in $\mathbf{U}$ and $\{y_1, \ldots, y_n\}$ be a basis in $\mathbf{V}$. Consider the set

$$\{x_1, \ldots, x_m, y_1, \ldots, y_n\}.$$

By Theorem 1.16, assertion (3), it follows that every element $z \in \mathbf{U} \oplus \mathbf{V}$ can be represented in a unique way in the form $u + v$, $u \in \mathbf{U}$, $v \in \mathbf{V}$. Since $u \in \mathbf{U}$ and $v \in \mathbf{V}$, we have

$$u = a_1 x_1 + \cdots + a_m x_m, \quad v = b_1 y_1 + \cdots + b_n y_n, \quad a_i, b_j \in \mathbf{F},$$

$i \in \{1, \ldots, m\}$, $j \in \{1, \ldots, n\}$. Hence,

$$z = u + v = a_1 x_1 + \cdots + a_m x_m + b_1 y_1 + \cdots + b_n y_n.$$

Consider

$$\alpha_1 x_1 + \cdots + \alpha_m x_m + \beta_1 y_1 + \cdots + \beta_n y_n = 0.$$

The uniqueness of the representation of 0 in the form of $u + v$ implies that

$$\alpha_1 x_1 + \cdots + \alpha_m x_m = \beta_1 y_1 + \cdots + \beta_n y_n = 0,$$

and hence the linear independence of $\{x_1, \ldots, x_m\}$ and the linear independence of $\{y_1, \ldots, y_n\}$ imply that

$$\alpha_1 = \cdots = \alpha_m = \beta_1 = \cdots = \beta_n = 0.$$

Consequently $\{x_1, \ldots, x_m, y_1, \ldots, y_n\}$ are linearly independent. This completes the proof. □

Theorem 1.18. *If* $\mathbf{E}$ *is a finite-dimensional vector space and if* $\{y_1, \ldots, y_m\}$ *is any set of linearly independent elements of* $\mathbf{E}$, *then unless* y*'s already form a basis. We can find elements* $y_{m+1}, \ldots, y_{m+p}$ *so that the totality of the* y, *that is,*

$$\{y_1, \ldots, y_m, y_{m+1}, \ldots, y_{m+p}\},$$

is a basis.

Proof. Let $\{x_1, \dots, x_n\}$ be a basis in **E**. Consider the set

$$S = \{y_1, \dots, y_m, x_1, \dots, x_n\}.$$

Since y's are linear combination of x's, the set S is linearly dependent. Hence, some element of S is a linear combination of the preceding ones. Because $\{y_1, \dots, y_m\}$ are linearly independent, this element may be different from y's. Let this element is x_i. We consider

$$S' = \{y_1, \dots, y_m, x_1, \dots, x_{i-1}, x_{i+1}, \dots, x_n\}.$$

If S' is linearly independent, then we are done. If it is not, we act as above while we reach a linearly independent set containing $y_1, \dots, y_m$, in terms of which we may express every element in **E**. This completes the proof. □

Theorem 1.19. *Let* **E** *be an* $n + m$*-dimensional vector space and* **U** *be its* n*-dimensional linear subspace of* **E**. *Then there exists an* m*-dimensional linear subspace* **V** *of* **E** *such that* $\mathbf{E} = \mathbf{U} \oplus \mathbf{V}$.

Proof. Let $\{x_1, \dots, x_n\}$ be a basis in **U**. Because **E** is $m + n$-dimensional vector space, using Theorem 1.18, there is a set $\{y_1, \dots, y_m\}$ such that $\{x_1, \dots, x_n, y_1, \dots, y_m\}$ is a basis of **E**. Let **V** be spanned by $\{y_1, \dots, y_m\}$. Then $\mathbf{E} = \mathbf{U} \oplus \mathbf{V}$. This completes the proof. □

Definition 1.15. Assume that **A** and **B** be two nonempty sets. The Cartesian product of **A** with **B**, denoted by $\mathbf{A} \times \mathbf{B}$, is defined to be the collection of all ordered pairs (a, b), where $a \in \mathbf{A}$ and $b \in \mathbf{B}$. We consider $(a, b) = (a'.b')$ if and only if $a = a'$ and $b = b'$. For a nonempty set **X**, we call a subset **R** of $\mathbf{X} \times \mathbf{X}$ a relation on **X** and write $x\mathbf{R}x'$ provided $(x, x') \in \mathbf{R}$. The relation **R** is said to be
1. reflexive provided $x\mathbf{R}x$ for all $x \in \mathbf{X}$,
2. symmetric provided $x\mathbf{R}x'$ if $x'\mathbf{R}x$,
3. transitive provided whenever $x\mathbf{R}x'$ and $x'\mathbf{R}x''$, we have $x\mathbf{R}x''$.

A relation **R** on a set **X** is called an equivalence relation provided it is reflexive, symmetric, and transitive. Sometimes, an equivalence relation is denoted by $\sim$.

Definition 1.16. Let **E** be a vector space and **V** be its linear subspace. We define an equivalence relation $\sim$ on **E** by stating that $x \sim y$ if $x - y \in \mathbf{V}$. The equivalence class of x is often denoted by

$$[x] = x + \mathbf{V}.$$

We have

$$[x] = \{x + y : y \in \mathbf{V}\}.$$

The quotient space $\mathbf{E}/\mathbf{V}$ is often denoted as $\mathbf{E}/\sim$ and defined as the set of all classes over $\mathbf{E}$ by $\sim$. Scalar multiplication and addition are defined on the equivalence classes by

$$\alpha[x] = [\alpha x], \quad \alpha \in \mathbf{F}, \quad x \in \mathbf{E}, \tag{1.19}$$

$$[x] + [y] = [x + y]. \tag{1.20}$$

Exercise 1.10. Prove that the operations (1.19) and (1.20) are well defined, i. e., do not depend on the choice of representatives.

Exercise 1.11. Prove that $\mathbf{E}/\mathbf{V}$ is a linear subspace of $\mathbf{E}$.

Theorem 1.20. *Let $m < n$ and $\mathbf{U}$ is an m-dimensional linear subspace of the n-dimensional vector space $\mathbf{E}$. Then $\mathbf{E}/\mathbf{U}$ has dimension $n - m$.*

Proof. By Theorem 1.19, it follows that there exists a linear subspace $\mathbf{V}$ of $\mathbf{E}$ such that $\mathbf{E} = \mathbf{U} \oplus \mathbf{V}$. Then we see that $\dim(\mathbf{V}) = n - m$ and $\mathbf{V}$ is isomorphic of $\mathbf{E}/\mathbf{U}$. Hence, by Corollary 1.7, it follows that $\dim(\mathbf{E}/\mathbf{U}) = n - m$. This completes the proof. □

1.2 Metric spaces

Definition 1.17. A metric space is an ordered pair $(\mathbf{M}, d)$, where $\mathbf{M}$ is a set and d is a metric on $\mathbf{M}$, i. e., a function $d : \mathbf{M}\times\mathbf{M} \longmapsto \mathbf{R}$ such that for any $x, y, z \in \mathbf{M}$ the following holds.
1. $d(x, y) \geq 0$ *non-negativity* or *separation axiom*,
2. $d(x, y) = 0 \iff x = y$ *identity of indiscernible*,
3. $d(x, y) = d(y, x)$ *symmetry*,
4. $d(x, z) \leq d(x, y) + d(y, z)$ *triangle inequality*.

The first condition follows from the other three. In fact, for any $x, y \in \mathbf{M}$, using the triangle inequality, we have

$$d(x, x) \leq d(x, y) + d(y, x).$$

Hence, using the symmetry, we have $d(x, y) = d(y, x)$ and

$$d(x, x) \leq 2d(x, y).$$

From this, using the identity of indiscernible, we get

$$2d(x, y) \geq 0 \quad \text{or} \quad d(x, y) \geq 0.$$

The function d is also called the *distance function* or simply the *distance*. When it is clear from the context what metric is used, d is omitted and one just writes $\mathbf{M}$ for a metric space.

Example 1.13. The real numbers with the distance function

$$d(x,y) = |x - y|$$

given by the absolute difference, is a metric space.

Example 1.14. The positive real numbers with the distance function

$$d(x,y) = \left|\log\left(\frac{y}{x}\right)\right|$$

is a metric space. In fact,

1. $d(x,y) \geq 0$ for any $x, y \in \mathbf{R}_+$.
2. $$\begin{aligned} d(x,y) = 0 \quad &\Longleftrightarrow \\ \left|\log\left(\frac{y}{x}\right)\right| = 0 \quad &\Longleftrightarrow \\ \log\left(\frac{y}{x}\right) = 0 \quad &\Longleftrightarrow \\ \frac{y}{x} = 1 \quad \text{or} \quad x = y, \quad & x, y \in \mathbf{R}_+. \end{aligned}$$
3. $$d(x,y) = \left|\log\left(\frac{y}{x}\right)\right| = \left|-\log\left(\frac{x}{y}\right)\right| = \left|\log\left(\frac{x}{y}\right)\right| = d(y,x), \quad x, y \in \mathbf{R}_+.$$
4. For $x, y, z \in \mathbf{R}_+$ we have $$\begin{aligned} d(x,z) &= \left|\log\left(\frac{z}{x}\right)\right| = \left|\log\left(\frac{y}{x}\cdot\frac{z}{y}\right)\right| = \left|\log\left(\frac{y}{x}\right) + \log\left(\frac{z}{y}\right)\right| \\ &\leq \left|\log\left(\frac{y}{x}\right)\right| + \left|\log\left(\frac{z}{y}\right)\right| = d(x,y) + d(y,z). \end{aligned}$$

Example 1.15. Let $\mathbf{M}$ be any nonempty set. Then $\mathbf{M}$ is a metric space with the distance function

$$d(x,y) = 0 \quad \text{if} \quad x = y \quad \text{and} \quad d(x,y) = 1 \quad \text{otherwise},$$

called the *discrete metric.*

Exercise 1.12. Prove that

$$d(x,y) = \max_{i\in\{1,\dots,n\}} |x_i - y_i|, \quad x = (x_1, \dots, x_n), \quad y = (y_1, \dots, y_n) \in \mathbf{E}_n,$$

is a metric on $\mathbf{E}_n$.

If $(\mathbf{M}, d)$ is a metric space and $\mathbf{X}$ is a subset of $\mathbf{M}$, then $(\mathbf{X}, d)$ becomes a metric space by restricting the domain of d to $\mathbf{X} \times \mathbf{X}$.

Definition 1.18. An element x of a metric space $\mathbf{M}$ is said to be the limit of a sequence $\{x_n\}_{n\in\mathbf{N}}$ of elements of $\mathbf{M}$, if $d(x_n, x) \to 0$ as $n \to \infty$. In this case, we write $x_n \to x$ as $n \to \infty$ or $\lim_{n\to\infty} x_n = x$.

Theorem 1.21. *If a sequence $\{x_n\}_{n\in\mathbf{N}}$ of elements of a metric space $\mathbf{M}$ converges to an element $x \in \mathbf{M}$, then every subsequence $\{x_{n_k}\}_{k\in\mathbf{N}}$ of the sequence $\{x_n\}_{n\in\mathbf{N}}$ also converges to the same limit.*

Proof. Let $\epsilon > 0$ be arbitrarily chosen. Then there exists $N = N(\epsilon) > 0$ such that

$$d(x_n, x) < \epsilon$$

for any $n > N$. In particular, when $n_k > N$ we get

$$d(x_{n_k}, x) < \epsilon,$$

which completes the proof. □

Theorem 1.22. *A sequence $\{x_n\}_{n\in\mathbf{N}}$ of elements of a metric space $\mathbf{M}$ can converge to at most one limit.*

Proof. Assume that $x_n \to x$ and $x_n \to y$ as $n \to \infty$, $x, y \in \mathbf{M}$. Let $\epsilon > 0$ be arbitrarily chosen. Then there exists $N = N(\epsilon) > 0$ such that

$$d(x_n, x) < \frac{\epsilon}{2} \quad \text{and} \quad d(x_n, y) < \frac{\epsilon}{2}$$

for any $n > N$. Hence, for any $n > N$, we get

$$d(x, y) \leq d(x, x_n) + d(x_n, y) < \frac{\epsilon}{2} + \frac{\epsilon}{2} = \epsilon.$$

Because $\epsilon > 0$ was arbitrarily chosen, we conclude that $x = y$. This completes the proof. □

Theorem 1.23. *If a sequence $\{x_n\}_{n\in\mathbf{N}}$ of elements of $\mathbf{M}$ converges to an element $x \in \mathbf{M}$, then the set $\{d(x_n, y) : n \in \mathbf{N}\}$ is bounded for every $y \in \mathbf{M}$.*

Proof. Let $y \in \mathbf{M}$ and $\epsilon > 0$ be arbitrarily chosen. Then there exists $N = N(\epsilon) > 1$, $N \in \mathbf{N}$, such that

$$d(x_n, x) < \frac{\epsilon}{2} \quad \text{for} \quad \text{any} \quad n > N.$$

Hence, for any $n > N$ we have

$$d(x_n, y) \leq d(x_n, x) + d(x, y) < \frac{\epsilon}{2} + d(x, y).$$

We set

$$K = \max\left\{d(x_1, y), \dots, d(x_N, y), \frac{\epsilon}{2} + d(x, y)\right\}.$$

Therefore

$$d(x_n, y) \le K \quad \text{for} \quad \text{any} \quad n \in \mathbf{N}.$$

This completes the proof. □

Definition 1.19. Let $\mathbf{M}$ be a metric space.

1. The set

$$B_r(a) = \{x \in \mathbf{M} : d(x, a) < r\}(B_r[a] = \{x \in \mathbf{M} : d(x, a) \le r\})$$

   is called an open (closed) ball with a center a and radius r.
2. Every open ball with center a point $x \in \mathbf{M}$ is called a neighborhood of the point x.
3. A set lying in an open ball is called bounded.
4. Let $\mathbf{X} \subseteq \mathbf{M}$. Then a point $a \in \mathbf{M}$ is called an accumulation (limit) point of $\mathbf{X}$ if

$$B_r(a) \cap (\mathbf{X} \setminus \{a\}) \neq \emptyset,$$

   for any $r > 0$. The set of all points of $\mathbf{X}$ plus the limit points of $\mathbf{X}$ is called the closure of $\mathbf{X}$ and it is denoted by $\overline{\mathbf{X}}$. Evidently, $\mathbf{X} \subseteq \overline{\mathbf{X}}$. The closure of the empty set we define as the empty set.

Theorem 1.24. *Let $\mathbf{M}$ be a metric space and $\mathbf{X}, \mathbf{Y} \subseteq \mathbf{M}$. Then*

1. $\overline{X \cup Y} = \overline{X} \cup \overline{Y}$,
2. $\overline{\mathbf{X} \cap \mathbf{Y}} \subseteq \overline{\mathbf{X}} \cap \overline{\mathbf{Y}}$.

Proof.

1. Let $a \in \overline{\mathbf{X} \cup \mathbf{Y}}$ and $r > 0$ be arbitrarily chosen.
   (a) Let $a \in X \cup Y$. Then

$$a \in X \quad \text{or} \quad a \in Y.$$

   Hence, $a \in \overline{X}$ or $a \in \overline{Y}$ and then $a \in \overline{X} \cup \overline{Y}$.
   (b) Let a is a limit point of $X \cup Y$. Then

$$B_r(a) \cap ((X \cup Y) \setminus \{a\}) \neq \emptyset.$$

   Let $y \in B_r(a) \cap ((X \cup Y) \setminus \{a\})$ be arbitrarily chosen. Then

$$y \neq a, \quad y \in X \cup Y \quad \text{and} \quad d(a, y) < r.$$

If $y \in X$, we get

$$y \neq a, \quad y \in \mathbf{X} \quad \text{and} \quad d(a,y) < r,$$

i. e., $y \in B_r(a) \cap (\mathbf{X} \setminus \{a\})$. From this, we conclude that

$$B_r(a) \cap (\mathbf{X} \setminus \{a\}) \neq \emptyset.$$

Because $r > 0$ was arbitrarily chosen, we obtain $a \in \overline{X}$. Hence, $a \in \overline{X} \cup \overline{Y}$. Similarly, if $y \in \mathbf{Y}$, we see that $a \in \overline{Y}$ and hence, $a \in \overline{X} \cup \overline{Y}$.

Since $a \in \overline{\mathbf{X} \cup \mathbf{Y}}$ was arbitrarily chosen and for it we see that it is an element of $\overline{X} \cup \overline{Y}$, we obtain

$$\overline{\mathbf{X} \cup \mathbf{Y}} \subseteq \overline{X} \cup \overline{Y}. \tag{1.21}$$

Let now $a \in \overline{\mathbf{X}} \cup \overline{\mathbf{Y}}$ and $r > 0$ be arbitrarily chosen. Then $a \in \overline{X}$ or $a \in \overline{Y}$.

(a) Suppose that $a \in \overline{X}$. If $a \in X$, then $a \in X \cup Y$ and $a \in \overline{X \cup Y}$. Assume that a is a limit point of X. Then

$$B_r(a) \cap (\mathbf{X} \setminus \{a\}) \neq \emptyset.$$

There exists $y \in B_r(a) \cap (\mathbf{X} \setminus \{a\})$. Hence,

$$y \neq a, \quad y \in \mathbf{X} \quad \text{and} \quad d(a,y) < r.$$

From this,

$$y \neq a, \quad y \in \mathbf{X} \cup \mathbf{Y} \quad \text{and} \quad d(a,y) < r,$$

i. e., $B_r(a) \cap ((X \cup Y) \setminus \{a\}) \neq \emptyset$. Since $r > 0$ was arbitrarily chosen, we conclude that $a \in \overline{\mathbf{X} \cup \mathbf{Y}}$.

(b) Let $a \in \overline{Y}$. As above, we get $a \in \overline{X \cup Y}$.

Because $a \in \overline{\mathbf{X}} \cup \overline{\mathbf{Y}}$ was arbitrarily chosen and for it we see that it is an element of $\overline{\mathbf{X} \cup \mathbf{Y}}$, we conclude that

$$\overline{X} \cup \overline{Y} \subseteq \overline{\mathbf{X} \cup \mathbf{Y}}.$$

From the previous relation and from (1.21) we obtain $\overline{X} \cup \overline{Y} = \overline{\mathbf{X} \cup \mathbf{Y}}$.

2. Let $a \in \overline{\mathbf{X} \cap \mathbf{Y}}$ and $r > 0$ be arbitrarily chosen and fixed.
   (a) Let $a \in X \cap Y$. Then $a \in X$ and $a \in Y$. Hence, $a \in \overline{X}$ and $a \in \overline{Y}$. Then $a \in \overline{X} \cap \overline{Y}$.
   (b) Assume that a is a limit point of $X \cap Y$. Then

$$B_r(a) \cap ((\mathbf{X} \cap \mathbf{Y}) \setminus \{a\}) \neq \emptyset.$$

Let $z \in B_r(a) \cap ((\mathbf{X} \cap \mathbf{Y}) \setminus \{a\})$. Hence,

$$z \in B_r(a), \quad z \in \mathbf{X} \cap \mathbf{Y} \quad \text{and} \quad z \neq a,$$

whereupon

$$z \in B_r(a), \quad z \in \mathbf{X}, \quad z \neq a \quad \text{and} \quad z \in B_r(a), \quad z \in Y, \quad z \neq a.$$

Since $r > 0$ was arbitrarily chosen, we see that $a \in \overline{\mathbf{X}}$ and $a \in \overline{\mathbf{Y}}$. Therefore $a \in \overline{\mathbf{X}} \cap \overline{\mathbf{Y}}$.

Because $a \in \overline{\mathbf{X} \cap \mathbf{Y}}$ was arbitrarily chosen and we have obtained $a \in \overline{\mathbf{X}} \cap \overline{\mathbf{Y}}$, we conclude the desired relation. This completes the proof. □

Definition 1.20. Let $\mathbf{M}$ be a metric space and $\mathbf{X}, \mathbf{Y} \subseteq \mathbf{M}$.
1. We say that the set $\mathbf{X}$ is closed if $\mathbf{X} = \overline{\mathbf{X}}$.
2. We say that the set $\mathbf{X}$ is open if $\mathbf{M} \setminus \mathbf{X}$ is closed.
3. We say that the set $\mathbf{X}$ is dense in $\mathbf{Y}$ if $\mathbf{Y} \subseteq \overline{\mathbf{X}}$.
4. The set $\mathbf{X}$ is said to be everywhere dense in $\mathbf{M}$ if $\mathbf{M} = \overline{\mathbf{X}}$.
5. The set $\mathbf{X}$ in $\mathbf{M}$ is said to be nowhere dense in $\mathbf{M}$ if every ball of $\mathbf{M}$ contains a ball free from points of $\mathbf{X}$.

Definition 1.21. Let $(\mathbf{L}, d_{\mathbf{L}})$ and $(\mathbf{M}, d_{\mathbf{M}})$ be metric spaces. We say that the function $f : \mathbf{L} \longmapsto \mathbf{M}$ is a continuous function at the point $x_0 \in \mathbf{L}$ if for every $\epsilon > 0$ there is a $\delta = \delta(\epsilon) > 0$ such that $d_{\mathbf{L}}(x, x_0) < \delta$, $x \in \mathbf{L}$, implies $d_{\mathbf{M}}(f(x), f(x_0)) < \epsilon$. We say that the function f is continuous on $\mathbf{L}$ if it is continuous at every point of $\mathbf{L}$.

1.3 Useful inequalities

We will give here some important inequalities. These inequalities play an important role in applied mathematics. Let $1 \leq p \leq \infty$, we denote by q the conjugate of p, i. e., $\frac{1}{p} + \frac{1}{q} = 1$.

Theorem 1.25 (Young's inequality). *Let $a, b > 0$, $p, q \in (1, \infty)$, $\frac{1}{p} + \frac{1}{q} = 1$. Then*

$$ab \leq \frac{a^p}{p} + \frac{b^q}{q}. \tag{1.22}$$

The equality holds if and only if $a^p = b^q$.

Proof. Because the mapping $x \longmapsto e^x$ is convex, we get

$$ab = e^{\log a + \log b} = e^{\frac{1}{p}\log a^p + \frac{1}{q}\log b^q} \leq \frac{1}{p} e^{\log a^p} + \frac{1}{q} e^{\log b^q} = \frac{a^p}{p} + \frac{b^q}{q}.$$

Now we will prove that the equality holds if and only if $a^p = b^q$. When $a = 0$ or $b = 0$ the assertion is evident. Assume that $a \neq 0$ and $b \neq 0$. We set $t = ab^{-\frac{q}{p}}$ or $a = tb^{\frac{q}{p}}$.

Then

$$\frac{a^p}{p} + \frac{b^q}{q} - ab = \frac{t^p b^q}{p} + \frac{b^q}{q} - tb^{\frac{q}{p}+1} = \frac{t^p b^q}{p} + \frac{b^q}{q} - tb^{q(\frac{1}{p}+\frac{1}{q})}$$
$$= \frac{t^p b^q}{p} + \frac{b^q}{q} - tb^q = b^q\left(\frac{t^p}{p} + \frac{1}{q} - t\right).$$

Hence,

$$\frac{a^p}{p} + \frac{b^q}{q} - ab = 0 \Longleftrightarrow$$
$$b^q\left(\frac{t^p}{p} + \frac{1}{q} - t\right) = 0 \Longleftrightarrow$$
$$\frac{t^p}{p} + \frac{1}{q} - t = 0 \Longleftrightarrow t = 1 \Longleftrightarrow a^p = b^q.$$

(because, if we take $f(t) = \frac{t^p}{p} + \frac{1}{q} - t$, we have $f(1) = 0$, $f'(t) = t^{p-1} - 1$, $f'(t) \geq 0$ for $t \geq 1$ and $f'(t) \leq 0$ for $t \in [0,1]$) This completes the proof. □

Theorem 1.26 (Young's inequality with ϵ). *Let $a, b, \epsilon > 0$, $p, q \in (1, \infty)$, $\frac{1}{p} + \frac{1}{q} = 1$. Then*

$$ab \leq \epsilon a^p + C(\epsilon) b^q, \quad C(\epsilon) = \frac{1}{q(\epsilon p)^{\frac{q}{p}}}.$$

The equality holds if and only if $a^p = \frac{b^q}{(\epsilon p)^q}$.

Proof. Let

$$a_1 = (\epsilon p)^{\frac{1}{p}} a, \quad b_1 = \frac{b}{(\epsilon p)^{\frac{1}{p}}}.$$

We apply Young's inequality (1.22) for a_1 and b_1 and we get

$$ab \leq \frac{\epsilon p a^p}{p} + \frac{1}{(\epsilon p)^{\frac{q}{p}} q} b^q = \epsilon a^p + C(\epsilon) b^q.$$

The equality holds if and only if $a_1^p = b_1^q$. This completes the proof. □

Theorem 1.27 (Hölder's inequality). *Let $\xi_j, \eta_j \in \mathbf{C}$, $j \in \{1, \ldots, l\}$, $p, q > 1$, $\frac{1}{p} + \frac{1}{q} = 1$. Then*

$$\sum_{k=1}^{l} |\xi_k \eta_k| \leq \left(\sum_{k=1}^{l} |\xi_k|^p\right)^{\frac{1}{p}} \left(\sum_{k=1}^{l} |\eta_k|^q\right)^{\frac{1}{q}}. \tag{1.23}$$

The equality holds if and only if $|\xi_j|^p(\sum_{k=1}^{l} |\eta_k|^q) = |\eta_j|^q(\sum_{k=1}^{l} |\xi_k|^p)$ for any $j \in \{1, \ldots, l\}$.

Proof. If $\sum_{k=1}^{l}|\xi_k|^p = 0$ ($\sum_{k=1}^{l}|\eta_k|^q = 0$), then $\xi_k = 0$ ($\eta_k = 0$) for any $k \in \{1,\dots,l\}$ and the assertion is evident. Assume that $\sum_{k=1}^{l}|\xi_k|^p \neq 0$ and $\sum_{k=1}^{l}|\eta_k|^q \neq 0$. We set

$$a = \frac{|\xi_j|}{(\sum_{k=1}^{l}|\xi_k|^p)^{\frac{1}{p}}}, \quad b = \frac{|\eta_j|}{(\sum_{k=1}^{l}|\eta_k|^q)^{\frac{1}{q}}}$$

for some $j \in \{1,\dots,l\}$. We apply Young's inequality. We obtain

$$\frac{|\xi_j\eta_j|}{(\sum_{k=1}^{l}|\xi_k|^p)^{\frac{1}{p}}(\sum_{k=1}^{l}|\eta_k|^q)^{\frac{1}{q}}} \leq \frac{|\xi_j|^p}{p\sum_{k=1}^{l}|\xi_k|^p} + \frac{|\eta_j|^q}{q\sum_{k=1}^{l}|\eta_k|^q},$$

whereupon

$$\begin{aligned}\sum_{j=1}^{l}\frac{|\xi_j\eta_j|}{(\sum_{k=1}^{l}|\xi_k|^p)^{\frac{1}{p}}(\sum_{k=1}^{l}|\eta_k|^q)^{\frac{1}{q}}} &\leq \sum_{j=1}^{l}\frac{|\xi_j|^p}{p\sum_{k=1}^{l}|\xi_k|^p} + \sum_{j=1}^{l}\frac{|\eta_j|^q}{q\sum_{k=1}^{l}|\eta_k|^q}\\ &= \frac{\sum_{j=1}^{l}|\xi_j|^p}{p\sum_{k=1}^{l}|\xi_k|^p} + \frac{\sum_{j=1}^{l}|\eta_j|^q}{q\sum_{k=1}^{l}|\eta_k|^q} = \frac{1}{p} + \frac{1}{q} = 1.\end{aligned}$$

From the previous inequality we get the inequality (1.23). This completes the proof. □

Theorem 1.28 (Minkowski's inequality). *Let $\xi_k, \eta_k \in \mathbf{C}$, $k \in \{1,\dots,l\}$. Then for every $p \geq 1$ we have*

$$\left(\sum_{k=1}^{l}|\xi_k + \eta_k|^p\right)^{\frac{1}{p}} \leq \left(\sum_{k=1}^{l}|\xi_k|^p\right)^{\frac{1}{p}} + \left(\sum_{k=1}^{l}|\eta_k|^p\right)^{\frac{1}{p}}.$$

Proof. If $\sum_{k=1}^{l}|\xi_k + \eta_k|^p = 0$ the assertion is evident. Assume that $\sum_{k=1}^{l}|\xi_k + \eta_k|^p \neq 0$.

1. Let $p = 1$. Then

$$\sum_{k=1}^{l}|\xi_k + \eta_k| \leq \sum_{k=1}^{l}(|\xi_k| + |\eta_k|) = \sum_{k=1}^{l}|\xi_k| + \sum_{k=1}^{l}|\eta_k|.$$

2. Let $p > 1$. We choose q to be the conjugate of p. Then $q(p-1) = p$ and using Hölder's inequality, we obtain

$$\begin{aligned}\sum_{k=1}^{l}|\xi_k + \eta_k|^p &= \sum_{k=1}^{l}|\xi_k + \eta_k|^{p-1}|\xi_k + \eta_k| \leq \sum_{k=1}^{l}|\xi_k + \eta_k|^{p-1}|\xi_k| + \sum_{k=1}^{l}|\xi_k + \eta_k|^{p-1}|\eta_k|\\ &\leq \left(\sum_{k=1}^{l}|\xi_k + \eta_k|^{(p-1)q}\right)^{\frac{1}{q}}\left(\sum_{k=1}^{l}|\xi_k|^p\right)^{\frac{1}{p}}\\ &\quad + \left(\sum_{k=1}^{l}|\xi_k + \eta_k|^{(p-1)q}\right)^{\frac{1}{q}}\left(\sum_{k=1}^{l}|\eta_k|^p\right)^{\frac{1}{p}}\end{aligned}$$

$$= \left(\sum_{k=1}^{l} |\xi_k + \eta_k|^p \right)^{\frac{1}{q}} \left(\left(\sum_{k=1}^{l} |\xi_k|^p \right)^{\frac{1}{p}} + \left(\sum_{k=1}^{l} |\eta_k|^p \right)^{\frac{1}{p}} \right),$$

whereupon

$$\left(\sum_{k=1}^{l} |\xi_k + \eta_k|^p \right)^{1-\frac{1}{q}} \leq \left(\sum_{k=1}^{l} |\xi_k|^p \right)^{\frac{1}{p}} + \left(\sum_{k=1}^{l} |\eta_k|^p \right)^{\frac{1}{p}},$$

which completes the proof. □

1.4 Complete spaces

Let **M** be a metric space.

Definition 1.22. A sequence $\{x_n\}_{n\in\mathbf{N}}$ of elements of **M** is said to be a fundamental sequence or a Cauchy sequence if for any $\epsilon > 0$ there is an $N = N(\epsilon) > 0$ such that $d(x_n, x_m) < \epsilon$ whenever $m, n \geq N$.

Theorem 1.29. *Let $\{x_n\}_{n\in\mathbf{N}}$ be a sequence of elements of the metric space* **M** *that converges to x_0. Then it is a Cauchy sequence.*

Proof. Let $\epsilon > 0$ be arbitrarily chosen. Since $\{x_n\}_{n\in\mathbf{N}}$ converges to x_0, there is an $N = N(\epsilon) > 0$ such that

$$d(x_n, x_0) < \frac{\epsilon}{2} \quad \text{whenever} \quad n \geq N.$$

Hence,

$$d(x_n, x_m) \leq d(x_n, x_0) + d(x_0, x_m) < \frac{\epsilon}{2} + \frac{\epsilon}{2} = \epsilon \quad \text{whenever} \quad n, m \geq N.$$

This completes the proof. □

Definition 1.23. The metric space **M** is said to be complete if every its Cauchy sequence is convergent to a limit point in **M**.

Example 1.16. Consider the space $\mathbf{C}([a,b])$, the space of all continuous functions on $[a,b]$. We provide this space with the metric

$$d(f,g) = \max_{t\in[a,b]} |f(t) - g(t)|, \quad f, g \in \mathbf{C}([a,b]).$$

Let $\{f_n\}_{n\in\mathbf{N}}$ be a Cauchy sequence of elements of $\mathbf{C}([a,b])$, i. e., $d(f_n, f_m) \to 0$ as $n, m \to \infty$. Then the sequence $\{f_n\}_{n\in\mathbf{N}}$ satisfies the Cauchy condition of uniform convergence on $[a,b]$. If f is the limit of this sequence, then f is continuous on $[a,b]$, i. e., $f \in \mathbf{C}([a,b])$. Therefore $\mathbf{C}([a,b])$ is a complete metric space.

Example 1.17. Consider the space **m**. In this space we define a metric as follows:

$$d(x,y) = \sup_{i\in\mathbf{N}} |\xi_i - \eta_i|, \quad x = \{\xi_l\}_{l\in\mathbf{N}}, y = \{\eta_l\}_{l\in\mathbf{N}} \in \mathbf{m}.$$

Let $\{x^n\}_{n\in\mathbf{N}}$ be a Cauchy sequence in **m**, i. e., $d(x^n, x^m) \to 0$ as $n, m \to \infty$. Let also, $x^n = \{\xi_l^n\}_{l\in\mathbf{N}}$. We take $\epsilon > 0$ arbitrarily. Then there is an $N = N(\epsilon) > 0$ such that

$$\sup_{l\in\mathbf{N}} |\xi_l^n - \xi_l^m| < \epsilon \quad \text{for any} \quad m, n \geq N. \tag{1.24}$$

Since $x^n \in \mathbf{m}$ for any $n \in \mathbf{N}$, then for any $n \in \mathbf{N}$ there is a $K_n > 0$ such that

$$|\xi_l^n| \leq K_n \quad \text{for any} \quad l \in \mathbf{N}. \tag{1.25}$$

By (1.24) we get

$$|\xi_l^n - \xi_l^m| < \epsilon \quad \text{for any} \quad m, n \geq N \tag{1.26}$$

and for any $l \in \mathbf{N}$. Therefore for any $l \in \mathbf{N}$ the sequence $\{\xi_l^n\}_{n\in\mathbf{N}}$ satisfies the Cauchy condition. Hence, for any $l \in \mathbf{N}$ the sequence $\{\xi_l^n\}_{n\in\mathbf{N}}$ converges to ξ_l. Let $m \to \infty$ in (1.26). Then

$$|\xi_l^n - \xi_l| < \epsilon \quad \text{for any} \quad n \geq N \tag{1.27}$$

and for any $l \in \mathbf{N}$. Hence, by (1.25), we get

$$|\xi_l| = |\xi_l^N - \xi_l - \xi_l^N| \leq |\xi_l^N - \xi_l| + |\xi_l^N| < \epsilon + K_N$$

for any $l \in \mathbf{N}$. Consequently the sequence $\{\xi_l\}_{l\in\mathbf{N}}$ is a bounded sequence. Let $x = \{\xi_l\}_{l\in\mathbf{N}}$. Then $x \in \mathbf{m}$. From (1.27) we obtain

$$\sup_{l\in\mathbf{N}} |\xi_l^n - \xi_l| < \epsilon \quad \text{for any} \quad n \geq N.$$

Since $\epsilon > 0$ was arbitrarily chosen, we conclude that $x^n \to x$ as $n \to \infty$. Therefore **m** is a complete metric space.

Example 1.18. Consider the set **Q** of rational numbers, in which a metric is defined by

$$d(r_1, r_2) = |r_1 - r_2|, \quad r_1, r_2 \in \mathbf{Q}.$$

Consider the sequence $\{(1 + \frac{1}{n})^n\}_{n\in\mathbf{N}}$. This sequence is a Cauchy sequence. Its limit is e, which is not a rational number. Therefore **Q** is not a complete metric space.

Exercise 1.13. For $p \geq 1$, $p < \infty$, with $\mathbf{l}_p$ we denote the set of all sequences $x = \{\xi_l\}_{l\in\mathbf{N}}$, $\xi_l \in \mathbf{C}$, $l \in \mathbf{N}$, such that

$$\sum_{l=1}^{\infty} |\xi_l|^p < \infty.$$

In $\mathbf{l}_p$ we define a metric by

$$d(x,y) = \left(\sum_{l=1}^{\infty} |\xi_l - \eta_l|^p \right)^{\frac{1}{p}}, \quad x = \{\xi_l\}_{l \in \mathbf{N}}, y = \{\eta_l\}_{l \in \mathbf{N}}. \tag{1.28}$$

1. Prove that (1.28) satisfies all axioms for a metric.
2. Prove that $\mathbf{l}_p$ is a complete metric space.

Exercise 1.14. For $k \in \mathbf{N}_0$ with $\mathbf{C}^k([a,b])$ we will denote the set of all k-times continuously differentiable functions on $[a,b]$. In $\mathbf{C}^k([a,b])$ we define a metric by

$$d(f,g) = \sum_{l=0}^{k} \max_{x \in [a,b]} |f^{(l)}(x) - g^{(l)}(x)|, \quad f, g \in \mathbf{C}^k([a,b]). \tag{1.29}$$

1. Prove that (1.29) satisfies all axioms for a metric.
2. Prove that $\mathbf{C}^k([a,b])$ is a complete metric space.

Below we will give analogs of Cantor's lemma for contracting intervals.

Theorem 1.30. *Consider the sequence of closed balls*

$$B_{r_1}[a_1] \supset B_{r_2}[a_2] \supset \cdots \supset B_{r_n}[a_n] \supset \cdots$$

in the complete metric space $\mathbf{M}$. *If* $r_n \to 0$ *as* $n \to \infty$, *then these balls have a unique common point.*

Proof. Consider the sequence $\{a_l\}_{l \in \mathbf{N}}$ of the centers of the considered sequence of closed balls. Let $p \in \mathbf{N}$ be arbitrarily chosen. Because $B_{r_{n+p}}[a_{n+p}] \subset B_{r_n}[a_n]$, we have $a_{n+p} \in B_{r_n}[a_n]$, whereupon

$$d(a_{n+p}, a_n) \le r_n.$$

Using that $r_n \to 0$ as $n \to \infty$, from the previous inequality we get $d(a_{n+p}, a_n) \to 0$ as $n \to \infty$. Since $p \in \mathbf{N}$ was arbitrarily chosen, we conclude that the sequence $\{a_l\}_{l \in \mathbf{N}}$ is a Cauchy sequence in the complete metric space $\mathbf{M}$. Therefore it is convergent and let its limit is a. We have $a \in \mathbf{M}$. Let $k \in \mathbf{N}$ be arbitrarily chosen and fixed. Then

$$a_k, a_{k+1}, \ldots \in B_{r_k}[a_k].$$

Hence, using the fact that $B_{r_k}[a_k]$ is a closed set and $a_n \to a$, as $n \to \infty$, we obtain $a \in B_{r_k}[a_k]$. Because $k \in \mathbf{N}$ was arbitrarily chosen and fixed, we conclude that $a \in B_{r_l}[a_l]$ for any $l \in \mathbf{N}$. Now we assume that $b \in \mathbf{M}$ is a common point of the considered sequence of closed balls such that $d(a,b) = \delta > 0$. Then

$$\delta = d(a,b) \le d(a,a_n) + d(a_n,b) \le r_n + r_n = 2r_n \quad \text{for} \quad \text{any} \quad n \in \mathbf{N},$$

which is a contradiction because $r_n \to 0$ as $n \to \infty$. This completes the proof. □

Definition 1.24. Let **X** be a bounded set in a metric space **M**. The diameter of **X** is defined by

$$\operatorname{diam}(X) = \sup_{x,y\in\mathbf{X}} d(x,y).$$

Using this definition, we can generalize Theorem 1.30 in the following manner.

Theorem 1.31. *Let*

$$B_{r_1}[a_1] \supset B_{r_2}[a_2] \supset \cdots \supset B_{r_n}[a_n] \supset \cdots$$

be a sequence of closed balls in the complete metric space **M** *whose diameters tend to zero. Then these balls have a unique common point.*

Proof. Since the diameters of $B_{r_n}[a_n]$ tend to zero as $n \to \infty$, we have $r_n \to 0$ as $n \to \infty$. Hence, by Theorem 1.30, we go to the desired result. This completes the proof. □

Theorem 1.32. *If in a metric space* **M** *any sequence of closed balls*

$$B_{r_1}[a_1] \supset B_{r_2}[a_2] \supset \cdots \supset B_{r_n}[a_n] \supset \cdots$$

whose diameters tend to zero has a nonempty intersection, then the space **M** *is complete.*

Proof. Let $\{x_n\}_{n\in\mathbf{N}}$ be arbitrarily chosen Cauchy sequence in **M**. Let also, $\epsilon \in (0, \frac{1}{2})$ be arbitrarily chosen. Then there is an $N = N(\epsilon) > 0$ such that

$$d(x_{n_k+p}, x_{n_k}) < \epsilon^k$$

for any $n_k > N$ and for any $p \in \mathbf{N}$. Consider the sequence of closed balls $B_{\epsilon^{k-1}}[x_{n_k}]$ for $n_k > N$ and $k \in \mathbf{N}$. Let $k \in \mathbf{N}$ be arbitrarily chosen and fixed, and $y \in B_{\epsilon^k}[x_{n_{k+1}}]$ be arbitrarily chosen. Then $d(x_{n_{k+1}}, y) < \epsilon^k$ and

$$\begin{aligned} d(x_{n_k}, y) &\le d(x_{n_{k+1}}, x_{n_k}) + d(x_{n_{k+1}}, y) < \epsilon^k + \epsilon^k \\ &= 2\epsilon^k < \epsilon^{k-1}. \end{aligned}$$

Therefore $y \in B_{\epsilon^{k-1}}[x_{n_k}]$. Because $y \in B_{\epsilon^k}[x_{n_{k+1}}]$ was arbitrarily chosen and we see that it is an element of $B_{\epsilon^{k-1}}[x_{n_k}]$, we conclude that

$$B_{\epsilon^{k-1}}[x_{n_k}] \supset B_{\epsilon^k}[x_{n_{k+1}}].$$

Since $k \in \mathbf{N}$ was arbitrarily chosen, we obtain

$$B_1[x_{n_1}] \supset B_\epsilon[x_{n_2}] \supset \cdots \supset B_{\epsilon^{k-1}}[x_{n_k}] \supset \cdots.$$

By the assumptions, we see that there is an $x_0 \in B_{\epsilon^{k-1}}[x_{n_k}]$ for any $k \in \mathbf{N}$. We have $x_{n_k} \to x_0$ as $k \to \infty$. For $n, n_k > N$, we get

$$d(x_n, x_0) \le d(x_n, x_{n_k}) + d(x_{n_k}, x_0) < \epsilon^k + \epsilon^{k-1}.$$

Because $\epsilon \in (0, \frac{1}{2})$ was arbitrarily chosen, we conclude that $x_n \to x_0$ as $n \to \infty$. This completes the proof. □

Definition 1.25. A set $\mathbf{M}$ is said to be of the first category if it can be written as a countable union of nowhere dense sets. Otherwise, it is said to be of the second category.

Theorem 1.33. *A nonempty complete metric space $\mathbf{M}$ is a set of the second category.*

Proof. Assume that $\mathbf{M}$ is a set of the first category. Then

$$\mathbf{M} = \bigcup_{n\in\mathbf{N}} \mathbf{X}_n,$$

where $\mathbf{X}_n$, $n \in \mathbf{N}$, are nowhere dense sets in $\mathbf{M}$. Let $\epsilon \in (0,1)$ be arbitrarily chosen. Then there is a ball $B_\epsilon[a_1] \subset \mathbf{M}$ that does not contain any point of $\mathbf{X}_1$. There is a ball $B_{\epsilon^2}[a_2] \subset B_\epsilon[a_1]$ that does not contain any point of $\mathbf{X}_2$, and so on. In this way we obtain the sequence of closed balls

$$B_\epsilon[a_1] \supset B_{\epsilon^2}[a_2] \supset \cdots \supset B_{\epsilon^n}[a_n] \supset \cdots.$$

Because $\epsilon^n \to 0$ as $n \to \infty$ and $\mathbf{M}$ is a complete metric space, there is $a \in B_{\epsilon^k}[a_k]$ for any $k \in \mathbf{N}$. We have $a \in \mathbf{M}$. On the other hand, $a \notin \mathbf{X}_k$ for any $k \in \mathbf{N}$. Hence, $a \notin \mathbf{M}$, which is a contradiction. This completes the proof. □

Definition 1.26. Let $(\mathbf{P}, d_\mathbf{P})$ and $(\mathbf{M}, d_\mathbf{M})$ be metric spaces. If there is a one-to-one correspondence f between $\mathbf{P}$ and $\mathbf{M}$, $f(\mathbf{P}) = \mathbf{M}$, and

$$d_\mathbf{P}(x,y) = d_\mathbf{M}(f(x), f(y)) \quad \text{for} \quad \text{any} \quad x,y \in \mathbf{P},$$

then the spaces $\mathbf{P}$ and $\mathbf{M}$ is said to be isometric.

Theorem 1.34 (The completion of metric spaces). *Let $\mathbf{M}$ be a metric space that is noncomplete. Then there is a complete metric space $\mathbf{P}$ such that it has a subset $\mathbf{L}$ which is everywhere dense in $\mathbf{P}$ and isometric to $\mathbf{M}$.*

Proof. Consider all sequences of elements of $\mathbf{M}$ which are Cauchy sequences. We associate every two Cauchy sequences $\{x_n\}_{n\in\mathbf{N}}$ and $\{x'_n\}_{n\in\mathbf{N}}$ to the same class if $d(x_n, x'_n) \to 0$ as $n \to \infty$. We consider this class $\tilde{x}$ as an element of a new set $\mathbf{P}$. Let $\tilde{x}$, $\tilde{y} \in \mathbf{P}$ and $\{x_n\}_{n\in\mathbf{N}} \in \tilde{x}$, $\{y_n\}_{n\in\mathbf{N}} \in \tilde{y}$. We have

$$d(x_n, y_n) \le d(x_n, x_m) + d(x_m, y_n) \le d(x_n, x_m) + d(x_m, y_m) + d(y_m, y_n).$$

Hence,

$$d(x_n, y_n) - d(x_m, y_m) \le d(x_n, x_m) + d(y_m, y_n). \tag{1.30}$$

If the indices m and n are interchanged, then we get

$$d(x_m, y_m) - d(x_n, y_n) \le d(x_n, x_m) + d(y_n, y_m).$$

From the previous inequality and from (1.30), we obtain

$$|d(x_n, y_n) - d(x_m, y_m)| \leq d(x_n, x_m) + d(y_n, y_m). \tag{1.31}$$

Since the sequences $\{x_n\}_{n\in\mathbf{N}}$ and $\{y_n\}_{n\in\mathbf{N}}$ are fundamental sequences, from the inequality (1.31), we conclude that the sequence $\{d(x_n, y_n)\}_{n\in\mathbf{N}}$ is a Cauchy sequence in **R**. Therefore there exists $\lim_{n\to\infty} d(x_n, y_n)$. Using this, we can define a metric in the space **P** as follows:

$$d(\tilde{x}, \tilde{y}) = \lim_{n\to\infty} d(x_n, y_n). \tag{1.32}$$

We will show that $d(\tilde{x}, \tilde{y})$ does not depend on the choice of the representatives $\{x_n\}_{n\in\mathbf{N}}$ and $\{y_n\}_{n\in\mathbf{N}}$ of the classes $\tilde{x}$ and $\tilde{y}$, respectively. Let $\{x'_n\}_{n\in\mathbf{N}} \in \tilde{x}$ and $\{y'_n\}_{n\in\mathbf{N}} \in \tilde{y}$. Then

$$d(x_n, y_n) \leq d(x_n, x'_n) + d(x'_n, y_n) \leq d(x_n, x'_n) + d(x'_n, y'_n) + d(y'_n, y_n).$$

From this,

$$\lim_{n\to\infty} d(x_n, y_n) \leq \lim_{n\to\infty} d(x_n, x'_n) + \lim_{n\to\infty} d(x'_n, y'_n) + \lim_{n\to\infty} d(y'_n, y_n) = \lim_{n\to\infty} d(x'_n, y'_n),$$

i. e.,

$$\lim_{n\to\infty} d(x_n, y_n) \leq \lim_{n\to\infty} d(x'_n, y'_n). \tag{1.33}$$

Also,

$$d(x'_n, y'_n) \leq d(x'_n, x_n) + d(x_n, y'_n) \leq d(x'_n, x_n) + d(x_n, y_n) + d(y_n, y'_n).$$

Hence,

$$\lim_{n\to\infty} d(x'_n, y'_n) \leq \lim_{n\to\infty} d(x'_n, x_n) + \lim_{n\to\infty} d(x_n, y_n) + \lim_{n\to\infty} d(y_n, y'_n) = \lim_{n\to\infty} d(x_n, y_n).$$

From the previous inequality and from (1.33), we obtain

$$\lim_{n\to\infty} d(x_n, y_n) = \lim_{n\to\infty} d(x'_n, y'_n).$$

Now we will show that (1.32) satisfies all axioms for a metric. Suppose that $\tilde{x}, \tilde{y}, \tilde{z} \in \mathbf{P}$ and $\{x_n\}_{n\in\mathbf{N}} \in \tilde{x}$, $\{y_n\}_{n\in\mathbf{N}} \in \tilde{y}$ and $\{z_n\}_{n\in\mathbf{N}} \in \tilde{z}$.

1. Since $d(x_n, y_n) \geq 0$ for any $n \in \mathbf{N}$, we have

$$d(\tilde{x}, \tilde{y}) \geq 0.$$

2. We have $d(\tilde{x}, \tilde{y}) = 0$ if and only if $\lim_{n\to\infty} d(x_n, y_n) = 0$ if and only if $\{x_n\}_{n\in\mathbf{N}}$ and $\{y_n\}_{n\in\mathbf{N}}$ belong to the same class, which is possible if and only if $\tilde{x} = \tilde{y}$.

3. We have

$$d(\tilde{x},\tilde{y}) = \lim_{n\to\infty} d(x_n,y_n) = \lim_{n\to\infty} d(y_n,x_n) = d(\tilde{y},\tilde{x}).$$

4. We have

$$d(\tilde{x},\tilde{y}) = \lim_{n\to\infty} d(x_n,y_n) \le \lim_{n\to\infty} d(x_n,z_n) + \lim_{n\to\infty} d(z_n,y_n) = d(\tilde{x},\tilde{z}) + d(\tilde{z},\tilde{y}).$$

Now we will prove that $\mathbf{P}$ is a complete metric space. Suppose that $\{\tilde{x}^n\}_{n\in\mathbf{N}}$ is a Cauchy sequence in $\mathbf{P}$. Then $d(\tilde{x}^n,\tilde{x}^m) \to 0$ as $m,n\to\infty$. Let $\{x_k^n\}_{k\in\mathbf{N}} \in \tilde{x}^n$. Since $\{x_k^n\}_{k\in\mathbf{N}}$ is a Cauchy sequence, there exists $k_n \in \mathbf{N}$ such that

$$d(x_p^n, x_{k_n}^n) \le \frac{1}{n} \quad \text{for any} \quad p > k_n. \tag{1.34}$$

Now we consider the sequence $\{x_{k_l}^l\}_{l\in\mathbf{N}}$. We have

$$d(x_{k_n}^n, x_{k_m}^m) \le d(x_{k_n}^n, x_p^n) + d(x_p^n, x_{k_m}^m) \le d(x_{k_n}^n, x_p^n) + d(x_p^n, x_p^m) + d(x_p^m, x_{k_m}^m). \tag{1.35}$$

Let $\epsilon > 0$ be arbitrarily chosen and fixed. Since $\{\tilde{x}^n\}_{n\in\mathbf{N}}$ is a Cauchy sequence in $\mathbf{P}$, we have $d(\tilde{x}^m,\tilde{x}^n) \to 0$ as $m,n\to\infty$. Therefore there exists $n_0 \in \mathbf{N}$ such that

$$d(\tilde{x}^n,\tilde{x}^m) = \lim_{p\to\infty} d(x_p^n, x_p^m) < \frac{\epsilon}{2}$$

whenever $m,n \ge n_0$. Hence, there is $p_1 \in \mathbf{N}$ so that

$$d(x_p^n, x_p^m) < \frac{\epsilon}{2} \tag{1.36}$$

whenever $m,n \ge n_0$ and $p \ge p_1$. If there is a need we enlarge n_0 so that $\frac{1}{n_0} < \frac{\epsilon}{4}$. Then, for $m,n \ge n_0$, $p > \max\{k_n, k_m, p_1\}$, using (1.34), (1.35) and (1.36), we get $\frac{1}{n}, \frac{1}{m} < \frac{\epsilon}{4}$ and

$$d(x_{k_n}^n, x_{k_m}^m) \le \frac{1}{n} + \frac{\epsilon}{2} + \frac{1}{m} < \frac{\epsilon}{4} + \frac{\epsilon}{2} + \frac{\epsilon}{4} = \epsilon. \tag{1.37}$$

Therefore the sequence $\{x_{k_n}^n\}_{n\in\mathbf{N}}$ is a Cauchy sequence. We have

$$d(\tilde{x}^n,\tilde{x}) = \lim_{p\to\infty} d(x_p^n, x_{k_p}^p) \le \lim_{p\to\infty} d(x_p^n, x_{k_n}^n) + \lim_{p\to\infty} d(x_{k_n}^n, x_{k_p}^p) \le \frac{1}{n} + \lim_{p\to\infty} d(x_{k_n}^n, x_{k_p}^p).$$

Hence, by (1.37), for $n \ge n_0$, we get

$$d(\tilde{x}^n,\tilde{x}) \le \frac{1}{n_0} + \epsilon \le \frac{\epsilon}{4} + \epsilon.$$

Consequently the sequence $\{\tilde{x}^n\}_{n\in\mathbf{N}}$ converges to $\tilde{x} \in \mathbf{P}$. Therefore $\mathbf{P}$ is a complete metric space.

Now we consider the set **L** of all sequences of the form $\{x,\dots,x,\dots\}$. We see that if $\{x,\dots,x,\dots\} \in \tilde{x}$ and $\{y,\dots,y,\dots\} \in \tilde{y}$, then $d(x,y) = d(\tilde{x},\tilde{y})$. Note that there is an one-to-one correspondence f between **L** and **M**, defined by

$$f(x) = \{x,\dots,x,\dots\}, \quad x \in \mathbf{M}.$$

Also, **L** and **M** are isometric. Now we will prove that **L** is everywhere dense in **P**. Let $\tilde{x} \in \mathbf{P}$ be the class containing the sequence $\{x_n\}_{n\in\mathbf{N}}$. We choose $n \in \mathbf{N}$ such that $d(x_n, x_m) < \epsilon$ for any $m > n$. We construct the sequence $\{x_n,\dots,x_n,\dots\}$ and denote by $\tilde{x}_\epsilon$ the class containing this sequence. We have $\tilde{x}_\epsilon \in \mathbf{L}$. Also,

$$d(\tilde{x},\tilde{x}_\epsilon) = \lim_{m\to\infty} d(x_m, x_n) \le \epsilon.$$

Therefore **L** is everywhere dense in **P**.

Now we will prove that the space **P** is uniquely defined to within isometry. Let **Y** be another complete metric space in which **M** is everywhere dense. Then every $\tilde{y} \in \mathbf{Y}$ is the limit of some sequence $\{x_n\}_{n\in\mathbf{N}} \subseteq \mathbf{M}$. Since $\{x_n\}_{n\in\mathbf{N}}$ is a Cauchy sequence, it defines some element $\tilde{x} \in \mathbf{P}$. We associate the element $\tilde{x}$ with the element $\tilde{y}$. Let now $\tilde{z} \in \mathbf{P}$ be given and $\{z_n\}_{n\in\mathbf{N}}$ be a fundamental sequence in the class $\tilde{z}$. Since $\{z_n\}_{n\in\mathbf{N}}$ belongs to **Y**, it defines some element $\tilde{\tilde{z}} \in \mathbf{Y}$. Associating this element with $\tilde{z}$, we obtain an one-to-one correspondence between the elements of the spaces **P** and **Y**. Also, we have

$$d(\tilde{x},\tilde{z}) = \lim_{n\to\infty} d(x_n, z_n) = d(\tilde{y},\tilde{\tilde{z}}),$$

i. e., the correspondence between **P** and **Y** is isometric. □

Definition 1.27. The space **P**, defined by Theorem 1.34, is called the completion of the space **M**.

Example 1.19. Let $\mathbf{P}([a,b])$ be the space of all polynomials defined on $[a,b]$. We provide this space with a metric

$$d(p,q) = \max_{t\in[a,b]} |p(t) - q(t)|, \quad p,q \in \mathbf{P}([a,b]).$$

We see that $\mathbf{P}([a,b])$ is not a complete metric space. Since $\mathbf{P}([a,b])$ is everywhere dense in $\mathbf{C}([a,b])$ and $\mathbf{C}([a,b])$ is complete, we see that $\mathbf{C}([a,b])$ is the completion of $\mathbf{P}([a,b])$.

1.5 Normed spaces

It is well known that the notion of norm is of fundamental importance in discussing linear topological spaces. We shall begin with the definition of the semi-norm.

Definition 1.28. A real-valued function ν defined on a vector space **E** is called a *semi-norm* on **E**, if the following conditions are satisfied:

1. $\nu(x + y) \le \nu(x) + \nu(y)$ for any $x, y \in \mathbf{E}$, *sub-additivity*.
2. $\nu(\alpha x) = |\alpha|\nu(x)$ for any $x \in \mathbf{E}$, *homogeneity*.
3. $\nu(x) \ge 0$ for any $x \in \mathbf{E}$, *non-negativity*.

Theorem 1.35. *If* $\mathbf{E}$ *is a real vector space and* $\nu : \mathbf{E} \longmapsto \mathbf{R}$ *is a semi-norm, then*

$$\nu(x - y) \ge |\nu(x) - \nu(y)|,$$

for any $x, y \in \mathbf{E}$.

Proof. We have

$$\nu(x) \le \nu(x - y) + \nu(y)$$

for any $x, y \in \mathbf{E}$,

$$\nu(x) - \nu(y) \le \nu(x - y) \tag{1.38}$$

for any $x, y \in \mathbf{E}$. Since

$$\nu(x - y) = |-1|\nu(y - x) \ge \nu(y) - \nu(x)$$

for any $x, y \in \mathbf{E}$, we have

$$-(\nu(x) - \nu(y)) \le \nu(x) - \nu(y) \tag{1.39}$$

for any $x, y \in \mathbf{E}$. The inequalities (1.38) and (1.39) give the desired inequality. □

Definition 1.29. A normed space is an ordered pair $(\mathbf{E}, \|\cdot\|)$, where $\mathbf{E}$ is a vector space over $\mathbf{F}$ and $\|\cdot\|$ is a norm on $\mathbf{E}$, i. e., a function $\|\cdot\| : \mathbf{E} \longmapsto \mathbf{R}$ such that for any $x, y, z \in \mathbf{E}$ the following hold:
1. $\|x\| \ge 0$ *non-negativity*,
2. $\|x\| = 0$ iff $x = 0$ *separate points*,
3. $\|\lambda x\| = |\lambda|\|x\|$ for any $\lambda \in \mathbf{F}$ *homogeneity of the norm*,
4. $\|x + y\| \le \|x\| + \|y\|$ *triangle inequality*.

Note that the first condition follows from the other three. To see this, we take $x \in \mathbf{E}$ arbitrarily. Then

$$\begin{aligned} \|0\| &= \|x + (-x)\| \le \|x\| + \|-x\| = \|x\| + \|(-1)x\| \\ &= \|x\| + |-1|\|x\| = 2\|x\|. \end{aligned}$$

Hence, using the second condition, we see that $\|x\| \ge 0$.

Example 1.20. In $\mathbf{E}_n$ we define a norm as follows:

$$\|x\| = \left(\sum_{l=1}^{n} |x_l|^2\right)^{\frac{1}{2}}, \quad x_l \in \mathbf{F}, \quad l \in \{1,\ldots,n\}, \quad x = (x_1,\ldots,x_n). \tag{1.40}$$

We will check that (1.40) satisfies all axioms for a norm. Take $x, y \in \mathbf{F}$, $x = (x_1,\ldots,x_n)$, $y = (y_1,\ldots,y_n)$, arbitrarily. Then

1. $\|x\| \geq 0$.
2. $\|x\| = 0$ iff $(\sum_{l=1}^{n} |x_l|^2)^{\frac{1}{2}} = 0$ iff $x_l = 0$ for any $l \in \{1,\ldots,n\}$.
3. $$\|\lambda x\| = \left(\sum_{l=1}^{n} |\lambda x_l|^2\right)^{\frac{1}{2}} = \left(\sum_{l=1}^{n} |\lambda|^2 |x_l|^2\right)^{\frac{1}{2}} = |\lambda| \left(\sum_{l=1}^{n} |x_l|^2\right)^{\frac{1}{2}} = |\lambda| \|x\|$$

   for any $\lambda \in \mathbf{F}$.
4. Applying Minkowski's inequality, we get

$$\|x+y\| = \left(\sum_{l=1}^{n} |x_l + y_l|^2\right)^{\frac{1}{2}} \leq \left(\sum_{l=1}^{n} |x_l|^2\right)^{\frac{1}{2}} + \left(\sum_{l=1}^{n} |y_l|^2\right)^{\frac{1}{2}} = \|x\| + \|y\|.$$

Example 1.21. In the space $\mathbf{C}^k([a,b])$ we define a norm

$$\|f\| = \sum_{l=0}^{k} \max_{t\in[a,b]} |f^{(l)}(t)|, \quad f \in \mathbf{C}^k([a,b]). \tag{1.41}$$

We will check that (1.41) satisfies all axioms for a norm. Let $f, g \in \mathbf{C}^k([a,b])$ and $\lambda \in \mathbf{F}$ be chosen arbitrarily. Then

1. $\|f\| \geq 0$.
2. $$0 = \|f\| \quad \text{iff} \quad \sum_{l=0}^{k} \max_{t\in[a,b]} |f^{(l)}(t)| = 0 \quad \text{iff}$$
   $$\max_{t\in[a,b]} |f^{(l)}(t)| = 0 \quad \text{for} \quad \text{any} \quad l \in \{0,\ldots,k\} \quad \text{iff} \quad f \equiv 0 \quad \text{on} \quad [a,b].$$
3. $$\begin{aligned}\|\lambda f\| &= \sum_{l=0}^{k} \max_{t\in[a,b]} |(\lambda f)^{(l)}(t)| = \sum_{l=0}^{k} \max_{t\in[a,b]} |\lambda f^{(l)}(t)| \\ &= |\lambda| \sum_{l=0}^{k} \max_{t\in[a,b]} |f^{(l)}(t)| = |\lambda| \|f\|.\end{aligned}$$

4.

$$
\begin{aligned}
\|f+g\| &= \sum_{l=0}^{k} \max_{t\in[a,b]} \left|(f+g)^{(l)}(t)\right| = \sum_{l=0}^{k} \max_{t\in[a,b]} \left|f^{(l)}(t)+g^{(l)}(t)\right| \\
&\leq \sum_{l=0}^{k} \max_{t\in[a,b]} \left|f^{(l)}(t)\right| + \sum_{l=0}^{k} \max_{t\in[a,b]} \left|g^{(l)}(t)\right| = \|f\| + \|g\|.
\end{aligned}
$$

Example 1.22. With $\mathbf{l}_p$, $p \geq 1$, we denote the set of all sequences $x = \{x_l\}_{l\in\mathbf{N}}$ for which $\sum_{l=1}^{\infty} |x_l|^p < \infty$. Note that $\mathbf{l}_p$ is a vector space. In $\mathbf{l}_p$, $p \geq 1$, $p < \infty$, we define

$$
\|x\| = \left(\sum_{l=1}^{\infty} |x_l|^p\right)^{\frac{1}{p}}, \quad x = \{x_l\}_{l\in\mathbf{N}} \in \mathbf{l}_p. \tag{1.42}
$$

We will check if (1.42) satisfies all axioms for a norm. Let $x, y \in \mathbf{l}_p$, $x = \{x_l\}_{l\in\mathbf{N}}$, $y = \{y_l\}_{l\in\mathbf{N}}$, $\lambda \in \mathbf{F}$, be arbitrarily chosen. Then

1. $\|x\| \geq 0$.
2.

$$
0 = \|x\| \quad \text{iff} \quad \left(\sum_{l=1}^{\infty} |x_l|^p\right)^{\frac{1}{p}} = 0 \quad \text{iff}
$$

$$
\sum_{l=1}^{\infty} |x_l|^p = 0 \quad \text{iff} \quad x_l = 0 \quad \text{for} \quad \text{any} \quad l \in \mathbf{N} \quad \text{iff} \quad x = 0.
$$

3.

$$
\|\lambda x\| = \left(\sum_{l=1}^{\infty} |\lambda x_l|^p\right)^{\frac{1}{p}} = \left(\sum_{l=1}^{\infty} |\lambda|^p |x_l|^p\right)^{\frac{1}{p}} = |\lambda| \left(\sum_{l=1}^{\infty} |x_l|^p\right)^{\frac{1}{p}} = |\lambda| \|x\|.
$$

4. Since for any $m \in \mathbf{N}$, using Minkowski's inequality, we have

$$
\left(\sum_{l=1}^{m} |x_l + y_l|^p\right)^{\frac{1}{p}} \leq \left(\sum_{l=1}^{m} |x_l|^p\right)^{\frac{1}{p}} + \left(\sum_{l=1}^{m} |y_l|^p\right)^{\frac{1}{p}},
$$

we conclude that

$$
\left(\sum_{l=1}^{\infty} |x_l + y_l|^p\right)^{\frac{1}{p}} \leq \left(\sum_{l=1}^{\infty} |x_l|^p\right)^{\frac{1}{p}} + \left(\sum_{l=1}^{\infty} |y_l|^p\right)^{\frac{1}{p}}.
$$

Therefore

$$
\|x+y\| = \left(\sum_{l=1}^{\infty} |x_l + y_l|^p\right)^{\frac{1}{p}} \leq \left(\sum_{l=1}^{\infty} |x_l|^p\right)^{\frac{1}{p}} + \left(\sum_{l=1}^{\infty} |y_l|^p\right)^{\frac{1}{p}} = \|x\| + \|y\|.
$$

Exercise 1.15. Check that

$$\|f\| = |f(a)| + |f'(a)| + \max_{t\in[a,b]} |f''(t)|, \quad f \in \mathbf{C}^2([a,b]),$$

satisfies all axioms for a norm.

Example 1.23. In $\mathbf{C}^1([a,b])$ we define

$$\|f\| = \max_{t\in[a,b]} |f'(t)|. \tag{1.43}$$

Since

$$\|f\| = 0 \quad \text{iff} \quad \max_{t\in[a,b]} |f'(t)| = 0 \quad \text{iff}$$
$$f'(t) = 0 \quad \text{for} \quad \text{any} \quad t \in [a,b] \quad \text{iff} \quad f \equiv \text{const} \quad \text{on} \quad [a,b],$$

(1.43) does not satisfy the axioms for a norm.

Exercise 1.16. Check if

$$\|f\| = \max_{t\in[a,b]} |f'(t)| + |f(b) - f(a)|$$

satisfies all axioms for a norm in $\mathbf{C}^1([a,b])$.

Answer. No.

Note that in a normed space, a metric can be defined by

$$d(x,y) = \|x - y\|.$$

It is evident that the defined metric satisfies all axioms for a metric.

Below we will suppose that $\mathbf{E}$ is a normed space with a norm $\|\cdot\|$.

Lemma 1.1. *For every $x, y \in \mathbf{E}$ the inequality*

$$\big|\|x\| - \|y\|\big| \le \|x - y\|$$

holds.

Proof. We have

$$\|x\| = \|x - y + y\| \le \|x - y\| + \|y\|.$$

Therefore

$$\|x\| - \|y\| \le \|x - y\|.$$

If we interchange the places of x and y in the previous inequality, we get

$$\|y\| - \|x\| \le \|y - x\| = \|x - y\|.$$

This completes the proof. □

Definition 1.30.

1. An element $x_0 \in \mathbf{E}$ will be called a limit of a sequence $\{x_n\}_{n\in\mathbf{N}} \subset \mathbf{E}$, if
$$\|x_n - x_0\| \to 0 \quad \text{as} \quad n \to \infty.$$
We will write $x_n \to x_0$ as $n \to \infty$ or $\lim_{n\to\infty} x_n = x_0$.
2. For $r > 0$ the set
$$S_r(x_0) = \{x \in \mathbf{E} : \|x - x_0\| < r\} (S_r[x_0] = \{x \in \mathbf{E} : \|x - x_0\| \le r\})$$
will be called an open (closed) ball with a center x_0 and radius r. Sometimes, we will say that $S_r(x_0)$ is a neighborhood of x_0.
3. A set $\mathbf{M} \subset \mathbf{E}$ is said to be bounded, if there exists a positive constant c such that $\|x\| \le c$ for any $x \in \mathbf{M}$.

Theorem 1.36. *Every convergent sequence in* $\mathbf{E}$ *is a bounded sequence.*

Proof. Let $\{x_n\}_{n\in\mathbf{N}}$ be a convergent sequence in $\mathbf{E}$ to the element $x_0 \in \mathbf{E}$. Let also, $\epsilon > 0$ be arbitrarily chosen and fixed. Then there exists an $N = N(\epsilon) \in \mathbf{N}$ such that
$$\|x_n - x_0\| < \epsilon$$
for any $n > N$, $n \in \mathbf{N}$. Hence, by Lemma 1.1, we conclude that
$$\|x_n\| - \|x_0\| < \epsilon \quad \text{or} \quad \|x_n\| < \epsilon + \|x_0\|$$
for any $n > N$, $n \in \mathbf{N}$. Let
$$c = \max\{\|x_1\|, \dots, \|x_N\|, \epsilon + \|x_0\|\}.$$
Then
$$\|x_n\| \le c$$
for any $n \in \mathbf{N}$. This completes the proof. □

Theorem 1.37. *Let* $\{x_n\}_{n\in\mathbf{N}}$ *be a convergent sequence in* $\mathbf{E}$ *to the element* $x_0 \in \mathbf{E}$.

1. *For any* $r > 0$ *there is an* $N = N(r) \in \mathbf{N}$ *such that* $x_n \in S_r(x_0)$ *for any* $n > N$.
2. *Every subsequence* $\{x_{n_k}\}_{k\in\mathbf{N}}$ *of the sequence* $\{x_n\}_{n\in\mathbf{N}}$ *is convergent to* x_0.
3. *If* $\{\lambda_n\}_{n\in\mathbf{N}} \subseteq \mathbf{F}$ *and* $\lambda_n \to \lambda_0$ *as* $n \to \infty$, $\lambda_0 \in \mathbf{F}$, *then* $\lambda_n x_n \to \lambda_0 x_0$ *as* $n \to \infty$.
4. *If* $\{y_n\}_{n\in\mathbf{N}} \subseteq \mathbf{E}$ *and* $y_n \to y_0$ *as* $n \to \infty$, $y_0 \in \mathbf{E}$, *then* $x_n + y_n \to x_0 + y_0$ *as* $n \to \infty$.
5. $\|x_n\| \to \|x_0\|$ *as* $n \to \infty$.
6. x_0 *is unique.*

Proof.

1. Let $r > 0$ be arbitrarily chosen and fixed. Then there is an $N = N(r) \in \mathbf{N}$ such that
$$\|x_n - x_0\| \le r$$
for any $n > N$, $n \in \mathbf{N}$, i. e., $x_n \in S_r(x_0)$ for any $n > N$, $n \in \mathbf{N}$.
2. Let $\epsilon > 0$ be arbitrarily chosen and fixed. Then there exists an $N = N(\epsilon) \in \mathbf{N}$ such that
$$\|x_n - x_0\| < \epsilon \tag{1.44}$$
for any $n > N$, $n \in \mathbf{N}$. Also, there is a $K = K(\epsilon) \in \mathbf{N}$ such that $n_k > N$ for any $k > K$, $k \in \mathbf{N}$. Hence, by (1.44), we get
$$\|x_{n_k} - x_0\| < \epsilon$$
for any $k > K$, $k \in \mathbf{N}$.
3. Since $x_n \to x_0$ and $\lambda_n \to \lambda_0$ as $n \to \infty$, there exist positive constants c_1 and c such that
$$|\lambda_n| \le c_1 \quad \text{and} \quad \|x_n\| \le c$$
for any $n \in \mathbf{N}$.
   (a) Let $\lambda_0 = 0$. Then
$$\|\lambda_n x_n\| = |\lambda_n|\|x_n\| \le c|\lambda_n| \to 0 \quad \text{as} \quad n \to \infty.$$
   (b) Let $\lambda_0 \neq 0$. Then
$$\begin{aligned}\|\lambda_n x_n - \lambda_0 x_0\| &= \|\lambda_n x_n - \lambda_n x_0 + \lambda_n x_0 - \lambda_0 x_0\| \le \|\lambda_n x_n - \lambda_n x_0\| + \|\lambda_n x_0 - \lambda_0 x_0\| \\ &= |\lambda_n|\|x_n - x_0\| + |\lambda_n - \lambda_0|\|x_0\| \\ &\le c_1\|x_n - x_0\| + |\lambda_n - \lambda_0|\|x_0\| \to 0 \quad \text{as} \quad n \to \infty.\end{aligned}$$
4. We have
$$\|x_n + y_n - x_0 - y_0\| = \|(x_n - x_0) + (y_n - y_0)\| \le \|x_n - x_0\| + \|y_n - y_0\| \to 0 \quad \text{as} \quad n \to \infty.$$
5. By Lemma 1.1, we get
$$\big|\|x_n\| - \|x_0\|\big| \le \|x_n - x_0\| \to 0 \quad \text{as} \quad n \to \infty.$$
6. Assume that there exists $y_0 \in \mathbf{E}$ such that $x_n \to y_0$ as $n \to \infty$. Then
$$\|y_0 - x_0\| = \|y_0 - x_n + x_n - x_0\| \le \|x_n - y_0\| + \|x_n - x_0\| \to 0 \quad \text{as} \quad n \to \infty.$$
Therefore $x_0 = y_0$.

This completes the proof. □

Definition 1.31. A set $\mathbf{M} \subset \mathbf{E}$ will be called open, if for every $x_0 \in \mathbf{M}$ there exists $r_0 > 0$ such that $S_{r_0}(x_0) \subset \mathbf{M}$.

Theorem 1.38. *Let $A_1, \dots, A_l \subset \mathbf{E}$ be open sets. Then $\bigcap_{k=1}^{l} A_k$ is an open set in $\mathbf{E}$.*

Proof. Let $x \in \bigcap_{k=1}^{l} A_k$ be arbitrarily chosen. Then $x \in A_k$ for any $k \in \{1, \dots, l\}$. Since A_k, $k \in \{1, \dots, l\}$, are open sets in $\mathbf{E}$, there are $r_k > 0$ so that $S_{r_k}(x) \subset A_k$. Let $r = \min_{1 \le k \le l} r_k$. Then $S_r(x) \subset A_k$ for any $k \in \{1, \dots, l\}$. Therefore $S_r(x) \subset \bigcap_{k=1}^{l} A_k$. This completes the proof. □

Theorem 1.39. *Let $\{A_k\}_{k \in \mathbf{N}}$ be open sets in $\mathbf{E}$. Then $\bigcup_{k \in \mathbf{N}} A_k$ is an open set in $\mathbf{E}$.*

Proof. Let $x \in \bigcup_{k \in \mathbf{N}} A_k$ be arbitrarily chosen and fixed. Then there is a $k_0 \in \mathbf{N}$ such that $x \in A_{k_0}$. Since A_{k_0} is an open set in $\mathbf{E}$, there is an $r_0 > 0$ such that $S_{r_0}(x) \subset A_{k_0}$. From this, $S_{r_0}(x) \subset \bigcup_{k \in \mathbf{N}} A_k$. This completes the proof. □

Definition 1.32. A point $a \in \mathbf{E}$ will be called a limit point for a set $\mathbf{M} \subset \mathbf{E}$ if for any $r > 0$ there is $x \in S_r(a) \cap M$, $x \neq a$.

Theorem 1.40. *A point $a \in \mathbf{E}$ is a limit point for the set $\mathbf{M} \subset \mathbf{E}$ if and only if there is a sequence $\{x_n\}_{n \in \mathbf{N}} \subset \mathbf{M}$ that converges to a and $x_n \neq a$ for any $n \in \mathbf{N}$.*

Proof.

1. Let $a \in \mathbf{E}$ be a limit point for the set $\mathbf{M}$. Then for any $n \in \mathbf{N}$ there are $x_n \in S_{\frac{1}{n}}(a) \cap \mathbf{M}$, $x_n \neq a$. In this way we get a sequence $\{x_n\}_{n \in \mathbf{N}}$ such that
$$\|x_n - a\| \to 0 \quad \text{as} \quad n \to \infty, \qquad x_n \neq a,$$
i. e., $x_n \to a$ as $n \to \infty$ and $x_n \neq a$.
2. Let there is a sequence $\{x_n\}_{n \in \mathbf{N}} \subset \mathbf{M}$ such that $x_n \neq a$ for any $n \in \mathbf{N}$ and $x_n \to a$ as $n \to \infty$. Hence, for any $r > 0$ there is an $N = N(r) \in \mathbf{N}$ such that $x_n \in S_r(a)$ for any $n > N$ and $x_n \neq a$.

This completes the proof. □

Definition 1.33. A set $\mathbf{M} \subset \mathbf{E}$ is said to be closed if it contains all its limit points.

Theorem 1.41. *Let $\mathbf{A}_1, \dots, \mathbf{A}_l$ be closed sets in $\mathbf{E}$. Then $\bigcup_{k=1}^{l} \mathbf{A}_k$ is a closed set in $\mathbf{E}$.*

Proof. Let $a \in \mathbf{E}$ be a limit point for $\bigcup_{k=1}^{l} \mathbf{A}_k$. Then there exists a sequence $\{x_n\}_{n \in \mathbf{N}} \subset \bigcup_{k=1}^{l} \mathbf{A}_k$ such that $x_n \to a$ as $n \to \infty$. Hence, there is an $m \in \{1, \dots, l\}$ and a subsequence $\{x_{n_s}\}_{s \in \mathbf{N}}$ of the sequence $\{x_n\}_{n \in \mathbf{N}}$ such that $\{x_{n_s}\}_{s \in \mathbf{N}} \subset \mathbf{A}_m$. We see that $x_{n_s} \to a$ as $s \to \infty$ and $x_n \neq a$. Hence, by Theorem 1.40, it follows that a is a limit point for $\mathbf{A}_m$. Because $\mathbf{A}_m$ is a closed set in $\mathbf{E}$, we conclude that $a \in \mathbf{A}_m$. Therefore $a \in \bigcup_{k=1}^{l} \mathbf{A}_k$ and $\bigcup_{k=1}^{l} \mathbf{A}_k$ is a closed set in $\mathbf{E}$. This completes the proof. □

Theorem 1.42. *Let $\{\mathbf{A}_k\}_{k \in \mathbf{N}}$ be closed sets in $\mathbf{E}$. Then $\bigcap_{k \in \mathbf{N}} \mathbf{A}_k$ is a closed set in $\mathbf{E}$.*

Proof. Let $a \in \mathbf{E}$ be a limit point for $\bigcap_{k\in\mathbf{N}} \mathbf{A}_k$. Then there exists a sequence $\{x_n\}_{n\in\mathbf{N}} \subset \bigcap_{k\in\mathbf{N}} \mathbf{A}_k$ such that $x_n \to a$ as $n \to \infty$. Hence, $\{x_n\}_{n\in\mathbf{N}} \subset \mathbf{A}_k$, $x_n \to a$ as $n \to \infty$ for any $k \in \mathbf{N}$. Therefore a is a limit point of $\mathbf{A}_k$ for any $k \in \mathbf{N}$. Because $\mathbf{A}_k$, $k \in \mathbf{N}$, are closed sets in $\mathbf{E}$, we have $a \in \mathbf{A}_k$ for any $k \in \mathbf{N}$. Therefore $a \in \bigcap_{k\in\mathbf{N}} A_k$ and $\bigcap_{k\in\mathbf{N}} A_k$ is a closed set in $\mathbf{E}$. This completes the proof. □

Definition 1.34. Let $\mathbf{M} \subset \mathbf{E}$.
1. The set $\mathbf{M}$ together with all of its limit points is called the closure of $\mathbf{M}$. It will be denoted by $\overline{\mathbf{M}}$.
2. The set $\mathbf{E} \setminus \mathbf{M}$ will be called the completion of the set $\mathbf{M}$ to $\mathbf{E}$.
3. A point $x_0 \in \mathbf{E}$ will be called an interior point for the set $\mathbf{M}$, if there is an $r > 0$ such that $S_r(x_0) \subset \mathbf{M}$.
4. A point $x_0 \in \mathbf{E}$ will be called an exterior point of the set $\mathbf{M}$, if there is an $r > 0$ such that $S_r(x_0) \cap \mathbf{M} = \emptyset$.
5. A point $x_0 \in \mathbf{E}$ will be called a boundary point of the set $\mathbf{M}$, if for every $r > 0$ we have
$$S_r(x_0) \cap \mathbf{M} \neq \emptyset \quad \text{and} \quad S_r(x_0) \cap (\mathbf{E} \setminus \mathbf{M}) \neq \emptyset.$$
6. The set of all boundary points of the set $\mathbf{M}$ will be called the boundary of the set $\mathbf{M}$ and it will be denoted by $\partial\mathbf{M}$.

Remark 1.1. Note that we have the following possibilities.
$$\partial\mathbf{M} \subset \mathbf{M} \quad \text{or} \quad \partial\mathbf{M} \cap \mathbf{M} = \emptyset \quad \text{or} \quad \partial\mathbf{M} \cap \mathbf{M} \neq \partial\mathbf{M}.$$

Definition 1.35. Two norms $\|\cdot\|_1$ and $\|\cdot\|_2$ in $\mathbf{E}$ will be called equivalent, if there are positive constants c_1 and c_2 such that
$$c_1\|x\|_2 \leq \|x\|_1 \leq c_2\|x\|_2$$
for any $x \in \mathbf{E}$. We will write $\|\cdot\|_1 \sim \|\cdot\|_2$.

Theorem 1.43. *In every finite-dimensional vector space every norms are equivalent.*

Proof. Let $\mathbf{U}$ be a finite-dimensional vector space. With $\{\phi_l\}_{l=1}^m$ we will denote a basis in $\mathbf{U}$. Then every $x \in \mathbf{U}$ has the following representation:
$$x = \sum_{k=1}^{m} \xi_k \phi_k, \quad \xi_k \in \mathbf{F}, \quad k \in \{1, \ldots, m\}.$$
In $\mathbf{U}$ we define a norm
$$\|x\| = \left(\sum_{l=1}^{m} |\xi_l|^2\right)^{\frac{1}{2}} \quad \text{for} \quad x \in \mathbf{U}. \tag{1.45}$$

We take an arbitrary norm $\|\cdot\|_1$ in **U**. Let

$$c_2 = \left(\sum_{l=1}^{m} \|\phi_l\|_1^2\right)^{\frac{1}{2}}.$$

Then, for $x = \sum_{l=1}^{m} \xi_l \phi_l$, $\xi_l \in \mathbf{F}$, $l \in \{1, \dots, m\}$, we have

$$\begin{aligned}\|x\|_1 &= \left\|\sum_{l=1}^{m} \xi_l \phi_l\right\|_1 \leq \sum_{l=1}^{m} |\xi_l| \|\phi_l\|_1 \\ &\leq \left(\sum_{l=1}^{m} |\xi_l|^2\right)^{\frac{1}{2}} \left(\sum_{l=1}^{m} \|\phi_l\|_1^2\right)^{\frac{1}{2}} = c_2 \|x\|,\end{aligned}$$

i. e.,

$$\|x\|_1 \leq c_2 \|x\|. \tag{1.46}$$

On the other hand, by Lemma 1.1 and (1.46), we get

$$\big| \|x\|_1 - \|y\|_1 \big| \leq \|x - y\|_1 \leq c_2 \|x - y\|$$

for any $x, y \in \mathbf{U}$. Therefore the function $\|\cdot\|_1$ is a continuous function in $\mathbf{E}_m$. Hence, there exists

$$c_1 = \inf_{\|x\|=1} \|x\|_1.$$

Consequently

$$\left\| \frac{x}{\|x\|} \right\|_1 \geq c_1 \quad \text{or} \quad \|x\|_1 \geq c_1 \|x\|.$$

This completes the proof. □

Exercise 1.17. Prove that (1.45) satisfies all axioms for a norm.

Theorem 1.44. *Let* **L** *be a linear subspace of* **E** *which is a closed set in* **E**. *Then*

$$\|l\|_{\mathbf{E}/\mathbf{L}} = \inf_{x \in l} \|x\|, \quad l \in \mathbf{E}/\mathbf{L}, \tag{1.47}$$

is a norm in **E**/**L**.

Proof. Firstly, we will prove that every $l \in \mathbf{E}/\mathbf{L}$ is a closed set. Let $\{x_n\}_{n\in\mathbf{N}}$ be a sequence of elements of l such that $x_n \to x_0$ as $n \to \infty$. We fix $m \in \mathbf{N}$ and consider $x_m - x_n$ for $n \in \mathbf{N}$. We have $x_m - x_n \in \mathbf{L}$ for any $n \in \mathbf{N}$ and $x_m - x_n \to x_m - x_0$ as $n \to \infty$. Hence, $x_m - x_0 \in \mathbf{L}$. Because $x_m \in l$, we see that $x_0 \in l$.

Let $l, m \in \mathbf{E}/\mathbf{L}$ be arbitrarily chosen.

1. $\|l\|_{\mathbf{E}/\mathbf{L}} \geq 0$.
2. We will prove that $\|l\|_{\mathbf{E}/\mathbf{L}} = 0$ iff $l = \mathbf{L}$.
   (a) Let $\|l\|_{\mathbf{E}/\mathbf{L}} = 0$. Then there exists a sequence $\{x_n\}_{n\in\mathbf{N}}$ of elements of l such that $x_n \to 0$ as $n \to \infty$. Because l is a closed set in $\mathbf{L}$, we obtain $0 \in l$ and hence $l = \mathbf{L}$.
   (b) Let $l = \mathbf{L}$. Then $0 \in l$ and $\|l\|_{\mathbf{E}/\mathbf{L}} = 0$.
3. Let $\lambda \in \mathbf{F}$ be arbitrarily chosen. Then

$$\lambda l = \{\lambda x : x \in l\}$$

and

$$\|\lambda l\|_{\mathbf{E}/\mathbf{L}} = \inf_{x\in l} \|\lambda x\| = |\lambda| \inf_{x\in l} \|x\| = |\lambda| \|l\|_{\mathbf{E}/\mathbf{L}}.$$

4. We have

$$\begin{aligned} \|l + m\|_{\mathbf{E}/\mathbf{L}} &= \inf_{x\in l+m} \|x\| \leq \inf_{\substack{x=x_1+x_2 \\ x_1\in l, x_2\in m}} (\|x_1 + x_2\|) \\ &\leq \inf_{\substack{x=x_1+x_2 \\ x_1\in l, x_2\in m}} (\|x_1\| + \|x_2\|) \leq \inf_{x_1\in l} \|x_1\| + \inf_{x_2\in m} \|x_2\| \\ &= \|l\|_{\mathbf{E}/\mathbf{L}} + \|m\|_{\mathbf{E}/\mathbf{L}}. \end{aligned}$$

This completes the proof. □

Theorem 1.45. *Let $\mathbf{L}$ be a closed linear subspace of $\mathbf{E}$. Then the sequence $\{l_n\}_{n\in\mathbf{N}}$ of elements of $\mathbf{E}/\mathbf{L}$ is convergent to l if and only if there exists a sequence $\{x_n\}_{n\in\mathbf{N}}$ of elements $x_n \in l_n$ such that $x_n \to x$ as $n \to \infty$, $x \in l$.*

Proof.
1. Let $\{l_n\}_{n\in\mathbf{N}}$ be a sequence of elements of $\mathbf{E}/\mathbf{L}$ that converges to l. Then we get

$$\|l_n - l\|_{\mathbf{E}/\mathbf{L}} \to 0 \quad \text{as} \quad n \to \infty,$$

that is

$$\|l_n - l\|_{\mathbf{E}/\mathbf{L}} = \epsilon_n, \quad \epsilon_n \to 0 \quad \text{as} \quad n \to \infty.$$

Hence, there exist $y_n \in l_n$ and $x \in l$ such that

$$\|y_n - x\| < 2\epsilon_n.$$

Let $x_0 \in l$ be arbitrarily chosen. Then

$$\|y_n - x\| = \|(y_n - x + x_0) - x_0\| < 2\epsilon_n.$$

Since $x_0, x \in l$, we have $x - x_0 \in \mathbf{L}$. Therefore

$$x_n = y_n - x + x_0 \in l_n.$$

Consequently for every $x_0 \in l$ there exists a sequence $\{x_n\}_{n\in\mathbf{N}}$, $x_n \in l_n$, such that $x_n \to x_0$ as $n \to \infty$.

2. Let there exists a sequence $\{x_n\}_{n\in\mathbf{N}}$, $x_n \in l_n$, such that $x_n \to x_0$ as $n \to \infty$, $x_0 \in l$. Then, using (1.47),

$$\|l_n - l\| \leq \|x_n - x_0\| \to 0 \quad \text{as} \quad n \to \infty.$$

This completes the proof. □

Definition 1.36. Let $\mathbf{L}$ be a linear subspace of $\mathbf{E}$. We define a distance from $x \in \mathbf{E}$ to $\mathbf{L}$ as follows:

$$\operatorname{dist}(x, \mathbf{L}) = \inf_{y\in\mathbf{L}} \|x - y\|.$$

By Definition 1.36, we get

1. $\operatorname{dist}(x, \mathbf{L}) \geq 0$,
2. for any $y \in \mathbf{L}$ we have

$$\operatorname{dist}(x, \mathbf{L}) \leq \|x - y\|,$$

3. for any $\epsilon > 0$ there exists $y_\epsilon \in \mathbf{L}$ such that

$$\|x - y_\epsilon\| < \epsilon + \operatorname{dist}(x, \mathbf{L}).$$

Theorem 1.46. *Let* $\mathbf{L}$ *be a closed linear subspace of* $\mathbf{E}$. *If* $x \notin \mathbf{L}$, *then* $\operatorname{dist}(x, \mathbf{L}) > 0$.

Proof. Assume that $\operatorname{dist}(x, \mathbf{L}) = 0$. Then there exists a sequence $\{y_n\}_{n\in\mathbf{N}}$ of elements of $\mathbf{L}$ such that

$$\|y_n - x\| < \frac{1}{n}$$

for any $n \in \mathbf{N}$. Since $\mathbf{L}$ is closed, we see that $x \in \mathbf{L}$, which is a contradiction. This completes the proof. □

Theorem 1.47. *Let* $\mathbf{L}$ *be a finite-dimensional linear subspace of* $\mathbf{E}$. *Then for any* $x \in \mathbf{E}$ *there exists* $x^* \in \mathbf{L}$ *such that*

$$\operatorname{dist}(x, \mathbf{L}) = \|x - x^*\|.$$

Proof. Suppose that $\mathbf{L}$ is m-dimensional.

1. If $x \in \mathbf{L}$, then $\operatorname{dist}(x, \mathbf{L}) = 0$ and $x = x^*$.

2. Let $x \notin \mathbf{L}$. Then $d = \mathrm{dist}(x, L) > 0$. We take $\{\phi_l\}_{l=1}^m$ to be a basis in $\mathbf{L}$. Then any $x \in \mathbf{L}$ can be represented in the following way:

$$y = \sum_{l=1}^{m} y_l \phi_l, \quad y_l \in \mathbf{F}, \quad l \in \{1, \dots, m\}.$$

We define a norm in $\mathbf{L}$ in the following way:

$$\|y\|_c = \left(\sum_{l=1}^{m} |y_l|^2 \right)^{\frac{1}{2}} \quad \text{for} \quad y = \sum_{l=1}^{m} y_l \phi_l \in \mathbf{L}.$$

Because $\mathbf{L}$ is finite-dimensional, all norms in $\mathbf{L}$ are equivalent. For a norm $\|\cdot\|$ in $\mathbf{L}$ there exist positive constants α and β such that

$$\alpha \|z\|_c \leq \|z\| \leq \beta \|z\|_c$$

for any $z \in \mathbf{L}$. We take

$$r = \frac{d + 1 + \|x\|}{\alpha}.$$

Let $y \in \mathbf{L}$ be arbitrarily chosen. If $\|y\|_c > r$, then

$$\|x - y\| \geq \|y\| - \|x\| \geq \alpha \|y\|_c - \|x\| > \alpha r - \|x\| = d + 1.$$

Therefore d is achieved for $\|y\|_c \leq r$. Since $\|y\|_c \leq r$ is a closed and a bounded set in $\mathbf{L}$, and $\|x - y\|$ is a continuous function on it, there exists $x^* \in \mathbf{L}$ such that

$$\inf_{\|y\|_c \leq r} \|x - y\| = \|x - x^*\|.$$

This completes the proof. □

Definition 1.37. The normed space $\mathbf{E}$ will be called a strongly normed space if the equality

$$\|x + y\| = \|x\| + \|y\|$$

holds if and only if $y = \lambda x$, $\lambda > 0$, $y, x \in \mathbf{E}$.

Theorem 1.48. *Let $\mathbf{E}$ be a strongly normed space and $\mathbf{L}$ be a finite-dimensional linear subspace of $\mathbf{E}$. If for $x \in \mathbf{E}$ there exists $x^* \in \mathbf{L}$ such that*

$$\|x - x^*\| = \inf_{y \in \mathbf{L}} \|x - y\|,$$

then x^* is unique.

Proof. If $\operatorname{dist}(x, \mathbf{L}) = 0$, then $x = x^*$. Suppose that $d = \operatorname{dist}(x, L) > 0$. Assume that there are $x_1^*, x_2^* \in \mathbf{L}$ such that

$$d = \|x - x_1^*\| = \|x - x_2^*\|.$$

Then

$$\begin{aligned}\left\|x - \frac{x_1^* + x_2^*}{2}\right\| &= \left\|\frac{x - x_1^*}{2} + \frac{x - x_2^*}{2}\right\| \le \frac{1}{2}\|x - x_1^*\| + \frac{1}{2}\|x - x_2^*\| \\ &= \frac{1}{2}d + \frac{1}{2}d = d.\end{aligned}$$

Hence,

$$\begin{aligned}\|2x - (x_1^* + x_2^*)\| = 2d &= \|(x - x_1^*) + (x - x_2^*)\| \\ &= \|x - x_1^*\| + \|x - x_2^*\|.\end{aligned}$$

Since $\mathbf{E}$ is a strongly normed space, we see that there exists $\lambda > 0$ such that

$$x - x_1^* = \lambda(x - x_2^*).$$

If $\lambda \ne 1$, then

$$x = \frac{1}{1 - \lambda}(x_1^* - \lambda x_2^*) \in \mathbf{L},$$

which is a contradiction. Therefore $\lambda = 1$ and $x_1^* = x_2^*$. This completes the proof. □

Lemma 1.2 (Riesz's lemma). *Let* $\mathbf{L}$ *be a closed linear subspace of* $\mathbf{E}$ *and* $\mathbf{L} \ne \mathbf{E}$. *Then for any* $\epsilon \in (0, 1)$ *there exists* $z_\epsilon \notin \mathbf{L}$, $\|z_\epsilon\| = 1$, *such that*

$$\operatorname{dist}(z_\epsilon, \mathbf{L}) > 1 - \epsilon.$$

Proof. Since $\mathbf{L} \ne \mathbf{E}$, there exists $x \in \mathbf{E}$ and $x \notin \mathbf{L}$. Let $d = \inf_{y \in \mathbf{L}} \|x - y\|$. We have $d > 0$. Then for any $\epsilon \in (0, 1)$ there exists $y_\epsilon \in \mathbf{L}$ such that

$$d \le \|y_\epsilon - x\| < \frac{d}{1 - \epsilon}.$$

Let

$$z_\epsilon = \frac{y_\epsilon - x}{\|y_\epsilon - x\|}.$$

We have $\|z_\epsilon\| = 1$. If we suppose that $z_\epsilon \in \mathbf{L}$, then $y_\epsilon - x \in \mathbf{L}$. Hence, $x \in \mathbf{L}$, which is a contradiction. Therefore $z_\epsilon \notin \mathbf{L}$. For $y \in \mathbf{L}$ we have

$$\begin{aligned}\|z_\epsilon - y\| &= \left\|\frac{y_\epsilon - x}{\|y_\epsilon - x\|} - y\right\| = \frac{\|x - (y_\epsilon - y\|y_\epsilon - x\|)\|}{\|y_\epsilon - x\|} \\ &\ge \frac{d}{\|y_\epsilon - x\|} > 1 - \epsilon.\end{aligned}$$

Therefore $\operatorname{dist}(z_\epsilon, \mathbf{L}) > 1 - \epsilon$. This completes the proof. □

1.6 Banach spaces

Definition 1.38. A normed vector space $\mathbf{E}$ that is complete in the sense of convergence in norm is called a Banach space.

Example 1.24. The space $\mathbf{E}_n$ is a Banach space with a norm

$$\|x\| = \left(\sum_{l=1}^{n} \xi_l^2\right)^{\frac{1}{2}}, \quad x = (\xi_1, \ldots, \xi_n) \in \mathbf{E}_n.$$

Example 1.25. The space $\mathbf{C}([a,b])$ is a Banach space with a norm

$$\|f\| = \max_{a \le t \le b} |f(t)|.$$

Example 1.26. The vector space $\mathbf{l}_p$, $p \in (1, \infty)$, is a Banach space with a norm

$$\|x\| = \left(\sum_{l=1}^{\infty} |x_l|^p\right)^{\frac{1}{p}}.$$

Theorem 1.49. *Let $\mathbf{E}$ be a Banach space and $\mathbf{L}$ be its closed linear subspace. Then $\mathbf{E}/\mathbf{L}$ is a Banach space.*

Proof. Let $\{l_n\}_{n\in\mathbf{N}}$ be a Cauchy sequence in $\mathbf{E}/\mathbf{L}$. We take $x_n \in l_n$ so that

$$\|x_n - x_m\| \le 2\|l_n - l_m\|_{\mathbf{E}/\mathbf{L}}.$$

In this way we get a Cauchy sequence $\{x_n\}_{n\in\mathbf{N}}$ of elements of $\mathbf{E}$. Because $\mathbf{E}$ is a Banach space, the sequence $\{x_n\}_{n\in\mathbf{N}}$ is convergent to an element $x \in \mathbf{E}$. Let l be the class containing x. Hence, by Theorem 1.45, we conclude that the sequence $\{l_n\}_{n\in\mathbf{N}}$ is convergent to l. Therefore $\mathbf{E}/\mathbf{L}$ is a Banach space. This completes the proof. □

Definition 1.39. Let $x_1, x_2, \ldots, x_n, \ldots$ be elements of a Banach space $\mathbf{E}$. An expression of the form $\sum_{l=1}^{\infty} x_l$ is called a series, made up of the elements of the space $\mathbf{E}$. Let $s_n = \sum_{l=1}^{n} x_l$. If the sequence $\{s_n\}_{n\in\mathbf{N}}$ converges, then $\sum_{l=1}^{\infty} x_l$ is said to be a convergent series.

Theorem 1.50. *Let $a_n \in \mathbf{F}$, $n \in \mathbf{N}$, and $\sum_{l=1}^{\infty} a_l$ be a convergent series. Let also, $\mathbf{E}$ be a Banach space and $x_n \in \mathbf{E}$, $\|x_n\| \le |a_n|$, $n \in \mathbf{N}$. Then $\sum_{n=1}^{\infty} x_n$ is a convergent series.*

Proof. For any $n, p \in \mathbf{N}$, we have

$$\|s_{n+p} - s_n\| = \left\|\sum_{l=n+1}^{n+p} x_l\right\| \le \sum_{l=n+1}^{n+p} \|x_l\| \le \sum_{l=n+1}^{n+p} |a_l|.$$

Therefore $\{s_n\}_{n\in\mathbf{N}}$ is a Cauchy sequence in $\mathbf{E}$. Because $\mathbf{E}$ is a Banach space, we conclude that the sequence $\{s_n\}_{n\in\mathbf{N}}$ is convergent. This completes the proof. □

1.7 Inner product spaces

Definition 1.40. An inner product space is an ordered pair $(\mathbf{E}, (\cdot,\cdot))$, where $\mathbf{E}$ is a vector space over $\mathbf{F}$ and $(\cdot,\cdot)$ is an inner product, i. e., a map $(\cdot,\cdot) : \mathbf{E}\times\mathbf{E} \longmapsto \mathbf{F}$ that satisfies the following axioms.

1. *Positive definiteness.*

$$(x,x) \geq 0 \quad \text{for} \quad \text{any} \quad x \in \mathbf{E},$$

   and

$$(x,x) = 0 \quad \text{iff} \quad x = 0.$$

2. *Linearity in the first argument.*

$$(\lambda x, y) = \lambda(x,y) \quad \text{and} \quad (x+y,z) = (x,z) + (y,z)$$

   for any $x,y,z \in \mathbf{E}$ and for any $\lambda \in \mathbf{F}$.

3. *Conjugate symmetry.*

$$(x,y) = \overline{(y,x)}$$

   for any $x,y \in \mathbf{E}$.

When it is clear from the context what inner product is used, $(\cdot,\cdot)$ is omitted and one just writes $\mathbf{E}$ for an inner product space.

Below we suppose that $\mathbf{E}$ is an inner product space.

By Definition 1.40 we get the following:

1. $(x,x) \in \mathbf{R}$ for any $x \in \mathbf{E}$ because

$$(x,x) = \overline{(x,x)}.$$

2. For any $\lambda \in \mathbf{R}$ and for any $x,y \in \mathbf{E}$ we have

$$(x,\lambda y) = \overline{(\lambda y, x)} = \overline{\lambda(y,x)} = \overline{\lambda}(x,y).$$

3. For any $x,y,z \in \mathbf{E}$ we have

$$(x,y+z) = \overline{(y+z,x)} = \overline{(y,x)+(z,x)} = \overline{(y,x)} + \overline{(z,x)} = (x,y) + (x,z).$$

4. For any $x,y \in \mathbf{E}$ we have

$$\begin{aligned}(x+y,x+y) &= (x,x+y) + (y,x+y) = (x,x) + (x,y) + (y,x) + (y,y)\\ &= (x,x) + (x,y) + \overline{(x,y)} + (y,y) = (x,x) + 2\operatorname{Re}(x,y) + (y,y).\end{aligned}$$

Example 1.27. In $\mathbf{E}_n$ we define the inner product as follows:

$$(x,y) = \sum_{l=1}^{n} x_l \overline{y_l}, \quad x = (x_1,\dots,x_n), \quad y = (y_1,\dots,y_n) \in \mathbf{E}_n. \tag{1.48}$$

We will check that (1.48) satisfies all axioms for inner product.

1. $$(x,x) = \sum_{l=1}^{n} x_l \overline{x_l} = \sum_{l=1}^{n} |x_l|^2 \geq 0$$

   for any $x \in \mathbf{E}_n$. Also,

   $$(x,x) = 0 \quad \text{iff} \quad \sum_{l=1}^{n} |x_l|^2 = 0 \quad \text{iff} \quad x_l = 0$$

   for any $l \in \{1,\dots,n\}$.
2. Let $\lambda \in \mathbf{F}$ and $x, y, z \in \mathbf{E}_n$,

   $$x = (x_1,\dots,x_n), \quad y = (y_1,\dots,y_n), \quad z = (z_1,\dots,z_n).$$

   Then

   $$(\lambda x, y) = \sum_{l=1}^{n} (\lambda x_l)\overline{y_l} = \lambda \sum_{l=1}^{n} x_l \overline{y_l} = \lambda (x,y).$$

   Also,

   $$(x+y,z) = \sum_{l=1}^{n} (x_l + y_l)\overline{z_l} = \sum_{l=1}^{n} x_l \overline{z_l} + \sum_{l=1}^{n} y_l \overline{z_l} = (x,z) + (y,z).$$
3. For $x, y \in \mathbf{E}_n$,

   $$x = (x_1,\dots,x_n), \quad y = (y_1,\dots,y_n),$$

   we have

   $$(x,y) = \sum_{l=1}^{n} x_l \overline{y_l} = \sum_{l=1}^{n} \overline{\overline{x_l} y_l} = \overline{\sum_{l=1}^{n} y_l \overline{x_l}} = \overline{(y,x)}.$$

Exercise 1.18. In $\mathbf{l}_2$ we define an inner product as follows:

$$(x,y) = \sum_{l=1}^{\infty} x_l \overline{y_l}, \quad x = \{x_l\}_{l \in \mathbf{N}}, \quad y = \{y_l\}_{l \in \mathbf{N}} \in \mathbf{l}_2. \tag{1.49}$$

Prove that (1.49) satisfies all axioms for an inner product.

In $\mathbf{E}$ we define a norm as follows:

$$\|x\| = \sqrt{(x,x)}, \quad x \in \mathbf{E}. \tag{1.50}$$

Theorem 1.51. *For any $x, y \in \mathbf{E}$ we have*

$$|(x,y)| \le \|x\|\|y\|.$$

Proof. If $x = 0$ and $y = 0$ the assertion is evident. Suppose that $x \neq 0$ or $y \neq 0$. Without loss of generality we assume that $y \neq 0$. By the axioms for an inner product we have

$$\begin{aligned} 0 &\le (x + \lambda y, x + \lambda y) = (x,x) + 2\operatorname{Re}(x, \lambda y) + (\lambda y, \lambda y) \\ &= \|x\|^2 + 2\operatorname{Re}(\overline{\lambda}(x,y)) + (\lambda y, \lambda y) = \|x\|^2 + 2\operatorname{Re}(\overline{\lambda}(x,y)) + |\lambda|^2\|y\|^2 \end{aligned} \tag{1.51}$$

for any $\lambda \in \mathbf{F}$. In particular, for $\lambda = -\frac{(x,y)}{\|y\|^2}$, we have

$$\begin{aligned} 0 &\le \|x\|^2 + 2\operatorname{Re}\left(-\frac{\overline{(x,y)}(x,y)}{\|y\|^2}\right) + \frac{|\overline{(x,y)}|^2}{\|y\|^4}\|y\|^2 \\ &= \|x\|^2 - 2\frac{|\overline{(x,y)}|^2}{\|y\|^2} + \frac{|\overline{(x,y)}|^2}{\|y\|^2} = \|x\|^2 - \frac{|\overline{(x,y)}|^2}{\|y\|^2}, \end{aligned}$$

whereupon we get the desired result. This completes the proof. □

Exercise 1.19. Prove that (1.50) satisfies all axioms for a norm.

Theorem 1.52. *Let $\mathbf{V}$ be a normed vector space over $\mathbf{C}$ with a norm $\|\cdot\|$ that satisfies*

$$\|x + y\|^2 + \|x - y\|^2 = 2(\|x\|^2 + \|y\|^2)$$

for any $x, y \in \mathbf{V}$. Then

$$(x,y) = \frac{1}{4}(\|x + y\|^2 - \|x - y\|^2) + \frac{i}{4}(\|x + iy\|^2 - \|x - iy\|^2) \tag{1.52}$$

is an inner product in $\mathbf{V}$.

Proof. We will prove that (1.52) satisfies all axioms for an inner product. Let $x, y, z \in \mathbf{V}$ and $\lambda \in \mathbf{C}$.

1.

$$(x,x) = \frac{1}{4}(\|2x\|^2) + \frac{i}{4}(\|x + ix\|^2 - \|x - ix\|^2) = \|x\|^2 \ge 0.$$

$(x,x) = 0$ iff $\|x\|^2 = 0$ iff $x = 0$.

2.

$$(x + z, y) = \frac{1}{4}(\|x + y + z\|^2 - \|x + z - y\|^2) + \frac{i}{4}(\|x + z + iy\|^2 - \|x + z - iy\|^2),$$

$$
\begin{aligned}
(x,y)+(z,y) &= \frac{1}{4}(\|x+y\|^2-\|x-y\|^2)+\frac{i}{4}(\|x+iy\|^2-\|x-iy\|^2)\\
&\quad+\frac{1}{4}(\|z+y\|^2-\|z-y\|^2)+\frac{i}{4}(\|z+iy\|^2-\|z-iy\|^2)\\
&= \frac{1}{4}(\|x+y\|^2+\|z+y\|^2-\|x-y\|^2-\|z-y\|^2)\\
&\quad+\frac{i}{4}(\|x+iy\|^2+\|z+iy\|^2-\|x-iy\|^2-\|z-iy\|^2)\\
&= \frac{1}{4}\left(\frac{1}{2}\|x+z+2y\|^2+\frac{1}{2}\|x-z\|^2-\frac{1}{2}\|x+z-2y\|^2-\frac{1}{2}\|x-z\|^2\right)\\
&\quad+\frac{i}{4}\left(\frac{1}{2}\|x+z+2iy\|^2+\frac{1}{2}\|x-z\|^2-\frac{1}{2}\|x+z-2iy\|^2-\frac{1}{2}\|x-z\|^2\right)\\
&= \frac{1}{8}(\|x+z+2y\|^2-\|x+z-2y\|^2)\\
&\quad+\frac{i}{8}(\|x+z+2iy\|^2-\|x+z-2iy\|^2)\\
&= \frac{1}{8}(\|x+z+y+y\|^2-\|x+z-y-y\|^2)\\
&\quad+\frac{i}{8}(\|x+z+iy+iy\|^2-\|x+z-iy-iy\|^2)\\
&= \frac{1}{8}(\|x+z+y+y\|^2+\|x+z\|^2-\|x+z\|^2-\|x+z-y-y\|^2)\\
&\quad+\frac{i}{8}(\|x+z+iy+iy\|^2+\|x+z\|^2-\|x+z\|^2-\|x+z-iy-iy\|^2)\\
&= \frac{1}{8}(2\|x+z+y\|^2+2\|y\|^2-2\|x+z-y\|^2-2\|y\|^2)\\
&\quad+\frac{i}{8}(2\|x+z+iy\|^2+2\|iy\|^2-2\|x+z-iy\|^2-2\|iy\|^2)\\
&= \frac{1}{4}(\|x+z+y\|^2-\|x+z-y\|^2)+\frac{i}{4}(\|x+z+iy\|^2-\|x+z-iy\|^2).
\end{aligned}
$$

Therefore

$$(x+z,y)=(x,y)+(z,y).$$

(a) Let $\lambda \in \mathbf{N}$.

i. $\lambda = 2$. Then

$$(2x,y)=(x+x,y)=(x,y)+(x,y)=2(x,y).$$

ii. Assume that

$$(\lambda x,y)=\lambda(x,y)$$

for some $\lambda \in \mathbf{N}$.

iii. We will prove that

$$((\lambda+1)x,y)=(\lambda+1)(x,y).$$

In fact,

$$((\lambda+1)x,y)=(\lambda x+x,y)=(\lambda x,y)+(x,y)=\lambda(x,y)+(x,y)=(\lambda+1)(x,y).$$

Therefore for any $\lambda\in\mathbf{N}$ we have

$$(\lambda x,y)=\lambda(x,y).$$

(b) Let $\lambda\in\mathbf{N}$ and $\frac{1}{\lambda}x=z$. Hence, $x=\lambda z$ and

$$(\lambda z,y)=\lambda(z,y).$$

Therefore

$$(x,y)=\lambda\left(\frac{1}{\lambda}x,y\right)$$

or

$$\left(\frac{1}{\lambda}x,y\right)=\frac{1}{\lambda}(x,y).$$

(c) We have

$$(0,y)=\frac{1}{4}(\|y\|^2-\|-y\|^2)+\frac{i}{4}(\|iy\|^2-\|-iy\|^2)=0.$$

On the other hand,

$$(0,y)=(x+(-x),y)=(x,y)+(-x,y).$$

Therefore

$$(-x,y)=-(x,y).$$

(d) Let $\lambda\in-\mathbf{N}=\{-1,-2,\dots\}$. Then $-\lambda\in\mathbf{N}$ and

$$(-\lambda x,y)=-(\lambda x,y)=-\lambda(x,y),$$

whereupon

$$(\lambda x,y)=\lambda(x,y).$$

(e) Let $\lambda = \frac{p}{q}$, where $p, q \in \mathbf{N}$. We have

$$(\lambda x, y) = \left(\frac{p}{q}x, y\right) = p\left(\frac{1}{q}x, y\right) = \frac{p}{q}(x, y) = \lambda(x, y).$$

(f) Let $\lambda = -\frac{p}{q}$, $p, q \in \mathbf{N}$. Then

$$(\lambda x, y) = \left(-\frac{p}{q}x, y\right) = -\left(\frac{p}{q}x, y\right) = -\frac{p}{q}(x, y) = \lambda(x, y).$$

(g) Let $\lambda \in \mathbf{R}$. Then there exists a sequence $\{\lambda_n\}_{n\in\mathbf{N}}$ such that $\lambda_n \in \mathbf{Q}$ for any $n \in \mathbf{N}$ and $\lambda_n \to \lambda$ as $n \to \infty$. We have

$$(\lambda_n x, y) = \lambda_n(x, y). \tag{1.53}$$

Also,

$$\lambda_n(x, y) \to \lambda(x, y) \quad \text{as} \quad n \to \infty. \tag{1.54}$$

On the other hand,

$$\begin{aligned}(\lambda_n x, y) &= \frac{1}{4}(\|\lambda_n x + y\|^2 - \|\lambda_n x - y\|^2)\\ &\quad + \frac{i}{4}(\|\lambda_n x + iy\|^2 - \|\lambda_n x - iy\|^2)\\ &\to \frac{1}{4}(\|\lambda x + y\|^2 - \|\lambda x - y\|^2)\\ &\quad + \frac{i}{4}(\|\lambda x + iy\|^2 - \|\lambda x - iy\|^2) = (\lambda x, y).\end{aligned}$$

Hence, by (1.53) and (1.54), we obtain

$$(\lambda x, y) = \lambda(x, y).$$

(h) We have

$$\begin{aligned}(ix, y) &= \frac{1}{4}(\|ix + y\|^2 - \|ix - y\|^2) + \frac{i}{4}(\|ix + iy\|^2 - \|ix - iy\|^2)\\ &= \frac{1}{4}(\|x - iy\|^2 - \|x + iy\|^2) + \frac{i}{4}(\|x + y\|^2 - \|x - y\|^2)\\ &= i\left(\frac{1}{4}(\|x + y\|^2 - \|x - y\|^2) + \frac{1}{4i}(\|x - iy\|^2 - \|x + iy\|^2)\right)\\ &= i\left(\frac{1}{4}(\|x + y\|^2 - \|x - y\|^2) + \frac{i}{4}(\|x + iy\|^2 - \|x - iy\|^2)\right) = i(x, y).\end{aligned}$$

(i) Let $\lambda = a + ib$, $a, b \in \mathbf{R}$. Then

$$\begin{aligned}(\lambda x, y) &= (ax + ibx, y) = (ax, y) + (ibx, y) = a(x, y) + i(bx, y) = a(x, y) + ib(x, y)\\ &= (a + ib)(x, y)\\ &= \lambda(x, y).\end{aligned}$$

3. We have

$$
\begin{aligned}
(x,y) &= \frac{1}{4}(\|x+y\|^2 - \|x-y\|^2) + \frac{i}{4}(\|x+iy\|^2 - \|x-iy\|^2) \\
&= \frac{1}{4}(\|y+x\|^2 - \|y-x\|^2) - \frac{i}{4}(\|y+ix\|^2 - \|y-ix\|^2) = \overline{(y,x)}.
\end{aligned}
$$

This completes the proof. □

Exercise 1.20. Let **E** be a normed vector space over **R** with a norm $\|\cdot\|$ such that

$$\|x+y\|^2 + \|x-y\|^2 = 2(\|x\|^2 + \|y\|^2).$$

Prove that

$$(x,y) = \frac{1}{2}(\|x+y\|^2 - \|x-y\|^2)$$

is an inner product in **E**.

Exercise 1.21. Let **E** be a normed vector space over **R** with a norm $\|\cdot\|$. Prove that

$$\|z-x\|^2 + \|z-y\|^2 = \frac{1}{2}\|x-y\|^2 + 2\left\|z - \frac{x+y}{2}\right\|^2$$

for any $x, y, z \in \mathbf{E}$.

Theorem 1.53. *The inner product in* **E** *is a continuous function with respect to norm convergence.*

Proof. Let $\{x_n\}_{n\in\mathbf{N}}$ and $\{y_n\}_{n\in\mathbf{N}}$ be two sequences in **E** such that

$$\|x_n - x\| \to 0 \quad \text{and} \quad \|y_n - y\| \to 0 \quad \text{as} \quad n \to \infty$$

for $x, y \in \mathbf{E}$. Note that there exists a positive constant M such that

$$\|x_n\| \le M \quad \text{and} \quad \|y_n\| \le M$$

for any $n \in \mathbf{N}$. Hence, by Theorem 1.51, we get

$$
\begin{aligned}
|(x_n,y_n) - (x,y)| &= |(x_n,y_n) - (x_n,y) + (x_n,y) - (x,y)| \\
&= |(x_n,y_n - y) + (x_n - x,y)| \le |(x_n,y_n - y)| + |(x_n - x,y)| \\
&\le \|x_n\|\|y_n - y\| + \|x_n - x\|\|y\| \le M\|y_n - y\| + \|y\|\|x_n - x\| \to 0
\end{aligned}
$$

as $n \to \infty$. This completes the proof. □

Theorem 1.54. *For any* $x, y \in \mathbf{E}$ *we have*

$$\|x+y\|^2 + \|x-y\|^2 = 2(\|x\|^2 + \|y\|^2).$$

Proof. We have

$$\|x+y\|^2 = \|x\|^2 + 2\,\mathrm{Re}((x,y)) + \|y\|^2,$$
$$\|x-y\|^2 = \|x\|^2 - 2\,\mathrm{Re}((x,y)) + \|y\|^2,$$

whereupon we get the desired result. This completes the proof. □

Definition 1.41.

1. Two elements $x, y \in \mathbf{E}$ will be called orthogonal if $(x,y) = 0$. We will write $x \perp y$.
2. An element $x \in \mathbf{E}$ will be called orthogonal to a linear subspace $\mathbf{L} \subset \mathbf{E}$ if it is orthogonal to every element of $\mathbf{L}$. We will write $x \perp \mathbf{L}$.
3. A system $\{e_l\}_{l\in\mathbf{N}}$ of elements of the space $\mathbf{E}$ is called orthonormal, if

$$(e_l, e_m) = \begin{cases} 1 & \text{if } l = m, \\ 0 & \text{if } l \neq m. \end{cases}$$

Theorem 1.55 (Schmidt orthogonalization system). *Any system of linearly independent elements $\{h_l\}_{l\in\mathbf{N}}$ can be converted into an orthonormal system.*

Proof. Because $\{h_l\}_{l\in\mathbf{N}}$ is a linearly independent system of elements, we have $h_l \neq 0$ for any $l \in \mathbf{N}$. We set

$$g_1 = \frac{h_1}{\|h_1\|}.$$

For $k = 2, \ldots$, we take

$$\widetilde{g}_k = h_k - c_1^k g_1 - \cdots - c_{k-1}^k g_{k-1},$$

where $g_m = \frac{\widetilde{g}_m}{\|\widetilde{g}_m\|}$ for $m \in \{1, \ldots, k-1\}$, c_l^k, $l \in \{1, \ldots, k-1\}$, are chosen so that

$$(\widetilde{g}_k, g_l) = 0 \quad \text{for} \quad l \in \{1, \ldots, k-1\}.$$

We have

$$0 = (\widetilde{g}_k, g_l) = (h_k, g_l) - c_l^k,$$

whereupon

$$c_l^k = (h_k, g_l), \quad l \in \{1, \ldots, k-1\}.$$

Then

$$g_k = \frac{\widetilde{g}_k}{\|\widetilde{g}_k\|}, \quad k \in \mathbf{N}.$$

This completes the proof. □

1.8 Hilbert spaces

The most important function spaces in modern physics and modern analysis, are known as Hilbert spaces. We give some important results on these spaces here.

Definition 1.42 (Hilbert space). An inner product space **H** will be called a Hilbert space if it is complete in the sense of the norm

$$\|\cdot\| = \sqrt{(\cdot,\cdot)}.$$

Example 1.28. The space $\mathbf{l}_2$ is a Hilbert space.

Below we will denote by **H** a Hilbert space.

Definition 1.43. A set **K** in a space **V** will be called convex if

$$\lambda x_1 + (1-\lambda)x_2 \in \mathbf{K}$$

for any $x_1, x_2 \in \mathbf{K}$ and for any $\lambda \in [0,1]$.

Theorem 1.56. *Let* **M** *be a closed convex subset of the Hilbert space* **H** *and* $x \notin \mathbf{M}$. *Then there exists a unique element* $y \in \mathbf{M}$ *such that*

$$\operatorname{dist}(x,\mathbf{M}) = |x-y\|.$$

Proof. By Theorem 1.46 we have $d = distd(x,\mathbf{M}) > 0$. For every $n \in \mathbf{N}$ there exists $y_n \in \mathbf{M}$ for which

$$d \le \|x-y_n\| < d + \frac{1}{n}. \tag{1.55}$$

We have

$$2\|x-y_n\|^2 + 2\|x-y_m\|^2 = \|y_n-y_m\|^2 + |2x-y_n-y_m\|^2, \quad n,m \in \mathbb{N}. \tag{1.56}$$

Since **M** is a convex subset of **H** we have

$$\frac{y_n+y_m}{2} \in \mathbf{M}.$$

Therefore

$$\|2x-y_n-y_m\|^2 = 4\left\|x-\frac{y_n+y_m}{2}\right\|^2 \ge 4d^2, \quad n,m \in \mathbb{N}.$$

Then, using (1.55) and (1.56), we obtain

$$\begin{aligned}\|y_n-y_m\|^2 &= 2\|x-y_n\|^2 + 2\|x-y_m\|^2 - 4\left\|x-\frac{y_n+y_m}{2}\right\|^2\\ &< 2\left(d+\frac{1}{n}\right)^2 + 2\left(d+\frac{1}{m}\right)^2 - 4d^2 \to 0 \quad \text{as} \quad m,n\to\infty.\end{aligned}$$

Therefore the sequence $\{y_n\}_{n\in\mathbf{N}}$ is a fundamental sequence in **M**. Since **M** is a closed subset of the Hilbert space **H**, we conclude that $\{y_n\}_{n\in\mathbf{N}}$ is convergent to $y \in \mathbf{M}$. Hence, by (1.55), we obtain

$$d = \|x - y\|.$$

Now we suppose that there are two elements $y_1, y_2 \in \mathbf{M}$ such that

$$d = \|x - y_1\| \quad \text{and} \quad d = \|x - y_2\|.$$

Since **M** is a convex subset of the Hilbert space **H**, we have

$$\frac{y_1 + y_2}{2} \in \mathbf{M} \quad \text{and} \quad \left\| x - \frac{y_1 + y_2}{2} \right\| \ge d.$$

Then

$$4d^2 = 2\|x - y_1\|^2 + 2\|x - y_2\|^2 = 4\left\| x - \frac{y_1 + y_2}{2} \right\|^2 + \|y_1 - y_2\|^2 \ge 4d^2 + \|y_1 - y_2\|^2,$$

whereupon

$$\|y_1 - y_2\| = 0.$$

Therefore $y_1 = y_2$. This completes the proof. □

Corollary 1.8. *Let* **M** *be a closed linear subspace of the Hilbert space* **H** *and* $x \in \mathbf{H} \setminus \mathbf{M}$. *Then there exists a unique* $y \in \mathbf{M}$ *such that*

$$\operatorname{dist}(x, M) = \|x - y\|.$$

Proof. Because **M** is a closed linear subspace of the Hilbert space **H**, it is a closed convex subset of the Hilbert space **H**. Hence, by Theorem 1.56, the desired result follows. □

Theorem 1.57. *Let* **M** *be a closed linear subspace of the Hilbert space* **H**, $x \in \mathbf{H} \setminus \mathbf{M}$ *and* $\|x - y\| = \operatorname{dist}(x, \mathbf{M})$ *for some* $y \in \mathbf{M}$. *Then* $x - y \perp \mathbf{M}$.

Proof. Let $\lambda \in \mathbf{F}$ be arbitrarily chosen and fixed. Then, for any $z \in \mathbf{M}$, $z \neq 0$, we have

$$\|x - y + \lambda z\| \ge \|x - y\|,$$

whereupon

$$\begin{aligned}
\|x - y\|^2 &\le \|x - y + \lambda z\|^2 \\
&= \|x - y\|^2 + 2\operatorname{Re}((x - y, \lambda z)) + \|\lambda z\|^2 = \|x - y\|^2 + 2\operatorname{Re}((x - y, \lambda z)) + |\lambda|^2\|z\|^2.
\end{aligned}$$

In particular, when $\lambda = -\frac{(x-y,z)}{\|z\|^2}$, we get

$$0 \leq -2\operatorname{Re}\left(\frac{\overline{(x-y,z)}(x-y,z)}{\|z\|^2}\right) + \frac{|(x-y,z)|^2}{\|z\|^2} = -\frac{|(x-y,z)|^2}{\|z\|^2}.$$

Therefore $(x-y,z) = 0$. Because $z \in \mathbf{M}$ was arbitrarily chosen, we conclude that $x - y \perp \mathbf{M}$. This completes the proof. □

Corollary 1.9. *Let* $\mathbf{M}$ *be a closed linear subspace of the Hilbert space* $\mathbf{H}$. *Then every* $x \in \mathbf{H}$ *can be represented in an unique way in the following manner:*

$$x = y + z,$$

where $y \in \mathbf{M}$, $z \in \mathbf{H} \setminus \mathbf{M}$ *and* $z \perp \mathbf{M}$.

Proof. If $x \in \mathbf{M}$, then $y = x$ and $z = 0$. Let $x \notin \mathbf{M}$. By Corollary 1.8 it follows that there exists a unique $y \in \mathbf{M}$ such that

$$\|x - y\| = \operatorname{dist}(x, M).$$

Let $z = x - y$. By Theorem 1.57 we have $z \perp \mathbf{M}$. This completes the proof. □

Definition 1.44. Let $\mathbf{M}$ be a closed linear subspace of the Hilbert space $\mathbf{H}$. The set of all elements of $\mathbf{H}$ which are orthogonal of $\mathbf{M}$ is called the orthogonal complement of $\mathbf{M}$. We will denote it by $\mathbf{M}^\perp$.

Theorem 1.58. *Let* $\mathbf{M}$ *be a closed linear subspace of the Hilbert space* $\mathbf{H}$. *Then* $\mathbf{M}^\perp$ *is a closed linear subspace of* $\mathbf{H}$.

Proof. Let $z_1, z_2 \in \mathbf{M}^\perp$ and $y \in \mathbf{M}$ be arbitrarily chosen. Then

$$(\lambda_1 z_1 + \lambda_2 z_2, y) = \lambda_1 (z_1, y) + \lambda_2 (z_2, y) = 0,$$

i. e., $\lambda_1 z_1 + \lambda_2 z_2 \in \mathbf{M}^\perp$. Note that $0 \in \mathbf{M}^\perp$. Hence, by Theorem 1.3, we conclude that $\mathbf{M}^\perp$ is a linear subspace of $\mathbf{H}$. Let $\{z_n\}_{n \in \mathbf{N}}$ be a convergent sequence of elements of $\mathbf{M}$ that converges to $z \in \mathbf{H}$. Then, using Theorem 1.53,

$$0 = (z_n, y) \to (z, y) \quad \text{as} \quad n \to \infty.$$

Therefore $(z, y) = 0$ and $z \in \mathbf{M}^\perp$. Consequently $\mathbf{M}^\perp$ is a closed subset of $\mathbf{H}$. This completes the proof. □

Theorem 1.59. *Let* $\mathbf{M}$ *be a linear subspace of the Hilbert space* $\mathbf{H}$. *Then* $\mathbf{M}$ *is dense in* $\mathbf{H}$ *if and only if* $\mathbf{M}^\perp = \{0\}$.

Proof.

1. Let $\mathbf{M}$ be dense in $\mathbf{H}$. Then $\overline{\mathbf{M}} = \mathbf{H}$. Assume that there exists $z_0 \in \mathbf{H}$ such that $z_0 \perp \mathbf{M}$. Let $y \in \mathbf{H}$ be arbitrarily chosen and fixed. Let $\{y_n\}_{n\in\mathbf{N}}$ be a sequence of elements of $\mathbf{M}$ such that $y_n \to y$ as $n \to \infty$. Then, using Theorem 1.53,
$$0 = (y_n, z_0) \to (y, z_0) \quad \text{as} \quad n \to \infty.$$
Therefore $(y, z_0) = 0$. Because $y \in \mathbf{H}$ was arbitrarily chosen, we can take $y = z_0$. Then $\|z_0\| = 0$. Therefore $z_0 = 0$.
2. Let $\mathbf{M}^\perp = \{0\}$. Assume that $\mathbf{M}$ is not dense in $\mathbf{H}$. Then there exists $x \in \mathbf{H}\backslash\overline{\mathbf{M}}$. Hence, by Corollary 1.9, we conclude that $x = y+z$, $y \in \overline{\mathbf{M}}$ and $z = \overline{\mathbf{M}}^\perp$. Note that $\overline{\mathbf{M}}^\perp = \mathbf{M}^\perp$. In fact, let $z_1 \in \overline{\mathbf{M}}^\perp$ be arbitrarily chosen and fixed. Then
$$(z_1, z_2) = 0 \tag{1.57}$$
for any $z_2 \in \overline{\mathbf{M}}$. In particular, (1.57) holds for any $z_2 \in \mathbf{M}$. Therefore $z_1 \in \mathbf{M}^\perp$. Since $z_1 \in \overline{\mathbf{M}}^\perp$ was arbitrarily chosen and for it we see that it is an element of $\mathbf{M}^\perp$, we conclude that
$$\overline{\mathbf{M}}^\perp \subseteq \mathbf{M}^\perp. \tag{1.58}$$
Let $z_3 \in \mathbf{M}^\perp$ be arbitrarily chosen and fixed. We take $z_4 \in \overline{\mathbf{M}}$ arbitrarily. Then there exists a sequence $\{z_4^n\}_{n\in\mathbf{N}}$ of elements of $\mathbf{M}$ such that
$$z_4^n \to z_4 \quad \text{as} \quad n \to \infty.$$
Since $z_3 \in \mathbf{M}^\perp$ and $z_4^n \in \mathbf{M}$, $n \in \mathbb{N}$, we have
$$(z_3, z_4^n) = 0 \quad \text{for} \quad \text{any} \quad n \in \mathbb{N}.$$
Hence, by Theorem 1.53, we get
$$0 = (z_3, z_4^n) \to (z_3, z_4),$$
i. e., $(z_3, z_4) = 0$. Because $z_4 \in \overline{\mathbf{M}}$ was arbitrarily chosen, we obtain $z_3 \perp \overline{\mathbf{M}}$. Therefore $z_3 \in \overline{\mathbf{M}}^\perp$. Since $z_3 \in \mathbf{M}^\perp$ was arbitrarily chosen and for it we see that it is an element of $\overline{\mathbf{M}}^\perp$, we obtain
$$\mathbf{M}^\perp \subseteq \overline{\mathbf{M}}^\perp.$$
From the previous relation and from (1.58), we get $\mathbf{M}^\perp = \overline{\mathbf{M}}^\perp$. Then $z = 0$ and $y = x \in \overline{\mathbf{M}}$, which is a contradiction. This completes the proof. □

Let $\mathbf{M}$ be a linear subspace of the Hilbert space $\mathbf{H}$, spanned by the orthonormal system $\{\phi_k\}_{k\in\mathbf{N}}$ and $x \in \mathbf{M}$.

Definition 1.45.
1. The series

$$\sum_{k=1}^{\infty} \alpha_k \phi_k, \qquad \alpha_k \in \mathbf{F},$$

   will be called a Fourier's series with respect to the orthonormal system $\{\phi_k\}_{k\in\mathbf{N}}$.
2. The numbers

$$\alpha_k = (x, \phi_k), \qquad k \in \mathbf{N},$$

   will be called the Fourier coefficients for the element x with respect to the orthonormal system $\{\phi_k\}_{k\in\mathbf{N}}$.

Let $\mathbf{M}_n = \operatorname{Span}\{\phi_1, \dots, \phi_n\}$ and $u_n = \sum_{k=1}^n c_k \phi_k \in \mathbf{M}_n$ for some $c_k \in \mathbf{F}$. Then

$$\left(x - \sum_{k=1}^n c_k \phi_k, x - \sum_{k=1}^n c_k \phi_k \right) = (x, x) - 2 \sum_{k=1}^n \operatorname{Re}(c_k (x, \phi_k)) + \sum_{k=1}^n |c_k|^2.$$

Let

$$\Delta_n = \|x - u_n\|.$$

Then

$$\Delta_n^2 = \|x\|^2 - 2 \sum_{k=1}^n \operatorname{Re}(\overline{c_k}(x, \phi_k)) + \sum_{k=1}^n |c_k|^2.$$

Hence,

$$\Delta_n^2 = \|x\|^2 - \sum_{k=1}^n \alpha_k \overline{c_k} - \sum_{k=1}^n \overline{\alpha_k} c_k + \sum_{k=1}^n |c_k|^2.$$

Note that

$$\begin{aligned} |\alpha_k - c_k|^2 &= (\alpha_k - c_k)(\overline{\alpha_k} - \overline{c_k}) \\ &= \alpha_k \overline{\alpha_k} - \alpha_k \overline{c_k} - c_k \overline{\alpha_k} + c_k \overline{c_k} = |\alpha_k|^2 - \alpha_k \overline{c_k} - c_k \overline{\alpha_k} + |c_k|^2, \end{aligned}$$

whereupon

$$-\alpha_k \overline{c_k} - c_k \overline{\alpha_k} + |c_k|^2 = |\alpha_k - c_k|^2 - |\alpha_k|^2.$$

Therefore

$$\Delta_n^2 = \|x\|^2 + \sum_{k=1}^n |\alpha_k - c_k|^2 - \sum_{k=1}^n |\alpha_k|^2.$$

Let

$$d_n = \operatorname{dist}(x, \mathbf{M}_n).$$

Then

$$\begin{aligned} d_n^2 &= \inf_{u_n \in \mathbf{M}_n} \|x - u_n\|^2 = \inf_{c_1,\ldots,c_n} \Delta_n^2 \\ &= \Delta_n^2|_{\alpha_1 = c_1,\ldots,\alpha_n = c_n} = \|x\|^2 - \sum_{k=1}^{n} |\alpha_k|^2 \end{aligned}$$

and

$$d_n = \left\| x - \sum_{k=1}^{n} \alpha_k \phi_k \right\|.$$

Theorem 1.60. *We have*

$$\sum_{k=1}^{n} |\alpha_k|^2 \le \|x\|^2 \quad \textit{for} \quad \textit{any} \quad n \in \mathbf{N}. \tag{1.59}$$

and $d_n < d_m$ for any $m < n$.

Proof. Since $d_n^2 \ge 0$ for any $n \in \mathbf{N}$, we get the inequality (1.59). For $m < n$ we have

$$d_m^2 = \|x\|^2 - \sum_{k=1}^{m} |\alpha_k|^2 > \|x\|^2 - \sum_{k=1}^{n} |\alpha_k|^2 = d_n^2,$$

whereupon $d_m > d_n$. This completes the proof. □

Theorem 1.61 (Bessel's inequality). *We have*

$$\sum_{k=1}^{\infty} |c_k|^2 \le \|x\|^2. \tag{1.60}$$

Proof. Since the inequality (1.59) holds for any $n \in \mathbf{N}$, we conclude that the series $\sum_{k=1}^{\infty} |\alpha_k|^2$ is convergent and (1.60) holds. This completes the proof. □

Definition 1.46 (Complete orthonormal system). An orthonormal system $\{\phi_k\}_{k \in \mathbf{N}}$ in the Hilbert space $\mathbf{H}$ is called a complete orthonormal system if for any $x \in \mathbf{H}$ the representation

$$x = \sum_{k=1}^{\infty} c_k \phi_k, \quad c_k = (x, \phi_k), \quad k \in \mathbf{N},$$

holds.

Theorem 1.62. *An orthonormal system* $\{\phi_k\}_{k\in\mathbf{N}}$ *in the Hilbert space* $\mathbf{H}$ *is complete if and only if*

$$\|x\|^2 = \sum_{k=1}^{\infty} |c_k|^2, \quad c_k = (x, \phi_k), \quad k \in \mathbf{N},$$

for any $x \in \mathbf{H}$.

Proof.

1. Let $\{\phi_k\}_{k\in\mathbf{N}}$ is a complete orthonormal system. Let also, $x \in \mathbf{H}$ be arbitrarily chosen and fixed. Then

$$x = \sum_{k=1}^{\infty} c_k\phi_k.$$

   Hence,

$$\|x\|^2 = (x, x) = \sum_{k=1}^{\infty} |c_k|^2.$$

2. Let $\|x\|^2 = \sum_{k=1}^{\infty} |c_k|^2$, $c_k = (x, \phi_k)$, $k \in \mathbf{N}$. Hence, using

$$\left\| x - \sum_{k=1}^{n} c_k\phi_k \right\|^2 = \|x\|^2 - \sum_{k=1}^{n} |c_k|^2$$

   for any $n \in \mathbf{N}$, we see that $x = \sum_{k=1}^{\infty} c_k\phi_k$. This completes the proof. □

Theorem 1.63. *Let* $\{\phi_k\}_{k\in\mathbf{N}}$ *be an orthonormal system in the Hilbert space* $\mathbf{H}$ *and* $\mathbf{M} = \operatorname{Span}\{\{\phi_k\}_{k\in\mathbf{N}}\}$. *Then* $\{\phi_k\}_{k\in\mathbf{N}}$ *is a complete orthonormal system if and only if* $\overline{\mathbf{M}} = \mathbf{H}$.

Proof.

1. Let $\{\phi_k\}_{k\in\mathbf{N}}$ is a complete orthonormal system in $\mathbf{H}$. Assume that $\overline{\mathbf{M}} \neq \mathbf{H}$. Then there exists $x_0 \in \mathbf{H} \setminus (\overline{\mathbf{M}})$, $x_0 \neq 0$. Hence, $(x_0, \phi_k) = 0$ for any $k \in \mathbf{N}$. Since $\{\phi_k\}_{k\in\mathbf{N}}$ is a complete orthonormal system, we have

$$x_0 = \sum_{k=1}^{\infty} (x_0, \phi_k)\phi_k = 0.$$

   This is a contradiction. Therefore $\overline{\mathbf{M}} = \mathbf{H}$.
2. Let $\overline{\mathbf{M}} = \mathbf{H}$ and $x \in \mathbf{H}$ be arbitrarily chosen and fixed. Then for any $\epsilon > 0$ there exists $x_\epsilon \in \mathbf{M}$ such that

$$\|x - x_\epsilon\| < \epsilon.$$

   Hence, for any $\epsilon > 0$ there exists $N = N(\epsilon) \in \mathbf{N}$ such that

$$x_\epsilon = \sum_{k=1}^{N} c_k\phi_k, \quad c_k = (x, \phi_k), \quad k \in \{1, \dots, N\}.$$

Therefore

$$\left\| x - \sum_{k=1}^{N} c_k \phi_k \right\| = \|x - x_\epsilon\| < \epsilon.$$

By Theorem 1.60, we obtain

$$\left\| x - \sum_{k=1}^{n} c_k \phi_k \right\| \le \left\| x - \sum_{k=1}^{N} c_k \phi_k \right\| = \|x - x_\epsilon\| < \epsilon$$

for any $n > N$. Therefore

$$x = \sum_{k=1}^{\infty} c_k \phi_k.$$

Since $x \in \mathbf{H}$ was arbitrarily chosen and fixed, we conclude that $\{\phi_k\}_{k\in\mathbf{N}}$ is a complete orthonormal system in $\mathbf{H}$. This completes the proof. □

1.9 Separable spaces

Definition 1.47 (Separable space). A vector space $\mathbf{X}$ is said to be separable if there is a sequence $\{x_n\}_{n\in\mathbf{N}}$ of elements of $\mathbf{X}$ such that for any $x \in \mathbf{X}$ there exists a subsequence $\{x_{n_k}\}_{k\in\mathbf{N}}$ of the sequence $\{x_n\}_{n\in\mathbf{N}}$ that converges to x. In other words, a vector space X is said to be separable if it contains a countable everywhere dense set.

Example 1.29. Consider $\mathbf{E}_n$. Let

$$\mathbf{E}_n^0 = \{(x_1, \ldots, x_n) : x_k \in \mathbf{Q}, k \in \{1, \ldots, n\}\}.$$

Then $\mathbf{E}_n^0$ is a countable everywhere dense set in $\mathbf{E}_n$. Consequently $\mathbf{E}_n$ is a separable space.

Example 1.30. Consider the space $\mathbf{C}([a,b])$. By $\widetilde{C}$ we will denote its linear subspace of all polynomials with rational coefficients. Then $\widetilde{C}$ is a countable everywhere dense set of the space $\mathbf{C}([a,b])$. Therefore $\mathbf{C}([a,b])$ is a separable space.

Theorem 1.64. *If in the Hilbert space $\mathbf{H}$ there exists a finite or a countable orthogonal basis $\{f_n\}$, then $\mathbf{H}$ is separable.*

Proof. The set $\{\alpha_n f_n : \alpha_n \in \mathbf{Q}\}$ is a countable everywhere dense set in $\mathbf{H}$. Therefore $\mathbf{H}$ is separable. This completes the proof. □

Theorem 1.65. *In every separable Hilbert space $\mathbf{H}$ there exists an orthogonal basis.*

Proof. Since $\mathbf{H}$ is a separable Hilbert space there exists a countable everywhere dense set $\mathbf{A}$. Let $e_1 \in \mathbf{A}$, $e_1 \neq 0$. Then we take $e_2 \in \mathbf{A} \setminus \{e_1\}$ so that e_1 and e_2 are linearly independent. Then we take $e_3 \in \mathbf{A} \setminus \{e_1, e_2\}$ such that e_1, e_2 and e_3 are linearly independent.

And so on. In this way we get a linearly independent system $\{e_n\}$. From this, using Theorem 1.55, we get a countable orthogonal basis in $\mathbf{H}$. This completes the proof. □

1.10 Advanced practical problems

Problem 1.1. Prove that $\mathbf{C}^k[a,b]$ is a vector space.

Problem 1.2. Let $\mathbf{D}[a,b]$ be the set of all differential operators

$$P(D) = \sum_{k=0}^{m} p_k(x)\frac{d^k}{dx^k}, \quad Q(D) = \sum_{k=0}^{m} q_k(x)\frac{d^k}{dx^k},$$

where $x \in [a,b]$, $p_k, q_k \in \mathbf{C}([a,b])$, $k \in \{1,\dots,m\}$, $\frac{d^0}{dx^0} = 1$, with operations

$$P(D) + Q(D) = \sum_{k=0}^{m} (p_k(x) + q_k(x))\frac{d^k}{dx^k},$$

$$\alpha P(D) = \sum_{k=0}^{m} (\alpha p_k(x))\frac{d^k}{dx^k}, \quad \alpha \in \mathbf{F}.$$

Prove that $\mathbf{D}[a,b]$ is a vector space.

Problem 1.3. Consider the set $E[a,b]$. Prove that
1. xe^x, $x^2e^{x^2}$ and $x^3e^{x^3}$ are linearly independent.
2. 0, $\sin x$, $\cos x$ and $\sin x + \cos x$ are linearly dependent.

Problem 1.4. Suppose that $\mathbf{A}$, $\mathbf{B}$ and $\mathbf{C}$ are linear subspaces of a vector space $\mathbf{E}$. Prove that

$$\mathbf{A} \cap (\mathbf{B} + (\mathbf{A} \cap \mathbf{C})) = (\mathbf{A} \cap \mathbf{B}) + (\mathbf{A} \cap \mathbf{C}).$$

Problem 1.5. Prove that

$$d(x,y) = \sum_{i=1}^{n} |x_i - y_i|, \quad x = (x_1,\dots,x_n), \quad y = (y_1,\dots,y_n) \in \mathbf{E}_n,$$

is a metric on $\mathbf{E}_n$.

Problem 1.6. We say that a function $f : [a,b] \longmapsto \mathbf{R}$ is a function of bounded variation if there exists a positive constant c such that for every partition $P = \{a = t_1 < \cdots < t_{n_p} = b\}$ of the interval $[a,b]$ we have

$$\sum_{l=1}^{n_p} |f(t_l) - f(t_{l-1})| < c.$$

Let $\mathcal{P}$ be the set of all partitions P of the interval $[a,b]$. If $f : [a,b] \longmapsto \mathbf{R}$ is a function of bounded variation, then its total variation is defined by

$$\bigvee_a^b f = \sup_{P\in\mathcal{P}} \sum_{l=1}^{n_p} |f(t_l) - f(t_{l-1})|.$$

The set of all functions of bounded variation on $[a,b]$ is denoted by $\mathbf{V}([a,b])$. In $\mathbf{V}([a,b])$ we define a metric by

$$d(f,g) = |f(a) - g(a)| + \bigvee_a^b (f-g), \quad f,g \in \mathbf{V}([a,b]). \tag{1.61}$$

Prove that (1.61) satisfies all axioms for a metric.

Problem 1.7. Check that

$$\|f\| = |f(a)| + |f(b)| + \max_{t\in[a,b]} |f''(t)|, \quad f \in \mathbf{C}^2([a,b]),$$

satisfies all axioms for a norm.

Problem 1.8. Check if

$$\|f\| = \max_{t\in[a,b]} |f(t)| + |f(b) - f(a)|$$

satisfies all axioms for a norm in $\mathbf{C}^1([a,b])$.

Answer. Yes.

Problem 1.9. Prove that the sequence

$$\left\{\frac{x^{n+1}}{n+1} - \frac{x^{n+2}}{n+2}\right\}$$

is convergent in $\mathbf{C}([0,1])$ and $\mathbf{C}^1([0,1])$.

Problem 1.10. Let $x_1, \ldots, x_n$ be an orthogonal system in a Hilbert space $\mathbf{H}$. Prove that

$$\left\|\sum_{k=1}^n x_k\right\|^2 = \sum_{k=1}^n \|x_k\|^2.$$

2 Lebesgue integration

Some preliminaries regarding sets and mappings that will be used throughout this book can be found in the appendix of this book.

2.1 Lebesgue outer measure. Measurable sets

Definition 2.1. Let **I** be a nonempty interval of real numbers. We define its length $l(\mathbf{I})$ to be ∞ if **I** is unbounded and if **I** is bounded, we define its length to be the difference of its end points.

Below with $\{I_k\}_{k\in\mathbf{N}}$ we will denote a countable collection of nonempty open, bounded intervals.

Definition 2.2. Let **A** be a set of real numbers. Then we define the outer measure of **A** as follows:

$$m^\star(A) = \inf_{\substack{\{\mathbf{I}_k\}_{k\in\mathbf{N}} \\ \mathbf{A}\subseteq\bigcup_{k=1}^\infty \mathbf{I}_k}} \left(\sum_{k=1}^\infty l(\mathbf{I}_k)\right).$$

By the definition, it follows that $m^\star(\emptyset) = 0$.

Theorem 2.1. *If $A \subseteq B$, then $m^*(A) \le m^*(B)$.*

Proof. We have

$$\begin{aligned} m^\star(A) &= \inf_{\substack{\{\mathbf{I}_k\}_{k\in\mathbf{N}} \\ \mathbf{A}\subseteq\bigcup_{k=1}^\infty \mathbf{I}_k}} \left(\sum_{k=1}^\infty l(\mathbf{I}_k)\right) \\ &\le \inf_{\substack{\{\mathbf{V}_k\}_{k\in\mathbf{N}} \\ \mathbf{A}\subseteq B\subseteq\bigcup_{k=1}^\infty \mathbf{V}_k}} \left(\sum_{k=1}^\infty l(\mathbf{V}_k)\right) = m^*(B). \end{aligned}$$

This completes the proof. □

Theorem 2.2. *The outer measure of an interval is its length.*

Proof.
1. Let $\mathbf{I} = [a, b]$. Let also $\epsilon > 0$ be arbitrarily chosen. Since

$$[a, b] \subset (a - \epsilon, b + \epsilon),$$

we have

$$m^\star([a, b]) \le l((a - \epsilon, b + \epsilon)) = b - a + 2\epsilon.$$

https://doi.org/10.1515/9783110657722-002

Because $\epsilon > 0$ was arbitrarily chosen, we conclude that

$$m^\star([a,b]) \leq b - a. \tag{2.1}$$

Let $\{\mathbf{V}_k\}_{k\in\mathbf{N}}$ be a collection of open intervals covering the interval $[a,b]$. By the Heine–Borel theorem, it follows that there exists a finite subcollection $\{\mathbf{I}_k\}_{k=1}^n$ of the collection $\{\mathbf{V}_k\}_{k\in\mathbf{N}}$ that covers $[a,b]$. There exists $l \in \{1,\ldots,n\}$ such that $a \in \mathbf{I}_l$. Let $\mathbf{I}_l = (a_1, b_1)$. We have $a_1 < a < b_1$. If $b_1 > b$, then

$$b - a \leq \sum_{k=1}^{n} l(\mathbf{I}_k). \tag{2.2}$$

Let $b_1 \in [a,b)$. Then there exists an interval (a_2,b_2) such that $b_1 \in (a_2,b_2)$. If $b_2 > b$, then (2.1) holds. If $b_2 < b$ we continue this process while it terminates. Then we obtain a subcollection $\{(a_k,b_k)\}_{k=1}^N$ of the collection $\{I_k\}_{k=1}^n$. We have

$$a_{k+1} < b_k \quad \text{for} \quad 1 \leq k \leq N-1$$

and since this process is terminated, we have $b_N > b$. Therefore (2.1) holds. By (2.1) and (2.2), we see that

$$m^\star([a,b]) = b - a.$$

2. Let $\mathbf{I}$ is any bounded interval. Then for any $\epsilon > 0$ there exist two closed bounded intervals $\mathbf{J}_1$ and $\mathbf{J}_2$ such that

$$\mathbf{J}_1 \leq \mathbf{I} \leq \mathbf{J}_2$$

and

$$l(\mathbf{I}) - \epsilon \leq l(\mathbf{J}_1) \leq l(\mathbf{J}_2) \leq l(\mathbf{I}) + \epsilon.$$

Hence,

$$l(\mathbf{I}) - \epsilon \leq m^\star(\mathbf{J}_1) \leq m^*(\mathbf{I}) \leq m^\star(\mathbf{J}_2) \leq l(\mathbf{I}) + \epsilon.$$

Because $\epsilon > 0$ was arbitrarily chosen, we see that

$$m^*(\mathbf{I}) = l(\mathbf{I}).$$

3. If $\mathbf{I}$ is unbounded interval, then for any $n \in \mathbf{N}$ there exists an interval $\mathbf{J} \subseteq \mathbf{I}$ such that $l(\mathbf{J}) = n$. Hence,

$$m^\star(\mathbf{I}) \geq m^\star(\mathbf{J}) = l(\mathbf{J}) = n$$

for any $n \in \mathbf{N}$. Therefore $m^\star(\mathbf{I}) = \infty$.

This completes the proof. □

Theorem 2.3. $m^\star(\{y\}) = 0$ *for any* $y \in \mathbf{R}$.

Proof. Let $\epsilon > 0$ be arbitrarily chosen and fixed. Then

$$\{y\} \subset (y-\epsilon, y+\epsilon).$$

Hence,

$$m^\star(\{y\}) \le 2\epsilon.$$

Because $\epsilon > 0$ was arbitrarily chosen, we conclude that $m^\star(\{y\}) = 0$. This completes the proof. □

Theorem 2.4. *For any set* $\mathbf{A}$ *and for any* $y \in \mathbf{R}$ *we have*

$$m^\star(\mathbf{A}+y) = m^\star(\mathbf{A}).$$

Proof. Note that any countable collection $\{I_k\}_{k\in\mathbf{N}}$ of open, bounded intervals covers $\mathbf{A}$ if and only if $\{I_k+y\}_{k\in\mathbf{N}}$ covers $\mathbf{A}+y$. Also, $m^\star(I_k+y) = m^\star(I_k)$ for any $k \in \mathbf{N}$. Therefore

$$\begin{aligned} m^\star(\mathbf{A}+y) &= \inf_{\substack{\{\mathbf{I}_k+y\}_{k\in\mathbf{N}}\\ \mathbf{A}+y\subseteq\bigcup_{k=1}^\infty \mathbf{I}_k+y}} \left(\sum_{k=1}^\infty l(\mathbf{I}_k+y)\right)\\ &= \inf_{\substack{\{\mathbf{I}_k\}_{k\in\mathbf{N}}\\ \mathbf{A}\subseteq\bigcup_{k=1}^\infty \mathbf{I}_k}} \left(\sum_{k=1}^\infty l(\mathbf{I}_k)\right) = m^\star(\mathbf{A}). \end{aligned}$$

This completes the proof. □

Theorem 2.5. *The outer measure is countable subadditive, i. e., if* $\{\mathbf{E}_k\}_{k\in\mathbf{N}}$ *is any countable collection of sets, disjoint or not, then*

$$m^\star\left(\bigcup_{k\in\mathbf{N}} \mathbf{E}_k\right) \le \sum_{k=1}^\infty m^\star(\mathbf{E}_k).$$

Proof. If there is a $k \in \mathbf{N}$ such that $m^\star(\mathbf{E}_k) = \infty$, then the assertion is evident. Suppose that $m^\star(\mathbf{E}_k) < \infty$ for any $k \in \mathbf{N}$. We take $\epsilon > 0$ arbitrarily. Then for any $k \in \mathbf{N}$ there exists a countable collection $\{\mathbf{I}_{k,m}\}_{m\in\mathbf{N}}$ of open, bounded intervals, such that

$$\sum_{m=1}^\infty l(\mathbf{I}_{k,m}) \le m^\star(\mathbf{E}_k) + \frac{\epsilon}{2^k}.$$

Note that $\{\mathbf{I}_{k,m}\}_{k,m\in\mathbf{N}}$ is a countable collection of open, bounded intervals for which

$$\bigcup_{k\in\mathbf{N}} \mathbf{E}_k \subseteq \bigcup_{k,m\in\mathbf{N}} \mathbf{I}_{k,m}.$$

Hence, by Definition 2.2, it follows that

$$m^{\star}\Big(\bigcup_{k\in\mathbf{N}}\mathbf{E}_k\Big)\le\sum_{k,m=1}^{\infty}l(\mathbf{I}_{k,m})=\sum_{k=1}^{\infty}\Big(\sum_{m=1}^{\infty}l(\mathbf{I}_{k,m})\Big)\le\sum_{k=1}^{\infty}\Big(m^{\star}(\mathbf{E}_k)+\frac{\epsilon}{2^k}\Big)$$
$$=\sum_{k=1}^{\infty}m^{\star}(\mathbf{E}_k)+\sum_{k=1}^{\infty}\frac{\epsilon}{2^k}=\sum_{k=1}^{\infty}m^{\star}(\mathbf{E}_k)+\epsilon.$$

Because $\epsilon > 0$ was arbitrarily chosen, we get to the desired result. This completes the proof. □

Corollary 2.1. *If $\{\mathbf{E}_k\}_{k=1}^{m}$ is any finite collection of sets, disjoint or not, then*

$$m^{\star}\Big(\bigcup_{k=1}^{m}\mathbf{E}_k\Big)\le\sum_{k=1}^{m}m^{\star}(\mathbf{E}_k).$$

Proof. We take $\mathbf{E}_k = \emptyset$ for $k > m$ and we apply Theorem 2.5. This completes the proof. □

Corollary 2.2. *The interval* $[0,1]$ *is not countable.*

Proof. Assume that $[0,1]$ is countable. Then there exists a sequence $\{y_k\}_{k\in\mathbf{N}}$ of elements of $[0,1]$ such that

$$[0,1]=\bigcup_{k\in\mathbf{N}}\{y_k\}.$$

Hence, using Theorems 2.3 and 2.5, it follows that

$$1=m^{\star}([0,1])=m^{\star}\Big(\bigcup_{k\in\mathbf{N}}\{y_k\}\Big)\le\sum_{k=1}^{\infty}m^{\star}(\{y_k\})=0,$$

which is a contradiction. □

Corollary 2.3. *The set of irrational numbers in the interval* $[0,1]$ *has outer measure* 1.

Proof. Let $\mathbf{A}$ be the set of irrational numbers in $[0,1]$ and B be the set of rational numbers in $[0,1]$. Then

$$[0,1]=\mathbf{A}\cup\mathbf{B}. \tag{2.3}$$

Since $\mathbf{A}\subset[0,1]$, using Theorem 2.1, we get

$$m^{\star}(\mathbf{A})\le m^{\star}([0,1])=1. \tag{2.4}$$

On the other hand, using Theorem 2.5, we obtain

$$1=m^{\star}([0,1])\le m^{\star}(\mathbf{A})+m^{\star}(\mathbf{B}). \tag{2.5}$$

Because **B** is countable, there exists a sequence $\{y_k\}_{k\in\mathbf{N}}$ of elements of **B** such that

$$\mathbf{B} = \bigcup_{k=1}^{\infty}\{y_k\}.$$

Then, using Theorems 2.3 and 2.5, we obtain

$$m^\star(\mathbf{B}) = m^\star\left(\bigcup_{k=1}^{\infty}\{y_k\}\right) \le \sum_{k=1}^{\infty} m^\star(\{y_k\}) = 0.$$

Hence, by (2.5), we find

$$1 \le m^\star(\mathbf{A}).$$

By the previous inequality and (2.4), we obtain

$$m^\star(\mathbf{A}) = 1.$$

This completes the proof. □

Theorem 2.6. *Let* $m^\star(\mathbf{A}) = 0$. *Then*

$$m^\star(\mathbf{A}\cup\mathbf{B}) = m^\star(\mathbf{B})$$

for any set **B**.

Proof. Since $\mathbf{B} \subseteq \mathbf{A}\cup\mathbf{B}$, we have

$$m^\star(\mathbf{B}) \le m^\star(\mathbf{A}\cup\mathbf{B}). \tag{2.6}$$

On the other hand,

$$m^\star(\mathbf{A}\cup\mathbf{B}) \le m^\star(\mathbf{A}) + m^\star(\mathbf{B}) = m^\star(\mathbf{B}).$$

Hence, by (2.6), we conclude the desired result. This completes the proof. □

Definition 2.3. Let **E** be a subset of **R**. With $\mathbf{E}^c$ we will denote the set

$$\mathbf{E}^c = \{x \in \mathbf{R} : x \notin \mathbf{E}\}.$$

The set $\mathbf{E}^c$ will be called the complement of the set **E** in **R**.

Definition 2.4. A set **E** is said to be measurable provided for any set **A**,

$$m^\star(\mathbf{A}) = m^\star(\mathbf{A}\cap\mathbf{E}) + m^\star(\mathbf{A}\cap\mathbf{E}^c).$$

Lemma 2.1. *For any sets* **A** *and* **E** *we have*

$$\mathbf{A} = (\mathbf{A}\cap\mathbf{E})\cup(\mathbf{A}\cap\mathbf{E}^c).$$

Proof. Let $x \in \mathbf{A}$ be arbitrarily chosen and fixed.

1. If $x \in \mathbf{E}$, then

$$x \in \mathbf{A} \cap \mathbf{E}$$

   and

$$x \in (\mathbf{A} \cap \mathbf{E}) \cup (\mathbf{A} \cap \mathbf{E}^c).$$

2. If $x \notin \mathbf{E}$, then $x \in \mathbf{E}^c$. Hence,

$$x \in \mathbf{A} \cap \mathbf{E}^c.$$

   Therefore

$$x \in (\mathbf{A} \cap \mathbf{E}) \cup (\mathbf{A} \cap \mathbf{E}^c).$$

Because $x \in \mathbf{A}$ was arbitrarily chosen and we see that it is an element of $(\mathbf{A}\cap\mathbf{E})\cup(\mathbf{A}\cap\mathbf{E}^c)$, we conclude that

$$\mathbf{A} \subseteq (\mathbf{A} \cap \mathbf{E}) \cup (\mathbf{A} \cap \mathbf{E}^c). \tag{2.7}$$

Let $x \in (\mathbf{A} \cap \mathbf{E}) \cup (\mathbf{A} \cap \mathbf{E}^c)$ be arbitrarily chosen and fixed. Then

$$x \in \mathbf{A} \cap \mathbf{E} \quad \text{or} \quad x \in \mathbf{A} \cap \mathbf{E}^c.$$

1. Let $x \in \mathbf{A} \cap \mathbf{E}$. Then $x \in \mathbf{A}$.
2. Let $x \in \mathbf{A} \cap \mathbf{E}^c$. Then $x \in \mathbf{A}$.

Because $x \in (\mathbf{A} \cap \mathbf{E}) \cup (\mathbf{A} \cap \mathbf{E}^c)$ was arbitrarily chosen and we see that it is an element of $\mathbf{A}$, we conclude that

$$(\mathbf{A} \cap \mathbf{E}) \cup (\mathbf{A} \cap \mathbf{E}^c) \subseteq \mathbf{A}.$$

From the previous relation and from (2.7), we get the desired result. This completes the proof. □

Remark 2.1. Using Lemma 2.1 and Corollary 2.1, we have

$$m^\star(\mathbf{A}) \le m^\star(\mathbf{A} \cap \mathbf{E}) + m^\star(\mathbf{A} \cap \mathbf{E}^c).$$

Therefore $\mathbf{E}$ is measurable if and only if

$$m^\star(\mathbf{A}) \ge m^\star(\mathbf{A} \cap \mathbf{E}) + m^\star(\mathbf{A} \cap \mathbf{E}^c). \tag{2.8}$$

This inequality trivially holds if $m^\star(\mathbf{A}) = \infty$. Thus it suffices to establish (2.8) for sets $\mathbf{A}$ that have finite outer measure.

Theorem 2.7. *If* $\mathbf{A}$ *is a measurable set and* $\mathbf{B}$ *is any set disjoint from* $\mathbf{A}$, *then*

$$m^\star(\mathbf{A} \cup \mathbf{B}) = m^\star(\mathbf{A}) + m^\star(\mathbf{B}).$$

Proof. Since $\mathbf{A}$ is a measurable set, using Definition 2.4, we get

$$m^\star(\mathbf{A} \cup \mathbf{B}) = m^\star((\mathbf{A} \cup \mathbf{B}) \cap \mathbf{A}) + m^\star((\mathbf{A} \cup \mathbf{B}) \cap \mathbf{A}^c) = m^\star(\mathbf{A}) + m^\star(\mathbf{B}).$$

This completes the proof. □

Theorem 2.8. *Any set of outer measure zero is measurable. In particular, any countable set is measurable.*

Proof. Let $\mathbf{E}$ be a set of outer measure zero and let $\mathbf{A}$ be any set. Since

$$\mathbf{A} \cap \mathbf{E} \subseteq \mathbf{E} \quad \text{and} \quad \mathbf{A} \cap \mathbf{E}^c \subseteq \mathbf{A},$$

we have

$$m^\star(\mathbf{A} \cap \mathbf{E}) \leq m^\star(\mathbf{E}) = 0 \quad \text{and} \quad m^\star(\mathbf{A} \cap \mathbf{E}^c) \leq m^\star(\mathbf{A}).$$

Thus

$$m^\star(\mathbf{A}) \geq m^\star(\mathbf{A} \cap \mathbf{E}^c) = 0 + m^\star(\mathbf{A} \cap \mathbf{E}^c) = m^\star(\mathbf{A} \cap \mathbf{E}) + m^\star(\mathbf{A} \cap \mathbf{E}^c)$$

and therefore $\mathbf{E}$ is measurable. Hence, using the fact that every countable set has outer measure zero, we conclude that every countable set is measurable. This completes the proof. □

Lemma 2.2. *For any sets* $\mathbf{A}$, $\mathbf{E}_1$ *and* $\mathbf{E}_2$ *we have*

$$(\mathbf{A} \cap \mathbf{E}_1^c) \cap \mathbf{E}_2^c = \mathbf{A} \cap (\mathbf{E}_1 \cup \mathbf{E}_2)^c \tag{2.9}$$

and

$$(\mathbf{A} \cap \mathbf{E}_1) \cup (\mathbf{A} \cap \mathbf{E}_1^c \cap \mathbf{E}_2) = \mathbf{A} \cap (\mathbf{E}_1 \cup \mathbf{E}_2). \tag{2.10}$$

Proof. Firstly, we will prove equation (2.9). Let $x \in (\mathbf{A} \cap \mathbf{E}_1^c) \cap \mathbf{E}_2^c$ be arbitrarily chosen and fixed. Then

$$x \in \mathbf{A} \cap \mathbf{E}_1^c \quad \text{and} \quad x \in \mathbf{E}_2^c.$$

Since $x \in \mathbf{A} \cap \mathbf{E}_1^c$, we get $x \in \mathbf{A}$ and $x \in \mathbf{E}_1^c$. From $x \in \mathbf{E}_1^c$ and $x \in \mathbf{E}_2^c$, we obtain $x \notin \mathbf{E}_1$ and $x \notin \mathbf{E}_2$. Therefore $x \notin \mathbf{E}_1 \cup \mathbf{E}_2$ and $x \in (\mathbf{E}_1 \cup \mathbf{E}_2)^c$. From $x \in \mathbf{A}$ and $x \in (\mathbf{E}_1 \cup \mathbf{E}_2)^c$, we conclude that $x \in \mathbf{A} \cap (\mathbf{E}_1 \cup \mathbf{E}_2)^c$. Because $x \in (\mathbf{A} \cap \mathbf{E}_1^c) \cap \mathbf{E}_2^c$ was arbitrarily chosen and we see that it is an element of the set $\mathbf{A} \cap (\mathbf{E}_1 \cup \mathbf{E}_2)^c$, we conclude that

$$(\mathbf{A} \cap \mathbf{E}_1^c) \cap \mathbf{E}_2^c \subseteq \mathbf{A} \cap (\mathbf{E}_1 \cup \mathbf{E}_2)^c. \tag{2.11}$$

Let $x \in \mathbf{A} \cap (\mathbf{E}_1 \cup \mathbf{E}_2)^c$ be arbitrarily chosen and fixed. Then $x \in \mathbf{A}$ and $x \in (\mathbf{E}_1 \cup \mathbf{E}_2)^c$. Hence, $x \notin \mathbf{E}_1 \cup \mathbf{E}_2$. Therefore $x \notin \mathbf{E}_1$ and $x \notin \mathbf{E}_2$. From this, it follows that $x \in \mathbf{E}_1^c$ and $x \in \mathbf{E}_2^c$. Since $x \in \mathbf{A}$ and $x \in \mathbf{E}_1^c$, we obtain $x \in \mathbf{A} \cap \mathbf{E}_1^c$. Because $x \in \mathbf{A} \cap \mathbf{E}_1^c$ and $x \in \mathbf{E}_2^c$, we see that $x \in (\mathbf{A} \cap \mathbf{E}_1^c) \cap \mathbf{E}_2^c$. Since $x \in \mathbf{A} \cap (\mathbf{E}_1 \cup \mathbf{E}_2)^c$ was arbitrarily chosen and we see that it is an element of $(\mathbf{A} \cap \mathbf{E}_1^c) \cap \mathbf{E}_2^c$, we conclude that

$$\mathbf{A} \cap (\mathbf{E}_1 \cup \mathbf{E}_2)^c \subseteq (\mathbf{A} \cap \mathbf{E}_1^c) \cap \mathbf{E}_2^c.$$

From the previous relation and from (2.11), we obtain equation (2.9). Now we will prove equation (2.10). Let $x \in (\mathbf{A} \cap \mathbf{E}_1) \cup (\mathbf{A} \cap \mathbf{E}_1^c \cap \mathbf{E}_2)$ be arbitrarily chosen and fixed. Then $x \in \mathbf{A} \cap \mathbf{E}_1$ or $x \in \mathbf{A} \cap \mathbf{E}_1^c \cap \mathbf{E}_2$.

1. Let $x \in \mathbf{A} \cap \mathbf{E}_1$. Then $x \in \mathbf{A}$ and $x \in \mathbf{E}_1$. Hence, $x \in \mathbf{E}_1 \cup \mathbf{E}_2$ and $x \in \mathbf{A} \cap (\mathbf{E}_1 \cup \mathbf{E}_2)$.
2. Let $x \in \mathbf{A} \cap \mathbf{E}_1^c \cap \mathbf{E}_2$. Then $x \in \mathbf{A}$, $x \in \mathbf{E}_1^c$ and $x \in \mathbf{E}_2$. Hence, $x \in \mathbf{A}$ and $x \in \mathbf{E}_1 \cup \mathbf{E}_2$. Therefore $x \in \mathbf{A} \cap (\mathbf{E}_1 \cup \mathbf{E}_2)$.

Because $x \in (\mathbf{A} \cap \mathbf{E}_1) \cup (\mathbf{A} \cap \mathbf{E}_1^c \cap \mathbf{E}_2)$ was arbitrarily chosen and we see that it is an element of $\mathbf{A} \cap (\mathbf{E}_1 \cup \mathbf{E}_2)$, we obtain

$$(\mathbf{A} \cap \mathbf{E}_1) \cup (\mathbf{A} \cap \mathbf{E}_1^c \cap \mathbf{E}_2) \subseteq \mathbf{A} \cap (\mathbf{E}_1 \cup \mathbf{E}_2). \tag{2.12}$$

Let $x \in \mathbf{A} \cap (\mathbf{E}_1 \cup \mathbf{E}_2)$ is arbitrarily chosen and fixed. Then $x \in \mathbf{A}$ and $x \in \mathbf{E}_1 \cup \mathbf{E}_2$.

1. Let $x \in \mathbf{E}_1$. Then $x \in \mathbf{A} \cap \mathbf{E}_1$ and
$$x \in (\mathbf{A} \cap \mathbf{E}_1) \cup (\mathbf{A} \cap \mathbf{E}_1^c \cap \mathbf{E}_2). \tag{2.13}$$
2. Let $x \in \mathbf{E}_2$. If $x \in \mathbf{E}_1$, then $x \in \mathbf{A} \cap \mathbf{E}_1$ and we get (2.13). If $x \notin \mathbf{E}_1$, then $x \in \mathbf{E}_1^c$ and $x \in \mathbf{A} \cap \mathbf{E}_1^c \cap \mathbf{E}_2$. Therefore we get (2.13).

Because $x \in \mathbf{A} \cap (\mathbf{E}_1 \cup \mathbf{E}_2)$ was arbitrarily chosen and we see that it is an element of $(\mathbf{A} \cap \mathbf{E}_1) \cup (\mathbf{A} \cap \mathbf{E}_1^c \cap \mathbf{E}_2)$, we obtain

$$\mathbf{A} \cap (\mathbf{E}_1 \cup \mathbf{E}_2) \subseteq (\mathbf{A} \cap \mathbf{E}_1) \cup (\mathbf{A} \cap \mathbf{E}_1^c \cap \mathbf{E}_2).$$

From the previous relation and from (2.12), we get equation (2.10). This completes the proof. □

Theorem 2.9. *The union of a finite collection of measurable sets is measurable.*

Proof. Let $\mathbf{E}_1$ and $\mathbf{E}_2$ be two measurable sets. Then for any set $\mathbf{A}$ we have

$$m^\star(\mathbf{A}) = m^\star(\mathbf{A} \cap \mathbf{E}_1) + m^\star(\mathbf{A} \cap \mathbf{E}_1^c)$$

and

$$m^\star(\mathbf{A} \cap \mathbf{E}_1^c) = m^\star(\mathbf{A} \cap \mathbf{E}_1^c \cap \mathbf{E}_2) + m^\star(\mathbf{A} \cap \mathbf{E}_1^c \cap \mathbf{E}_2^c).$$

Therefore

$$m^\star(\mathbf{A}) = m^\star(\mathbf{A} \cap \mathbf{E}_1) + m^\star(\mathbf{A} \cap \mathbf{E}_1^c \cap \mathbf{E}_2) + m^\star(\mathbf{A} \cap \mathbf{E}_1^c \cap \mathbf{E}_2^c).$$

Hence, by (2.9), we obtain

$$\begin{aligned} m^\star(\mathbf{A}) &= m^\star(\mathbf{A} \cap \mathbf{E}_1) + m^\star(\mathbf{A} \cap (\mathbf{E}_1 \cup \mathbf{E}_2)^c) \\ &\quad + m^\star(\mathbf{A} \cap \mathbf{E}_1^c \cap \mathbf{E}_2) \geq m^\star(\mathbf{A} \cap (\mathbf{E}_1 \cup \mathbf{E}_2)) + m^\star(\mathbf{A} \cap (\mathbf{E}_1 \cup \mathbf{E}_2)^c). \end{aligned}$$

Therefore $\mathbf{E}_1 \cup \mathbf{E}_2$ is a measurable set. Assume that $\bigcup_{k=1}^n \mathbf{E}_k$ is a measurable set, where $\mathbf{E}_k$, $k \in \{1, \dots, n\}$, are measurable sets. Let $\mathbf{E}_{n+1}$ be a measurable set. Then, as above, $(\bigcup_{k=1}^n \mathbf{E}_k) \cup \mathbf{E}_{n+1}$ is a measurable set. Therefore $\bigcup_{k=1}^{n+1} \mathbf{E}_k$ is a measurable set. This completes the proof. □

Theorem 2.10. *Let* $\mathbf{A}$ *be any set and* $\{\mathbf{E}_k\}_{k=1}^n$ *be a finite disjoint collection of measurable sets. Then*

$$m^\star\left(\mathbf{A} \cap \left(\bigcup_{k=1}^n \mathbf{E}_k\right)\right) = \sum_{k=1}^n m^\star(\mathbf{A} \cap \mathbf{E}_k).$$

In particular,

$$m^\star\left(\bigcup_{k=1}^n \mathbf{E}_k\right) = \sum_{k=1}^n m^\star(\mathbf{E}_k).$$

Proof. We will use induction.

1. Let $n = 1$. Then the assertion is evident.
2. Assume that the assertion is valid for some $n \in \mathbf{N}$.
3. We will prove the assertion for $n + 1$. Because $\{\mathbf{E}_k\}_{k=1}^{n+1}$ is a disjoint collection, we have

$$\mathbf{A} \cap \left(\bigcup_{k=1}^{n+1} \mathbf{E}_k\right) \cap \mathbf{E}_{n+1} = \mathbf{A} \cap \mathbf{E}_{n+1}$$

and

$$\mathbf{A} \cap \left(\bigcup_{k=1}^{n+1} \mathbf{E}_k\right) \cap \mathbf{E}_{n+1}^c = \mathbf{A} \cap \left(\bigcup_{k=1}^{n} \mathbf{E}_k\right).$$

Hence, by the measurability of $\mathbf{E}_{n+1}$, we get

$$\begin{aligned} m^\star\left(\mathbf{A} \cap \left(\bigcup_{k=1}^{n+1} \mathbf{E}_k\right)\right) &= m^\star\left(\mathbf{A} \cap \left(\bigcup_{k=1}^{n+1} \mathbf{E}_k\right) \cap \mathbf{E}_{n+1}\right) + m^\star\left(\mathbf{A} \cap \left(\bigcup_{k=1}^{n+1} \mathbf{E}_k\right) \cap \mathbf{E}_{n+1}^c\right) \\ &= m^\star(\mathbf{A} \cap \mathbf{E}_{n+1}) + m^\star\left(\mathbf{A} \cap \left(\bigcup_{k=1}^{n} \mathbf{E}_k\right)\right) \end{aligned}$$

$$= m^\star(\mathbf{A} \cap \mathbf{E}_{n+1}) + \sum_{k=1}^{n} m^\star(\mathbf{A} \cap \mathbf{E}_k)$$
$$= \sum_{k=1}^{n+1} m^\star(\mathbf{A} \cap \mathbf{E}_k).$$

This completes the proof. □

Lemma 2.3. *For any sets* **A**, **B** *and* **C** *we have*

$$\mathbf{A} \setminus \mathbf{B} = \mathbf{A} \cap \mathbf{B}^c, \tag{2.14}$$
$$\mathbf{A} \cap (\mathbf{B} \setminus \mathbf{C}) \subseteq \mathbf{A} \cap (\mathbf{B} \cap \mathbf{C}^c), \tag{2.15}$$
$$(\mathbf{A} \cap \mathbf{B})^c = \mathbf{A}^c \cup \mathbf{B}^c. \tag{2.16}$$

Proof.

1. We will prove (2.14). Let $x \in \mathbf{A} \setminus \mathbf{B}$ be arbitrarily chosen and fixed. Then $x \in \mathbf{A}$ and $x \notin \mathbf{B}$. From $x \notin \mathbf{B}$, it follows that $x \in \mathbf{B}^c$. By $x \in \mathbf{A}$ and $x \in \mathbf{B}^c$, we obtain $x \in \mathbf{A} \cap \mathbf{B}^c$. Because $x \in \mathbf{A} \setminus \mathbf{B}$ was arbitrarily chosen and we see that it is an element of $\mathbf{A} \cap \mathbf{B}^c$, we get
$$\mathbf{A} \setminus \mathbf{B} \subseteq \mathbf{A} \cap \mathbf{B}^c. \tag{2.17}$$
Let $x \in \mathbf{A} \cap \mathbf{B}^c$ be arbitrarily chosen. Then $x \in \mathbf{A}$ and $x \in \mathbf{B}^c$. From $x \in \mathbf{B}^c$ it follows that $x \notin \mathbf{B}$. From $x \in \mathbf{A}$ and $x \notin \mathbf{B}$, we get $x \in \mathbf{A} \setminus \mathbf{B}$. Because $x \in \mathbf{A} \cap \mathbf{B}^c$ was arbitrarily chosen and fixed and we see that it is an element of the set $\mathbf{A} \setminus \mathbf{B}$, we obtain
$$\mathbf{A} \cap \mathbf{B}^c \subseteq \mathbf{A} \setminus \mathbf{B}.$$
From the previous relation and from (2.17), we obtain (2.14).
2. We will prove (2.15). Let $x \in \mathbf{A} \cap (\mathbf{B} \setminus \mathbf{C})$ is arbitrarily chosen. Then $x \in \mathbf{A}$ and $x \in \mathbf{B} \setminus \mathbf{C}$. From $x \in \mathbf{B} \setminus \mathbf{C}$, it follows that $x \in \mathbf{B}$ and $x \notin \mathbf{C}$. By $x \notin \mathbf{C}$, we get $x \in \mathbf{C}^c$. From $x \in \mathbf{A}$, $x \in \mathbf{B}$ and $x \in \mathbf{C}^c$, we see that $x \in \mathbf{A} \cap (\mathbf{B} \cap \mathbf{C}^c)$. Because $x \in \mathbf{A} \cap (\mathbf{B} \setminus \mathbf{C})$ was arbitrarily chosen, we conclude to equation (2.15).
3. We will prove (2.16). Let $x \in (\mathbf{A} \cap \mathbf{B})^c$ be arbitrarily chosen and fixed. Then $x \notin \mathbf{A} \cap \mathbf{B}$. Hence, $x \notin \mathbf{A}$ or $x \notin \mathbf{B}$.
   (a) If $x \notin \mathbf{A}$, then $x \in \mathbf{A}^c$ and $x \in \mathbf{A}^c \cup \mathbf{B}^c$.
   (b) If $x \notin \mathbf{B}$, then $x \in \mathbf{B}^c$ and $x \in \mathbf{A}^c \cup \mathbf{B}^c$.

   Because $x \in (\mathbf{A} \cap \mathbf{B})^c$ was arbitrarily chosen and we see that it is an element of $\mathbf{A}^c \cup \mathbf{B}^c$, we obtain
$$(\mathbf{A} \cap \mathbf{B})^c \subseteq \mathbf{A}^c \cup \mathbf{B}^c. \tag{2.18}$$
Let $x \in \mathbf{A}^c \cup \mathbf{B}^c$ be arbitrarily chosen and fixed. Then $x \in \mathbf{A}^c$ or $x \in \mathbf{B}^c$. Hence, $x \notin \mathbf{A}$ or $x \notin \mathbf{B}$. Therefore $x \notin \mathbf{A} \cap \mathbf{B}$ and $x \in (\mathbf{A} \cap \mathbf{B})^c$. Because $x \in \mathbf{A}^c \cup \mathbf{B}^c$ was arbitrarily

chosen and we see that it is an element of $(\mathbf{A} \cap \mathbf{B})^c$, we obtain

$$\mathbf{A}^c \cup \mathbf{B}^c \subseteq (\mathbf{A} \cap \mathbf{B})^c.$$

From the previous relation and from (2.18), we get equation (2.16). This completes the proof. □

Theorem 2.11. *Let $\mathbf{E}_1$ and $\mathbf{E}_2$ be measurable sets. Then $\mathbf{E}_1 \setminus \mathbf{E}_2$ is a measurable set.*

Proof. Let $\mathbf{A}$ be any set. Since $\mathbf{E}_1$ is a measurable set, then $\mathbf{E}_1^c$ is a measurable set. Hence, by Theorem 2.9, we see that $\mathbf{E}_1^c \cup \mathbf{E}_2$ is a measurable set. Then

$$m^\star(\mathbf{A}) \geq m^\star(\mathbf{A} \cap (\mathbf{E}_1^c \cup \mathbf{E}_2)) + m^\star(\mathbf{A} \cap (\mathbf{E}_1^c \cup \mathbf{E}_2)^c). \tag{2.19}$$

Using (2.14) and (2.16), we have

$$(\mathbf{E}_1 \setminus \mathbf{E}_2)^c = (\mathbf{E}_1 \cap \mathbf{E}_2^c)^c = \mathbf{E}_1^c \cup \mathbf{E}_2.$$

Therefore

$$\mathbf{A} \cap (\mathbf{E}_1 \setminus \mathbf{E}_2)^c = \mathbf{A} \cap (\mathbf{E}_1^c \cup \mathbf{E}_2)$$

and

$$m^\star(\mathbf{A} \cap (\mathbf{E}_1 \setminus \mathbf{E}_2)^c) = m^\star(\mathbf{A} \cap (\mathbf{E}_1^c \cup \mathbf{E}_2)). \tag{2.20}$$

By (2.15) and (2.16), we get

$$\mathbf{A} \cap (\mathbf{E}_1 \setminus \mathbf{E}_2) \subseteq \mathbf{A} \cap (\mathbf{E}_1 \cap \mathbf{E}_2^c) = \mathbf{A} \cap (\mathbf{E}_1^c \cup \mathbf{E}_2)^c.$$

Hence, by Theorem 2.1, we obtain

$$m^\star(\mathbf{A} \cap (\mathbf{E}_1 \setminus \mathbf{E}_2)) \leq m^\star(\mathbf{A} \cap (\mathbf{E}_1^c \cup \mathbf{E}_2)^c).$$

From the previous inequality and from (2.19) and (2.20), we arrive at

$$\begin{aligned} m^\star(\mathbf{A} \cap (\mathbf{E}_1 \setminus \mathbf{E}_2)) + m^\star(\mathbf{A} \cap (\mathbf{E}_1 \setminus \mathbf{E}_2)^c) &\leq m^\star(\mathbf{A} \cap (\mathbf{E}_1^c \cup \mathbf{E}_2)^c) \\ &\quad + m^\star(\mathbf{A} \cap (\mathbf{E}_1^c \cup \mathbf{E}_2)) \leq m^\star(\mathbf{A}). \end{aligned}$$

Because $\mathbf{A}$ was arbitrarily chosen and fixed, we conclude that $\mathbf{E}_1 \setminus \mathbf{E}_2$ is a measurable set. This completes the proof. □

Theorem 2.12. *The union of a countable collection of sets is also the union of a countable disjoint collection of sets. In particular, the union of a countable collection of measurable sets is also the union of a countable disjoint collection of measurable sets.*

Proof. Let $\mathbf{E} = \bigcup_{k=1}^{\infty} \mathbf{E}'_k$. We set

$$\mathbf{E}_1 = \mathbf{E}'_1,$$
$$\mathbf{E}_k = \mathbf{E}'_k \setminus \bigcup_{l=1}^{k-1} \mathbf{E}'_l.$$

Suppose that $x \in \mathbf{E}$ is arbitrarily chosen and fixed. Assume that $x \notin \mathbf{E}_k$ for any $k \in \mathbf{N}$. Then $x \notin \mathbf{E}'_k$ for any $k \in \mathbf{N}$ and $x \notin \mathbf{E} = \bigcup_{k=1}^{\infty} \mathbf{E}'_k$, which is a contradiction. Therefore $x \in \mathbf{E}_k$ for some $k \in \mathbf{N}$ and hence, $x \in \bigcup_{k=1}^{\infty} \mathbf{E}_k$. Because $x \in \mathbf{E}$ was arbitrarily chosen and we see that it is an element of $\bigcup_{k=1}^{\infty} \mathbf{E}_k$, we get

$$\mathbf{E} \subseteq \bigcup_{k=1}^{\infty} \mathbf{E}_k. \tag{2.21}$$

Let now $x \in \bigcup_{k=1}^{\infty} \mathbf{E}_k$ is arbitrarily chosen. Suppose that $x \notin \mathbf{E}'_k$ for any $k \in \mathbf{N}$. Hence, $x \notin \mathbf{E}_k$ for any $k \in \mathbf{N}$ and $x \notin \bigcup_{k=1}^{\infty} \mathbf{E}_k$. This is a contradiction. Therefore $x \in \mathbf{E}'_k$ for some $k \in \mathbf{N}$ and $x \in \mathbf{E} = \bigcup_{k=1}^{\infty} \mathbf{E}'_k$. Because $x \in \bigcup_{k=1}^{\infty} \mathbf{E}_k$ was arbitrarily chosen and we see that it is an element of $\mathbf{E}$, we conclude that

$$\bigcup_{k=1}^{\infty} \mathbf{E}_k \subseteq \mathbf{E}.$$

From the previous relation and from (2.21), we obtain $\mathbf{E} = \bigcup_{k=1}^{\infty} \mathbf{E}_k$. Suppose that there is $x \in \mathbf{E}$ such that $x \in \mathbf{E}_k$ and $x \in \mathbf{E}_l$ for some $k > l$. From the definition of the set $\mathbf{E}_k$ it follows that $x \notin \bigcup_{m=1}^{k-1} \mathbf{E}_m$. Therefore $x \notin \mathbf{E}_m$ for any $m \in \{1, \dots, k-1\}$. In particular, $x \notin \mathbf{E}_l$, which is a contradiction. Consequently $\{\mathbf{E}_k\}_{k \in \mathbf{N}}$ is a collection of disjoint sets.

Now we suppose that $\mathbf{E}'_k$, $k \in \mathbf{N}$, are measurable sets. Then for $k > 2$, using Theorem 2.9, we see that $\bigcup_{l=1}^{k-1} \mathbf{E}'_l$ is a measurable set. Hence, by Theorem 2.11, the sets $\mathbf{E}_k$, $k \in \mathbf{N}$, are measurable sets. This completes the proof. □

Theorem 2.13. *The union of countable collection of measurable sets is measurable.*

Proof. Let $\mathbf{E} = \bigcup_{k=1}^{\infty} \mathbf{E}'_k$ be a countable collection of measurable sets. By Theorem 2.12, it follows that there is a countable disjoint collection $\{\mathbf{E}_k\}_{k \in \mathbf{N}}$ of measurable sets such that $\mathbf{E} = \bigcup_{k=1}^{\infty} \mathbf{E}_k$. Let $\mathbf{A}$ be any set, $n \in \mathbf{N}$ be arbitrarily chosen and

$$\mathbf{F}_n = \bigcup_{k=1}^{n} \mathbf{E}_k.$$

Then by Theorem 2.9 we see that $\mathbf{F}_n$ is a measurable set. Also, $\mathbf{F}_n \subseteq \mathbf{E}$ and $\mathbf{F}_n^c \supseteq \mathbf{E}^c$. Hence, by Theorem 2.1, we get

$$m^\star(\mathbf{A} \cap \mathbf{E}^c) \le m^\star(\mathbf{A} \cap \mathbf{F}_n^c).$$

Then, using the fact that $\mathbf{F}_n$ is a measurable set, we obtain

$$\begin{aligned} m^\star(\mathbf{A}) &= m^\star(\mathbf{A} \cap \mathbf{F}_n) + m^\star(\mathbf{A} \cap \mathbf{F}_n^c) \\ &\geq m^\star(\mathbf{A} \cap \mathbf{F}_n) + m^\star(\mathbf{A} \cap \mathbf{E}^c). \end{aligned} \tag{2.22}$$

By Theorem 2.10, we obtain

$$m^\star(\mathbf{A} \cap \mathbf{F}_n) = m^\star\left(\mathbf{A} \cap \left(\bigcup_{k=1}^{n} \mathbf{E}_k\right)\right) = \sum_{k=1}^{n} m^\star(\mathbf{A} \cap \mathbf{E}_k).$$

Hence, by (2.22), we get

$$m^\star(\mathbf{A}) \geq \sum_{k=1}^{n} m^\star(\mathbf{A} \cap \mathbf{E}_k) + m^\star(\mathbf{A} \cap \mathbf{E}^c).$$

Because $n \in \mathbf{N}$ was arbitrarily chosen, from the previous inequality we find

$$m^\star(\mathbf{A}) \geq \sum_{k=1}^{\infty} m^\star(\mathbf{A} \cap \mathbf{E}_k) + m^\star(\mathbf{A} \cap \mathbf{E}^c). \tag{2.23}$$

From Theorem 2.5, we have

$$\begin{aligned} m^\star(\mathbf{A} \cap \mathbf{E}) &= m^\star\left(\mathbf{A} \cap \left(\bigcup_{k=1}^{\infty} \mathbf{E}_k\right)\right) \leq m^\star\left(\bigcup_{k=1}^{\infty} (\mathbf{A} \cap \mathbf{E}_k)\right) \\ &\leq \sum_{k=1}^{\infty} m^\star(\mathbf{A} \cap \mathbf{E}_k). \end{aligned}$$

Hence, by (2.23), we find

$$m^\star(\mathbf{A}) \geq m^\star(\mathbf{A} \cap \mathbf{E}) + m^\star(\mathbf{A} \cap \mathbf{E}^c).$$

This completes the proof. □

Definition 2.5. A collection $\mathcal{F}$ of subsets of $\mathbf{R}$ is called σ-algebra if
1. it contains $\mathbf{R}$,
2. $\mathbf{A} \in \mathcal{F}$, then $\mathbf{A}^c \in \mathcal{F}$,
3. $\mathbf{A}_k \in \mathcal{F}$, $k \in \mathbf{N}$, then $\bigcup_{k=1}^{\infty} \mathbf{A}_k \in \mathcal{F}$.

Remark 2.2. Let $\mathcal{F}$ be a σ-algebra. Let also $\mathbf{A}_k \in \mathcal{F}$, $k \in \mathbf{N}$. Then $\mathbf{A}_k^c \in \mathcal{F}$, $\bigcup_{k=1}^{\infty} \mathbf{A}_k^c \in \mathcal{F}$. Hence,

$$\left(\bigcup_{k=1}^{\infty} \mathbf{A}_k^c\right)^c = \bigcap_{k=1}^{\infty} \mathbf{A}_k \in \mathcal{F}.$$

Therefore every σ-algebra is closed with respect to the formation of countable intersections.

Remark 2.3. By Theorems 2.11 and 2.13, it follows that the set $\mathcal{M}$ of all measurable sets is a σ-algebra.

Theorem 2.14. *If a σ-algebra $\mathcal{F}$ of subsets of* $\mathbf{R}$ *contains intervals of the form* (a, ∞), *then it contains all intervals.*

Proof. Since $(a, \infty) \in \mathcal{F}$ and $\mathcal{F}$ is a σ-algebra, we have

$$(-\infty, a] = (a, \infty)^c \in \mathcal{F}.$$

Because

$$(a, b]^c = (-\infty, a] \cup (b, \infty)$$

and $(-\infty, a]$, $(b, \infty) \in \mathcal{F}$ and $\mathcal{F}$ is closed with respect to the formation of finite unions, we conclude that $(a, b]^c \in \mathcal{F}$. Hence, using the fact that $\mathcal{F}$ is closed with respect to complements, we see that $(a, b] \in \mathcal{F}$. Now we consider $[a, \infty)$. We will prove that

$$[a, \infty) = \bigcap_{n=1}^{\infty} \left(a - \frac{1}{n}, \infty\right). \tag{2.24}$$

Let $x \in [a, \infty)$ be arbitrarily chosen and fixed. Then $x \in (a - \frac{1}{n}, \infty)$ for any $n \in \mathbf{N}$. Therefore $x \in \bigcap_{n=1}^{\infty}(a - \frac{1}{n}, \infty)$. Since $x \in [a, \infty)$ was arbitrarily chosen and we see that it is an element of $\bigcap_{n=1}^{\infty}(a - \frac{1}{n}, \infty)$, we obtain the relation

$$[a, \infty) \subseteq \bigcap_{n=1}^{\infty} \left(a - \frac{1}{n}, \infty\right). \tag{2.25}$$

Let now $x \in \bigcap_{n=1}^{\infty}(a - \frac{1}{n}, \infty)$ be arbitrarily chosen and fixed. Then

$$a - x < \frac{1}{n}$$

for any $n \in \mathbf{N}$. Hence, $a - x \leq 0$, i. e., $x \in [a, \infty)$. Because $x \in \bigcap_{n=1}^{\infty}(a - \frac{1}{n}, \infty)$ was arbitrarily chosen and we see that it is an element of $[a, \infty)$, we conclude that

$$\bigcap_{n=1}^{\infty} \left(a - \frac{1}{n}, \infty\right) \subseteq [a, \infty).$$

From the previous relation and from (2.25), we obtain equation (2.24). Since

$$\left(a - \frac{1}{n}, \infty\right) \in \mathcal{F}$$

for any $n \in \mathbf{N}$ and $\mathcal{F}$ is closed with respect to the formation of countable intersections, we obtain $[a, \infty) \in \mathcal{F}$. Hence, using the fact that $\mathcal{F}$ is closed with respect to the complements, we find

$$(-\infty, a) = [a, \infty)^c \in \mathcal{F}.$$

Note that

$$[a,b) = \left((-\infty,a) \cup [b,\infty)\right)^c.$$

Because $(-\infty,a)$, $[b,\infty) \in \mathcal{F}$ and $\mathcal{F}$ is closed with respect to the formation of finite unions and complements, we conclude that $[a,b) \in \mathcal{F}$. Also,

$$[a,b] = (-\infty,b] \cap [a,\infty).$$

Since $(-\infty,b]$, $[a,\infty) \in \mathcal{F}$ and $\mathcal{F}$ is closed with respect to the formation of finite intersections, we obtain $[a,b] \in \mathcal{F}$. This completes the proof. □

Theorem 2.15. *Every interval is measurable.*

Proof. We will prove that every interval of the form (a,∞) is measurable. Let $\mathbf{A}$ be a set such that $a \notin \mathbf{A}$. Let also $\{\mathbf{I}_k\}_{k\in\mathbf{N}}$ be any collection of open, bounded intervals that covers $\mathbf{A}$. We define

$$\begin{aligned} \mathbf{A}_1 &= \mathbf{A} \cap (-\infty,a), \quad & \mathbf{A}_2 &= \mathbf{A} \cap (a,\infty), \\ \mathbf{I}'_k &= \mathbf{I}_k \cap (-\infty,a), \quad & \mathbf{I}''_k &= \mathbf{I}_k \cap (a,\infty) \end{aligned}$$

for each index k. Then

$$l(\mathbf{I}_k) = l(\mathbf{I}'_k) + l(\mathbf{I}''_k)$$

for each index k. Note that $\{\mathbf{I}'_k\}_{k\in\mathbf{N}}$ and $\{\mathbf{I}''_k\}_{k\in\mathbf{N}}$ are countable collections of open, bounded intervals that cover $\mathbf{A}_1$ and $\mathbf{A}_2$, respectively. Hence, using the definition for the outer measure, we get

$$m^\star(\mathbf{A}_1) \le \sum_{k=1}^{\infty} l(\mathbf{I}'_k) \quad \text{and} \quad m^\star(\mathbf{A}_2) \le \sum_{k=1}^{\infty} l(\mathbf{I}''_k).$$

Therefore

$$\begin{aligned} m^\star(\mathbf{A}_1) + m^\star(\mathbf{A}_2) &\le \sum_{k=1}^{\infty} l(\mathbf{I}'_k) + \sum_{k=1}^{\infty} l(\mathbf{I}''_k) \\ &= \sum_{k=1}^{\infty} \left(l(\mathbf{I}'_k) + l(\mathbf{I}''_k)\right) = \sum_{k=1}^{\infty} l(\mathbf{I}_k), \end{aligned}$$

i. e.,

$$m^\star(\mathbf{A}_1) + m^\star(\mathbf{A}_2) \le \sum_{k=1}^{\infty} l(\mathbf{I}_k).$$

Because $\{\mathbf{I}_k\}_{k\in\mathbf{N}}$ was arbitrarily chosen, from the previous inequality we get

$$m^\star(\mathbf{A} \cap (a,\infty)) + m^\star(\mathbf{A} \cap (a,\infty)^c) \le m^\star(\mathbf{A}). \tag{2.26}$$

Let now $a \in \mathbf{A}$. Then we consider the set $\mathbf{B} = \mathbf{A} \setminus \{a\}$. As above,

$$m^\star(\mathbf{B} \cap (a, \infty)) + m^\star(\mathbf{B} \cap (a, \infty)^c) \le m^\star(\mathbf{B}). \tag{2.27}$$

By Theorem 2.4, it follows that

$$\begin{aligned} m^\star(\mathbf{B}) &= m^\star(\mathbf{A}), \\ m^\star(\mathbf{B} \cap (a, \infty)) &= m^\star(\mathbf{A} \cap (a, \infty)), \\ m^\star(\mathbf{B} \cap (a, \infty)^c) &= m^\star(\mathbf{A} \cap (a, \infty)^c). \end{aligned}$$

Hence, by (2.27), we get (2.26). Therefore (a, ∞) is measurable. Because the measurable sets are σ-algebra and every interval of the form (a, ∞) is measurable, using Theorem 2.14, we conclude that every interval is measurable. This completes the proof. □

Theorem 2.16. *The translate of a measurable set is a measurable set.*

Proof. Let $\mathbf{E}$ be a measurable set. Let also $\mathbf{A}$ be any set and $y \in \mathbf{R}$. Since $\mathbf{E}$ is measurable, using Theorem 2.4, we have

$$\begin{aligned} m^\star(\mathbf{A}) = m^\star(\mathbf{A} - y) &= m^\star((\mathbf{A} - y) \cap \mathbf{E}) + m^\star((\mathbf{A} - y) \cap \mathbf{E}^c) \\ &= m^\star(\mathbf{A} \cap (\mathbf{E} + y)) + m^\star(\mathbf{A} \cap (\mathbf{E} + y)^c). \end{aligned}$$

Therefore $\mathbf{E} + y$ is a measurable set. This completes the proof. □

Theorem 2.17. *The intersection of an arbitrary nonempty collection of σ-algebras on $\mathbf{R}$ is a σ-algebra on $\mathbf{R}$.*

Proof. Let $\mathcal{A}_\alpha$, $\alpha \in \mathbf{I}$, are σ-algebras on $\mathbf{R}$. Here $\mathbf{I}$ is an index set. Let also

$$\mathcal{A} = \bigcap_{\alpha \in \mathbf{I}} \mathcal{A}_\alpha.$$

1. Since $\mathcal{A}_\alpha$, $\alpha \in \mathbf{I}$, are σ-algebras on $\mathbf{R}$, we have $\mathbf{R} \in \mathcal{A}_\alpha$ for any $\alpha \in \mathbf{I}$. Therefore $\mathbf{R} \in \mathcal{A} = \bigcap_{\alpha \in \mathbf{I}} \mathcal{A}_\alpha$.
2. Let $\mathbf{A} \in \mathcal{A}$ be any set. Then $\mathbf{A} \in \mathcal{A}_\alpha$ for any $\alpha \in \mathbf{I}$. Because $\mathcal{A}_\alpha$, $\alpha \in \mathbf{I}$, are σ-algebras, we have $\mathbf{A}^c \in \mathcal{A}_\alpha$ for any $\alpha \in \mathbf{I}$. Consequently $\mathbf{A}^c \in \mathcal{A}$.
3. Let $\{\mathbf{A}_n\}_{n \in \mathbf{N}} \subseteq \mathcal{A}$. Then $\{\mathbf{A}_n\}_{n \in \mathbf{N}} \subseteq \mathcal{A}_\alpha$ for any $\alpha \in \mathbf{I}$. Because $\mathcal{A}_\alpha$, $\alpha \in \mathbf{N}$, are σ-algebras, we obtain $\bigcup_{n=1}^\infty \mathbf{A}_n \in \mathcal{A}_\alpha$ for any $\alpha \in \mathbf{I}$. Therefore $\bigcup_{n=1}^\infty \mathbf{A}_n \in \mathcal{A}$.

Consequently $\mathcal{A}$ is a σ-algebra. This completes the proof. □

Definition 2.6. The intersection of all σ-algebras of subsets of $\mathbf{R}$ that contain the open sets is a σ-algebra called the Borel σ-algebra. It will be denoted by $\mathcal{B}(\mathbf{R})$. The members of $\mathcal{B}(\mathbf{R})$ are called Borel sets.

The Borel σ-algebra is contained in every σ-algebra that contains all open sets. Since the measurable sets are a σ-algebra containing all open sets, every Borel set is measurable.

Definition 2.7. A set is said to be a $\mathbf{G}_\delta$ set provided it is the intersection of a countable collection of open sets.

Exercise 2.1. Prove that the translate of a $\mathbf{G}_\delta$ set is a $\mathbf{G}_\delta$ set.

Definition 2.8. A set is said to be a $\mathbf{F}_\sigma$ set provided it is the union of a countable collection of closed sets.

Exercise 2.2. Prove that the translate of an $\mathbf{F}_\sigma$ set is a $\mathbf{F}_\sigma$ set.

Theorem 2.18. *Every open set is measurable.*

Proof. Note that every open set is a disjoint union of a countable collection of open intervals. Hence, from Theorems 2.15 and 2.13, it follows that every open set is measurable. This completes the proof. □

Exercise 2.3. Prove that every $\mathbf{F}_\sigma$ set and every $\mathbf{G}_\delta$ set is measurable.

Theorem 2.19. *Let* $\mathbf{E}$ *be any set of real numbers. Then* $\mathbf{E}$ *is measurable if and only if for each* $\epsilon > 0$ *there is an open set* $\mathbf{O}$*, containing* $\mathbf{E}$*, so that* $m^\star(\mathbf{O} \setminus \mathbf{E}) < \epsilon$.

Proof.

1. Suppose that $\mathbf{E}$ is measurable. We take $\epsilon > 0$ arbitrarily.
   (a) Let $m^\star(\mathbf{E}) < \infty$. By the definition of the outer measure, it follows that there is a countable collection of open intervals $\{\mathbf{I}_k\}_{k\in\mathbf{N}}$ which covers $\mathbf{E}$ and

$$\sum_{k=1}^{\infty} l(\mathbf{I}_k) < m^\star(\mathbf{E}) + \epsilon.$$

   We define

$$\mathbf{O} = \bigcup_{k=1}^{\infty} \mathbf{I}_k.$$

   Then $\mathbf{O}$ is an open set that contains $\mathbf{E}$. Hence, by Theorem 2.5, we obtain

$$m^\star(\mathbf{O}) = m^\star\left(\bigcup_{k=1}^{\infty} \mathbf{I}_k\right) \le \sum_{k=1}^{\infty} m^\star(\mathbf{I}_k) \le \sum_{k=1}^{\infty} l(\mathbf{I}_k) < m^\star(\mathbf{E}) + \epsilon.$$

   Therefore

$$m^\star(\mathbf{O}) - m^\star(\mathbf{E}) < \epsilon.$$

   Because

$$\mathbf{O} = (\mathbf{O} \setminus \mathbf{E}) \cup \mathbf{E} \quad \text{and} \quad (\mathbf{O} \setminus \mathbf{E}) \cap \mathbf{E} = \emptyset,$$

   using Theorem 2.7, we get

$$m^\star(\mathbf{O}) = m^\star(\mathbf{O} \setminus \mathbf{E}) + m^\star(\mathbf{E}).$$

Consequently

$$m^\star(\mathbf{O} \setminus \mathbf{E}) = m^\star(\mathbf{O}) - m^\star(\mathbf{E}) < \epsilon.$$

(b) Let $m^\star(\mathbf{E}) = \infty$. Then $\mathbf{E}$ may be expressed as the disjoint union of a countable collection $\{\mathbf{E}_k\}_{k\in\mathbf{N}}$ of measurable sets, each of which has finite outer measure. By the first case, for each $k \in \mathbf{N}$ there is an open set $\mathbf{O}_k$ containing $\mathbf{E}_k$ such that

$$m^\star(\mathbf{O}_k \setminus \mathbf{E}_k) < \frac{\epsilon}{2^k}.$$

Let

$$\mathbf{O} = \bigcup_{k=1}^{\infty} \mathbf{O}_k.$$

Then $\mathbf{O}$ is an open set that contains $\mathbf{E}$ and

$$\begin{aligned} m^\star(\mathbf{O} \setminus \mathbf{E}) &= m^\star\left(\left(\bigcup_{k=1}^{\infty} \mathbf{O}_k\right) \setminus \mathbf{E}\right) \leq m^\star\left(\bigcup_{k=1}^{\infty} (\mathbf{O}_k \setminus \mathbf{E}_k)\right) \\ &\leq \sum_{k=1}^{\infty} m^\star(\mathbf{O}_k \setminus \mathbf{E}_k) < \sum_{k=1}^{\infty} \frac{\epsilon}{2^k} = \epsilon. \end{aligned}$$

2. Let for each $\epsilon > 0$ there is an open set $\mathbf{O}$ containing $\mathbf{E}$ such that $m^\star(\mathbf{O} \setminus \mathbf{E}) < \epsilon$. Let $\mathbf{A}$ be any set. Then, since $\mathbf{O}$ is an open set, using Theorem 2.18, it is measurable. Therefore

$$m^\star(\mathbf{A}) \geq m^\star(\mathbf{A} \cap \mathbf{O}) + m^\star(\mathbf{A} \cap \mathbf{O}^c). \tag{2.28}$$

Using $\mathbf{A} \cap \mathbf{E} \subseteq \mathbf{A} \cap \mathbf{O}$ and Theorem 2.1, we have

$$m^\star(\mathbf{A} \cap \mathbf{E}) \leq m^\star(\mathbf{A} \cap \mathbf{O}). \tag{2.29}$$

Now we will prove that

$$\mathbf{E} = \mathbf{O} \cap (\mathbf{O} \setminus \mathbf{E})^c. \tag{2.30}$$

Let $x \in \mathbf{E}$ be arbitrarily chosen. Then $x \in \mathbf{O}$ and $x \notin \mathbf{O} \setminus \mathbf{E}$. Therefore $x \in (\mathbf{O} \setminus \mathbf{E})^c$. Because $x \in \mathbf{O}$ and $x \in (\mathbf{O} \setminus \mathbf{E})^c$, we get $x \in \mathbf{O} \cap (\mathbf{O} \setminus \mathbf{E})^c$. Since $x \in \mathbf{E}$ was arbitrarily chosen and we see that it is an element of the set $\mathbf{O} \cap (\mathbf{O} \setminus \mathbf{E})^c$, we conclude that

$$\mathbf{E} \subseteq \mathbf{O} \cap (\mathbf{O} \setminus \mathbf{E})^c. \tag{2.31}$$

Let $x \in \mathbf{O} \cap (\mathbf{O} \setminus \mathbf{E})^c$ be arbitrarily chosen and fixed. Then $x \in \mathbf{O}$ and $x \in (\mathbf{O} \setminus \mathbf{E})^c$. Hence, $x \notin \mathbf{O} \setminus \mathbf{E}$ and $x \in \mathbf{E}$. Because $x \in \mathbf{O} \cap (\mathbf{O} \setminus \mathbf{E})^c$ was arbitrarily chosen and we see that it is an element of $\mathbf{E}$, we obtain

$$\mathbf{O} \cap (\mathbf{O} \setminus \mathbf{E})^c \subseteq \mathbf{E}.$$

From the previous relation and from (2.31), we get equation (2.30). By (2.30), we find

$$\mathbf{E}^c = (\mathbf{O} \cap (\mathbf{O} \setminus \mathbf{E})^c)^c = \mathbf{O}^c \cup (\mathbf{O} \setminus \mathbf{E}).$$

Now we will prove

$$\mathbf{A} \cap \mathbf{E}^c \subseteq (\mathbf{A} \cap \mathbf{O}^c) \cup (\mathbf{O} \setminus \mathbf{E}). \tag{2.32}$$

Let $x \in \mathbf{A} \cap \mathbf{E}^c$ be arbitrarily chosen and fixed. Then $x \in \mathbf{A}$ and $x \in \mathbf{E}^c$. Hence, $x \notin \mathbf{E}$.
(a) If $x \in \mathbf{O}^c$, then $x \in \mathbf{A} \cap \mathbf{O}^c$ and $x \in (\mathbf{A} \cap \mathbf{O}^c) \cup (\mathbf{O} \setminus \mathbf{E})$.
(b) If $x \notin \mathbf{O}^c$, then $x \in \mathbf{O}$ and hence, $x \in \mathbf{O} \setminus \mathbf{E}$ and $x \in (\mathbf{A} \cap \mathbf{O}^c) \cup (\mathbf{O} \setminus \mathbf{E})$.
Because $x \in \mathbf{A} \cap \mathbf{E}^c$ was arbitrarily chosen and we see that it is an element of $(\mathbf{A} \cap \mathbf{O}^c) \cup (\mathbf{O} \setminus \mathbf{E})$, we get equation (2.32). By (2.32), we find

$$m^*(\mathbf{A} \cap \mathbf{E}^c) \le m^*((\mathbf{A} \cap \mathbf{O}^c) \cup (\mathbf{O} \setminus \mathbf{E})) \le m^*(\mathbf{A} \cap \mathbf{O}^c) + m^*(\mathbf{O} \setminus \mathbf{E}) < \epsilon + m^*(\mathbf{A} \cap \mathbf{O}^c).$$

From the previous inequality and from (2.29) and (2.28), we obtain

$$m^*(\mathbf{A} \cap \mathbf{E}) + m^*(\mathbf{A} \cap \mathbf{E}^c) \le m^*(\mathbf{A} \cap \mathbf{O}) + m^*(\mathbf{A} \cap \mathbf{O}^c) + \epsilon \le m^*(\mathbf{A}) + \epsilon.$$

Because $\epsilon > 0$ was arbitrarily chosen, we conclude that

$$m^*(\mathbf{A}) \ge m^*(\mathbf{A} \cap \mathbf{E}) + m^*(\mathbf{A} \cap \mathbf{E}^c),$$

i. e., **E** is measurable. This completes the proof. □

Theorem 2.20. *Let* **E** *be any set of real numbers. Then* **E** *is measurable if and only if there is a* $\mathbf{G}_\delta$ *set* **G**, *containing* **E**, *such that* $m^*(\mathbf{G} \setminus \mathbf{E}) = 0$.

Proof.

1. Let **E** be measurable. By Theorem 2.19, it follows that for any $k \in \mathbf{N}$ there is an open set $\mathbf{O}_k$, containing **E**, such that

$$m^*(\mathbf{O}_k \setminus \mathbf{E}) < \frac{1}{k}.$$

Let $\mathbf{G} = \bigcap_{k=1}^{\infty} \mathbf{O}_k$. Then **G** is a $\mathbf{G}_\delta$ set containing **E**. Also,

$$\mathbf{G} \setminus \mathbf{E} \subseteq \bigcap_{k=1}^{\infty} (\mathbf{O}_k \setminus \mathbf{E}). \tag{2.33}$$

Really, let $x \in \mathbf{G} \setminus \mathbf{E}$ be arbitrarily chosen. Then $x \in \mathbf{G}$ and $x \notin \mathbf{E}$. By the definition of the set **G**, it follows that $x \in \mathbf{O}_k$ for any $k \in \mathbf{N}$. Therefore $x \in \mathbf{O}_k \setminus \mathbf{E}$ for any $k \in \mathbf{N}$. Hence, $x \in \bigcap_{k=1}^{\infty} (\mathbf{O}_k \setminus \mathbf{E})$. Because $x \in \mathbf{G} \setminus \mathbf{E}$ was arbitrarily chosen and we see that it is an element of the set $\bigcap_{k=1}^{\infty}(\mathbf{O}_k \setminus \mathbf{E})$, we get equation (2.33). By (2.33) and Theorem 2.1, we find

$$m^\star(\mathbf{G} \setminus \mathbf{E}) \leq m^\star\left(\bigcap_{k=1}^{\infty}(\mathbf{O}_k \setminus \mathbf{E})\right) \leq m^\star(\mathbf{O}_k \setminus \mathbf{E}) < \frac{1}{k}$$

for any $k \in \mathbf{N}$. Therefore $m^\star(\mathbf{G} \setminus \mathbf{E}) = 0$.

2. Let there is a $\mathbf{G}_\delta$ set $\mathbf{G}$ containing $\mathbf{E}$ such that $m^\star(\mathbf{G} \setminus \mathbf{E}) = 0$. Since any set of outer measure zero is measurable, we have $\mathbf{G} \setminus \mathbf{E} \in \mathcal{M}$. Because $\mathcal{M}$ is a σ- algebra, we have $(\mathbf{G} \setminus \mathbf{E})^c \in \mathcal{M}$ and $\mathbf{G} \cap (\mathbf{G} \setminus \mathbf{E})^c \in \mathcal{M}$. From this, using $\mathbf{E} = \mathbf{G} \cap (\mathbf{G} \setminus \mathbf{E})^c$ (which one can prove as in (2.30)), we conclude that $\mathbf{E}$ is measurable. This completes the proof. □

Theorem 2.21. *Let $\mathbf{E}$ be any set of real numbers. Then $\mathbf{E}$ is measurable if and only if for each $\epsilon > 0$ there is a closed set $\mathbf{B}$ contained in $\mathbf{E}$ such that $m^\star(\mathbf{E} \setminus \mathbf{B}) < \epsilon$.*

Proof.

1. Let $\mathbf{E}$ be a measurable set. Then $\mathbf{E}^c$ is a measurable set. We take $\epsilon > 0$ arbitrarily. By Theorem 2.19, it follows that there is an open set $\mathbf{O}$ containing $\mathbf{E}^c$ such that

$$m^\star(\mathbf{O} \setminus \mathbf{E}^c) < \epsilon. \tag{2.34}$$

Note that $\mathbf{O}^c$ is a closed set and $\mathbf{O}^c \subset \mathbf{E}$. Now we will prove that

$$\mathbf{E} \setminus \mathbf{O}^c = \mathbf{O} \setminus \mathbf{E}^c. \tag{2.35}$$

Really, let $x \in \mathbf{E} \setminus \mathbf{O}^c$ be arbitrarily chosen. Then $x \in \mathbf{E}$ and $x \notin \mathbf{O}^c$. Hence, $x \in \mathbf{O}$ and $x \notin \mathbf{E}^c$. Therefore $x \in \mathbf{O} \setminus \mathbf{E}^c$. Since $x \in \mathbf{E} \setminus \mathbf{O}^c$ was arbitrarily chosen and we see that it is an element of $\mathbf{O} \setminus \mathbf{E}^c$, and we obtain

$$\mathbf{E} \setminus \mathbf{O}^c \subseteq \mathbf{O} \setminus \mathbf{E}^c. \tag{2.36}$$

Let now $x \in \mathbf{O} \setminus \mathbf{E}^c$ be arbitrarily chosen. Then $x \in \mathbf{O}$ and $x \notin \mathbf{E}^c$. Hence, $x \notin \mathbf{O}^c$ and $x \in \mathbf{E}$. Therefore $x \in \mathbf{E} \setminus \mathbf{O}^c$. Because $x \in \mathbf{O} \setminus \mathbf{E}^c$ was arbitrarily chosen and we see that it is an element of the set $\mathbf{E} \setminus \mathbf{O}^c$, we get

$$\mathbf{O} \setminus \mathbf{E}^c \subseteq \mathbf{E} \setminus \mathbf{O}^c.$$

From the previous relation and from (2.36), we obtain equation (2.35). By (2.35), Theorem 2.1 and (2.34), we find

$$m^\star(\mathbf{E} \setminus \mathbf{O}^c) = m^\star(\mathbf{O} \setminus \mathbf{E}^c) < \epsilon.$$

We set $\mathbf{B} = \mathbf{O}^c$.

2. Let for any $\epsilon > 0$ there is a closed set $\mathbf{B}$, contained in $\mathbf{E}$, such that $m^\star(\mathbf{E}\setminus\mathbf{B}) < \epsilon$. By (2.35), we obtain $m^\star(\mathbf{B}^c \setminus \mathbf{E}^c) < \epsilon$. Note that $\mathbf{B}^c$ is an open set and $\mathbf{E}^c \subseteq \mathbf{B}^c$. Hence, by Theorem 2.19, we conclude that $\mathbf{E}^c$ is measurable, i. e., $\mathbf{E}^c \in \mathcal{M}$. Because $\mathcal{M}$ is a σ-algebra, we see that $\mathbf{E} \in \mathcal{M}$, i. e., $\mathbf{E}$ is a measurable set. This completes the proof. □

Theorem 2.22. *Let* $\mathbf{E}$ *be any set of real numbers. Then* $\mathbf{E}$ *is measurable if and only if there is an* F_σ *set* $\mathbf{B}$, *contained in* $\mathbf{E}$, *such that* $m^\star(\mathbf{E} \setminus \mathbf{B}) = 0$.

Proof. By Theorem 2.20, it follows that the set $\mathbf{E}^c$ is measurable if and only if there is a $\mathbf{G}_\delta$ set $\mathbf{A}$, containing $\mathbf{E}^c$, such that $m^\star(\mathbf{A}\setminus\mathbf{E}^c) = 0$. Hence, using (2.35) and using the fact that the complement of a $\mathbf{G}_\delta$ set is an $\mathbf{F}_\sigma$ set, we get the desired result. This completes the proof. □

Theorem 2.23. *Let* $\mathbf{E}$ *be a measurable set of finite outer measure. Then for each* $\epsilon > 0$ *there is a finite disjoint collection of open intervals* $\{\mathbf{I}_k\}_{k=1}^n$ *for which if* $\mathbf{O} = \bigcup_{k=1}^n \mathbf{I}_k$, *then*

$$m^\star(\mathbf{E} \setminus \mathbf{O}) + m^\star(\mathbf{O} \setminus \mathbf{E}) < \epsilon.$$

Proof. By Theorem 2.19, it follows that for each $\epsilon > 0$ there is an open set $\mathbf{B}$, containing $\mathbf{E}$, such that $m^\star(\mathbf{B} \setminus \mathbf{E}) < \frac{\epsilon}{2}$. Since every open set is measurable, we see that $\mathbf{B}$ is measurable. Also, we have

$$m^\star(\mathbf{B}) \le m^\star(\mathbf{E}) + m^\star(\mathbf{B} \setminus \mathbf{E}) < m^\star(\mathbf{E}) + \frac{\epsilon}{2}.$$

Therefore $\mathbf{B}$ is a measurable set of finite outer measure. Because every open set of real numbers is a disjoint union of a countable collection of open intervals, there is a disjoint countable collection of open intervals $\{\mathbf{I}_k\}_{k\in\mathbf{N}}$ for which

$$\mathbf{B} = \bigcup_{k=1}^{\infty} \mathbf{I}_k, \quad \mathbf{I}_k \cap \mathbf{I}_l = \emptyset, \quad k \neq l, \quad k, l \in \mathbf{N}.$$

Since for any $n \in \mathbf{N}$ we have $\bigcup_{k=1}^n \mathbf{I}_k \subseteq \mathbf{B}$, using Theorem 2.1, we get

$$\sum_{k=1}^{n} l(\mathbf{I}_k) = \sum_{k=1}^{n} m^\star(\mathbf{I}_k) = m^\star\left(\bigcup_{k=1}^{n} \mathbf{I}_k\right) \le m^\star(\mathbf{B}) < \infty.$$

Therefore

$$\sum_{k=1}^{\infty} l(\mathbf{I}_k) < \infty.$$

From this, there is an $n \in \mathbf{N}$ such that

$$\sum_{k=n+1}^{\infty} l(\mathbf{I}_k) < \frac{\epsilon}{2}.$$

We define

$$\mathbf{O} = \bigcup_{k=1}^{n} \mathbf{I}_k.$$

Using that $\mathbf{O} \setminus \mathbf{E} \subseteq \mathbf{B} \setminus \mathbf{E}$ and Theorem 2.1, we obtain

$$m^\star(\mathbf{O} \setminus \mathbf{E}) \le m^\star(\mathbf{B} \setminus \mathbf{E}) < \frac{\epsilon}{2}.$$

On the other hand, $\mathbf{E} \subseteq \mathbf{B}$ and

$$\mathbf{E} \setminus \mathbf{O} \subseteq \mathbf{B} \setminus \mathbf{O} = \bigcup_{k=n+1}^{\infty} \mathbf{I}_k.$$

By the definition of the outer measure, we get

$$m^\star(\mathbf{E} \setminus \mathbf{O}) \le \sum_{k=n+1}^{\infty} l(\mathbf{I}_k) < \frac{\epsilon}{2}.$$

Thus,

$$m^\star(\mathbf{O} \setminus \mathbf{E}) + m^\star(\mathbf{E} \setminus \mathbf{O}) < \frac{\epsilon}{2} + \frac{\epsilon}{2} = \epsilon.$$

This completes the proof. □

Exercise 2.4. Show that a set $\mathbf{E}$ is measurable if and only if for each $\epsilon > 0$, there is a closed set $\mathbf{A}$ and an open set $\mathbf{B}$ for which

$$\mathbf{A} \subseteq \mathbf{E} \subseteq \mathbf{B} \quad \text{and} \quad m^\star(\mathbf{B} \setminus \mathbf{A}) < \epsilon.$$

Solution. By Theorems 2.19 and 2.21, it follows that the set $\mathbf{E}$ is measurable if and only if for each $\epsilon > 0$ there are a closed set $\mathbf{A}$ and an open set $\mathbf{B}$ such that

$$\mathbf{A} \subseteq \mathbf{E} \subseteq \mathbf{B}$$

and

$$m^\star(\mathbf{E} \setminus \mathbf{A}) < \frac{\epsilon}{2}, \quad m^\star(\mathbf{B} \setminus \mathbf{E}) < \frac{\epsilon}{2}. \tag{2.37}$$

Now we will prove

$$\mathbf{B} \setminus \mathbf{A} = (\mathbf{B} \setminus \mathbf{E}) \cup (\mathbf{E} \setminus \mathbf{A}). \tag{2.38}$$

Let $x \in \mathbf{B} \setminus \mathbf{A}$ be arbitrarily chosen. Then $x \in \mathbf{B}$ and $x \notin \mathbf{A}$.

1. If $x \in \mathbf{E}$, then $x \in \mathbf{E} \setminus \mathbf{A}$ and $x \in (\mathbf{B} \setminus \mathbf{E}) \cup (\mathbf{E} \setminus \mathbf{A})$.
2. If $x \notin \mathbf{E}$, then $x \in \mathbf{B} \setminus \mathbf{E}$ and $x \in (\mathbf{B} \setminus \mathbf{E}) \cup (\mathbf{E} \setminus \mathbf{A})$.

Because $x \in \mathbf{B} \setminus \mathbf{A}$ was arbitrarily chosen and we see that it is an element of $(\mathbf{B} \setminus \mathbf{E}) \cup (\mathbf{E} \setminus \mathbf{A})$, we get

$$\mathbf{B} \setminus \mathbf{A} \subseteq (\mathbf{B} \setminus \mathbf{E}) \cup (\mathbf{E} \setminus \mathbf{A}). \tag{2.39}$$

Let now $x \in (\mathbf{B} \setminus \mathbf{E}) \cup (\mathbf{E} \setminus \mathbf{A})$ be arbitrarily chosen. Then $x \in \mathbf{B} \setminus \mathbf{E}$ or $x \in \mathbf{E} \setminus \mathbf{A}$.

1. Let $x \in \mathbf{B} \setminus \mathbf{E}$. Then $x \in \mathbf{B}$ and $x \notin \mathbf{E}$. Hence, $x \notin \mathbf{A}$ and $x \in \mathbf{B}$. Therefore $x \in \mathbf{B} \setminus \mathbf{A}$.
2. Let $x \in \mathbf{E} \setminus \mathbf{A}$. Then $x \in \mathbf{E}$ and $x \notin \mathbf{A}$. Hence, $x \in \mathbf{B}$ and $x \notin \mathbf{A}$. Therefore $x \in \mathbf{B} \setminus \mathbf{A}$.

Because $x \in (\mathbf{B} \setminus \mathbf{E}) \cup (\mathbf{E} \setminus \mathbf{A})$ was arbitrarily chosen and we see that it is an element of $\mathbf{B} \setminus \mathbf{A}$, we conclude that

$$(\mathbf{B} \setminus \mathbf{E}) \cup (\mathbf{E} \setminus \mathbf{A}) \subseteq \mathbf{B} \setminus \mathbf{A}.$$

From the previous relation and from (2.39), we get equation (2.38). By (2.38) and Theorem 2.1, we obtain

$$m^\star(\mathbf{B} \setminus \mathbf{A}) = m^\star((\mathbf{B} \setminus \mathbf{E}) \cup (\mathbf{E} \setminus \mathbf{A})) \le m^\star(\mathbf{B} \setminus \mathbf{E}) + m^\star(\mathbf{E} \setminus \mathbf{A}) < \frac{\epsilon}{2} + \frac{\epsilon}{2} = \epsilon.$$

This completes the proof.

Exercise 2.5. Let $\mathbf{E}$ has finite outer measure. Show that there is an F_σ set $\mathbf{A}$ and a G_δ set $\mathbf{B}$ such that

$$\mathbf{A} \subseteq \mathbf{E} \subseteq \mathbf{B} \quad \text{and} \quad m^\star(\mathbf{A}) = m^\star(\mathbf{E}) = m^\star(\mathbf{B}).$$

2.2 The Lebesgue measure. The Borel–Cantelli lemma

Definition 2.9. The restriction of the set function outer measure to the class of measurable sets is called Lebesgue measure. It is denoted by m. In other words, if the set $\mathbf{E}$ is measurable, its Lebesgue measure, $m(\mathbf{E})$, is defined by

$$m(\mathbf{E}) = m^\star(\mathbf{E}).$$

Theorem 2.24. *The Lebesgue measure is countably additive, i. e., if $\{\mathbf{E}_k\}_{k\in\mathbf{N}}$ is a countable disjoint collection of measurable sets, then its union $\bigcup_{k=1}^\infty \mathbf{E}_k$ is also measurable and*

$$m\left(\bigcup_{k=1}^\infty \mathbf{E}_k\right) = \sum_{k=1}^\infty m(\mathbf{E}_k).$$

Proof. By Theorem 2.13, it follows that $\bigcup_{k=1}^\infty \mathbf{E}_k$ is a measurable set. By Theorem 2.5, we get

$$m\left(\bigcup_{k=1}^\infty \mathbf{E}_k\right) \le \sum_{k=1}^\infty m(\mathbf{E}_k). \tag{2.40}$$

On the other hand, using Theorem 2.10, we obtain

$$m\left(\bigcup_{k=1}^{n} \mathbf{E}_k\right) = \sum_{k=1}^{n} m(\mathbf{E}_k). \tag{2.41}$$

Since $\bigcup_{k=1}^{n} \mathbf{E}_k \subseteq \bigcup_{k=1}^{\infty} \mathbf{E}_k$ for any $n \in \mathbf{N}$, using Theorem 2.1, we obtain

$$m\left(\bigcup_{k=1}^{n} \mathbf{E}_k\right) \le m\left(\bigcup_{k=1}^{\infty} \mathbf{E}_k\right)$$

for any $n \in \mathbf{N}$. Hence, by (2.41), we find

$$m\left(\bigcup_{k=1}^{\infty} \mathbf{E}_k\right) \ge \sum_{k=1}^{n} m(\mathbf{E}_k)$$

for any $n \in \mathbf{N}$. Therefore

$$m\left(\bigcup_{k=1}^{\infty} \mathbf{E}_k\right) \ge \sum_{k=1}^{\infty} m(\mathbf{E}_k).$$

From the previous inequality and from (2.40), we get the desired result. This completes the proof. □

Remark 2.4. The set function Lebesgue measure, defined on the σ-algebra of Lebesgue measurable sets, assigns the length to any interval, is translation invariant, and is countable additive.

Definition 2.10. A countable collection of sets $\{\mathbf{E}_k\}_{k\in\mathbf{N}}$ is said to be:
1. ascending provided for each $k \in \mathbf{N}$, $\mathbf{E}_k \subseteq \mathbf{E}_{k+1}$;
2. descending provided for each $k \in \mathbf{N}$, $\mathbf{E}_{k+1} \subseteq \mathbf{E}_k$.

Theorem 2.25 (The continuity of measure). *The Lebesgue measure possesses the following properties.*
1. *If $\{\mathbf{E}_k\}_{k\in\mathbf{N}}$ is an ascendent collection of measurable sets, then*

$$m\left(\bigcup_{k=1}^{\infty} \mathbf{E}_k\right) = \lim_{k\to\infty} m(\mathbf{E}_k). \tag{2.42}$$

2. *If $\{\mathbf{E}_k\}_{k\in\mathbf{N}}$ is a descendent collection of measurable sets and $m(\mathbf{E}_1) < \infty$, then*

$$m\left(\bigcap_{k=1}^{\infty} \mathbf{E}_k\right) = \lim_{k\to\infty} m(\mathbf{E}_k). \tag{2.43}$$

Proof.

1. Suppose that there is an index l such that $m(\mathbf{E}_l) = \infty$. Then, using

$$m\left(\bigcup_{k=1}^{\infty} \mathbf{E}_k\right) \geq m(\mathbf{E}_l) \quad \text{and} \quad m(\mathbf{E}_p) \geq m(\mathbf{E}_l)$$

for any $p \geq l$, $p \in \mathbf{N}$, we get

$$m\left(\bigcup_{k=1}^{\infty} \mathbf{E}_k\right) = \infty \quad \text{and} \quad m(\mathbf{E}_p) = \infty$$

for any $p \geq l$, $p \in \mathbf{N}$. Therefore (2.42) holds. Now we assume that $m(\mathbf{E}_l) < \infty$ for any $l \in \mathbf{N}$. We set $A_0 = \emptyset$ and define the sets

$$\mathbf{A}_k = \mathbf{E}_k \setminus \mathbf{E}_{k-1}, \quad k \in \mathbf{N}.$$

We will prove that $\{\mathbf{A}_k\}_{k \in \mathbf{N}}$ is a disjoint collection. Assume that there are $k, l \in \mathbf{N}$, $k > l$, such that there is an $x \in \mathbf{A}_k \cap \mathbf{A}_l$. By the definition of the sets $\mathbf{A}_k$, we obtain $x \notin \mathbf{E}_{k-1}$. Since $\{\mathbf{E}_k\}_{k \in \mathbf{N}}$ is an ascendent collection, we conclude that $x \notin \mathbf{E}_p$ for any $p \leq k-1$, $p \in \mathbf{N}$. Therefore $x \notin \mathbf{A}_l$. This is a contradiction. Consequently $\{\mathbf{A}_k\}_{k \in \mathbf{N}}$ is a disjoint collection. Now we will prove that

$$\bigcup_{k=1}^{\infty} \mathbf{A}_k = \bigcup_{k=1}^{\infty} \mathbf{E}_k. \tag{2.44}$$

Let $x \in \bigcup_{k=1}^{\infty} \mathbf{A}_k$ be arbitrarily chosen. Then there is an $l \in \mathbf{N}$ such that $x \in \mathbf{A}_l = \mathbf{E}_l \setminus \mathbf{E}_{l-1}$. Hence, $x \in \mathbf{E}_l$ and $x \in \bigcup_{k=1}^{\infty} \mathbf{E}_k$. Because $x \in \bigcup_{k=1}^{\infty} \mathbf{A}_k$ was arbitrarily chosen and we see that it is an element of $\bigcup_{k=1}^{\infty} \mathbf{E}_k$, we conclude that

$$\bigcup_{k=1}^{\infty} \mathbf{A}_k \subseteq \bigcup_{k=1}^{\infty} \mathbf{E}_k. \tag{2.45}$$

Let now $x \in \bigcup_{k=1}^{\infty} \mathbf{E}_k$ be arbitrarily chosen. Then there is an $l \in \mathbf{N}$ such that $x \in \mathbf{E}_l$. Assume that $x \notin \mathbf{A}_k$ for any $k \in \mathbf{N}$. Then $x \notin \mathbf{E}_k$ for any $k \in \mathbf{N}$. In particular, $x \notin \mathbf{E}_l$. This is a contradiction. Therefore there is a $p \in \mathbf{N}$ such that $x \in \mathbf{A}_p$. Hence, $x \in \bigcup_{k=1}^{\infty} \mathbf{A}_k$. Because $x \in \bigcup_{k=1}^{\infty} \mathbf{E}_k$ was arbitrarily chosen and we see that it is an element of $\bigcup_{k=1}^{\infty} \mathbf{A}_k$, we obtain

$$\bigcup_{k=1}^{\infty} \mathbf{E}_k \subseteq \bigcup_{k=1}^{\infty} \mathbf{A}_k.$$

From the previous relation and from (2.45), we get equation (2.44). By the countable additivity of m, we have

$$m\left(\bigcup_{k=1}^{\infty} \mathbf{E}_k\right) = m\left(\bigcup_{k=1}^{\infty} \mathbf{A}_k\right) = \sum_{k=1}^{\infty} m(\mathbf{A}_k) = \sum_{k=1}^{\infty} m(\mathbf{E}_k \setminus \mathbf{E}_{k-1})$$

$$= \sum_{k=1}^{\infty}(m(\mathbf{E}_k) - m(\mathbf{E}_{k-1})) = \lim_{n\to\infty}\sum_{k=1}^{n}(m(\mathbf{E}_k) - m(\mathbf{E}_{k-1}))$$
$$= \lim_{n\to\infty}(m(\mathbf{E}_n) - m(\emptyset)) = \lim_{n\to\infty} m(\mathbf{E}_n).$$

2. Let

$$\mathbf{B}_k = \mathbf{E}_1 \setminus \mathbf{E}_k, \quad k \in \mathbf{N}.$$

Since $\{\mathbf{E}_k\}_{k\in\mathbf{N}}$ is a descendent collection, we see that $\{\mathbf{B}_k\}_{k\in\mathbf{N}}$ is an ascendent collection. By (2.42), we get

$$m\left(\bigcup_{k=1}^{\infty}\mathbf{B}_k\right) = \lim_{n\to\infty} m(\mathbf{B}_n). \tag{2.46}$$

Now we will prove that

$$\bigcup_{k=1}^{\infty}\mathbf{B}_k = \mathbf{E}_1 \setminus \left(\bigcap_{k=1}^{\infty}\mathbf{E}_k\right). \tag{2.47}$$

Let $x \in \bigcup_{k=1}^{\infty}\mathbf{B}_k$ be arbitrarily chosen. Then there is an $l \in \mathbf{N}$ such that $x \in \mathbf{B}_l = \mathbf{E}_1 \setminus \mathbf{E}_l$. Hence, $x \in \mathbf{E}_1$ and $x \notin \mathbf{E}_l$. Therefore $x \notin \bigcap_{k=1}^{\infty}\mathbf{E}_k$. From this, $x \in \mathbf{E}_1 \setminus (\bigcap_{k=1}^{\infty}\mathbf{E}_k)$. Because $x \in \bigcup_{k=1}^{\infty}\mathbf{B}_k$ was arbitrarily chosen and we see that it is an element of $\mathbf{E}_1 \setminus (\bigcap_{k=1}^{\infty}\mathbf{E}_k)$, we conclude that

$$\bigcup_{k=1}^{\infty}\mathbf{B}_k \subseteq \mathbf{E}_1 \setminus \left(\bigcap_{k=1}^{\infty}\mathbf{E}_k\right). \tag{2.48}$$

Let $x \in \mathbf{E}_1 \setminus (\bigcap_{k=1}^{\infty}\mathbf{E}_k)$ be arbitrarily chosen. Then $x \in \mathbf{E}_1$ and $x \notin \bigcap_{k=1}^{\infty}\mathbf{E}_k$. Hence, there is an $l \in \mathbf{N}$, $l \neq 1$, so that $x \notin \mathbf{E}_l$. Therefore $x \in \mathbf{B}_l$ and $x \in \bigcup_{k=1}^{\infty}\mathbf{B}_k$. Since $x \in \mathbf{E}_1 \setminus (\bigcap_{k=1}^{\infty}\mathbf{E}_k)$ was arbitrarily chosen and we see that it is an element of the set $\bigcup_{k=1}^{\infty}\mathbf{B}_k$, we conclude that

$$\mathbf{E}_1 \setminus \left(\bigcap_{k=1}^{\infty}\mathbf{E}_k\right) \subseteq \bigcup_{k=1}^{\infty}\mathbf{B}_k.$$

From the previous relation and from (2.48), we get equation (2.47). By equation (2.47), we find

$$m\left(\bigcup_{k=1}^{\infty}\mathbf{B}_k\right) = m\left(\mathbf{E}_1 \setminus \left(\bigcap_{k=1}^{\infty}\mathbf{E}_k\right)\right) = m(\mathbf{E}_1) - m\left(\bigcap_{k=1}^{\infty}\mathbf{E}_k\right).$$

Hence, by (2.46), we obtain

$$m(\mathbf{E}_1) - m\left(\bigcap_{k=1}^{\infty}\mathbf{E}_k\right) = \lim_{n\to\infty} m(\mathbf{B}_n)$$

$$= \lim_{n\to\infty} m(\mathbf{E}_1 \setminus \mathbf{E}_n) = \lim_{n\to\infty} (m(\mathbf{E}_1) - m(\mathbf{E}_n))$$
$$= m(\mathbf{E}_1) - \lim_{n\to\infty} m(\mathbf{E}_n),$$

whereupon we obtain (2.43). This completes the proof. □

Definition 2.11. For a measurable set $\mathbf{E}$, we say that a property holds almost everywhere on $\mathbf{E}$, or it holds for almost all $x \in \mathbf{E}$, provided there is a subset $\mathbf{E}_0$ of $\mathbf{E}$ for which $m(\mathbf{E}_0) = 0$ and the property holds for all $x \in \mathbf{E} \setminus \mathbf{E}_0$.

Lemma 2.4 (Borel–Cantelli lemma). *Let $\{\mathbf{E}_k\}_{k\in\mathbf{N}}$ be a countable collection of measurable sets for which $\sum_{k=1}^{\infty} m(\mathbf{E}_k) < \infty$. Then almost all $x \in \mathbf{R}$ belong to at most finitely many of the $\mathbf{E}_k$.*

Proof. By Theorem 2.5, we have

$$m\left(\bigcup_{k=1}^{\infty} \mathbf{E}_k\right) \le \sum_{k=1}^{\infty} m(\mathbf{E}_k) < \infty.$$

Note that $\{\bigcup_{k=n}^{\infty} \mathbf{E}_k\}_{n\in\mathbf{N}}$ is a descendent collection. Hence, by (2.43), we obtain

$$m\left(\bigcap_{n=1}^{\infty}\left(\bigcup_{k=n}^{\infty} \mathbf{E}_k\right)\right) = \lim_{n\to\infty} m\left(\bigcup_{k=n}^{\infty} \mathbf{E}_k\right) \le \lim_{n\to\infty} \sum_{k=n}^{\infty} m(\mathbf{E}_k) = 0.$$

Therefore almost all $x \in \mathbf{R}$ fail to belong to $\bigcap_{n=1}^{\infty}(\bigcup_{k=n}^{\infty} \mathbf{E}_k)$ and therefore belong to at most finitely many $\mathbf{E}_k$. This completes the proof. □

Exercise 2.6. Show that if $\mathbf{E}_1$ and $\mathbf{E}_2$ are measurable, then

$$m(\mathbf{E}_1 \cup \mathbf{E}_2) + m(\mathbf{E}_1 \cap \mathbf{E}_2) = m(\mathbf{E}_1) + m(\mathbf{E}_2).$$

2.3 Nonmeasurable sets

Theorem 2.26. *Let $\mathbf{E}$ be a bounded measurable set of real numbers. Suppose that there is a bounded countable infinite set Λ of real numbers for which the collection of translates of $\mathbf{E}$, $\{\lambda + \mathbf{E}\}_{\lambda\in\Lambda}$, is disjoint. Then $m(\mathbf{E}) = 0$.*

Proof. Assume that $m(\mathbf{E}) > 0$. Because $\mathbf{E}$ is a bounded set, we have $m(\mathbf{E}) < \infty$. Since the translate of a measurable set is a measurable set, we see that $\lambda + \mathbf{E}$ is a measurable set and $m(\mathbf{E}) = m(\lambda + \mathbf{E})$ for any $\lambda \in \Lambda$. By Theorem 2.24, it follows that

$$m\left(\bigcup_{\lambda\in\Lambda}(\lambda + \mathbf{E})\right) = \sum_{\lambda\in\Lambda} m(\lambda + \mathbf{E}) = \sum_{\lambda\in\Lambda} m(\mathbf{E}). \tag{2.49}$$

Because $\mathbf{E}$ and Λ are bounded sets, we see that $\bigcup_{\lambda\in\Lambda}(\lambda + \mathbf{E})$ is a bounded set. Hence,

$$m\left(\bigcup_{\lambda\in\Lambda}(\lambda + \mathbf{E})\right) < \infty. \tag{2.50}$$

Since $m(\mathbf{E}) < \infty$ and Λ is countable infinite, we get $\sum_{\lambda \in \Lambda} m(\mathbf{E}) = \infty$. Hence, by (2.49), we obtain $m(\bigcup_{\lambda \in \Lambda} (\lambda + \mathbf{E})) = \infty$. This contradicts (2.50). Therefore $m(\mathbf{E}) = 0$. This completes the proof. □

Definition 2.12. For any nonempty set $\mathbf{E}$ of real numbers, we say that two points x and y of $\mathbf{E}$ are rationally equivalent provided their difference $x - y$ belongs to the set of rational numbers $\mathbf{Q}$. We write $x \sim_{\mathbf{Q}} y$.

Proposition 2.1. *The relation $\sim_{\mathbf{Q}}$ is an equivalence relation.*

Proof. Let $x, y, z \in \mathbf{E}$.
1. Since $0 \in \mathbf{Q}$, we have $x - x \in \mathbf{Q}$ and $x \sim_{\mathbf{Q}} x$.
2. Let $x \sim_{\mathbf{Q}} y$. Then $x - y \in \mathbf{Q}$. Hence, $y - x \in \mathbf{Q}$ and $y \sim_{\mathbf{Q}} x$.
3. Let $x \sim_{\mathbf{Q}} y$ and $y \sim_{\mathbf{Q}} z$. Then $x - y \in \mathbf{Q}$ and $y - z \in \mathbf{Q}$. Hence,

$$x - z = (x - y) + (y - z) \in \mathbf{Q}.$$

Therefore $x \sim_{\mathbf{Q}} z$. This completes the proof. □

For the rational equivalence relation, there is the disjoint decomposition of $\mathbf{E}$ into the collection of equivalence classes.

Definition 2.13. Let $\mathcal{F}$ be a nonempty family of nonempty sets. A choice function f on $\mathcal{F}$ is a function $f : \mathcal{F} \longmapsto \bigcup_{\mathbf{F} \in \mathcal{F}} \mathbf{F}$ with the property that for each set $\mathbf{F}$ in $\mathcal{F}$, $f(\mathbf{F})$ is a member of $\mathbf{F}$.

Zermelo's axiom of choice. Let $\mathcal{F}$ be a nonempty collection of nonempty sets. Then there is a choice function on $\mathcal{F}$.

Definition 2.14. By a choice set for the rational equivalence relation on $\mathbf{E}$ we mean a set $\mathcal{C}_{\mathbf{E}}$ consisting of exactly one member of each equivalence class.

By Zermelo's axiom of choice there are such choice sets. A choice set $\mathcal{C}_{\mathbf{E}}$ has the following properties.
1. The difference of two points of $\mathcal{C}_{\mathbf{E}}$ is not rational.
2. For each point $x \in \mathbf{E}$ there is a $c \in \mathcal{C}_{\mathbf{E}}$ such that $x = c + q$ with q a rational number.

Note that for any set $\Lambda \subseteq \mathbf{Q}$ we have $\{\lambda + \mathcal{C}_{\mathbf{E}}\}_{\lambda \in \Lambda}$ is disjoint.

Theorem 2.27 (Vitali's theorem). *Any set $\mathbf{E}$ of real numbers with positive outer measure contains a subset that fails to be measurable.*

Proof. Suppose that $\mathbf{E}$ is bounded. Let $\mathcal{C}_{\mathbf{E}}$ be any choice set for the rational equivalence relation on $\mathbf{E}$. Assume that $\mathcal{C}_{\mathbf{E}}$ is measurable. Let Λ_0 be any bounded, countable infinite set of rational numbers. By Theorem 2.26, we get $m(\mathcal{C}_{\mathbf{E}}) = 0$. Hence,

$$m\Big(\bigcup_{\lambda \in \Lambda_0} (\lambda + \mathcal{C}_{\mathbf{E}})\Big) = \sum_{\lambda \in \Lambda_0} m(\lambda + \mathcal{C}_{\mathbf{E}}) = 0.$$

Since $\mathbf{E}$ is bounded, there is a $b > 0$ such that $\mathbf{E} \subset [-b, b]$. We take

$$\Lambda_0 = [-2b, 2b] \cap \mathbf{Q}.$$

Then Λ_0 is bounded and countably infinite. We will prove that

$$\mathbf{E} \subseteq \bigcup_{\lambda \in [-2b,2b] \cap \mathbf{Q}} (\lambda + \mathcal{C}_{\mathbf{E}}). \tag{2.51}$$

Let $x \in \mathbf{E}$ be arbitrarily chosen. Then there is $c \in \mathcal{C}_{\mathbf{E}}$ such that $x = c + q$ with q a rational number. Since $\mathbf{E} \subset [-b, b]$, we have $x, c \in [-b, b]$. Therefore $q \in [-2b, 2b]$ and $x \in q + \mathcal{C}_{\mathbf{E}}$. Hence, $x \in \bigcup_{\lambda \in [-2b,2b] \cap \mathbf{Q}} (\lambda + \mathcal{C}_{\mathbf{E}})$. Because $x \in \mathbf{E}$ was arbitrarily chosen and we see that it is an element of the set $\bigcup_{\lambda \in [-2b,2b] \cap \mathbf{Q}} (\lambda + \mathcal{C}_{\mathbf{E}})$, we obtain equation (2.51). By (2.51), we get

$$m(\mathbf{E}) \leq \sum_{\lambda \in [-2b,2b] \cap \mathbf{Q}} m(\lambda + \mathcal{C}_{\mathbf{E}}) = 0,$$

which is a contradiction. Therefore $\mathcal{C}_{\mathbf{E}}$ is not measurable. If $\mathbf{E}$ is unbounded, then it can be represented as an union of bounded sets $\mathbf{E}_k$. Then we take a such set $\mathbf{E}_k$ and as above we prove that $\mathcal{C}_{\mathbf{E}_k}$ is not measurable. This completes the proof. □

Theorem 2.28. *There are disjoint sets* $\mathbf{A}$ *and* $\mathbf{B}$ *such that*

$$m^\star(\mathbf{A} \cup \mathbf{B}) < m^\star(\mathbf{A}) + m^\star(\mathbf{B}).$$

Proof. Assume that for every disjoint sets $\mathbf{A}$ and $\mathbf{B}$ we have

$$m^\star(\mathbf{A} \cup \mathbf{B}) = m^\star(\mathbf{A}) + m^\star(\mathbf{B}). \tag{2.52}$$

Let the set $\mathbf{E}$ be arbitrarily chosen set of real numbers. Then for any set $\mathbf{A}$ of real numbers we have

$$\mathbf{A} = (\mathbf{A} \cap \mathbf{E}) \cup (\mathbf{A} \cap \mathbf{E}^c) \quad \text{and} \quad (\mathbf{A} \cap \mathbf{E}) \cap (\mathbf{A} \cap \mathbf{E}^c) = \emptyset.$$

Hence, by (2.52), we get

$$m^\star(\mathbf{A}) = m^\star((\mathbf{A} \cap \mathbf{E}) \cup (\mathbf{A} \cap \mathbf{E}^c)) = m^\star(\mathbf{A} \cap \mathbf{E}) + m^\star(\mathbf{A} \cap \mathbf{E}^c).$$

Therefore $\mathbf{E}$ is a measurable set. Because $\mathbf{E}$ was arbitrarily chosen set of real numbers, we conclude that every set of real numbers is measurable. This contradicts Theorem 2.27. This completes the proof. □

2.4 The Cantor set. The Cantor–Lebesgue function

Consider the interval $\mathbf{I} = [0,1]$. We subdivide the interval $\mathbf{I}$ into three intervals, each of length $\frac{1}{3}$, and we remove the interior of the middle interval, i. e., we remove the interval $(\frac{1}{3}, \frac{2}{3})$ from the interval $[0,1]$. We set

$$\mathbf{C}_1 = \left[0, \frac{1}{3}\right] \cup \left[\frac{2}{3}, 1\right].$$

We now repeat this operation "open middle one-third removal" on each of the two intervals in $\mathbf{C}_1$ to obtain a closed set, which is the union of 2^2 closed intervals, each of length $\frac{1}{3^2}$,

$$\mathbf{C}_2 = \left[0, \frac{1}{9}\right] \cup \left[\frac{2}{9}, \frac{1}{3}\right] \cup \left[\frac{2}{3}, \frac{7}{9}\right] \cup \left[\frac{8}{9}, 1\right].$$

We now repeat the above operation "open middle one-third removal" on each of the four intervals of $\mathbf{C}_2$ to obtain a closed set $\mathbf{C}_3$, which is the union of 2^3 closed intervals, each of length $\frac{1}{3^3}$. We continue this operation countably many times to obtain the countable collection of sets $\{\mathbf{C}_k\}_{k\in\mathbf{N}}$.

Definition 2.15. We define the Cantor set $\mathbb{C}$ by

$$\mathbb{C} = \bigcap_{k=1}^{\infty} \mathbf{C}_k.$$

The collection $\{\mathbf{C}_k\}_{k\in\mathbf{N}}$ has the following properties.

1. It is a descending sequence of closed sets.
2. For each $k \in \mathbf{N}$, the set $\mathbf{C}_k$ is the disjoint union of 2^k closed intervals, each of length $\frac{1}{3^k}$.

Theorem 2.29. *The Cantor set $\mathbb{C}$ is a closed, uncountable set of measure zero.*

Proof. Since $\mathbf{C}_k$ are closed sets for any $k \in \mathbb{N}$ and the intersection of closed sets is a closed set, we conclude that $\mathbb{C}$ is a closed set. By Theorem 2.18 we see that every open set is measurable. Since the collection of the measurable sets is a σ-algebra, we conclude that every closed set is measurable. Therefore $\mathbf{C}_k$ are measurable sets for any $k \in \mathbf{N}$. Because the countable intersection of measurable sets is a measurable set, we obtain $\mathbb{C}$ is a measurable set. Also, $\mathbb{C} \subseteq \mathbf{C}_k$ for any $k \in \mathbf{N}$. Therefore

$$m(\mathbb{C}) \leq m(\mathbf{C}_k) = \left(\frac{2}{3}\right)^k$$

for any $k \in \mathbf{N}$. Consequently $m(\mathbb{C}) = 0$. Now we suppose that $\mathbb{C}$ is countable. Let $\{c_k\}_{k\in\mathbf{N}}$ be an enumeration of $\mathbb{C}$. One of the two disjoint Cantor intervals whose union is $\mathbf{C}_1$ does not contain the point c_1. We denote it by $\mathbf{F}_1$. One of the two disjoint Cantor intervals in $\mathbf{C}_2$ whose union is $\mathbf{F}_1$ does not contain the point c_2. We denote it by $\mathbf{F}_2$. Continuing in this way, we get a countable collection of sets $\{\mathbf{F}_k\}_{k\in\mathbf{N}}$ such that

1. $\mathbf{F}_k$ is closed for any $k \in \mathbf{N}$,
2. $\mathbf{F}_k \subseteq \mathbf{C}_k$ for any $k \in \mathbf{N}$,
3. $\mathbf{F}_{k+1} \subseteq \mathbf{F}_k$ for any $k \in \mathbf{N}$,
4. $c_k \notin \mathbf{F}_k$ for any $k \in \mathbf{N}$.

From this and from the nested set theorem (see the appendix), it follows that $\bigcap_{k=1}^{\infty} \mathbf{F}_k$ is not empty. Let $x \in \bigcap_{k=1}^{\infty} \mathbf{F}_k$. From the construction of the collection $\{\mathbf{F}_k\}_{k\in\mathbf{N}}$, we have

$$\bigcap_{k=1}^{\infty} \mathbf{F}_k \subseteq \bigcap_{k=1}^{\infty} \mathbf{C}_k = \mathbb{C}.$$

Therefore $x \in \mathbb{C}$. Hence, there is an $l \in \mathbf{N}$ such that $x = c_l$. Because $x \in \mathbf{F}_k$ for any $k \in \mathbf{N}$, we see that $x \in \mathbf{F}_l$. Hence, $c_l \in \mathbf{F}_l$. This is a contradiction. Consequently $\mathbb{C}$ is uncountable. This completes the proof. □

Now we define the sets

$$\mathbf{O}_k = [0,1] \setminus \mathbf{C}_k, \quad k \in \mathbf{N},$$
$$\mathbf{O} = \bigcup_{k=1}^{\infty} \mathbf{O}_k.$$

We have

$$\mathbf{O}_1 = \left(\frac{1}{3}, \frac{2}{3}\right),$$
$$\mathbf{O}_2 = \left(\frac{1}{9}, \frac{2}{9}\right) \cup \left(\frac{1}{3}, \frac{2}{3}\right) \cup \left(\frac{7}{9}, \frac{8}{9}\right),$$
$$\mathbf{O}_3 = \left(\frac{1}{27}, \frac{2}{27}\right) \cup \left(\frac{1}{9}, \frac{2}{9}\right) \cup \left(\frac{7}{27}, \frac{8}{27}\right) \cup \left(\frac{1}{3}, \frac{2}{3}\right) \cup \left(\frac{19}{27}, \frac{20}{27}\right)$$
$$\cup \left(\frac{7}{9}, \frac{8}{9}\right) \cup \left(\frac{25}{27}, \frac{26}{27}\right).$$

Now we will prove that

$$\mathbb{C} = [0,1] \setminus \mathbf{O}. \tag{2.53}$$

Let $x \in \mathbb{C}$ be arbitrarily chosen. Then $x \in \mathbf{C}_k$ for any $k \in \mathbf{N}$. Hence, $x \notin \mathbf{O}_k$ for any $k \in \mathbf{N}$. Therefore $x \in [0,1]$ and $x \notin \mathbf{O}$. Consequently $x \in [0,1] \setminus \mathbf{O}$. Because $x \in \mathbb{C}$ was arbitrarily chosen and we see that it is an element of $[0,1] \setminus \mathbf{O}$, we obtain

$$\mathbb{C} \subseteq [0,1] \setminus \mathbf{O}. \tag{2.54}$$

Let $x \in [0,1] \setminus \mathbf{O}$ be arbitrarily chosen. Then $x \in [0,1]$ and $x \notin \mathbf{O}$. Hence, $x \notin \mathbf{O}_k$ for any $k \in \mathbf{N}$. Therefore $x \in \mathbf{C}_k$ for any $k \in \mathbf{N}$. From this, $x \in \mathbf{C}$. Because $x \in [0,1] \setminus \mathbf{O}$ was arbitrarily chosen and we see that it is an element of $\mathbf{C}$, we obtain the relation

$$[0,1] \setminus \mathbf{O} \subseteq \mathbb{C}.$$

From the previous relation and from equation (2.54), we get equation (2.53).

Definition 2.16 (The Cantor–Lebesgue function). Fix a natural number k. Define ϕ on $\mathbf{O}_k$ to be the increasing function on $\mathbf{O}_k$ which is a constant on each of its $2^k - 1$ intervals and takes the $2^k - 1$ values

$$\frac{1}{2^k}, \quad \frac{2}{2^k}, \quad \frac{3}{2^k}, \quad \dots, \quad \frac{2^k - 1}{2^k}.$$

We extend ϕ to all $[0,1]$ by defining it on $\mathbf{C}$ as follows:

$$\phi(0) = 0 \quad \text{and} \quad \phi(x) = \sup\{\phi(t) : t \in \mathbf{O} \cap [0,x)\} \quad \text{if} \quad x \in \mathbf{C} \setminus \{0\}.$$

Thus, on $\mathbf{O}_1$ we have

$$\phi(x) = \frac{1}{2} \quad \text{for} \quad x \in \left(\frac{1}{3}, \frac{2}{3}\right),$$

on $\mathbf{O}_2$ we have

$$\phi(x) = \begin{cases} \frac{1}{4} & \text{if } x \in (\frac{1}{9}, \frac{2}{9}), \\ \frac{1}{2} & \text{if } x \in (\frac{1}{3}, \frac{2}{3}), \\ \frac{3}{4} & \text{if } x \in (\frac{7}{9}, \frac{8}{9}), \end{cases}$$

on $\mathbf{O}_3$ we have

$$\phi(x) = \begin{cases} \frac{1}{8} & \text{if } x \in (\frac{1}{27}, \frac{2}{27}), \\ \frac{1}{4} & \text{if } x \in (\frac{1}{9}, \frac{2}{9}), \\ \frac{3}{8} & \text{if } x \in (\frac{7}{27}, \frac{8}{27}), \\ \frac{1}{2} & \text{if } x \in (\frac{1}{3}, \frac{2}{3}), \\ \frac{5}{8} & \text{if } x \in (\frac{19}{27}, \frac{20}{27}), \\ \frac{3}{4} & \text{if } x \in (\frac{7}{9}, \frac{8}{9}), \\ \frac{7}{8} & \text{if } x \in (\frac{25}{27}, \frac{26}{27}), \end{cases}$$

and so on.

Theorem 2.30. *The Cantor–Lebesgue function ϕ is an increasing continuous function that maps $[0,1]$ onto $[0,1]$. Its derivative exists on $\mathbf{O}$ and*

$$\phi' = 0 \quad \textit{on} \quad \mathbf{O} \quad \textit{while} \quad m(\mathbf{O}) = 1.$$

Proof. Since $m(\mathbb{C}) = 0$, we get

$$1 = m([0,1]) = m(\mathbb{C} \cup \mathbf{O}) = m(\mathbb{C}) + m(\mathbf{O}) = m(\mathbf{O}).$$

By the definition, ϕ is increasing on $\mathbf{O}$ and $\phi(x) \geq 0$ for any $x \in [0,1]$. Let $x_1, x_2 \in \mathbb{C}$, $x_1 \geq x_2 > 0$. Then

$$\mathbf{O} \cap [0, x_2) \subseteq \mathbf{O} \cap [0, x_1)$$

and

$$\phi(x_1) = \sup\{\phi(t) : t \in \mathbf{O} \cap [0, x_1)\} \geq \sup\{\phi(t) : t \in \mathbf{O} \cap [0, x_2)\} = \phi(x_2).$$

Therefore ϕ is increasing on $\mathbb{C}$ and it is increasing on $[0,1]$. Let now $x \in \mathbf{O}$ be arbitrarily chosen. Then x belongs to an open interval on which it is a constant. Therefore ϕ is continuous at x. Because $x \in \mathbf{O}$ was arbitrarily chosen, we conclude that ϕ is continuous on $\mathbf{O}$. Now we take $x_0 \in \mathbb{C}$, $x_0 \neq 0, 1$. Then x_0 is not a member of the $2^k - 1$ intervals removed in the first k stages of the removal process, whose union is denoted by $\mathbf{O}_k$. We take k large enough so that x_0 lies between two consecutive intervals (a_{1k}, a_k) and (b_{1k}, b_k), $a_k < b_{1k}$, in $\mathbf{O}_k$. By the definition of ϕ on $\mathbf{O}_k$, it follows that

$$\phi(b_k) - \phi(a_k) = \frac{1}{2^k}, \quad a_k < x_0 < b_{1k}.$$

Hence, if $x_1 < x_0$, $x_1 \in (a_k, b_{1k})$, using the fact that ϕ is increasing,

$$0 \leq \phi(x_0) - \phi(x_1) \leq \phi(b_k) - \phi(a_k) = \frac{1}{2^k}. \tag{2.55}$$

If $x_1 > x_0$, $x_1 \in (a_k, b_{1k})$, using the fact that ϕ is increasing, we have

$$0 \leq \phi(x_1) - \phi(x_0) \leq \phi(b_k) - \phi(a_k) = \frac{1}{2^k}.$$

By the previous inequalities and the inequalities (2.55), we conclude that ϕ is continuous at x_0. Because $x_0 \in \mathbb{C}$ was arbitrarily chosen, we see that ϕ is continuous on $\mathbb{C}$. Therefore ϕ is continuous on $[0,1]$. Because $\phi(0) = 0$ and $\phi(1) = 1$, and ϕ is increasing and continuous on $[0,1]$, we conclude that ϕ maps $[0,1]$ onto $[0,1]$. Since ϕ is a constant on each of the intervals removed at any stage of the removal process, its derivative exists and it is 0 at each point of $\mathbf{O}$. This completes the proof. □

Theorem 2.31. *Let ϕ be the Cantor–Lebesgue function and define the function ψ as follows:*

$$\psi(x) = \phi(x) + x, \quad x \in [0,1].$$

Then:
1. *ψ is strictly increasing continuous function that maps $[0,1]$ onto $[0,2]$,*
2. *ψ maps the Cantor set $\mathbb{C}$ onto a measurable set of positive measure,*
3. *ψ maps a measurable set, subset of the Cantor set $\mathbb{C}$, onto a nonmeasurable set.*

Proof.
1. Because ψ is a sum of two continuous functions on $[0,1]$, we conclude that ψ is continuous on $[0,1]$. Let $x_1, x_2 \in [0,1]$, $x_1 > x_2$. Then $\phi(x_1) \geq \phi(x_2)$ and

$$\psi(x_1) = \phi(x_1) + x_1 \geq \phi(x_2) + x_1 > \phi(x_2) + x_2 = \psi(x_2).$$

Consequently ψ is strictly increasing on $[0,1]$. Since $\psi(0) = 0$, $\psi(1) = 2$ and ψ is strictly increasing continuous function on $[0,1]$, we conclude that ψ maps $[0,1]$ onto $[0,2]$.

2. We have $[0,1] = \mathbb{C} \cup \mathbf{O}$. We will prove that

$$[0,2] = \psi(\mathbf{C}) \cup \psi(\mathbf{O}). \tag{2.56}$$

Let $y \in [0,2]$ be arbitrarily chosen. Since $\psi([0,1]) = [0,2]$, there is an $x \in [0,1]$ such that $y = \psi(x)$. If $x \in \mathbf{O}$, then $y \in \psi(\mathbf{O})$ and $y \in \psi(\mathbb{C}) \cup \psi(\mathbf{O})$. If $x \in \mathbb{C}$, then $y \in \psi(\mathbb{C})$ and $y \in \psi(\mathbb{C}) \cup \psi(\mathbf{O})$. Because $y \in [0,2]$ was arbitrarily chosen and we see that it is an element of $\psi(\mathbb{C}) \cup \psi(\mathbf{O})$, we obtain the relation

$$[0,2] \subseteq \psi(\mathbb{C}) \cup \psi(\mathbf{O}). \tag{2.57}$$

Let $y \in \psi(\mathbb{C}) \cup \psi(\mathbf{O})$ be arbitrarily chosen. If $y \in \psi(\mathbb{C})$, then there is an $x \in \mathbb{C}$ such that $y = \psi(x)$. Because $\mathbb{C} \subset [0,1]$, we get $x \in [0,1]$ and $y \in \psi([0,1]) = [0,2]$. If $y \in \psi(\mathbf{O})$, then there is an $x \in \mathbf{O}$ so that $y = \psi(x)$. Since $\mathbf{O} \subset [0,1]$, we obtain $y \in \psi([0,1]) = [0,2]$. Because $y \in \psi(\mathbb{C}) \cup \psi(\mathbf{O})$ was arbitrarily chosen and we see that it is an element of $[0,2]$, we get the relation

$$\psi(\mathbb{C}) \cup \psi(\mathbf{O}) \subseteq [0,2].$$

From the previous relation and from equation (2.57), we obtain equation (2.56). Assume that there is an $y \in \psi(\mathbf{O})$ and $y \in \psi(\mathbb{C})$. Then there exist $x_1 \in \mathbf{O}$ and $x_2 \in \mathbf{C}$ such that

$$y = \psi(x_1) = \psi(x_2).$$

This is a contradiction because $x_1 \neq x_2$ and ψ is a strictly increasing function on $[0,1]$. Consequently $\psi(\mathbb{C}) \cap \psi(\mathbf{O}) = \emptyset$. Since ψ is strictly increasing continuous on $[0,1]$ it has a continuous inverse. Therefore $\psi(\mathbb{C})$ is closed and $\psi(\mathbf{O})$ is open. Hence, $\psi(\mathbb{C})$ and $\psi(\mathbf{O})$ are measurable. Let $\{\mathbf{I}_k\}_{k\in\mathbf{N}}$ be an enumeration of the collection of the intervals that are removed in the Cantor removal process. We have $\mathbf{O} = \bigcup_{k=1}^{\infty} \mathbf{I}_k$. Since ψ is one-to-one, the collection $\{\psi(\mathbf{I}_k)\}_{k\in\mathbf{N}}$ is disjoint. Because ϕ is a constant on each $\mathbf{I}_k$, we have

$$m(\psi(\mathbf{I}_k)) = m(\mathbf{I}_k) = l(\mathbf{I}_k).$$

By Theorem 2.24, we get

$$m(\psi(\mathbf{O})) = \sum_{k=1}^{\infty} m(\psi(\mathbf{I}_k)) = \sum_{k=1}^{\infty} l(\mathbf{I}_k) = m(\mathbf{O}).$$

Because $m(\mathbf{O}) = 1$, we obtain $m(\psi(\mathbf{O})) = 1$. Hence, by (2.56), we find

$$2 = m([0,2]) = m(\psi(\mathbf{O}) \cup \psi(\mathbb{C})) = m(\psi(\mathbf{O})) + m(\psi(\mathbb{C})) = 1 + m(\psi(\mathbb{C})).$$

From this, $m(\psi(\mathbb{C})) = 1$.

3. From Theorem 2.27, it follows that there is a nonmeasurable set $\mathbf{W} \subset \psi(\mathbb{C})$. Because $\psi^{-1}(\mathbf{W}) \subset \mathbb{C}$ and $m(\mathbf{C}) = 0$, we see that $\psi^{-1}(\mathbf{W})$ is measurable and has measure zero. This completes the proof. □

Exercise 2.7. Prove that a strictly increasing continuous function, defined on an interval, maps Borel sets onto Borel sets.

Theorem 2.32. *There is a measurable set, a subset of the Cantor set, that is not a Borel set.*

Proof. We take the function ψ described in Theorem 2.31. By Theorem 2.31, ψ maps a measurable set $\mathbf{A}$ onto a nonmeasurable set. If we assume that A is a Borel set, then, using Exercise 2.7, we see that $\psi(\mathbf{A})$ is a Borel set, which is a measurable set. Therefore $\mathbf{A}$ is not a Borel set. This completes the proof. □

2.5 Lebesgue measurable functions

Lemma 2.5. *Let* $\mathbf{E}$ *be a measurable set and* f *be a function defined on* $\mathbf{E}$. *Then for any* $c \in \mathbf{R}$ *the following statements are equivalent.*
1. *For each* $c \in \mathbf{R}$, *the set* $\{x \in \mathbf{E} : f(x) > c\}$ *is measurable.*
2. *For each* $c \in \mathbf{R}$, *the set* $\{x \in \mathbf{E} : f(x) \geq c\}$ *is measurable.*
3. *For each* $c \in \mathbf{R}$, *the set* $\{x \in \mathbf{E} : f(x) < c\}$ *is measurable.*
4. *For each* $c \in \mathbf{R}$, *the set* $\{x \in \mathbf{E} : f(x) \leq c\}$ *is measurable.*

Proof. Since the complement in $\mathbf{E}$ of a measurable subset of $\mathbf{E}$ is measurable and (1) and (4), (2) and (3) are complementary in $\mathbf{E}$, we see that (1) and (4), (2) and (3) are equivalent. Now we assume (1). Let $c \in \mathbf{R}$ be arbitrarily chosen. We will prove that

$$\{x \in \mathbf{E} : f(x) \geq c\} = \bigcap_{k=1}^{\infty} \left\{x \in \mathbf{E} : f(x) > c - \frac{1}{k}\right\}. \tag{2.58}$$

Let $y \in \{x \in \mathbf{E} : f(x) \geq c\}$ be arbitrarily chosen. Then $y \in \mathbf{E}$ and $f(y) \geq c$. Hence, $f(y) > c - \frac{1}{k}$ for any $k \in \mathbf{N}$. Therefore $y \in \{x \in \mathbf{E} : f(x) > c - \frac{1}{k}\}$ for any $k \in \mathbf{N}$ and $y \in \bigcap_{k=1}^{\infty}\{x \in \mathbf{E} : f(x) > c - \frac{1}{k}\}$. Because $y \in \{x \in \mathbf{E} : f(x) \geq c\}$ was arbitrarily chosen and we see that it is an element of $\bigcap_{k=1}^{\infty}\{x \in \mathbf{E} : f(x) > c - \frac{1}{k}\}$, we conclude that

$$\{x \in \mathbf{E} : f(x) \geq c\} \subseteq \bigcap_{k=1}^{\infty} \left\{x \in \mathbf{E} : f(x) > c - \frac{1}{k}\right\}. \tag{2.59}$$

Let now $y \in \bigcap_{k=1}^{\infty}\{x \in \mathbf{E} : f(x) > c - \frac{1}{k}\}$ be arbitrarily chosen. Then

$$y \in \left\{x \in \mathbf{E} : f(x) > c - \frac{1}{k}\right\}$$

for any $k \in \mathbf{N}$. Hence, $y \in \mathbf{E}$ and $f(y) > c - \frac{1}{k}$ for any $k \in \mathbf{N}$. From this, $f(y) \geq c$ and $y \in \{x \in \mathbf{E} : f(x) \geq c\}$. Because $y \in \bigcap_{k=1}^{\infty}\{x \in \mathbf{E} : f(x) > c - \frac{1}{k}\}$ was arbitrarily chosen and we see that it is an element of $\{x \in \mathbf{E} : f(x) \geq c\}$, we find

$$\bigcap_{k=1}^{\infty}\left\{x \in \mathbf{E} : f(x) > c - \frac{1}{k}\right\} \subseteq \{x \in \mathbf{E} : f(x) \geq c\}.$$

From the previous relation and from equation (2.59), we obtain equation (2.58). We have $\{x \in \mathbf{E} : f(x) > c - \frac{1}{k}\}$ are measurable sets for any $k \in \mathbf{N}$. Since the intersection of countable collection of measurable sets is a measurable set, we see that $\bigcap_{k=1}^{\infty}\{x \in \mathbf{E} : f(x) > c - \frac{1}{k}\}$ is a measurable set. Hence, by (2.58), we conclude that $\{x \in \mathbf{E} : f(x) \geq c\}$ is a measurable set. Now we assume (2). Let $c \in \mathbf{R}$ be arbitrarily chosen. We will prove

$$\{x \in \mathbf{E} : f(x) > c\} = \bigcup_{k=1}^{\infty}\left\{x \in \mathbf{E} : f(x) \geq c + \frac{1}{k}\right\}. \tag{2.60}$$

Let $y \in \{x \in \mathbf{E} : f(x) > c\}$ be arbitrarily chosen. Then $y \in \mathbf{E}$ and $f(y) > c$. Hence, there is an $l \in \mathbf{N}$ such that $f(y) \geq c + \frac{1}{l}$. Therefore $y \in \{x \in \mathbf{E} : f(x) \geq c + \frac{1}{l}\}$, and from this $y \in \bigcup_{k=1}^{\infty}\{x \in \mathbf{E} : f(x) \geq c + \frac{1}{k}\}$. Because $y \in \{x \in \mathbf{E} : f(x) > c\}$ was arbitrarily chosen and we see that it is an element of $\bigcup_{k=1}^{\infty}\{x \in \mathbf{E} : f(x) \geq c + \frac{1}{k}\}$, we get the relation

$$\{x \in \mathbf{E} : f(x) > c\} \subseteq \bigcup_{k=1}^{\infty}\left\{x \in \mathbf{E} : f(x) \geq c + \frac{1}{k}\right\}. \tag{2.61}$$

Let now $y \in \bigcup_{k=1}^{\infty}\{x \in \mathbf{E} : f(x) \geq c + \frac{1}{k}\}$ be arbitrarily chosen. Then there is an $l \in \mathbf{N}$ such that $y \in \{x \in \mathbf{E} : f(x) \geq c + \frac{1}{l}\}$. Hence, $y \in \mathbf{E}$ and $f(y) \geq c + \frac{1}{l} > c$. Therefore $y \in \{x \in \mathbf{E} : f(x) > c\}$. Because $y \in \bigcup_{k=1}^{\infty}\{x \in \mathbf{E} : f(x) > c + \frac{1}{k}\}$ was arbitrarily chosen and we see that it is an element of the set $\{x \in \mathbf{E} : f(x) > c\}$, we obtain the relation

$$\bigcup_{k=1}^{\infty}\left\{x \in \mathbf{E} : f(x) \geq c + \frac{1}{k}\right\} \subseteq \{x \in \mathbf{E} : f(x) > c\}.$$

From the previous relation and from (2.61), we get equation (2.60). We see that $\{x \in \mathbf{E} : f(x) \geq c + \frac{1}{k}\}$ are measurable sets for any $k \in \mathbf{N}$. Since the union of countable collection of measurable sets is a measurable set, we see that the set $\bigcup_{k=1}^{\infty}\{x \in \mathbf{E} : f(x) \geq c + \frac{1}{k}\}$ is a measurable set. Hence, by (2.60), we conclude that $\{x \in \mathbf{E} : f(x) > c\}$ is a measurable set. This completes the proof. □

Exercise 2.8. Prove that

1.

$$\{x \in \mathbf{E} : f(x) = c\} = \{x \in \mathbf{E} : f(x) \geq c\} \cap \{x \in \mathbf{E} : f(x) \leq c\}, \tag{2.62}$$

2.

$$\{x \in \mathbf{E} : f(x) = \infty\} = \bigcap_{k=1}^{\infty}\{x \in \mathbf{E} : f(x) > k\}, \tag{2.63}$$

3.

$$\{x \in \mathbf{E} : f(x) = -\infty\} = \bigcap_{k=1}^{\infty} \{x \in \mathbf{E} : f(x) < -k\}, \tag{2.64}$$

for any $c \in \mathbf{R}$. Here $\mathbf{E}$ is a set and f is a function defined on $\mathbf{E}$.

Lemma 2.6. *Let $\mathbf{E}$ be a measurable set and f be a function defined on $\mathbf{E}$. Assume that one of the sets in Lemma 2.5 is measurable. Then for each extended real number c, the set $\{x \in \mathbf{E} : f(x) = c\}$ is a measurable set.*

Proof.

1. Let $-\infty < c < \infty$ be arbitrarily chosen. By Lemma 2.5, we see that the sets $\{x \in \mathbf{E} : f(x) \geq c\}$ and $\{x \in \mathbf{E} : f(x) \leq c\}$ are measurable sets. Because the intersection of a finite collection of measurable sets is a measurable set, we see that the set $\{x \in \mathbf{E} : f(x) \geq c\} \cap \{x \in \mathbf{E} : f(x) \leq c\}$ is a measurable set. Hence, by (2.62), we conclude that the set $\{x \in \mathbf{E} : f(x) = c\}$ is a measurable set.
2. Let $c = \infty$. By Lemma 2.5, we see that the sets $\{x \in \mathbf{E} : f(x) > k\}$ are measurable sets for any $k \in \mathbf{N}$. Since the intersection of a countable collection of measurable sets is a measurable set, we see that the set $\bigcap_{k=1}^{\infty} \{x \in \mathbf{E} : f(x) > k\}$ is a measurable set. Hence, by (2.63), we conclude that the set $\{x \in \mathbf{E} : f(x) = \infty\}$ is a measurable set.
3. Let $c = -\infty$. By Lemma 2.5, we see that the sets $\{x \in \mathbf{E} : f(x) < -k\}$ are measurable sets for any $k \in \mathbf{N}$. Because the intersection of a countable collection of measurable sets is a measurable set, we see that the set $\bigcap_{k=1}^{\infty} \{x \in \mathbf{E} : f(x) < -k\}$ is a measurable set. From this and from (2.64), we conclude that the set $\{x \in \mathbf{E} : f(x) = -\infty\}$ is a measurable set. This completes the proof. □

Definition 2.17. Let $\mathbf{E}$ be a measurable set. An extended real-valued function f defined on $\mathbf{E}$ is said to be Lebesgue measurable, or simply measurable, if it satisfies one of the statements of Lemma 2.5.

Theorem 2.33. *Let a function f be defined on a measurable set $\mathbf{O}$. Then f is measurable if and only if for any open set $\mathbf{O}$ the set $f^{-1}(\mathbf{O}) = \{x \in \mathbf{E} : f(x) \in \mathbf{O}\}$ is measurable.*

Proof.

1. Let f be measurable. Let also $\mathbf{O}$ be any open set. Then it can be represented as an union of countable collection of open bounded intervals $\{\mathbf{I}_k\}_{k \in \mathbf{N}}$ each of which is in the form $\mathbf{I}_k = \mathbf{A}_k \cap \mathbf{B}_k$, where $\mathbf{A}_k = (-\infty, a_k)$ and $B_k = (b_k, \infty)$. We will prove that

$$f^{-1}(\mathbf{O}) = \bigcup_{k=1}^{\infty} (f^{-1}(\mathbf{A}_k) \cap f^{-1}(\mathbf{B}_k)). \tag{2.65}$$

Let $x \in f^{-1}(\mathbf{O})$ be arbitrarily chosen. Then $x \in \mathbf{E}$ and $f(x) \in \mathbf{O}$. Hence, there is an $l \in \mathbf{N}$ such that $f(x) \in \mathbf{I}_l = \mathbf{A}_l \cap \mathbf{B}_l$. From this, it follows that $f(x) \in \mathbf{A}_l$ and $f(x) \in \mathbf{B}_l$.

Therefore $x \in f^{-1}(\mathbf{A}_l)$ and $x \in f^{-1}(\mathbf{B}_l)$. Then we get the relations $x \in f^{-1}(\mathbf{A}_l) \cap f^{-1}(\mathbf{B}_l)$ and $x \in \bigcup_{k=1}^{\infty}(f^{-1}(\mathbf{A}_k) \cap f^{-1}(\mathbf{B}_k))$. Because $x \in f^{-1}(\mathbf{O})$ was arbitrarily chosen and we see that it is an element of the set $\bigcup_{k=1}^{\infty}(f^{-1}(\mathbf{A}_k) \cap f^{-1}(\mathbf{B}_k))$, we conclude that

$$f^{-1}(\mathbf{O}) \subseteq \bigcup_{k=1}^{\infty}(f^{-1}(\mathbf{A}_k) \cap f^{-1}(\mathbf{B}_k)). \tag{2.66}$$

Let now $x \in \bigcup_{k=1}^{\infty}(f^{-1}(\mathbf{A}_k) \cap f^{-1}(\mathbf{B}_k))$ be arbitrarily chosen. Then there is an $l \in \mathbf{N}$ such that $x \in f^{-1}(\mathbf{A}_l) \cap f^{-1}(\mathbf{B}_l)$. Hence, $x \in f^{-1}(\mathbf{A}_l)$ and $x \in f^{-1}(\mathbf{B}_l)$. Then $f(x) \in \mathbf{A}_l$ and $f(x) \in \mathbf{B}_l$. From this, $f(x) \in \mathbf{I}_l = \mathbf{A}_l \cap \mathbf{B}_l \subseteq \mathbf{O}$. Consequently $x \in f^{-1}(\mathbf{O})$. Because $x \in \bigcup_{k=1}^{\infty}(f^{-1}(\mathbf{A}_k) \cap f^{-1}(\mathbf{B}_k))$ was arbitrarily chosen and we see that it is an element of $f^{-1}(\mathbf{O})$, we get the relation

$$\bigcup_{k=1}^{\infty}(f^{-1}(\mathbf{A}_k) \cap f^{-1}(\mathbf{B}_k)) \subseteq f^{-1}(\mathbf{O}).$$

From the previous relation and from equation (2.66), we have equation (2.65). Since f is measurable, we see that $f^{-1}(\mathbf{A}_l)$ and $f^{-1}(\mathbf{B}_l)$ are measurable sets. Hence, $f^{-1}(\mathbf{A}_l) \cap f^{-1}(\mathbf{B}_l)$ is a measurable set and $\bigcup_{l=1}^{\infty}(f^{-1}(\mathbf{A}_l) \cap f^{-1}(\mathbf{B}_l))$ is a measurable set. From this and from equation (2.65), we conclude that $f^{-1}(\mathbf{O})$ is a measurable set.

2. Let for any open set $\mathbf{O}$ the set $f^{-1}(\mathbf{O})$ is a measurable set. Since (c, ∞) is an open set for any $c \in \mathbf{R}$, we see that $f^{-1}((c, \infty))$ is a measurable set. Therefore the function f is measurable. This completes the proof. □

Theorem 2.34. *Let* $\mathbf{E}$ *be a measurable set and* $f : \mathbf{E} \longmapsto \mathbf{R}$ *be a continuous function. Then* f *is measurable.*

Proof. Let $\mathbf{O}$ be any open set. Since $f : \mathbf{E} \longmapsto \mathbf{R}$ is a continuous function, there is an open set $\mathbf{U}$ such that

$$f^{-1}(\mathbf{O}) = \mathbf{E} \cap \mathbf{U}.$$

Because $\mathbf{U}$ is an open set, it is measurable. Since the intersection of two measurable sets is a measurable set, we see that $\mathbf{E} \cap \mathbf{U}$ is a measurable set, or the set $f^{-1}(\mathbf{O})$ is a measurable set. Hence, by Theorem 2.33, we conclude that f is a measurable function. This completes the proof. □

Exercise 2.9. Let $\mathbf{E}$ be any set of real numbers and f, g, f_l, $l \in \{1, \ldots, n\}$, be extended real-valued functions on $\mathbf{E}$. Prove

1.

$$\{x \in \mathbf{E} : g(x) > c\} = \{x \in \mathbf{A} : g(x) > c\} \cup (\{x \in \mathbf{E} : f(x) > c\} \cap (\mathbf{E} \setminus \mathbf{A})), \tag{2.67}$$

$\mathbf{A} = \{x \in \mathbf{E} : f(x) \neq g(x)\}$, for any $c \in \mathbf{R}$,

2.

$$\{x \in \mathbf{E} : f(x) > c\} = \{x \in \mathbf{D} : f(x) > c\} \cup \{x \in \mathbf{E} \setminus \mathbf{D} : f(x) > c\}, \tag{2.68}$$

$D \subseteq \mathbf{E}$, for any $c \in \mathbf{R}$,

3.

$$\{x \in \mathbf{D} : f(x) > c\} = \mathbf{D} \cap \{x \in \mathbf{E} : f(x) > c\}, \tag{2.69}$$

$D \subseteq \mathbf{E}$, for any $c \in \mathbf{R}$,

4.

$$\{x \in \mathbf{E} : |f(x)| > c\} = \{x \in \mathbf{E} : f(x) > c\} \cup \{x \in \mathbf{E} : f(x) < -c\} \tag{2.70}$$

for any $c \geq 0$,

5.

$$\{x \in \mathbf{E} : \alpha f(x) > c\} = \left\{x \in \mathbf{E} : f(x) > \frac{c}{\alpha}\right\} \tag{2.71}$$

for any $c \in \mathbf{R}$ and for any $\alpha > 0$,

6.

$$\{x \in \mathbf{E} : \alpha f(x) > c\} = \left\{x \in \mathbf{E} : f(x) < \frac{c}{\alpha}\right\} \tag{2.72}$$

for any $c \in \mathbf{R}$ and for any $\alpha < 0$,

7.

$$\{x \in \mathbf{E} : f(x) + g(x) < c\} = \bigcup_{q \in \mathbf{Q}} (\{x \in \mathbf{E} : g(x) < c - q\} \cap \{x \in \mathbf{E} : f(x) < q\}) \tag{2.73}$$

for any $c \in \mathbf{R}$,

8.

$$\{x \in \mathbf{E} : (f(x))^2 > c\} = \{x \in \mathbf{E} : f(x) > \sqrt{c}\} \cup \{x \in \mathbf{E} : f(x) < -\sqrt{c}\} \tag{2.74}$$

for any $c \in \mathbf{R}$, $c > 0$,

9.

$$\{x \in \mathbf{E} : \max\{f_1(x), \dots, f_n(x)\} > c\} = \bigcup_{k=1}^{n} \{x \in \mathbf{E} : f_k(x) > c\} \tag{2.75}$$

for any $c \in \mathbf{R}$,

10.

$$\{x \in \mathbf{E} : \min\{f_1(x), \dots, f_n(x)\} > c\} = \bigcap_{k=1}^{n} \{x \in \mathbf{E} : f_k(x) > c\} \tag{2.76}$$

for any $c \in \mathbf{R}$.

Theorem 2.35. *Let* $\mathbf{E}$ *be a measurable set and* f, g *be extended real-valued functions on* $\mathbf{E}$. *If* f *is measurable on* $\mathbf{E}$ *and* $f = g$ *a. e. on* $\mathbf{E}$, *then* g *is measurable on* $\mathbf{E}$.

Proof. Let $c \in \mathbf{R}$ be arbitrarily chosen. Let also $\mathbf{A} = \{x \in \mathbf{E} : f(x) \neq g(x)\}$. Then $m(\mathbf{A}) = 0$. Hence, by Theorem 2.8, we see that the set $\mathbf{A}$ is a measurable set. Because every subset of a set of measure zero is measurable, we see that the set $\{x \in \mathbf{A} : g(x) > c\}$ is a measurable set. Since $\mathbf{E}$ and $\mathbf{A}$ are measurable sets, using Theorem 2.11, we see that the set $\mathbf{E}\setminus\mathbf{A}$ is a measurable set. Because f is a measurable function on $\mathbf{E}$, we conclude that the set $\{x \in \mathbf{E} : f(x) > c\}$ is a measurable set. Now, using the fact that the intersection of two measurable sets is a measurable set, we see that the set $\{x \in \mathbf{E} : f(x) > c\} \cap (\mathbf{E}\setminus\mathbf{A})$ is a measurable set. Since the union of two measurable sets is a measurable set, we see that the set

$$\{x \in \mathbf{A} : g(x) > c\} \cup (\{x \in \mathbf{E} : f(x) > c\} \cap (\mathbf{E} \setminus \mathbf{A}))$$

is a measurable set. Hence, by (2.67), we conclude that the set $\{x \in \mathbf{E} : g(x) > c\}$ is a measurable set. Because $c \in \mathbf{R}$ was arbitrarily chosen, we see that the function g is a measurable function on $\mathbf{E}$. This completes the proof. □

Theorem 2.36. *Let* $\mathbf{E}$ *and* $\mathbf{D}$, $\mathbf{D} \subseteq \mathbf{E}$, *be measurable sets and* f *be an extended real-valued function on* $\mathbf{E}$. *Then* f *is measurable on* $\mathbf{E}$ *if and only if its restrictions to* $\mathbf{D}$ *and* $\mathbf{E} \setminus \mathbf{D}$ *are measurable.*

Proof.

1. Let the restrictions of f to $\mathbf{D}$ and $\mathbf{E} \setminus \mathbf{D}$ be measurable. We take $c \in \mathbf{R}$ arbitrarily. Then the sets

$$\{x \in \mathbf{D} : f(x) > c\} \quad \text{and} \quad \{x \in \mathbf{E} \setminus \mathbf{D} : f(x) > c\}$$

   are measurable sets. Because the union of two measurable sets is a measurable set, we see that the set

$$\{x \in \mathbf{D} : f(x) > c\} \cup \{x \in \mathbf{E} \setminus \mathbf{D} : f(x) > c\}$$

   is a measurable set. Hence, by (2.68), the set $\{x \in \mathbf{E} : f(x) > c\}$ is a measurable set. Because $c \in \mathbf{R}$ was arbitrarily chosen, we conclude that the function f is measurable on $\mathbf{E}$.
2. Now we suppose that the function f is a measurable function on $\mathbf{E}$. We take $c > 0$ arbitrarily. Then the set $\{x \in \mathbf{E} : f(x) > c\}$ is a measurable set. Because $\mathbf{D}$ is a measurable set and the intersection of two measurable sets is a measurable set, we conclude that the set $\mathbf{D} \cap \{x \in \mathbf{E} : f(x) > c\}$ is a measurable set. From this and from (2.69), we see that the set $\{x \in \mathbf{D} : f(x) > c\}$ is a measurable set. As above, one can prove that the set $\{x \in \mathbf{E} \setminus \mathbf{D} : f(x) > c\}$ is a measurable set. Because $c \in \mathbf{R}$ was arbitrarily chosen, we see that the restrictions of f to $\mathbf{D}$ and $\mathbf{E} \setminus \mathbf{D}$ are measurable. This completes the proof. □

Theorem 2.37. *Let* $\mathbf{E}$ *be a measurable set and* f *be an extended real-valued measurable function on* $\mathbf{E}$. *Then* $|f|$ *is a measurable function on* $\mathbf{E}$.

Proof. Let $c \in \mathbf{R}$ be arbitrarily chosen.

1. Let $c < 0$. Then $\mathbf{E} = \{x \in \mathbf{E} : |f(x)| > c\}$ and $\{x \in \mathbf{E} : |f(x)| > c\}$ is a measurable set.
2. Let $c \geq 0$. Since f is measurable on $\mathbf{E}$, the sets

$$\{x \in \mathbf{E} : f(x) > c\} \quad \text{and} \quad \{x \in \mathbf{E} : f(x) < -c\}$$

are measurable sets. Hence, using the fact that the union of two measurable sets is a measurable set, we see that the set

$$\{x \in \mathbf{E} : f(x) > c\} \cup \{x \in \mathbf{E} : f(x) < -c\}$$

is a measurable set. From this and from (2.70), we see that the set $\{x \in \mathbf{E} : |f(x)| > c\}$ is a measurable set. Because $c \in \mathbf{R}$ was arbitrarily chosen, we conclude that the function $|f|$ is a measurable function on $\mathbf{E}$. This completes the proof. □

Theorem 2.38. *Let* $\mathbf{E}$ *be a measurable set of measure zero. Then every extended real-valued function on* $\mathbf{E}$ *is measurable.*

Proof. Let f be an extended real-valued function on $\mathbf{E}$. We take $c \in \mathbf{R}$ arbitrarily. Then $\{x \in \mathbf{E} : f(x) > c\} \subseteq \mathbf{E}$. Hence, using $m(\mathbf{E}) = 0$, we see that the set $\{x \in \mathbf{E} : f(x) > c\}$ is a measurable set. Because $c \in \mathbf{R}$ was arbitrarily chosen, we conclude that f is measurable on $\mathbf{E}$. This completes the proof. □

Theorem 2.39. *Let* $\mathbf{E}$ *be a measurable set and* f *be an extended real-valued measurable function on* $\mathbf{E}$. *Then* αf *is a measurable function on* $\mathbf{E}$ *for any* $\alpha \in \mathbf{R}$.

Proof. Let $c \in \mathbf{R}$ be arbitrarily chosen. We take $\alpha \in \mathbf{R}$ arbitrarily.

1. Let $\alpha = 0$. Then

$$\{x \in \mathbf{E} : 0 > c\} = \begin{cases} \mathbf{E} & \text{if } c < 0, \\ \emptyset & \text{if } c \geq 0. \end{cases}$$

Since $\mathbf{E}$ and $\emptyset$ are measurable sets, we see that the set $\{\mathbf{E} : 0 > c\}$ is a measurable set.
2. Let $\alpha > 0$. Since f is a measurable function on $\mathbf{E}$, we see that the set $\{x \in \mathbf{E} : f(x) > \frac{c}{\alpha}\}$ is a measurable set. Hence, by (2.71), we see that the set $\{x \in \mathbf{E} : \alpha f(x) > c\}$ is a measurable set.
3. Let $\alpha < 0$. Because f is a measurable set, the set $\{x \in \mathbf{E} : f(x) < \frac{c}{\alpha}\}$ is a measurable set. From this and from (2.72), we see that the set $\{x \in \mathbf{E} : \alpha f(x) > c\}$ is a measurable set.

Because $c \in \mathbf{R}$ was arbitrarily chosen and we see that the set $\{x \in \mathbf{E} : \alpha f(x) > c\}$ is a measurable set, we conclude that the function αf is a measurable function on $\mathbf{E}$. This completes the proof. □

Theorem 2.40. *Let* $\mathbf{E}$ *be a measurable set and* f, g *be extended real-valued measurable functions on* $\mathbf{E}$. *Then* $f + g$ *is a measurable function on* $\mathbf{E}$.

Proof. Let $c \in \mathbf{R}$ be arbitrarily chosen. Since f and g are measurable functions on $\mathbf{E}$, we see that the sets

$$\{x \in \mathbf{E} : g(x) < c - q\} \quad \text{and} \quad \{x \in \mathbf{E} : f(x) < q\}$$

are measurable sets for any $q \in \mathbf{Q}$. Because the intersection of two measurable sets is a measurable set, we see that the set

$$\{x \in \mathbf{E} : g(x) < c - q\} \cap \{x \in \mathbf{E} : f(x) < q\}$$

is a measurable set for any $q \in \mathbf{Q}$. Since the union of a countable collection of measurable sets is a measurable set, we see that the set

$$\bigcup_{q \in \mathbf{Q}} (\{x \in \mathbf{E} : g(x) < c - q\} \cap \{x \in \mathbf{E} : f(x) < q\})$$

is a measurable set. Hence, by (2.73), we see that the set $\{x \in \mathbf{E} : f(x) + g(x) < c\}$ is a measurable set. Because $c \in \mathbf{R}$ was arbitrarily chosen, we conclude that $f + g$ is a measurable function on $\mathbf{E}$. This completes the proof. □

Exercise 2.10. Let $\mathbf{E}$ be a measurable set and f, g be extended real-valued measurable functions on $\mathbf{E}$. Prove that $\alpha f + \beta g$ is a measurable function on $\mathbf{E}$ for any $\alpha, \beta \in \mathbf{R}$.

Corollary 2.4. *Let* $\mathbf{E}$ *be a measurable set and* f_l, $l \in \{1, \ldots, n\}$, *be extended real-valued measurable functions on* $\mathbf{E}$. *Then* $f_1 + \cdots + f_n$ *is a measurable function on* $\mathbf{E}$.

Proof. If $n = 2$, then the assertion follows immediately from Theorem 2.40. Let $n > 2$. Since f_1 and f_2 are measurable functions on $\mathbf{E}$, by Theorem 2.40, it follows that $f_1 + f_2$ is a measurable function on $\mathbf{E}$. Because $f_1 + f_2$ and f_3 are measurable functions on $\mathbf{E}$, by Theorem 2.40, it follows that $f_1 + f_2 + f_3$ is a measurable function on $\mathbf{E}$. And so on, $f_1 + \cdots + f_n$ is a measurable function on $\mathbf{E}$. This completes the proof. □

Theorem 2.41. *Let* $\mathbf{E}$ *be a measurable set and* f *be an extended real-valued measurable function on* $\mathbf{E}$. *Then* f^2 *is a measurable function on* $\mathbf{E}$.

Proof. Let $c \in \mathbf{R}$ be arbitrarily chosen.

1. Let $c < 0$. Then

$$\{x \in \mathbf{E} : (f(x))^2 > c\} = \mathbf{E}$$

   and hence, $\{x \in \mathbf{E} : (f(x))^2 > c\}$ is a measurable set.
2. Let $c \geq 0$. Since f is a measurable function on $\mathbf{E}$, the sets

$$\{x \in \mathbf{E} : f(x) > \sqrt{c}\} \quad \text{and} \quad \{x \in \mathbf{E} : f(x) < -\sqrt{c}\}$$

are measurable sets. Hence,

$$\{x \in \mathbf{E} : f(x) > \sqrt{c}\} \cup \{x \in \mathbf{E} : f(x) < -\sqrt{c}\}$$

is a measurable set. From this and from (2.74), we see that the set

$$\{x \in \mathbf{E} : (f(x))^2 > c\}$$

is a measurable set.

Because $c \in \mathbf{R}$ was arbitrarily chosen, we conclude that f is a measurable function on $\mathbf{E}$. This completes the proof. □

Theorem 2.42. *Let* E *be a measurable set and* f, g *be extended real-valued measurable functions on* $\mathbf{E}$. *Then* fg *is a measurable function on* $\mathbf{E}$.

Proof. Note that

$$fg = \frac{1}{4}((f+g)^2 - (f-g)^2). \tag{2.77}$$

Since f and g are measurable functions on $\mathbf{E}$, using Theorem 2.40, we see that $f+g$ is a measurable function on $\mathbf{E}$. Hence, by Theorem 2.41, it follows that $(f+g)^2$ is a measurable function on $\mathbf{E}$. Because g is a measurable function on $\mathbf{E}$, using Theorem 2.39, we see that $-g$ is a measurable function on $\mathbf{E}$. From this and from Theorem 2.40, we see that $f-g$ is a measurable function on $\mathbf{E}$. Hence, by Theorem 2.41, it follows that $(f-g)^2$ is a measurable function on $\mathbf{E}$. Since $(f-g)^2$ is a measurable function on $\mathbf{E}$, using Theorem 2.39, it follows that $-(f-g)^2$ is a measurable function on $\mathbf{E}$. Because $(f+g)^2$ and $-(f-g)^2$ are measurable functions, by Theorem 2.40, we conclude that $(f+g)^2 - (f-g)^2$ is a measurable function on $\mathbf{E}$. From this and from Theorem 2.39, using (2.77), we see that the function fg is a measurable function on $\mathbf{E}$. This completes the proof. □

Theorem 2.43. *Let* $\mathbf{E}$ *be a measurable set and* $\{f_k\}_{k=1}^n$ *be a finite collection of measurable functions on* $\mathbf{E}$. *Then* $\max\{f_1, \dots, f_n\}$ *and* $\min\{f_1, \dots, f_n\}$ *are measurable functions on* $\mathbf{E}$.

Proof. Let $c \in \mathbf{R}$ be arbitrarily chosen. Because f_k, $k \in \{1, \dots, n\}$, are measurable functions on $\mathbf{E}$, the sets

$$\{x \in \mathbf{E} : f_k(x) > c\}$$

are measurable sets for any $k \in \{1, \dots, n\}$. Hence,

$$\bigcup_{k=1}^{n} \{x \in \mathbf{E} : f_k(x) > c\} \quad \text{and} \quad \bigcap_{k=1}^{n} \{x \in \mathbf{E} : f_k(x) > c\}$$

are measurable sets. Hence, by (2.75), (2.76), we conclude that the sets

$$\{x \in \mathbf{E} : \max\{f_1(x), \ldots, f_n(x)\} > c\} \quad \text{and} \quad \{x \in \mathbf{E} : \min\{f_1(x), \ldots, f_n(x)\} > c\}$$

are measurable sets. Because $c \in \mathbf{R}$ was arbitrarily chosen, we conclude that the functions $\max\{f_1, \ldots, f_n\}$ and $\min\{f_1, \ldots, f_n\}$ are measurable functions on **E**. This completes the proof. □

For a function f, defined on **E**, we define

$$f^+(x) = \max\{f(x), 0\}, \quad f^-(x) = \max\{-f(x), 0\}, \quad x \in \mathbf{E}.$$

Exercise 2.11. Let **E** be a measurable set and f be an extended real-valued measurable function on **E**. Prove that f^+ and f^- are measurable functions on **E**.

Exercise 2.12. Let **E** be a measurable set and f be an extended real-valued measurable function on **E**. Prove that $|f|^p$ is a measurable function on **E** for each $p > 0$.

Theorem 2.44. *Let* **E** *be a measurable set, g be an extended real-valued measurable function on* **E** *and f be an extended real-valued continuous function on* **R**. *Then the composition $f \circ g$ is a measurable function on* **E**.

Proof. Let **O** be an open set. We will prove that

$$(f \circ g)^{-1}(\mathbf{O}) = g^{-1}(f^{-1}(\mathbf{O})). \tag{2.78}$$

Let $x \in (f \circ g)^{-1}(\mathbf{O})$ be arbitrarily chosen. Then $x \in \mathbf{E}$ and $f(g(x)) \in \mathbf{O}$. Since $g(x) \in \mathbf{R}$ and $f(g(x)) \in \mathbf{O}$, we see that $g(x) \in f^{-1}(\mathbf{O})$. Because $x \in \mathbf{E}$ and $g(x) \in f^{-1}(\mathbf{O})$, we obtain $x \in g^{-1}(f^{-1}(\mathbf{O}))$. Since $x \in (f \circ g)^{-1}(\mathbf{O})$ was arbitrarily chosen and we see that it is an element of $g^{-1}(f^{-1}(\mathbf{O}))$, we obtain the relation

$$(f \circ g)^{-1}(\mathbf{O}) \subseteq g^{-1}(f^{-1}(\mathbf{O})). \tag{2.79}$$

Let $x \in g^{-1}(f^{-1}(\mathbf{O}))$ be arbitrarily chosen. Then $x \in \mathbf{E}$ and $g(x) \in f^{-1}(\mathbf{O})$. Hence, $g(x) \in \mathbf{R}$ and $f(g(x)) \in \mathbf{O}$. Since $x \in \mathbf{E}$ and $f(g(x)) \in \mathbf{O}$, we conclude that $x \in (f \circ g)^{-1}(\mathbf{O})$. Because $x \in g^{-1}(f^{-1}(\mathbf{O}))$ was arbitrarily chosen and we see that it is an element of $(f \circ g)^{-1}(\mathbf{O})$, we obtain the relation

$$(f \circ g)^{-1}(\mathbf{O}) \subseteq (f \circ g)^{-1}(\mathbf{O}).$$

From the previous relation and from (2.79), we obtain equation (2.78). Since f is a continuous function on **R** and **O** is an open set, we see that the set $f^{-1}(\mathbf{O})$ is an open set. Because g is a measurable function on **E**, using Theorem 2.33, we conclude that $g^{-1}(f^{-1}(\mathbf{O}))$ is a measurable set. Hence, by equation (2.78), we see that $(f \circ g)^{-1}(\mathbf{O})$ is a measurable set. From this and from Theorem 2.33, we see that $f \circ g$ is a measurable function on **E**. This completes the proof. □

Exercise 2.13. Let $\mathbf{E}$ be a measurable set. If $f \equiv \text{const}$ on $\mathbf{E}$, prove that f is a measurable function on $\mathbf{E}$.

Definition 2.18. Let $\mathbf{E}$ be any set of real numbers and $\mathbf{A} \subseteq \mathbf{E}$. Let also $\{f_n\}_{n\in\mathbf{N}}$ be a sequence of extended real-valued functions with common domain $\mathbf{E}$ and f be an extended real-valued function on $\mathbf{A}$.

1. We say that the sequence $\{f_n\}_{n\in\mathbf{N}}$ converges to f pointwise on $\mathbf{A}$ provided
$$\lim_{n\to\infty} f_n(x) = f(x) \quad \text{for} \quad \text{all} \quad x \in \mathbf{A}.$$
2. We say that the sequence $\{f_n\}_{n\in\mathbf{N}}$ converges to f pointwise a. e. on $\mathbf{A}$ provided it converges to f pointwise on $\mathbf{A} \setminus \mathbf{B}$, where $\mathbf{B} \subseteq \mathbf{A}$ and $m(\mathbf{B}) = 0$.
3. We say that the sequence $\{f_n\}_{n\in\mathbf{N}}$ converges uniformly to f on $\mathbf{A}$ provided for each $\epsilon > 0$, there is an $N \in \mathbf{N}$ such that
$$|f_n - f| < \epsilon \quad \text{on} \quad \mathbf{A} \quad \text{for} \quad \text{all} \quad n \geq N.$$

Theorem 2.45. *Let $\mathbf{E}$ be a measurable set of real numbers, $\{f_n\}_{n\in\mathbf{N}}$ be a sequence of extended real-valued measurable functions on $\mathbf{E}$ that converges pointwise a. e. on $\mathbf{E}$ to the extended real-valued function f, defined on $\mathbf{E}$. Then f is measurable on $\mathbf{E}$.*

Proof. Let $\mathbf{A} \subseteq \mathbf{E}$ be such that the sequence $\{f_n\}_{n\in\mathbf{N}}$ converges pointwise to f on $\mathbf{A}$ and $m(\mathbf{E} \setminus \mathbf{A}) = 0$. Hence, f is measurable on $\mathbf{E} \setminus \mathbf{A}$. By Theorem 2.36, it follows that the functions f_j are measurable functions on $\mathbf{A}$ and $\mathbf{E} \setminus \mathbf{A}$ for any $j \in \mathbf{N}$. We will prove that f is measurable on $\mathbf{A}$. Let $c \in \mathbf{R}$ be arbitrarily chosen. We will show that

$$\{x \in \mathbf{A} : f(x) < c\} = \bigcup_{1\leq k,n<\infty} \left(\bigcap_{j=k}^{\infty} \left\{ x \in \mathbf{A} : f_j(x) < c - \frac{1}{n} \right\} \right). \tag{2.80}$$

Let $y \in \{x \in \mathbf{A} : f(x) < c\}$ be arbitrarily chosen. Because $f_j(y) \to f(y)$, as $j \to \infty$, and $f(y) < c$, there are $n, k \in \mathbf{N}$ such that for any $j \geq k$ we have $f_j(y) < c - \frac{1}{n}$, i. e., $y \in \{x \in \mathbf{A} : f_j(x) < c - \frac{1}{n}\}$ for any $j \geq k$. Therefore

$$y \in \bigcap_{j=k}^{\infty} \left\{ x \in \mathbf{A} : f_j(x) < c - \frac{1}{n} \right\}$$

and from this

$$y \in \bigcup_{1\leq k,n<\infty} \left(\bigcap_{j=k}^{\infty} \left\{ x \in \mathbf{A} : f_j(x) < c - \frac{1}{n} \right\} \right).$$

Because $y \in \{x \in \mathbf{A} : f(x) < c\}$ was arbitrarily chosen and we see that it is an element of $\bigcup_{1\leq k,n<\infty}(\bigcap_{j=k}^{\infty}\{x \in \mathbf{A} : f_j(x) < c - \frac{1}{n}\})$, we obtain the relation

$$\{x \in \mathbf{A} : f(x) < c\} \subseteq \bigcup_{1\leq k,n<\infty} \left(\bigcap_{j=k}^{\infty} \left\{ x \in \mathbf{A} : f_j(x) < c - \frac{1}{n} \right\} \right). \tag{2.81}$$

Let now $y \in \bigcup_{1\le k,n<\infty}(\bigcap_{j=k}^{\infty}\{x \in \mathbf{A} : f_j(x) < c - \frac{1}{n}\})$ be arbitrarily chosen. Then there are $k, n \in \mathbf{N}$ such that $y \in \{x \in \mathbf{A} : f_j(x) < c - \frac{1}{n}\}$ for any $j \ge k$. Hence,

$$f_j(y) < c - \frac{1}{n}$$

for any $j \ge k$. Assume that $f(y) \ge c$. Then

$$f(y) - f_j(y) > \frac{1}{n}$$

for any $j \ge k$. This is a contradiction. Therefore $f(y) < c$ and $y \in \{x \in \mathbf{A} : f(x) < c\}$. Because $y \in \bigcup_{1\le k,n<\infty}(\bigcap_{j=k}^{\infty}\{x \in \mathbf{A} : f_j(x) < c - \frac{1}{n}\})$ was arbitrarily chosen and we see that it is an element of $\{x \in \mathbf{A} : f(x) < c\}$, we conclude that

$$\bigcup_{1\le k,n<\infty}\left(\bigcap_{j=k}^{\infty}\left\{x \in \mathbf{A} : f_j(x) < c - \frac{1}{n}\right\}\right) \subseteq \{x \in \mathbf{A} : f(x) < c\}.$$

From the previous relation and from equation (2.81), we get equation (2.80). Since f_j are measurable functions on $\mathbf{A}$ for any $j \in \mathbf{N}$, we see that the sets $\{x \in \mathbf{A} : f_j(x) < c - \frac{1}{n}\}$ are measurable sets for any $j, n \in \mathbf{N}$. From this, the sets

$$\bigcap_{j=k}^{\infty}\left\{x \in \mathbf{A} : f_j(x) < c - \frac{1}{n}\right\}$$

are measurable sets for any $k, n \in \mathbf{N}$. Hence,

$$\bigcup_{1\le k,n<\infty}\left(\bigcap_{j=k}^{\infty}\left\{x \in \mathbf{A} : f_j(x) < c - \frac{1}{n}\right\}\right)$$

is a measurable set. From this and from equation (2.80), we conclude that the set $\{x \in \mathbf{A} : f(x) < c\}$ is a measurable set. Because $c \in \mathbf{R}$ was arbitrarily chosen, we conclude that the function f is a measurable function on $\mathbf{A}$. Hence, by Theorem 2.36, we see that the function f is a measurable function on $\mathbf{E}$. This completes the proof. □

Definition 2.19. Let $\mathbf{E}$ be any set of real numbers. The characteristic function of $\mathbf{E}$, $\kappa_{\mathbf{E}}$, is the function defined by

$$\kappa_{\mathbf{E}} = \begin{cases} 1 & \text{if } x \in \mathbf{E}, \\ 0 & \text{if } x \notin \mathbf{E}. \end{cases}$$

Note that if $\mathbf{E}$ is a measurable set, then its characteristic function is a measurable function on $\mathbf{E}$.

Exercise 2.14. Let $\mathbf{A}$ and $\mathbf{B}$ be any sets of real numbers. Prove

1. $\kappa_{\mathbf{A}\cap\mathbf{B}} = \kappa_{\mathbf{A}}\kappa_{\mathbf{B}}$.

2. $\kappa_{\mathbf{A}\cup\mathbf{B}} = \kappa_{\mathbf{A}} + \kappa_{\mathbf{B}} - \kappa_{\mathbf{A}}\kappa_{\mathbf{B}}$.
3. $\kappa_{\mathbf{R}\setminus\mathbf{A}} = 1 - \kappa_{\mathbf{A}}$.

Definition 2.20. A real-valued function ϕ defined on a measurable set is called simple provided it is measurable and takes only a finite number of values.

If $\mathbf{E}$ is a measurable set and ϕ is a simple function on $\mathbf{E}$ and takes the distinct values $c_1, \ldots, c_n$, then

$$\phi = \sum_{k=1}^{n} c_k \kappa_{\mathbf{E}_k} \quad \text{on} \quad \mathbf{E}, \tag{2.82}$$

where $\mathbf{E}_k = \{x \in \mathbf{E} : \phi(x) = c_k\}$, $k \in \{1, \ldots, n\}$.

Definition 2.21. Equation (2.82) of ϕ as a linear combination of characteristic functions is called the canonical representation of the simple function ϕ.

Theorem 2.46. *Let $\mathbf{E}$ be a measurable set and f be a measurable bounded real-valued function on $\mathbf{E}$. Then for each $\epsilon > 0$, there are simple functions ϕ_ϵ and ψ_ϵ defined on $\mathbf{E}$ such that*

$$\phi_\epsilon \le f \le \psi_\epsilon \quad \textit{and} \quad 0 \le \psi_\epsilon - \phi_\epsilon < \epsilon \quad \textit{on} \quad \mathbf{E}.$$

Proof. We take $\epsilon > 0$ arbitrarily. Let (c, d) be an open, bounded interval such that $f(\mathbf{E}) \subset (c, d)$. Consider a partition

$$c = y_0 < y_1 < \cdots < y_{n-1} < y_n = d$$

of the closed bounded interval $[c, d]$ such that $y_k - y_{k-1} < \epsilon$ for any $k \in \{1, \ldots, n\}$. Let

$$\mathbf{I}_k = [y_{k-1}, y_k), \quad \mathbf{E}_k = f^{-1}(\mathbf{I}_k), \quad 1 \le k \le n.$$

Since f is a measurable function on $\mathbf{E}$, we see that the sets $\mathbf{E}_k$ are measurable sets for any $k \in \{1, \ldots, n\}$. Define the simple functions ϕ_ϵ and ψ_ϵ on $\mathbf{E}$ as follows:

$$\phi_\epsilon = \sum_{k=1}^{n} y_{k-1}\kappa_{\mathbf{E}_k} \quad \text{and} \quad \psi_\epsilon = \sum_{k=1}^{n} y_k \kappa_{\mathbf{E}_k}.$$

Let $x \in \mathbf{E}$ be arbitrarily chosen. Since $f(\mathbf{E}) \subset (c, d)$, we have $f(x) \in (c, d)$. Hence, there is a $k \in \{1, \ldots, n\}$ such that $f(x) \in \mathbf{I}_k$. Therefore

$$\phi_\epsilon(x) = y_{k-1}, \quad \psi_\epsilon(x) = y_k$$

and

$$\phi_\epsilon(x) \le f(x) < \psi_\epsilon(x), \quad \psi_\epsilon(x) - \phi_\epsilon(x) = y_k - y_{k-1} < \epsilon.$$

This completes the proof. □

Theorem 2.47 (The simple approximation theorem). *An extended real-valued function f on a measurable set $\mathbf{E}$ is measurable if and only if there is a sequence $\{\phi_n\}_{n\in\mathbf{N}}$ of simple functions on $\mathbf{E}$ which converges pointwise on $\mathbf{E}$ to f and*

$$|\phi_n| \le |f| \quad \text{on} \quad \mathbf{E} \quad \text{for} \quad \text{all} \quad n \in \mathbf{N}.$$

If f is nonnegative, we can choose $\{\phi_n\}_{n\in\mathbf{N}}$ to be increasing.

Proof.

1. Let $\{\phi_n\}_{n\in\mathbf{N}}$ be a sequence of simple functions on $\mathbf{E}$ that converges pointwise on $\mathbf{E}$ to f and $|\phi_n| \le |f|$ on $\mathbf{E}$ for all $n \in \mathbf{N}$. Because ϕ_n are measurable functions on $\mathbf{E}$ for each $n \in \mathbf{N}$, using Theorem 2.45, we see that f is a measurable function on $\mathbf{E}$.
2. Let f is a measurable function on $\mathbf{E}$. Assume that $f \ge 0$ on $\mathbf{E}$. Take $n \in \mathbf{N}$ and define

$$\mathbf{E}_n = \{x \in \mathbf{E} : f(x) \le n\}.$$

Because f is measurable on $\mathbf{E}$, we have $\mathbf{E}_n$ are measurable sets for each $n \in \mathbf{N}$. By Theorem 2.36, it follows that the restriction of f to $\mathbf{E}_n$ is a nonnegative bounded measurable function. We apply Theorem 2.46 to the restriction of f to $\mathbf{E}$ with the choice $\epsilon = \frac{1}{n}$. In other words, we may choose simple functions ϕ_n and ψ_n defined on $\mathbf{E}_n$ such that

$$0 \le \phi_n \le f \le \psi_n \quad \text{on} \quad \mathbf{E}_n \quad \text{and} \quad 0 \le \psi_n - \phi_n < \frac{1}{n} \quad \text{on} \quad \mathbf{E}_n.$$

Note that

$$0 \le f - \phi_n \le \psi_n - \phi_n < \frac{1}{n} \quad \text{on} \quad \mathbf{E}_n.$$

We extend ϕ_n to all $\mathbf{E}$ by setting $\phi_n(x) = n$ if $f(x) > n$, $n \in \mathbf{N}$. The function ϕ_n is a simple function defined on $\mathbf{E}$ and $0 \le \phi_n \le f$ on $\mathbf{E}$, $n \in \mathbf{N}$. Let $x \in \mathbf{E}$ be arbitrarily chosen.

(a) Assume that $f(x)$ is finite. Then there exists an $N \in \mathbf{N}$ such that $f(x) < N$. Then

$$0 \le f(x) - \phi_n(x) < \frac{1}{n} \quad \text{for} \quad \text{all} \quad n > N.$$

Therefore $\phi_n(x) \to f(x)$, as $n \to \infty$.

(b) Assume that $f(x) = \infty$. Then $\phi_n(x) = n$ for all $n \in \mathbf{N}$. Therefore $\phi_n(x) \to f(x)$, as $n \to \infty$. By replacing ϕ_n, $n \in \mathbf{N}$, with $\max\{\phi_1, \dots, \phi_n\}$, we see that the sequence $\{\phi_n\}_{n\in\mathbf{N}}$ is increasing.

In the general case, if f is not nonnegative, we take the representation $f = f^+ - f^-$ on $\mathbf{E}$ and apply the above for f^+ and f^-. This completes the proof. □

Exercise 2.15. Let $\mathbf{E}$ be a measurable set of real numbers and f be a bounded measurable function on $\mathbf{E}$. Prove that there are sequences $\{\phi_n\}_{n\in\mathbf{N}}$ and $\{\psi_n\}_{n\in\mathbf{N}}$ of simple functions on $\mathbf{E}$ such that $\{\phi_n\}_{n\in\mathbf{N}}$ is increasing and $\{\psi_n\}_{n\in\mathbf{N}}$ is decreasing and each of these sequences converges to f uniformly on $\mathbf{E}$.

Theorem 2.48. *Assume that* $\mathbf{E}$ *is a measurable set with a finite measure and* $\{f_n\}_{n\in\mathbf{N}}$ *be a sequence of bounded measurable functions on* $\mathbf{E}$ *that converges poinwise on* $\mathbf{E}$ *to the real-valued function* f. *Then for each* $\epsilon > 0$ *and* $\delta > 0$, *there are a measurable subset* $\mathbf{A}$ *of* $\mathbf{E}$ *and an index* N *for which*

$$|f_n - f| < \epsilon \quad \text{on} \quad \mathbf{A} \quad \text{for} \quad \text{all} \quad n \geq N \quad \text{and} \quad m(\mathbf{E} \setminus \mathbf{A}) < \delta.$$

Proof. By Theorem 2.45, it follows that f is a measurable function on $\mathbf{E}$. Hence, using Theorem 2.39, the function $-f$ is a measurable function on $\mathbf{E}$. From Theorem 2.40, it follows that $f_n - f$ are measurable functions on $\mathbf{E}$ for each $n \in \mathbf{N}$. From this and from Theorem 2.37, we see that $|f_n - f|$ are measurable functions on $\mathbf{E}$ for all $n \in \mathbf{N}$. We take $\epsilon > 0$ and $\delta > 0$ arbitrarily. Note that the sets

$$\mathbf{E}_n = \{x \in \mathbf{E} : |f(x) - f_k(x)| < \epsilon \quad \text{for} \quad \text{all} \quad k \geq n\}$$

are measurable sets for all $n \in \mathbf{N}$. Then the sequence $\{\mathbf{E}_n\}_{n\in\mathbf{N}}$ is an ascending collection of measurable sets and since $\{f_n\}_{n\in\mathbf{N}}$ converges pointwise to f on $\mathbf{E}$, we have $\mathbf{E} = \bigcup_{n=1}^{\infty} \mathbf{E}_n$. Hence, by Theorem 2.25, we get

$$m(\mathbf{E}) = \lim_{n\to\infty} m(\mathbf{E}_n).$$

Since $m(\mathbf{E}) < \infty$, we can choose an index N such that

$$m(\mathbf{E}_N) > m(\mathbf{E}) - \delta.$$

Let $\mathbf{A} = \mathbf{E}_N$. Then

$$m(\mathbf{E} \setminus \mathbf{A}) = m(\mathbf{E}) - m(\mathbf{E}_N) < m(\mathbf{E}) - m(\mathbf{E}) + \delta = \delta.$$

This completes the proof. □

Theorem 2.49 (Egoroff's theorem). *Assume that* $\mathbf{E}$ *is a measurable set with finite measure and* $\{f_n\}_{n\in\mathbf{N}}$ *be a sequence of measurable functions on* $\mathbf{E}$ *that converges pointwise on* $\mathbf{E}$ *to the real-valued function* f. *Then for each* $\epsilon > 0$, *there is a closed set* $\mathbf{A}$ *contained in* $\mathbf{E}$ *for which* $f_n \to f$, *as* $n \to \infty$, *uniformly on* $\mathbf{A}$ *and* $m(\mathbf{E} \setminus \mathbf{A}) < \epsilon$.

Proof. Take $\epsilon > 0$ arbitrarily. By Theorem 2.48, it follows that for each $n \in \mathbf{N}$ there is a measurable subset $\mathbf{A}_n$ of $\mathbf{E}$ and an index $N(n)$ such that

$$|f_l - f| < \frac{1}{n} \quad \text{on} \quad \mathbf{A}_n \quad \text{for} \quad \text{all} \quad l \geq N \quad \text{and} \quad m(\mathbf{E} \setminus \mathbf{A}_n) < \frac{\epsilon}{2^{n+1}}.$$

Define the set

$$\widetilde{\mathbf{A}} = \bigcap_{n=1}^{\infty} \mathbf{A}_n.$$

Then, using Theorem 2.5, we get

$$m(\mathbf{E}\setminus\widetilde{\mathbf{A}}) = m\left(\mathbf{E}\setminus\left(\bigcap_{n=1}^{\infty}\mathbf{A}_n\right)\right) = m\left(\bigcup_{n=1}^{\infty}(\mathbf{E}\setminus\mathbf{A}_n)\right)$$
$$\leq \sum_{n=1}^{\infty} m(\mathbf{E}\setminus\mathbf{A}_n) < \sum_{n=1}^{\infty}\frac{\epsilon}{2^{n+1}} = \frac{\epsilon}{2}.$$

Now we choose an index n_0 such that $\frac{1}{n_0} < \epsilon$. Then

$$|f_l - f| < \frac{1}{n_0} \quad \text{on} \quad \mathbf{A}_{n_0} \quad \text{for} \quad \text{all} \quad l \geq N(n_0).$$

Since $\widetilde{\mathbf{A}} \subseteq \mathbf{A}_{n_0}$ and $\frac{1}{n_0} < \epsilon$, we get

$$|f_l - f| < \epsilon \quad \text{on} \quad \widetilde{\mathbf{A}} \quad \text{for} \quad \text{all} \quad l \geq N(n_0).$$

Therefore $\{f_n\}_{n\in\mathbf{N}}$ converges uniformly to f on $\widetilde{\mathbf{A}}$ and $m(\mathbf{E}\setminus\widetilde{\mathbf{A}}) < \frac{\epsilon}{2}$. By Theorem 2.21, it follows that there is a closed set $\mathbf{A}$ contained in $\widetilde{\mathbf{A}}$ such that $m(\widetilde{\mathbf{A}}\setminus\mathbf{A}) < \frac{\epsilon}{2}$. Thus, $\{f_n\}_{n\in\mathbf{N}}$ converges uniformly to f on $\mathbf{A}$ and

$$m(\mathbf{E}\setminus\mathbf{A}) = m(\mathbf{E}\setminus\widetilde{\mathbf{A}}) + m(\widetilde{\mathbf{A}}\setminus\mathbf{A}) < \frac{\epsilon}{2} + \frac{\epsilon}{2} = \epsilon.$$

This completes the proof. □

Theorem 2.50. *Let* $\mathbf{E}$ *be a measurable set,* f *be a simple function defined on* $\mathbf{E}$. *Then for each* $\epsilon > 0$, *there is a continuous function* g *on* $\mathbb{R}$ *and a closed set* $\mathbf{A}$ *contained in* $\mathbf{E}$ *for which*

$$f = g \quad \text{on} \quad \mathbf{A} \quad \text{and} \quad m(\mathbf{E}\setminus\mathbf{A}) < \epsilon.$$

Proof. Let $a_1, \ldots, a_n$ be the finite numbers of distinct values taken by f, and let them be taken on the sets $\mathbf{E}_1, \ldots, \mathbf{E}_n$, respectively. Since the a_k are distinct, the collection $\{\mathbf{E}_k\}_{k=1}^{n}$ is disjoint. By Theorem 2.21, there are closed sets $A_1, \ldots, A_n$ such that

$$\mathbf{A}_k \subseteq \mathbf{E}_k \quad \text{and} \quad m(\mathbf{E}_k\setminus\mathbf{A}_k) < \frac{\epsilon}{n}$$

for any $k \in \{1, \ldots, n\}$. Then the set $\mathbf{A} = \bigcup_{k=1}^{n}\mathbf{A}_k$ is a closed set and since $\{\mathbf{A}_k\}_{k=1}^{n}$ is disjoint, we get

$$m(\mathbf{E}\setminus\mathbf{A}) = m\left(\bigcup_{k=1}^{n}(\mathbf{E}\setminus\mathbf{A}_k)\right) = \sum_{k=1}^{n} m(\mathbf{E}\setminus\mathbf{A}_k) < \sum_{k=1}^{n}\frac{\epsilon}{n} = \epsilon.$$

Define g on $\mathbf{A}$ to take the values a_k on $\mathbf{A}_k$, $k \in \{1, \ldots, n\}$. Since the collection $\{\mathbf{A}_k\}_{k=1}^{n}$ is disjoint, the function g is properly defined and continuous. We extend g from a continuous function on $\mathbf{A}$ to a continuous function on all $\mathbf{R}$. This completes the proof. □

Theorem 2.51 (Luzin's theorem). *Let* $\mathbf{E}$ *be a measurable set and* f *be a real-valued measurable function on* $\mathbf{E}$. *Then for each* $\epsilon > 0$, *there is a continuous function* g *on* $\mathbf{R}$ *and a closed set* $\mathbf{A}$ *contained in* $\mathbf{E}$ *for which*

$$f = g \quad on \quad \mathbf{A} \quad and \quad m(\mathbf{E} \setminus \mathbf{A}) < \epsilon.$$

Proof.

1. Let $m(\mathbf{E}) < \infty$. By Theorem 2.47, there is a sequence $\{f_n\}_{n\in\mathbf{N}}$ of simple functions defined on $\mathbf{E}$ that converges to f pointwise on $\mathbf{E}$. Take $n \in \mathbf{N}$ arbitrarily. By Theorem 2.50, it follows that there is a continuous function g_n on $\mathbf{R}$ and a closed set $\mathbf{A}_n$ contained in $\mathbf{E}$ for which

$$g_n = f_n \quad \text{on} \quad \mathbf{A}_n \quad \text{and} \quad m(\mathbf{E} \setminus \mathbf{A}_n) < \frac{\epsilon}{2^{n+1}}.$$

By Egoroff's theorem, there is a closed set $\mathbf{A}_0$ contained in $\mathbf{E}$ such that $\{f_n\}_{n\in\mathbf{N}}$ converges to f uniformly on $\mathbf{A}_0$ and $m(\mathbf{E} \setminus \mathbf{A}_0) < \frac{\epsilon}{2}$. Define $\mathbf{A} = \bigcap_{n=0}^{\infty} \mathbf{A}_n$. Then

$$\begin{aligned} m(\mathbf{E} \setminus \mathbf{A}) &= m\left((\mathbf{E} \setminus \mathbf{A}_0) \cup \left(\bigcup_{n=1}^{\infty} (\mathbf{E} \setminus \mathbf{A}_n) \right) \right) \le m(\mathbf{E} \setminus \mathbf{A}_0) + m\left(\bigcup_{n=1}^{\infty} (\mathbf{E} \setminus \mathbf{A}_n) \right) \\ &\le m(\mathbf{E} \setminus \mathbf{A}_0) + \sum_{n=1}^{\infty} m(\mathbf{E} \setminus \mathbf{A}_n) < \frac{\epsilon}{2} + \sum_{n=1}^{\infty} \frac{\epsilon}{2^{n+1}} = \frac{\epsilon}{2} + \frac{\epsilon}{2} = \epsilon. \end{aligned}$$

The set $\mathbf{A}$ is closed. Each f_n is continuous on $\mathbf{A}$ because $\mathbf{A} \subseteq \mathbf{A}_n$ and $f_n = g_n$ on $\mathbf{A}_n$. Since the sequence $\{f_n\}_{n\in\mathbf{N}}$ converges uniformly to f on $\mathbf{A} \subseteq \mathbf{A}_0$, we conclude that the restriction of f to $\mathbf{A}$ is continuous on $\mathbf{A}$. Let g be the restriction of f to $\mathbf{A}$. We extend g from a continuous function on $\mathbf{A}$ to a continuous function on all $\mathbf{R}$.

2. Let $m(\mathbf{E}) = \infty$. Then $\mathbf{E}$ can be expressed as the disjoint union of a countable collection $\{\mathbf{E}_k\}_{k\in\mathbf{N}}$ of measurable sets with finite measure. For each $\mathbf{E}_k$, as in the previous case, we construct a continuous function g^k on $\mathbf{E}_k$ and a closed set $\mathbf{A}_k \subseteq \mathbf{E}_k$ such that

$$f = g^k \quad \text{on} \quad \mathbf{A}_k \quad \text{and} \quad m(\mathbf{E}_k \setminus \mathbf{A}_k) < \frac{\epsilon}{2^{k+1}}.$$

The collection $\{\mathbf{A}_k\}_{k\in\mathbf{N}}$ is disjoint. Let $\mathbf{A} = \bigcup_{k\in\mathbf{N}} \mathbf{A}_k$. Then $\mathbf{A}$ is closed. We construct a continuous function g on $\mathbf{A}$ such that its restrictions on $\mathbf{A}_k$ coincide with g_k. We extend g from a continuous function on $\mathbf{A}$ to a continuous function on all $\mathbf{R}$. Note that

$$\mathbf{E} \setminus \mathbf{A} \subseteq \bigcup_{k=1}^{\infty} (\mathbf{E}_k \setminus \mathbf{A}_k)$$

and

$$m(\mathbf{E} \setminus \mathbf{A}) \le m\left(\bigcup_{k=1}^{\infty} (\mathbf{E}_k \setminus \mathbf{A}_k) \right) \le \sum_{k=1}^{\infty} m(\mathbf{E}_k \setminus \mathbf{A}_k) < \sum_{k=1}^{\infty} \frac{\epsilon}{2^{k+1}} < \epsilon.$$

This completes the proof. □

2.6 The Riemann integral

In this section we recall some definitions to the Riemann integral.

Definition 2.22. Let f be a bounded real-valued function defined upon the closed bounded interval $[a, b]$. Let also $P = \{x_0, x_1, \dots, x_n\}$ be a partition of the interval $[a, b]$, i. e.,

$$a = x_0 < x_1 < \cdots < x_n = b.$$

Define the lower and upper Darboux sums for f with respect to P, respectively, by

$$L(f, P) = \sum_{i=1}^{n} m_i(x_i - x_{i-1})$$

and

$$U(f, P) = \sum_{i=1}^{n} M_i(x_i - x_{i-1}),$$

where

$$m_i = \inf\{f(x) : x_{i-1} < x < x_i\}, \quad M_i = \sup\{f(x) : x_{i-1} < x < x_i\}, \quad i \in \{1, \dots, n\}.$$

Definition 2.23. We define the lower and upper Riemann integrals of f over $[a, b]$, respectively, by

$$(R)\underline{\int_a^b} f = \sup\{L(f, P) : P \quad \text{is} \quad \text{a} \quad \text{partition} \quad \text{of} \quad [a, b]\},$$
$$(R)\overline{\int_a^b} f = \inf\{U(f, P) : P \quad \text{is} \quad \text{a} \quad \text{partition} \quad \text{of} \quad [a, b]\}.$$

Note that the lower and upper Riemann integrals are finite. The upper Riemann integral is at least as large the lower Riemann integral, and if the two are equal we say that f is Riemann integrable over $[a, b]$ and call the common value the Riemann integral of f over $[a, b]$. We denote

$$(R)\int_a^b f.$$

Definition 2.24. A real-valued function ψ defined on $[a, b]$ is called a step function if there is a partition $P = \{x_0, x_1, \dots, x_n\}$ of $[a, b]$ and numbers $c_1, \dots, c_n$, such that

$$\psi(x) = c_i \quad \text{if} \quad x_{i-1} < x < x_i, \quad i \in \{1, \dots, n\}.$$

Suppose that ψ is a step function on $[a,b]$. Observe that

$$L(\psi,P)=\sum_{i=1}^{n} c_i(x_i-x_{i-1})=U(\psi,P)$$

and

$$(R)\int_a^b \psi=\sum_{i=1}^{n} c_i(x_i-x_{i-1}).$$

Therefore we may reformulate the definition of the lower and upper Riemann integrals as follows.

Definition 2.25.

$$(R)\underline{\int_a^b} f=\sup\left\{(R)\int_a^b \phi : \phi \quad \text{is a step function,} \quad \phi\le f \quad \text{on} \quad [a,b]\right\},$$

$$(R)\overline{\int_a^b} f=\inf\left\{(R)\int_a^b \psi : \psi \quad \text{is a step function,} \quad \psi\ge f \quad \text{on} \quad [a,b]\right\}.$$

2.7 Lebesgue integration

2.7.1 The Lebesgue integral of a bounded measurable function over a set of finite measure

Suppose that $\mathbf{E}$ is a measurable set of finite measure.

Definition 2.26. Let ϕ be a simple function on $\mathbf{E}$,

$$\phi=\sum_{i=1}^{n} a_i\kappa_{\mathbf{E}_i},\quad \mathbf{E}_i=\phi^{-1}(a_i)=\{x\in\mathbf{E}:\phi(x)=a_i\},$$

$a_i\ne a_j$, $\mathbf{E}_i\cap\mathbf{E}_j=\emptyset$, $i,j\in\{1,\dots,n\}$. We define the integral of ϕ over $\mathbf{E}$ by

$$\int_{\mathbf{E}}\phi=\sum_{i=1}^{n} a_i m(\mathbf{E}_i).$$

Theorem 2.52. *Let $\{\mathbf{E}_i\}_{i=1}^n$ be a finite collection of disjoint measurable subsets of $\mathbf{E}$. Let also $a_i\in\mathbf{R}$, $i\in\{1,\dots,n\}$, and*

$$\phi=\sum_{i=1}^{n} a_i\kappa_{\mathbf{E}_i}.$$

Then

$$\int_{\mathbf{E}} \phi = \sum_{i=1}^{n} a_i m(\mathbf{E}_i).$$

Proof.

1. If a_i, $i \in \{1, \ldots, n\}$, are distinct, we get the desired result using Definition 2.26.
2. Assume that a_i, $i \in \{1, \ldots, n\}$, are not distinct. Let $\{b_1, \ldots, b_m\}$ be the distinct values taken by ϕ. We define the sets

   $$B_j = \{x \in \mathbf{E} : \phi(x) = b_j\}, \quad j \in \{1, \ldots, m\}.$$

   Hence, by Definition 2.26,

   $$\int_{\mathbf{E}} \phi = \sum_{i=1}^{m} b_i m(\mathbf{B}_i). \tag{2.83}$$

   For $j \in \{1, \ldots, m\}$, with $\mathbf{I}_j$ we denote the set of indices $i \in \{1, \ldots, n\}$, such that $a_i = b_j$. Then $\{1, \ldots, n\} = \bigcup_{j=1}^{m} \mathbf{I}_j$ and $\mathbf{I}_k \cap \mathbf{I}_l = \emptyset$, $k \neq l$, $k, l \in \{1, \ldots, m\}$. From the finite additivity of the measure, we have

   $$m(\mathbf{B}_j) = m\Big(\bigcup_{i \in \mathbf{I}_j} \mathbf{E}_i\Big) = \sum_{i \in \mathbf{I}_j} m(\mathbf{E}_i).$$

   Hence, using (2.83), we find

   $$\sum_{i=1}^{n} a_i m(\mathbf{E}_i) = \sum_{j=1}^{m} \Big(\sum_{i \in \mathbf{I}_j} a_i m(\mathbf{E}_i)\Big) = \sum_{j=1}^{m} b_j \Big(\sum_{i \in \mathbf{I}_j} m(\mathbf{E}_i)\Big) = \sum_{j=1}^{m} b_j m(\mathbf{B}_j) = \int_{\mathbf{E}} \phi.$$

   This completes the proof. □

Theorem 2.53. *Let ϕ be a simple function on $\mathbf{E}$ and $\phi \geq 0$ on $\mathbf{E}$. Then $\int_{\mathbf{E}} \phi \geq 0$.*

Proof. Since ϕ is a simple function on $\mathbf{E}$, there is a finite disjoint collection $\{\mathbf{E}_i\}_{i=1}^{m}$ of subsets of $\mathbf{E}$ such that $\mathbf{E} = \bigcup_{j=1}^{m} \mathbf{E}_j$ and

$$\phi = \sum_{j=1}^{m} a_j \kappa_{\mathbf{E}_j},$$

for some $a_j \in \mathbf{R}$, $j \in \{1, \ldots, m\}$. Since $\phi \geq 0$ on $\mathbf{E}$, we have $a_j \geq 0$ for any $j \in \{1, \ldots, m\}$. Hence,

$$\int_{\mathbf{E}} \phi = \sum_{j=1}^{m} a_j m(\mathbf{E}_j) \geq 0.$$

This completes the proof. □

Theorem 2.54. *Let ϕ and ψ be simple functions on* $\mathbf{E}$*. Then for any $\alpha, \beta \in \mathbf{R}$, we have*

$$\int_{\mathbf{E}} (\alpha\phi + \beta\psi) = \alpha \int_{\mathbf{E}} \phi + \beta \int_{\mathbf{E}} \psi.$$

Moreover, if $\phi \leq \psi$, then

$$\int_{\mathbf{E}} \phi \leq \int_{\mathbf{E}} \psi.$$

Proof. Since both ϕ and ψ take finite values, there is a finite disjoint collection $\{\mathbf{E}_j\}_{j=1}^{m}$ of subsets of $\mathbf{E}$ such that $\mathbf{E} = \bigcup_{j=1}^{m} \mathbf{E}_j$ and

$$\phi = \sum_{j=1}^{m} a_j \kappa_{\mathbf{E}_j} \quad \text{and} \quad \psi = \sum_{j=1}^{m} b_j \kappa_{\mathbf{E}_j}$$

for some $a_j, b_j \in \mathbf{R}$, $j \in \{1, \dots, m\}$. Then

$$\begin{aligned}
\alpha\phi &= \alpha \sum_{j=1}^{m} a_j \kappa_{\mathbf{E}_j} = \sum_{j=1}^{m} (\alpha a_j) \kappa_{\mathbf{E}_j}, \\
\beta\psi &= \beta \sum_{j=1}^{m} b_j \kappa_{\mathbf{E}_j} = \sum_{j=1}^{m} (\beta b_j) \kappa_{\mathbf{E}_j}, \\
\alpha\phi + \beta\psi &= \sum_{j=1}^{m} (\alpha a_j) \kappa_{\mathbf{E}_j} + \sum_{j=1}^{m} (\beta b_j) \kappa_{\mathbf{E}_j} = \sum_{j=1}^{m} (\alpha a_j + \beta b_j) \kappa_{\mathbf{E}_j}, \\
\int_{\mathbf{E}} \phi &= \sum_{j=1}^{m} a_j m(\mathbf{E}_j), \\
\int_{\mathbf{E}} \psi &= \sum_{j=1}^{m} b_j m(\mathbf{E}_j), \\
\int_{\mathbf{E}} (\alpha\phi + \beta\psi) &= \sum_{j=1}^{m} (\alpha a_j + \beta b_j) m(\mathbf{E}_j), \\
\alpha \int_{\mathbf{E}} \phi &= \alpha \sum_{j=1}^{m} a_j m(\mathbf{E}_j) = \sum_{j=1}^{m} (\alpha a_j) m(\mathbf{E}_j), \\
\beta \int_{\mathbf{E}} \psi &= \beta \sum_{j=1}^{m} b_j m(\mathbf{E}_j) = \sum_{j=1}^{m} (\beta b_j) m(\mathbf{E}_j), \\
\alpha \int_{\mathbf{E}} \phi + \beta \int_{\mathbf{E}} \psi &= \sum_{j=1}^{m} (\alpha a_j) m(\mathbf{E}_j) + \sum_{j=1}^{m} (\beta b_j) m(\mathbf{E}_j) = \sum_{j=1}^{m} (\alpha a_j + \beta b_j) m(\mathbf{E}_j).
\end{aligned}$$

Therefore

$$\int_{\mathbf{E}} (\alpha\phi + \beta\psi) = \alpha \int_{\mathbf{E}} \phi + \beta \int_{\mathbf{E}} \psi.$$

Let now $\phi \le \psi$ on $\mathbf{E}$. Then $\psi - \phi \ge 0$ on $\mathbf{E}$. By Theorem 2.53, we get

$$\int_{\mathbf{E}} (\psi - \phi) \ge 0.$$

Hence,

$$\int_{\mathbf{E}} \psi - \int_{\mathbf{E}} \phi = \int_{\mathbf{E}} (\psi - \phi) \ge 0.$$

This completes the proof. □

Definition 2.27. Let f be a bounded real-valued function defined on $\mathbf{E}$. We define the lower and upper Lebesgue integral, respectively, to be

$$\int_{\mathbf{E}}^{l} f = \sup\left\{\int_{\mathbf{E}} \phi : \phi \quad \text{is} \quad \text{simple} \quad \text{and} \quad \phi \le f \quad \text{on} \quad \mathbf{E}\right\},$$

$$\int_{\mathbf{E}}^{u} f = \inf\left\{\int_{\mathbf{E}} \psi : \psi \quad \text{is} \quad \text{simple} \quad \text{and} \quad f \le \psi \quad \text{on} \quad \mathbf{E}\right\}.$$

Since f is a bounded function on $\mathbf{E}$, there is a constant M such that $f \le M$ on $\mathbf{E}$. Let ϕ is a simple function on $\mathbf{E}$ such that $\phi \le f$ on $\mathbf{E}$. Then $\phi \le M$ on $\mathbf{E}$ and $M - \phi$ is a simple function on $\mathbf{E}$. By Theorem 2.54, we get

$$0 \le \int_{\mathbf{E}} (M - \phi) = \int_{\mathbf{E}} M - \int_{\mathbf{E}} \phi = Mm(\mathbf{E}) - \int_{\mathbf{E}} \phi,$$

whereupon

$$\int_{\mathbf{E}} \phi \le Mm(\mathbf{E}).$$

Therefore

$$\int_{\mathbf{E}}^{l} f \le Mm(\mathbf{E}).$$

Let now ϕ and ψ be simple functions on $\mathbf{E}$ such that $\phi \le f \le \psi$ on $\mathbf{E}$. Hence, by Theorem 2.54, we find

$$\int_{\mathbf{E}} \phi \le \int_{\mathbf{E}} \psi.$$

Consequently

$$\int_{\mathbf{E}}^{l} f \le \int_{\mathbf{E}}^{u} f.$$

Definition 2.28. A bounded real-valued function f on $\mathbf{E}$ is said to be Lebesgue integrable provided its upper and lower Lebesgue integrals are equal. The common value of the upper and lower integrals is called the Lebesgue integral, or simply the integral, of f over $\mathbf{E}$. It is denoted by $(L)\int_{\mathbf{E}} f$ or simply $\int_{\mathbf{E}} f$.

Theorem 2.55. *Let f be a bounded function defined on a closed, bounded interval $\mathbf{I} = [a, b]$. If f is Riemann integrable over $\mathbf{I}$, then it is Lebesgue integrable over $\mathbf{I}$ and the two integrals are equal.*

Proof. Since f is Riemann integrable over $\mathbf{I}$, we have

$$\begin{aligned}(R)\int_{\mathbf{I}} f &= \sup\left\{(R)\int_{\mathbf{I}} \phi : \phi \quad \text{is a step function,} \quad \phi \le f\right\}\\ &= \inf\left\{(R)\int_{\mathbf{I}} \psi : \psi \quad \text{is a step function,} \quad f \le \psi\right\}.\end{aligned}$$

Because every step function is a simple function and every simple function on a bounded closed interval is a step function, and $(R)\int_{\mathbf{I}} \phi = (L)\int_{\mathbf{I}} \phi$ for any simple function ϕ on $\mathbf{I}$, we get

$$\begin{aligned}&\sup\left\{(R)\int_{\mathbf{I}} \phi : \phi \quad \text{is a step function,} \quad \phi \le f\right\}\\ &= \sup\left\{(L)\int_{\mathbf{I}} \phi : \phi \quad \text{is a simple function,} \quad \phi \le f\right\}\\ &= \inf\left\{(R)\int_{\mathbf{I}} \psi : \psi \quad \text{is a step function,} \quad f \le \psi\right\}\\ &= \inf\left\{(L)\int_{\mathbf{I}} \psi : \psi \quad \text{is a step function,} \quad f \le \psi\right\}.\end{aligned}$$

Hence, f is Lebesgue integrable over $\mathbf{I}$ and

$$(L)\int_{\mathbf{I}} f = (R)\int_{\mathbf{I}} f.$$

This completes the proof. □

Theorem 2.56. *Let f be a bounded measurable function on $\mathbf{E}$. Then f is integrable over $\mathbf{E}$.*

Proof. By Theorem 2.46, it follows that for any $n \in \mathbf{N}$ there are simple functions ϕ_n and ψ_n on $\mathbf{E}$ such that

$$\phi_n \le f \le \psi_n$$

and $\psi_n - \phi_n \le \frac{1}{n}$ on **E**. Hence, by Theorem 2.54, it follows that

$$0 \le \int_{\mathbf{E}} \psi_n - \int_{\mathbf{E}} \phi_n = \int_{\mathbf{E}} (\psi_n - \phi_n) \le \frac{1}{n} m(\mathbf{E})$$

for any $n \in \mathbf{N}$. From this,

$$\begin{aligned} 0 &\le \inf\left\{\int_{\mathbf{E}} \psi : \psi \quad \text{simple}, \psi \ge f\right\} - \sup\left\{\int_{\mathbf{E}} \phi : \phi \quad \text{simple}, \phi \le f\right\} \\ &\le \int_{\mathbf{E}} \psi_n - \int_{\mathbf{E}} \phi_n \le \frac{1}{n} m(\mathbf{E}) \end{aligned}$$

for any $n \in \mathbf{N}$. Therefore the upper and lower integrals of f over **E** are equal. Consequently the function f is integrable over **E**. This completes the proof. □

Theorem 2.57. *Let f be a bounded measurable function on* **E** *and $m(\mathbf{E}) = 0$. Then*

$$\int_{\mathbf{E}} f = 0.$$

Proof. Let ϕ be an arbitrary simple function on **E** with canonical representation

$$\phi = \sum_{i=1}^{n} a_i \kappa_{\mathbf{E}_i},$$

where $\mathbf{E} = \bigcup_{i=1}^{n} \mathbf{E}_i$, $\mathbf{E}_i \cap \mathbf{E}_j = \emptyset$, $a_i \in \mathbf{R}$, $a_i \neq a_j$, $i \neq j$, $i, j \in \{1, \dots, n\}$. Then $m(\mathbf{E}_i) = 0$, $i \in \{1, \dots, n\}$, and

$$\int_{\mathbf{E}} \phi = \sum_{i=1}^{n} a_i m(\mathbf{E}_i) = 0.$$

Hence,

$$\int_{\mathbf{E}} f = \sup\left\{\int_{\mathbf{E}} \phi : \phi \quad \text{is} \quad \text{simple}, \quad \phi \le f\right\} = 0.$$

This completes the proof. □

Theorem 2.58. *Let f and g be bounded measurable functions on* **E**. *Then, for any $\alpha, \beta \in$* **R**, *we have*

$$\int_{\mathbf{E}} (\alpha f + \beta g) = \alpha \int_{\mathbf{E}} f + \beta \int_{\mathbf{E}} g.$$

Moreover, if $f \ge g$ on **E**, *we have*

$$\int_{\mathbf{E}} f \ge \int_{\mathbf{E}} g.$$

Proof. Suppose that $\alpha > 0$. Using that the Lebesgue integral is equal to the upper Lebesgue integral, we have

$$\int_{\mathbf{E}} (\alpha f) = \inf_{\psi \geq \alpha f} \int_{\mathbf{E}} \psi = \alpha \inf_{\frac{\psi}{\alpha} \geq f} \int_{\mathbf{E}} \frac{\psi}{\alpha} = \alpha \int_{\mathbf{E}} f.$$

Let $\alpha < 0$. Then, using the fact that the Lebesgue integral is equal to the upper and lower Lebesgue integrals, we get

$$\int_{\mathbf{E}} (\alpha f) = \inf_{\psi \geq \alpha f} \int_{\mathbf{E}} \psi = \alpha \sup_{\frac{\psi}{\alpha} \leq f} \int_{\mathbf{E}} \frac{\psi}{\alpha} = \alpha \int_{\mathbf{E}} f.$$

Now we will prove that

$$\int_{\mathbf{E}} (f + g) = \int_{\mathbf{E}} f + \int_{\mathbf{E}} g. \tag{2.84}$$

Let ϕ_1, ϕ_2, ψ_1 and ψ_2 be simple functions on $\mathbf{E}$ such that

$$\phi_1 \leq f \leq \psi_1 \quad \text{and} \quad \phi_2 \leq g \leq \psi_2 \quad \text{on} \quad \mathbf{E}.$$

Then $\phi_1 + \phi_2$ and $\psi_1 + \psi_2$ are simple functions on $\mathbf{E}$ and

$$\phi_1 + \phi_2 \leq f + g \leq \psi_1 + \psi_2 \quad \text{on} \quad \mathbf{E}.$$

Hence, by Theorem 2.54, we get

$$\int_{\mathbf{E}} (f + g) \leq \int_{\mathbf{E}} (\psi_1 + \psi_2) = \int_{\mathbf{E}} \psi_1 + \int_{\mathbf{E}} \psi_2.$$

Then

$$\int_{\mathbf{E}} (f + g) \leq \inf_{\psi_1 \geq f} \int_{\mathbf{E}} \psi_1 + \inf_{\psi_2 \geq g} \int_{\mathbf{E}} \psi_2 = \int_{\mathbf{E}} f + \int_{\mathbf{E}} g. \tag{2.85}$$

On the other hand,

$$\int_{\mathbf{E}} (f + g) \geq \int_{\mathbf{E}} (\phi_1 + \phi_2) = \int_{\mathbf{E}} \phi_1 + \int_{\mathbf{E}} \phi_2.$$

Then

$$\int_{\mathbf{E}} (f + g) \geq \sup_{\phi_1 \leq f} \int_{\mathbf{E}} \phi_1 + \sup_{\phi_2 \leq g} \int_{\mathbf{E}} \phi_2 = \int_{\mathbf{E}} f + \int_{\mathbf{E}} g.$$

From the previous inequality and from (2.85), we get (2.84). Hence, since αf and βg are bounded measurable functions on $\mathbf{E}$ for any $\alpha, \beta \in \mathbf{R}$, we get

$$\int_{\mathbf{E}} (\alpha f + \beta g) = \int_{\mathbf{E}} (\alpha f) + \int_{\mathbf{E}} (\beta g) = \alpha \int_{\mathbf{E}} f + \beta \int_{\mathbf{E}} g.$$

Now we assume that $f \geq g$ on $\mathbf{E}$. Then $h = f - g \geq 0$ on $\mathbf{E}$. Since 0 is a simple function on $\mathbf{E}$, we obtain

$$\int_{\mathbf{E}} h \geq 0.$$

From this,

$$0 \leq \int_{\mathbf{E}} (f - g) = \int_{\mathbf{E}} f - \int_{\mathbf{E}} g.$$

This completes the proof. □

Theorem 2.59. *Let f be a bounded measurable function on $\mathbf{E}$. Then for any measurable subset $\mathbf{A}$ of $\mathbf{E}$ we have*

$$\int_{\mathbf{E}} f \kappa_{\mathbf{A}} = \int_{\mathbf{A}} f.$$

Proof. Note that $f\kappa_{\mathbf{A}}$ is a bounded measurable function on $\mathbf{E}$ and $\mathbf{A}$. Also, f is a bounded measurable function on $\mathbf{A}$. Then, by Theorem 2.56, we conclude that $f\kappa_{\mathbf{A}}$ is integrable on $\mathbf{E}$ and $\mathbf{A}$, and f is integrable on $\mathbf{A}$. Let ϕ be any simple function on $\mathbf{E}$ with the canonical representation

$$\phi = \sum_{i=1}^{n} a_i \kappa_{\mathbf{E}_i}, \quad a_i \in \mathbf{R}, \quad a_i \neq a_j, \quad i,j \in \{1, \ldots, n\},$$

$\mathbf{E} = \bigcup_{i=1}^{n} \mathbf{E}_i$, $\mathbf{E}_i \cap \mathbf{E}_j = \emptyset$, $i \neq j$, $i,j \in \{1, \ldots, n\}$. Then

$$\phi \kappa_{\mathbf{A}} = \sum_{i=1}^{n} a_i \kappa_{\mathbf{E}_i} \kappa_{\mathbf{A}} = \sum_{i=1}^{n} a_i \kappa_{\mathbf{E}_i \cap \mathbf{A}} + 0 \kappa_{\mathbf{E} \setminus (\bigcup_{i=1}^{n} (\mathbf{E}_i \cap \mathbf{A}))}. \tag{2.86}$$

Hence, $\phi \kappa_{\mathbf{A}}$ is a simple function on $\mathbf{E}$ with the canonical representation (2.86). Then, using Definition 2.26, we obtain

$$\int_{\mathbf{E}} \phi \kappa_{\mathbf{A}} = \sum_{i=1}^{n} a_i m(\mathbf{E}_i \cap \mathbf{A}). \tag{2.87}$$

Note that $\mathbf{A} = \bigcup_{i=1}^{n}(\mathbf{E}_i \cap \mathbf{A})$, $(\mathbf{E}_i \cap \mathbf{A}) \cap (\mathbf{E}_j \cap \mathbf{A}) = \emptyset$. Hence, (2.86) is the canonical representation of the simple function $\phi\kappa_{\mathbf{A}}$ on $\mathbf{A}$. Then, using Definition 2.26 and using the fact that $\phi\kappa_{\mathbf{A}}$ is the restriction of ϕ on $\mathbf{A}$, we get

$$\int_{\mathbf{A}} \phi = \int_{\mathbf{A}} \phi\kappa_{\mathbf{A}} = \sum_{i=1}^{n} a_i m(\mathbf{E}_i \cap \mathbf{A}).$$

From this and from (2.87), we obtain

$$\int_{\mathbf{E}} \phi\kappa_{\mathbf{A}} = \int_{\mathbf{A}} \phi.$$

Next,

$$\begin{aligned}\int_{\mathbf{E}} f\kappa_{\mathbf{A}} &= \sup\left\{\int_{\mathbf{E}} \phi : \phi \quad \text{simple}, \phi \le f\kappa_{\mathbf{A}}\right\} = \sup\left\{\int_{\mathbf{E}} \phi\kappa_{\mathbf{A}} : \phi \quad \text{simple}, \phi \le f\kappa_{\mathbf{A}}\right\} \\ &= \sup\left\{\int_{\mathbf{A}} \phi : \phi \quad \text{simple}, \phi \le f\kappa_{\mathbf{A}}\right\} = \int_{\mathbf{A}} f\kappa_{\mathbf{A}} = \int_{\mathbf{A}} f.\end{aligned}$$

This completes the proof. □

Theorem 2.60. *Let f be a bounded measurable function on* $\mathbf{E}$. *Suppose that* $\mathbf{A}$ *and* $\mathbf{B}$ *are disjoint measurable subsets of* $\mathbf{E}$. *Then*

$$\int_{\mathbf{A}\cup\mathbf{B}} f = \int_{\mathbf{A}} f + \int_{\mathbf{B}} f.$$

Proof. Note that $f\kappa_{\mathbf{A}\cup\mathbf{B}}$, $f\kappa_{\mathbf{A}}$ and $f\kappa_{\mathbf{B}}$ are bounded measurable functions on $\mathbf{E}$. Since $\mathbf{A}$ and $\mathbf{B}$ are disjoint, we have

$$f\kappa_{\mathbf{A}\cup\mathbf{B}} = f\kappa_{\mathbf{A}} + f\kappa_{\mathbf{B}}.$$

Hence, by Theorems 2.58 and 2.59, we get

$$\begin{aligned}\int_{\mathbf{A}\cup\mathbf{B}} f &= \int_{\mathbf{A}\cup\mathbf{B}} f\kappa_{\mathbf{A}\cup\mathbf{B}} = \int_{\mathbf{A}\cup\mathbf{B}} (f\kappa_{\mathbf{A}} + f\kappa_{\mathbf{B}}) = \int_{\mathbf{A}\cup\mathbf{B}} f\kappa_{\mathbf{A}} + \int_{\mathbf{A}\cup\mathbf{B}} f\kappa_{\mathbf{B}} \\ &= \int_{\mathbf{A}} f\kappa_{\mathbf{A}} + \int_{\mathbf{B}} f\kappa_{\mathbf{B}} = \int_{\mathbf{A}} f + \int_{\mathbf{B}} f.\end{aligned}$$

This completes the proof. □

Theorem 2.61. *Let f be a bounded measurable function on* $\mathbf{E}$. *Then*

$$\left|\int_{\mathbf{E}} f\right| \le \int_{\mathbf{E}} |f|.$$

Proof. By Theorem 2.37, we see that $|f|$ is a measurable function on **E**. Since f is bounded on **E**, the function $|f|$ is a bounded measurable function on **E**. By Theorem 2.56, it follows that it is integrable over **E**. Note that

$$-f \le |f| \le f \quad \text{on} \quad \mathbf{E}.$$

Hence, by Theorem 2.58, we obtain

$$-\int_{\mathbf{E}} f \le \int_{\mathbf{E}} |f| \le \int_{\mathbf{E}} f.$$

This completes the proof. □

Theorem 2.62. *Let $\{f_n\}_{n\in\mathbf{N}}$ be a sequence of bounded measurable functions on* **E**. *If $f_n \to f$, as $n \to \infty$, uniformly on* **E**, *then*

$$\lim_{n\to\infty} \int_{\mathbf{E}} f_n = \int_{\mathbf{E}} f.$$

Proof. Let $\epsilon > 0$ be arbitrarily chosen. Since $\{f_n\}_{n\in\mathbf{N}}$ is uniformly convergent to f on **E**, there is an $N \in \mathbf{N}$ such that

$$|f - f_n| < \frac{\epsilon}{m(\mathbf{E})} \quad \text{on} \quad \mathbf{E} \tag{2.88}$$

for any $n \ge N$. In particular,

$$|f - f_N| < \frac{\epsilon}{m(\mathbf{E})} \quad \text{on} \quad \mathbf{E}.$$

Hence,

$$-\frac{\epsilon}{m(\mathbf{E})} < f - f_N < \frac{\epsilon}{m(\mathbf{E})} \quad \text{on} \quad \mathbf{E},$$

or

$$f_N - \frac{\epsilon}{m(\mathbf{E})} < f < f_N + \frac{\epsilon}{m(\mathbf{E})} \quad \text{on} \quad \mathbf{E}.$$

From the previous inequalities, since f_N is bounded on **E**, we conclude that the function f is a bounded function on **E**. Since the sequence $\{f_n\}_{n\in\mathbf{N}}$ is uniformly convergent to f on **E**, it is pointwise convergent to f on **E**. Because f_n are measurable functions on **E** for any $n \in \mathbf{N}$, using Theorem 2.45, we conclude that f is measurable on **E**. Therefore f is a bounded measurable function on **E**. By Theorem 2.56, it follows that f is integrable on **E**. Hence, by Theorem 2.54, Theorem 2.61 and (2.88), we get

$$\left|\int_{\mathbf{E}} f_n - \int_{\mathbf{E}} f\right| = \left|\int_{\mathbf{E}} (f_n - f)\right| \le \int_{\mathbf{E}} |f_n - f| < \int_{\mathbf{E}} \frac{\epsilon}{m(\mathbf{E})} = \epsilon$$

for any $n \ge N$. This completes the proof. □

Theorem 2.63 (The bounded convergence theorem). *Let* $\{f_n\}_{n\in\mathbf{N}}$ *be an uniformly bounded sequence of measurable functions on* $\mathbf{E}$. *If* $f_n \to f$, *as* $n \to \infty$, *pointwise on* $\mathbf{E}$, *then*

$$\lim_{n\to\infty} \int_{\mathbf{E}} f_n = \int_{\mathbf{E}} f.$$

Proof. Let M be a positive constant such that $|f_n| \leq M$ on $\mathbf{E}$ for any $n \in \mathbf{N}$. Then $|f| \leq M$ on $\mathbf{E}$ and

$$\begin{aligned} |f_n - f| &\leq |f_n| + |f| \\ &\leq M + M = 2M \quad \text{on} \quad \mathbf{E} \end{aligned}$$

for any $n \in \mathbf{N}$. We take $\epsilon > 0$ arbitrarily. By Egoroff's theorem, Theorem 2.49, it follows that there exists a closed subset $\mathbf{A}$ of $\mathbf{E}$ such that $f_n \to f$, as $n \to \infty$, uniformly on $\mathbf{A}$ and $m(\mathbf{E} \setminus \mathbf{A}) < \frac{\epsilon}{4M}$. Hence, there is an index $N \in \mathbf{N}$ such that

$$|f_n - f| < \frac{\epsilon}{2m(\mathbf{E})} \quad \text{on} \quad \mathbf{A}$$

for any $n \geq N$. Then, using Theorems 2.58, 2.61, and 2.60, we get

$$\begin{aligned} \left| \int_{\mathbf{E}} f_n - \int_{\mathbf{E}} f \right| &= \left| \int_{\mathbf{E}} (f_n - f) \right| \leq \int_{\mathbf{E}} |f_n - f| \\ &= \int_{\mathbf{A}} |f_n - f| + \int_{\mathbf{E}\setminus\mathbf{A}} |f_n - f| \\ &< \frac{\epsilon}{2m(\mathbf{E})} m(\mathbf{A}) + 2Mm(\mathbf{E} \setminus \mathbf{A}) < \frac{\epsilon}{2} + \frac{\epsilon}{2} = \epsilon \end{aligned}$$

for any $n \geq N$. This completes the proof. □

2.7.2 The Lebesgue integral of a measurable nonnegative function

Suppose that $\mathbf{E}$ is a measurable set.

Definition 2.29. A measurable function on $\mathbf{E}$ is said to vanish outside a set of finite measure if there is a subset $\mathbf{E}_0$ of $\mathbf{E}$ for which $m(\mathbf{E}_0) < \infty$ and $f \equiv 0$ on $\mathbf{E} \setminus \mathbf{E}_0$. We say that a measurable function f on $\mathbf{E}$ which vanishes outside a set of finite measure has finite support and define its support to be

$$\{x \in \mathbf{E} : f(x) \neq 0\}.$$

Definition 2.30. Let f be a bounded measurable function on $\mathbf{E}$ that has finite support $\mathbf{E}_0$, $\mathbf{E}_0 \subseteq \mathbf{E}$. We define its integral over $\mathbf{E}$ by

$$\int_{\mathbf{E}} f = \int_{\mathbf{E}_0} f.$$

The set of all bounded measurable functions on $\mathbf{E}$ which have finite support will be denoted by $\mathcal{E}$.

Definition 2.31. For a nonnegative measurable function f on $\mathbf{E}$ we define the integral of f over $\mathbf{E}$ by

$$\int_{\mathbf{E}} f = \sup\left\{\int_{\mathbf{E}} \phi : \phi \in \mathcal{E}, 0 \le \phi \le f \quad \text{on} \quad \mathbf{E}\right\}.$$

Theorem 2.64 (Chebychev's inequality). *Let f be a nonnegative measurable function on $\mathbf{E}$. Then for any $\lambda > 0$ we have*

$$m(\mathbf{E}_\lambda) \le \frac{1}{\lambda}\int_{\mathbf{E}} f, \tag{2.89}$$

where $\mathbf{E}_\lambda = \{x \in \mathbf{E} : f(x) \ge \lambda\}$.

Proof.

1. Let $m(\mathbf{E}_\lambda) = \infty$. Define the sets $\mathbf{E}_{\lambda,n} = \mathbf{E}_\lambda \cap [-n, n]$ and the functions $\psi_n = \lambda\kappa_{\mathbf{E}_{\lambda,n}}$ for any $n \in \mathbf{N}$. Then ψ_n are bounded measurable functions on $\mathbf{E}$ with finite support $\mathbf{E}_{\lambda,n}$ for any $n \in \mathbf{N}$. Hence, by Definition 2.30, we get

$$\int_{\mathbf{E}} \psi_n = \int_{\mathbf{E}_{\lambda,n}} \lambda = \lambda m(\mathbf{E}_{\lambda,n})$$

for any $n \in \mathbf{N}$. Note that $0 \le \psi_n \le f$ on $\mathbf{E}$ for any $n \in \mathbf{N}$. From this and from Definition 2.31, it follows

$$\int_{\mathbf{E}} \psi_n \le \int_{\mathbf{E}} f$$

for any $n \in \mathbf{N}$. Also,

$$\mathbf{E}_\lambda = \bigcup_{n=1}^{\infty} \mathbf{E}_{\lambda,n} \quad \text{and} \quad \mathbf{E}_{\lambda,n} \subseteq \mathbf{E}_{\lambda,m}$$

for any $m, n \in \mathbf{N}$, $n \le m$. Hence, by Theorem 2.25, it follows that

$$\infty = \lambda m(\mathbf{E}_\lambda) = \lambda \lim_{n\to\infty} m(\mathbf{E}_{\lambda,n}) = \lim_{n\to\infty} (\lambda m(\mathbf{E}_{\lambda,n})) = \lim_{n\to\infty} \int_{\mathbf{E}} \psi_n \le \int_{\mathbf{E}} f.$$

Therefore $\int_{\mathbf{E}} f = \infty$. The equality (2.89) holds since both sides equal to ∞.

2. Let $m(\mathbf{E}_\lambda) < \infty$. Define $\phi = \lambda\kappa_{\mathbf{E}_\lambda}$. Then ϕ is a bounded measurable function on $\mathbf{E}$ of finite support $\mathbf{E}_\lambda$ and $0 \le \phi \le f$. Hence, by Definition 2.31, we get

$$\lambda m(\mathbf{E}_\lambda) = \int_{\mathbf{E}} \phi \le \int_{\mathbf{E}} f.$$

This completes the proof. □

Exercise 2.16. Let f be any function on a set $\mathbf{A}$. Prove

$$\{x \in \mathbf{A} : f(x) > 0\} = \bigcup_{n=1}^{\infty} \left\{x \in \mathbf{A} : f(x) \ge \frac{1}{n}\right\}.$$

Theorem 2.65. *Let f be a nonnegative measurable function on $\mathbf{E}$. Then $\int_{\mathbf{E}} f = 0$ if and only if $f = 0$ a. e. on $\mathbf{E}$.*

Proof.

1. Let $\int_{\mathbf{E}} f = 0$. Then for each $n \in \mathbf{N}$, using Chebychev's inequality, we get

$$m\left(\left\{x \in \mathbf{E} : f(x) \ge \frac{1}{n}\right\}\right) \le n \int_{\mathbf{E}} f = 0$$

for any $n \in \mathbf{N}$. Using Exercise 2.16 and Theorem 2.5, we get

$$m(\{x \in \mathbf{E} : f(x) > 0\}) \le \sum_{n=1}^{\infty} m\left(\left\{x \in \mathbf{E} : f(x) \ge \frac{1}{n}\right\}\right) = 0.$$

Because f is nonnegative on $\mathbf{E}$, we conclude that $f = 0$ a. e. on $\mathbf{E}$.

2. Let $f = 0$ a. e. on $\mathbf{E}$. Let ϕ be a simple function on $\mathbf{E}$ and ψ be a bounded measurable function on $\mathbf{E}$ of finite support such that $0 \le \phi \le \psi \le f$ on $\mathbf{E}$. Then $\phi = 0$ a. e. on $\mathbf{E}$ and $\int_{\mathbf{E}} \phi = 0$. Since this holds for all such functions ϕ, we get $\int_{\mathbf{E}} \psi = 0$. Because this holds for all such functions ψ, we obtain $\int_{\mathbf{E}} f = 0$. This completes the proof. □

Theorem 2.66. *Let f and g be any nonnegative measurable functions on $\mathbf{E}$. Then for any $\alpha > 0$ and $\beta > 0$ we have*

$$\int_{\mathbf{E}} (\alpha f + \beta g) = \alpha \int_{\mathbf{E}} f + \beta \int_{\mathbf{E}} g.$$

Moreover, if $f \le g$ on $\mathbf{E}$, then

$$\int_{\mathbf{E}} f \le \int_{\mathbf{E}} g.$$

Proof. Let $\alpha > 0$ and ϕ be a bounded measurable function on $\mathbf{E}$ of finite support such that $0 \le \phi \le f$ on $\mathbf{E}$. Then $0 \le \alpha\phi \le \alpha f$ on $\mathbf{E}$. Hence, using Theorem 2.58, we get

$$\alpha \int_{\mathbf{E}} \phi = \int_{\mathbf{E}} (\alpha\phi).$$

Because this holds for any such ϕ, we get

$$\alpha \int_{\mathbf{E}} f = \int_{\mathbf{E}} (\alpha f). \tag{2.90}$$

Now we will prove that

$$\int_{\mathbf{E}} (f+g) = \int_{\mathbf{E}} f + \int_{\mathbf{E}} g. \tag{2.91}$$

Let ϕ and ψ be bounded measurable functions on $\mathbf{E}$ of finite support such that $0 \le \phi \le f$ and $0 \le \psi \le g$ on $\mathbf{E}$. Then $0 \le \phi + \psi \le f + g$ on $\mathbf{E}$ and using Theorem 2.58, we get

$$\int_{\mathbf{E}} (f+g) \ge \int_{\mathbf{E}} (\phi + \psi) = \int_{\mathbf{E}} \phi + \int_{\mathbf{E}} \psi.$$

Hence,

$$\int_{\mathbf{E}} (f+g) \ge \int_{\mathbf{E}} f + \int_{\mathbf{E}} g. \tag{2.92}$$

Let h be a bounded measurable function on $\mathbf{E}$ of finite support such that $0 \le h \le f + g$ on $\mathbf{E}$. We define the functions

$$l = \min\{f, h\} \quad \text{and} \quad k = h - l \quad \text{on} \quad \mathbf{E}.$$

Let $x \in \mathbf{E}$ be arbitrarily chosen.
1. If $h(x) \le f(x)$, then $l(x) = h(x)$ and
$$k(x) = 0 \le g(x).$$
2. Let $h(x) > f(x)$. Then $l(x) = f(x)$ and
$$k(x) = h(x) - f(x) \le f(x) + g(x) - f(x) = g(x).$$

Consequently $k \le g$ on $\mathbf{E}$. Because h is a bounded measurable function on $\mathbf{E}$ of finite support, we see that l is a bounded measurable function on $\mathbf{E}$ of finite support. From this, k is a bounded measurable function on $\mathbf{E}$ of finite support. We have

$$0 \le l \le f \quad \text{and} \quad 0 \le k \le g \quad \text{on} \quad \mathbf{E}.$$

Hence, by Theorem 2.58, we get

$$\int_{\mathbf{E}} h = \int_{\mathbf{E}} l + \int_{\mathbf{E}} k \leq \int_{\mathbf{E}} f + \int_{\mathbf{E}} g.$$

Because the previous inequality holds for all such h, we obtain

$$\int_{\mathbf{E}} (f+g) \leq \int_{\mathbf{E}} f + \int_{\mathbf{E}} g.$$

From the previous inequality and from (2.92), we get (2.91). Since αf and βg are nonnegative measurable functions on $\mathbf{E}$, using (2.91) and (2.90), we get

$$\int_{\mathbf{E}} (\alpha f + \beta g) = \int_{\mathbf{E}} (\alpha f) + \int_{\mathbf{E}} (\beta g) = \alpha \int_{\mathbf{E}} f + \beta \int_{\mathbf{E}} g.$$

Let now $f \leq g$ on $\mathbf{E}$. Let ϕ be a bounded measurable function on $\mathbf{E}$ of finite support such that $0 \leq \phi \leq f$ on $\mathbf{E}$. Then $\phi \leq g$ on $\mathbf{E}$ and

$$\int_{\mathbf{E}} \phi \leq \int_{\mathbf{E}} g.$$

Since the previous inequality holds for any such ϕ, we conclude that

$$\int_{\mathbf{E}} f \leq \int_{\mathbf{E}} g.$$

This completes the proof. □

Theorem 2.67. *Let f be a nonnegative measurable function on $\mathbf{E}$. If $\mathbf{A}$ and $\mathbf{B}$ are disjoint measurable subsets of $\mathbf{E}$, then*

$$\int_{\mathbf{A}\cup\mathbf{B}} f = \int_{\mathbf{A}} f + \int_{\mathbf{B}} f. \tag{2.93}$$

In particular, if $\mathbf{B}$ is a subset of $\mathbf{E}$ of measure zero, then

$$\int_{\mathbf{E}} f = \int_{\mathbf{E}\setminus\mathbf{B}} f. \tag{2.94}$$

Proof. Let $\phi \in \mathcal{E}$ be arbitrarily chosen such that $0 \leq \phi \leq f$ and its support is $\mathbf{E}_0 \subseteq \mathbf{A}\cup\mathbf{B}$. Let $\mathbf{E}_0 = \mathbf{E}_1 \cup \mathbf{E}_2$, $\mathbf{E}_1 \cap \mathbf{E}_2 = \emptyset$, $\mathbf{E}_1 \subseteq \mathbf{A}$, $\mathbf{E}_2 \subseteq \mathbf{B}$. Then, using Theorem 2.58, we get

$$\int_{\mathbf{A}\cup\mathbf{B}} \phi = \int_{\mathbf{E}_0} \phi = \int_{\mathbf{E}_1\cup\mathbf{E}_2} \phi = \int_{\mathbf{E}_1} \phi + \int_{\mathbf{E}_2} \phi = \int_{\mathbf{A}} \phi + \int_{\mathbf{B}} \phi.$$

Hence, using Definition 2.30, we obtain (2.93). Now we will prove (2.94). Since for any $\phi \in \mathcal{E}$ for which $0 \le \phi \le f$ we have $\int_{\mathbf{B}} \phi = 0$ and

$$\int_{\mathbf{E}} \phi = \int_{(\mathbf{E}\setminus\mathbf{B})\cup\mathbf{B}} \phi = \int_{(\mathbf{E}\setminus\mathbf{B})} \phi + \int_{\mathbf{B}} \phi = \int_{(\mathbf{E}\setminus\mathbf{B})} \phi.$$

Because $\phi \in \mathcal{E}$, $0 \le \phi \le f$, was arbitrarily chosen, we get (2.94). This completes the proof. □

Lemma 2.7 (Fatou's lemma). *Let $\{f_n\}_{n\in\mathbf{N}}$ be a sequence of nonnegative measurable functions on* **E**. *If $f_n \to f$, as $n \to \infty$, pointwise a. e. on* **E**, *then*

$$\int_{\mathbf{E}} f \le \liminf \int_{\mathbf{E}} f_n.$$

Proof. By Theorem 2.45, it follows that f is measurable on **E**. Let $\mathbf{E}_0$ be such that $f_n \to f$, as $n \to \infty$, pointwise on $\mathbf{E}_0$ and $m(\mathbf{E} \setminus \mathbf{E}_0) = 0$. Let h be any bounded measurable function on **E** of finite support such that $0 \le h \le f$ on **E**. Let also M be a positive constant such that $0 \le h \le M$ on **E**. Define the set

$$\mathbf{E}_1 = \{x \in \mathbf{E} : h(x) \neq 0\}.$$

Then $m(\mathbf{E}_1) < \infty$. For any $n \in \mathbf{N}$ we define the function h_n on **E** by

$$h_n = \min\{h, f_n\} \quad \text{on} \quad \mathbf{E}.$$

Since h and f_n are measurable functions on **E**, by Theorem 2.43, it follows that the functions h_n are measurable on $\mathbf{E}_0$ for any $n \in \mathbf{N}$. Also,

$$0 \le h_n \le M \quad \text{and} \quad h_n = 0 \quad \text{on} \quad \mathbf{E} \setminus \mathbf{E}_1.$$

Furthermore, for each $x \in \mathbf{E}_0$ we have $h(x) \le f(x)$ and $h_n(x) \to h(x)$ as $n \to \infty$. Hence, by Theorem 2.63, we get

$$\lim_{n\to\infty} \int_{\mathbf{E}} h_n = \lim_{n\to\infty} \int_{\mathbf{E}_0} h_n = \int_{\mathbf{E}_0} h.$$

However, for each $n \in \mathbf{N}$ we have $h_n \le f_n$ on $\mathbf{E}_0$ and using Definition 2.31, we have

$$\int_{\mathbf{E}_0} h_n \le \int_{\mathbf{E}_0} f_n.$$

Thus,

$$\int_{\mathbf{E}} h = \liminf \int_{\mathbf{E}_0} h_n \le \liminf \int_{\mathbf{E}_0} f_n = \liminf \int_{\mathbf{E}} f_n.$$

The previous inequality is valid for all such h. Therefore

$$\int_{\mathbf{E}} f \le \liminf \int_{\mathbf{E}} f_n.$$

This completes the proof. □

Theorem 2.68 (The monotone convergence theorem). *Let $\{f_n\}_{n\in\mathbf{N}}$ be an increasing sequence of nonnegative measurable functions on $\mathbf{E}$. If $f_n \to f$, as $n \to \infty$, a. e. on $\mathbf{E}$, then*

$$\lim_{n\to\infty} \int_{\mathbf{E}} f_n = \int_{\mathbf{E}} f.$$

Proof. By Theorem 2.45, it follows that f is measurable on $\mathbf{E}$. By Fatou's lemma, we have

$$\int_{\mathbf{E}} f \le \liminf \int_{\mathbf{E}} f_n. \tag{2.95}$$

On the other hand, using the fact that $\{f_n\}_{n\in\mathbf{N}}$ is an increasing sequence, we see that $f_n \le f$ a. e. on $\mathbf{E}$ for each $n \in \mathbf{N}$. Hence, by Theorem 2.66, we get

$$\int_{\mathbf{E}} f_n \le \int_{\mathbf{E}} f$$

for any $n \in \mathbf{N}$ and

$$\limsup \int_{\mathbf{E}} f_n \le \int_{\mathbf{E}} f.$$

From the previous inequality and from (2.95), we get the desired result. This completes the proof. □

Corollary 2.5. *Let $\{f_n\}_{n\in\mathbf{N}}$ be a sequence of nonnegative measurable functions on $\mathbf{E}$. If*

$$f = \sum_{n=1}^{\infty} f_n \quad a.e. \quad on \quad \mathbf{E},$$

then

$$\int_{\mathbf{E}} f = \sum_{n=1}^{\infty} \int_{\mathbf{E}} f_n.$$

Proof. Let

$$S_m = \sum_{n=1}^{m} f_n, \quad m \in \mathbf{N}.$$

Then $\{S_n\}_{n\in\mathbf{N}}$ is an increasing sequence of measurable functions on $\mathbf{E}$ and $S_n \to f$, as $n \to \infty$, a. e. on $\mathbf{E}$. By Theorem 2.45, we see that f is measurable on $\mathbf{E}$. By the monotone convergence theorem, it follows that

$$\lim_{n\to\infty} \int_{\mathbf{E}} S_n = \int_{\mathbf{E}} f,$$

whereupon we get the desired result. This completes the proof. □

Definition 2.32. A nonnegative measurable function f on $\mathbf{E}$ is said to be integrable over $\mathbf{E}$ if $\int_{\mathbf{E}} f < \infty$.

Theorem 2.69. *Let the nonnegative measurable function f be integrable over $\mathbf{E}$. Then f is finite a. e. on $\mathbf{E}$.*

Proof. Note that

$$\{x \in \mathbf{E} : f(x) = \infty\} \subseteq \{x \in \mathbf{E} : f(x) \geq n\}$$

for any $n \in \mathbf{N}$. Hence, by Chebychev's inequality (Theorem 2.64), we get

$$m(\{x \in \mathbf{E} : f(x) = \infty\}) \leq m(\{x \in \mathbf{E} : f(x) \geq n\}) \leq \frac{1}{n} \int_{\mathbf{E}} f \to 0 \quad \text{as} \quad n \to \infty,$$

because $0 \leq \int_{\mathbf{E}} f < \infty$. Consequently $m(\{x \in \mathbf{E} : f(x) = \infty\}) = 0$. This completes the proof. □

Lemma 2.8 (Beppo Levi's lemma). *Let $\{f_n\}_{n\in\mathbf{N}}$ be an increasing sequence of nonnegative measurable functions on $\mathbf{E}$. If the sequence $\{\int_{\mathbf{E}} f_n\}_{n\in\mathbf{N}}$ is bounded, then $\{f_n\}_{n\in\mathbf{N}}$ converges pointwise on $\mathbf{E}$ to a measurable function f that is finite a. e. on $\mathbf{E}$ and*

$$\lim_{n\to\infty} \int_{\mathbf{E}} f_n = \int_{\mathbf{E}} f < \infty.$$

Proof. There exists an extended real-valued nonnegative function f on $\mathbf{E}$ such that

$$f(x) = \lim_{n\to\infty} f_n(x), \quad x \in \mathbf{E}.$$

By Theorem 2.45, we see that f is measurable on $\mathbf{E}$. By the Monotone Convergence Theorem, it follows that

$$\lim_{n\to\infty} \int_{\mathbf{E}} f_n = \int_{\mathbf{E}} f.$$

Since the sequence $\{\int_{\mathbf{E}} f_n\}_{n\in\mathbf{N}}$ is a bounded sequence, we see that $\int_{\mathbf{E}} f < \infty$. From this and from Theorem 2.69, we see that f is finite a. e. on $\mathbf{E}$. This completes the proof. □

Theorem 2.70. *Let f be a nonnegative measurable function on* **E** *and* $\mathbf{A} \subseteq \mathbf{E}$ *be a measurable set. Then*

$$\int_{\mathbf{E}} f\kappa_{\mathbf{A}} = \int_{\mathbf{A}} f.$$

Proof. Let ϕ be any measurable function of finite support on **E** such that $0 \le \phi \le f$. Then $\phi\kappa_{\mathbf{A}}$ is a measurable function of finite support on **E** and $0 \le \phi\kappa_{\mathbf{A}} \le f\kappa_{\mathbf{A}}$. Then, using Theorem 2.59, we get

$$\int_{\mathbf{E}} \phi\kappa_{\mathbf{A}} = \int_{\mathbf{A}} \phi.$$

Because this holds for any such ϕ, we conclude that

$$\int_{\mathbf{E}} f\kappa_{\mathbf{A}} = \int_{\mathbf{A}} f.$$

This completes the proof. □

2.7.3 The general Lebesgue integral

Suppose that **E** is a measurable set.

Theorem 2.71. *Let f be a measurable function on* **E**. *Then f^+ and f^- are integrable over* **E** *if and only if $|f|$ is integrable over* **E**.

Proof.
1. Assume that f^+ and f^- are integrable over **E**. Then

$$f^+ \ge 0, \quad f^- \ge 0 \quad \text{on} \quad \mathbf{E},$$

and

$$\int_{\mathbf{E}} f^+ < \infty, \quad \int_{\mathbf{E}} f^- < \infty.$$

Hence, using $|f| = f^+ + f^-$, we get

$$\int_{\mathbf{E}} |f| = \int_{\mathbf{E}} f^+ + \int_{\mathbf{E}} f^- < \infty.$$

2. Assume that $|f|$ is integrable over **E**. We have

$$0 \le f^+ \le |f|, \quad 0 \le f^- \le |f| \quad \text{on} \quad \mathbf{E}.$$

Hence, by Theorem 2.66, it follows that

$$0 \le \int_{\mathbf{E}} f^+ \le \int_{\mathbf{E}} |f| < \infty,$$
$$0 \le \int_{\mathbf{E}} f^- \le \int_{\mathbf{E}} |f| < \infty.$$

This completes the proof. □

Definition 2.33. A measurable function f on $\mathbf{E}$ is said to be integrable over $\mathbf{E}$ provided $|f|$ is integrable over $\mathbf{E}$. In this case we define

$$\int_{\mathbf{E}} f = \int_{\mathbf{E}} f^+ - \int_{\mathbf{E}} f^-.$$

Theorem 2.72. *Let f be a measurable function on $\mathbf{E}$. Suppose that g is a nonnegative function on $\mathbf{E}$ that is integrable over $\mathbf{E}$ and*

$$|f| \le g \quad on \quad \mathbf{E}.$$

Then f is integrable over $\mathbf{E}$ and

$$\left| \int_{\mathbf{E}} f \right| \le \int_{\mathbf{E}} |f|.$$

Proof. By Theorem 2.66, we get

$$\int_{\mathbf{E}} |f| \le \int_{\mathbf{E}} g.$$

Hence, using the fact that g is integrable over $\mathbf{E}$, we conclude that $|f|$ is integrable over $\mathbf{E}$. By Definition 2.33, it follows that f is integrable over $\mathbf{E}$. Also, using Theorem 2.66 and the triangle inequality for real numbers, we obtain

$$\begin{aligned}\left| \int_{\mathbf{E}} f \right| &= \left| \int_{\mathbf{E}} f^+ - \int_{\mathbf{E}} f^- \right| \le \left| \int_{\mathbf{E}} f^+ \right| + \left| \int_{\mathbf{E}} f^- \right| = \int_{\mathbf{E}} f^+ + \int_{\mathbf{E}} f^- \\ &= \int_{\mathbf{E}} (f^+ + f^-) = \int_{\mathbf{E}} |f|.\end{aligned}$$

This completes the proof. □

Theorem 2.73. *Let the functions f and g are integrable over $\mathbf{E}$. Then for any $\alpha, \beta \in \mathbf{R}$, the function $\alpha f + \beta g$ is integrable over $\mathbf{E}$ and*

$$\int_{\mathbf{E}} (\alpha f + \beta g) = \alpha \int_{\mathbf{E}} f + \beta \int_{\mathbf{E}} g.$$

Moreover, if $f \leq g$ on $\mathbf{E}$, *then*

$$\int_{\mathbf{E}} f \leq \int_{\mathbf{E}} g.$$

Proof. Let $\alpha > 0$. Then

$$(\alpha f)^+ = \alpha f^+, \quad (\alpha f)^- = \alpha f^-.$$

Hence, by Theorem 2.66, we obtain

$$\begin{aligned}\int_{\mathbf{E}} (\alpha f) &= \int_{\mathbf{E}} ((\alpha f)^+ - (\alpha f)^-) = \int_{\mathbf{E}} (\alpha f)^+ - \int_{\mathbf{E}} (\alpha f)^- = \int_{\mathbf{E}} (\alpha f^+) - \int_{\mathbf{E}} (\alpha f^-) \\ &= \alpha \int_{\mathbf{E}} f^+ - \alpha \int_{\mathbf{E}} f^- = \alpha\Big(\int_{\mathbf{E}} f^+ + \int_{\mathbf{E}} (-f^-)\Big) = \alpha \int_{\mathbf{E}} (f^+ - f^-) = \alpha \int_{\mathbf{E}} f.\end{aligned}$$

Let $\alpha < 0$. Then

$$(\alpha f)^+ = -\alpha f^-, \quad (\alpha f)^- = -\alpha f^+.$$

Hence, by Theorem 2.66, we obtain

$$\begin{aligned}\int_{\mathbf{E}} (\alpha f) &= \int_{\mathbf{E}} ((\alpha f)^+ - (\alpha f)^-) = \int_{\mathbf{E}} (\alpha f)^+ - \int_{\mathbf{E}} (\alpha f)^- \\ &= \int_{\mathbf{E}} (-\alpha f^-) - \int_{\mathbf{E}} (-\alpha f^+) = \alpha \int_{\mathbf{E}} (-f^-) + \alpha \int_{\mathbf{E}} f^+ \\ &= \alpha\Big(\int_{\mathbf{E}} (-f^-) + \int_{\mathbf{E}} f^+\Big) = \alpha \int_{\mathbf{E}} (f^+ - f^-) = \alpha \int_{\mathbf{E}} f.\end{aligned}$$

Note that

$$(f+g)^+ - (f+g)^- = f + g = f^+ - f^- + g^+ - g^-.$$

Hence,

$$(f+g)^+ + f^- + g^- = (f+g)^- + f^+ + g^+,$$

and

$$\int_{\mathbf{E}} ((f+g)^+ + f^- + g^-) = \int_{\mathbf{E}} ((f+g)^- + f^+ + g^+). \tag{2.96}$$

By Theorem 2.66, we have

$$\int_{\mathbf{E}} ((f+g)^+ + f^- + g^-) = \int_{\mathbf{E}} (f+g)^+ + \int_{\mathbf{E}} f^- + \int_{\mathbf{E}} g^- \tag{2.97}$$

and

$$\int_{\mathbf{E}}((f+g)^- + f^+ + g^+) = \int_{\mathbf{E}}(f+g)^- + \int_{\mathbf{E}} f^+ + \int_{\mathbf{E}} g^+. \tag{2.98}$$

From (2.96), (2.97) and (2.98), we obtain

$$\int_{\mathbf{E}}(f+g)^+ + \int_{\mathbf{E}} f^- + \int_{\mathbf{E}} g^- = \int_{\mathbf{E}}(f+g)^- + \int_{\mathbf{E}} f^+ + \int_{\mathbf{E}} g^+.$$

Hence,

$$\int_{\mathbf{E}}(f+g)^+ - \int_{\mathbf{E}}(f+g)^- = \int_{\mathbf{E}} f^+ - \int_{\mathbf{E}} f^- + \int_{\mathbf{E}} g^+ - \int_{\mathbf{E}} g^-,$$

or

$$\int_{\mathbf{E}}((f+g)^+ - (f+g)^-) = \int_{\mathbf{E}}(f^+ - f^-) + \int_{\mathbf{E}}(g^+ - g^-),$$

or

$$\int_{\mathbf{E}}(f+g) = \int_{\mathbf{E}} f + \int_{\mathbf{E}} g. \tag{2.99}$$

Because αf and βg are integrable over $\mathbf{E}$ for any $\alpha, \beta \in \mathbf{R}$, using (2.99), we obtain

$$\int_{\mathbf{E}}(\alpha f + \beta g) = \int_{\mathbf{E}}(\alpha f) + \int_{\mathbf{E}}(\beta g) = \alpha \int_{\mathbf{E}} f + \beta \int_{\mathbf{E}} g.$$

Now we suppose that $f \leq g$ on $\mathbf{E}$. Define $h = g - f$ on $\mathbf{E}$. Then $h \geq 0$ on $\mathbf{E}$. By Theorem 2.66, we get

$$0 \leq \int_{\mathbf{E}} h = \int_{\mathbf{E}}(g-f) = \int_{\mathbf{E}} g - \int_{\mathbf{E}} f.$$

This completes the proof. □

Theorem 2.74. *Let f be integrable over $\mathbf{E}$ and $\mathbf{A} \subseteq \mathbf{E}$ be measurable. Then*

$$\int_{\mathbf{E}} f\kappa_{\mathbf{A}} = \int_{\mathbf{A}} f.$$

Proof. By Theorem 2.70, we have

$$\int_{\mathbf{E}} f^+\kappa_{\mathbf{A}} = \int_{\mathbf{A}} f^+ \quad \text{and} \quad \int_{\mathbf{E}} f^-\kappa_{\mathbf{A}} = \int_{\mathbf{A}} f^-.$$

Hence, using Theorem 2.73, we get

$$\int_{\mathbf{E}} f\kappa_{\mathbf{A}} = \int_{\mathbf{E}} (f^+ - f^-)\kappa_{\mathbf{A}} = \int_{\mathbf{E}} (f^+\kappa_{\mathbf{A}} - f^-\kappa_{\mathbf{A}})$$
$$= \int_{\mathbf{E}} f^+\kappa_{\mathbf{A}} - \int_{\mathbf{E}} f^-\kappa_{\mathbf{A}} = \int_{\mathbf{A}} f^+ - \int_{\mathbf{A}} f^- = \int_{\mathbf{A}} (f^+ - f^-) = \int_{\mathbf{A}} f.$$

This completes the proof. □

Theorem 2.75. *Let f be integrable over $\mathbf{E}$. Assume that $\mathbf{A}$ and $\mathbf{B}$ are disjoint measurable subsets of $\mathbf{E}$. Then*

$$\int_{\mathbf{A}\cup\mathbf{B}} f = \int_{\mathbf{A}} f + \int_{\mathbf{B}} f.$$

Proof. We have

$$|f\kappa_{\mathbf{A}}| \le |f| \quad \text{and} \quad |f\kappa_{\mathbf{B}}| \le |f| \quad \text{on} \quad \mathbf{E}.$$

Then, using Theorem 2.66, we get

$$\int_{\mathbf{E}} |f\kappa_{\mathbf{A}}| \le \int_{\mathbf{E}} |f| \quad \text{and} \quad \int_{\mathbf{E}} |f\kappa_{\mathbf{B}}| \le \int_{\mathbf{E}} |f|.$$

Hence, $f\kappa_{\mathbf{A}}$ and $f\kappa_{\mathbf{B}}$ are integrable over $\mathbf{E}$. Because $\mathbf{A}$ and $\mathbf{B}$ are disjoint, we have

$$f\kappa_{\mathbf{A}\cup\mathbf{B}} = f\kappa_{\mathbf{A}} + f\kappa_{\mathbf{B}}.$$

From this, $f\kappa_{\mathbf{A}\cup\mathbf{B}}$ is integrable over $\mathbf{E}$ and, using Theorem 2.74, we obtain

$$\int_{\mathbf{A}\cup\mathbf{B}} f = \int_{\mathbf{A}\cup\mathbf{B}} f\kappa_{\mathbf{A}\cup\mathbf{B}} = \int_{\mathbf{A}\cup\mathbf{B}} (f\kappa_{\mathbf{A}} + f\kappa_{\mathbf{B}})$$
$$= \int_{\mathbf{A}\cup\mathbf{B}} f\kappa_{\mathbf{A}} + \int_{\mathbf{A}\cup\mathbf{B}} f\kappa_{\mathbf{B}} = \int_{\mathbf{A}} f\kappa_{\mathbf{A}} + \int_{\mathbf{B}} f\kappa_{\mathbf{B}} = \int_{\mathbf{A}} f + \int_{\mathbf{B}} f.$$

This completes the proof. □

Theorem 2.76. *Let f be integrable over $\mathbf{E}$. Then f is finite a. e. on $\mathbf{E}$ and*

$$\int_{\mathbf{E}} f = \int_{\mathbf{E}_0} f \quad \textit{if} \quad \mathbf{E}_0 \subseteq \mathbf{E} \quad \textit{and} \quad m(\mathbf{E} \setminus \mathbf{E}_0) = 0.$$

Proof. By Theorem 2.69, it follows that f^+ and f^- are finite a. e. on $\mathbf{E}$. Hence, $f = f^+ - f^-$ is finite a. e. on $\mathbf{E}$. Let $\mathbf{E}_0$ be a measurable subset of $\mathbf{E}$ such that $m(\mathbf{E} \setminus \mathbf{E}_0) = 0$. Hence, by Theorem 2.67, we obtain

$$\begin{aligned}\int_{\mathbf{E}} f &= \int_{\mathbf{E}} (f^+ - f^-) = \int_{\mathbf{E}} f^+ - \int_{\mathbf{E}} f^- = \int_{\mathbf{E}_0} f^+ - \int_{\mathbf{E}_0} f^- \\ &= \int_{\mathbf{E}_0} (f^+ - f^-) = \int_{\mathbf{E}_0} f.\end{aligned}$$

This completes the proof. □

Theorem 2.77 (The Lebesgue dominated convergence theorem). *Let $\{f_n\}_{n\in\mathbf{N}}$ be a sequence of measurable functions on* $\mathbf{E}$. *Let g be a nonnegative function on* $\mathbf{E}$ *that is integrable over* $\mathbf{E}$ *and $|f_n| \le g$ on* $\mathbf{E}$ *for all $n \in \mathbf{N}$. If $f_n \to f$ pointwise a. e. on* $\mathbf{E}$, *then f is integrable over* $\mathbf{E}$ *and* $\lim_{n\to\infty} \int_{\mathbf{E}} f_n = \int_{\mathbf{E}} f$.

Proof. Since $|f_n| \le g$ on $\mathbf{E}$ for all $n \in \mathbf{N}$, we see that $|f| \le g$ a. e. on $\mathbf{E}$. There exists a measurable set $\mathbf{E}_0 \subseteq \mathbf{E}$ such that $|f| \le g$ on $\mathbf{E}_0$ and $m(\mathbf{E} \setminus \mathbf{E}_0) = 0$. By Theorem 2.72, it follows that f_n are integrable over $\mathbf{E}$ for all $n \in \mathbf{N}$ and f is integrable over $\mathbf{E}_0$. Because $m(\mathbf{E} \setminus \mathbf{E}_0) = 0$, the function f is integrable over $\mathbf{E}$. By Theorem 2.76, it follows that f_n, f and g are finite a. e. on $\mathbf{E}$. Therefore by possible excising from $\mathbf{E}$ a countable collection of sets of measure zero and using the countable additivity of the Lebesgue measure, we can suppose that f_n, f and g are finite on $\mathbf{E}$. The functions $g - f$, $g - f_n$ are properly defined, nonnegative and measurable on $\mathbf{E}$. Also, $\{g - f_n\}_{n\in\mathbf{N}}$ and $\{g + f_n\}_{n\in\mathbf{N}}$ converge pointwise a. e. on $\mathbf{E}$ to $g - f$ and $g + f$, respectively. By Fatou's lemma, we get

$$\int_{\mathbf{E}} (g - f) \le \liminf \int_{\mathbf{E}} (g - f_n) \quad \text{and} \quad \int_{\mathbf{E}} (g + f) \le \liminf \int_{\mathbf{E}} (g + f_n).$$

On the other hand,

$$\int_{\mathbf{E}} g - \int_{\mathbf{E}} f = \int_{\mathbf{E}} (g - f) \le \liminf \int_{\mathbf{E}} (g - f_n) = \int_{\mathbf{E}} g - \limsup \int_{\mathbf{E}} f_n.$$

Therefore

$$\limsup \int_{\mathbf{E}} f_n \le \int_{\mathbf{E}} f. \tag{2.100}$$

Also,

$$\int_{\mathbf{E}} g + \int_{\mathbf{E}} f = \int_{\mathbf{E}} (g + f) \le \liminf \int_{\mathbf{E}} (g + f_n) = \int_{\mathbf{E}} g + \liminf \int_{\mathbf{E}} f_n.$$

Hence,

$$\int_{\mathbf{E}} f \le \liminf \int_{\mathbf{E}} f_n.$$

By the previous inequality and (2.100), we get

$$\int_{\mathbf{E}} f = \lim_{n\to\infty} \int_{\mathbf{E}} f_n.$$

This completes the proof. □

Theorem 2.78. *Let $\{f_n\}_{n\in\mathbf{N}}$ be a sequence of measurable functions on $\mathbf{E}$ that converges pointwise a. e. on $\mathbf{E}$ to f. Suppose that there is a sequence $\{g_n\}_{n\in\mathbf{N}}$ of nonnegative measurable functions on $\mathbf{E}$ that converges pointwise a. e. on $\mathbf{E}$ to g and $|f_n| \le g_n$ on $\mathbf{E}$ for all $n \in \mathbf{N}$. If $\lim_{n\to\infty} \int_{\mathbf{E}} g_n = \int_{\mathbf{E}} g < \infty$, then $\lim_{n\to\infty} \int_{\mathbf{E}} f_n = \int_{\mathbf{E}} f$.*

Proof. Since $\lim_{n\to\infty} \int_{\mathbf{E}} g_n = \int_{\mathbf{E}} g < \infty$, we see that g is integrable over $\mathbf{E}$ and there is an $N \in \mathbf{N}$ such that g_n are integrable over $\mathbf{E}$ for any $n > N$. By Theorem 2.72, it follows that f_n are integrable over $\mathbf{E}$ for any $n \ge N$. Because $|f_n| \le g_n$ on $\mathbf{E}$ for any $n \in \mathbf{N}$, we conclude that $|f| \le g$ a. e. on $\mathbf{E}$. Therefore f is integrable over $\mathbf{E}$. Note that the sequences $\{f_n+g_n\}_{n\in\mathbf{N}}$ and $\{g_n-f_n\}_{n\in\mathbf{N}}$ converge pointwise a. e. on $\mathbf{E}$ to $f+g$ and $g-f$, respectively. Hence, by Fatou's lemma, we get

$$\begin{aligned}\int_{\mathbf{E}} f + \int_{\mathbf{E}} g = \int_{\mathbf{E}} (f+g) &\le \liminf \int_{\mathbf{E}} (f_n+g_n)\\ &= \liminf \int_{\mathbf{E}} f_n + \liminf \int_{\mathbf{E}} g_n = \liminf \int_{\mathbf{E}} f_n + \int_{\mathbf{E}} g,\end{aligned}$$

whereupon

$$\int_{\mathbf{E}} f \le \liminf \int_{\mathbf{E}} f_n. \tag{2.101}$$

Also,

$$\begin{aligned}\int_{\mathbf{E}} g - \int_{\mathbf{E}} f = \int_{\mathbf{E}} (g-f) &\le \liminf \int_{\mathbf{E}} (g_n-f_n)\\ &= \liminf \int_{\mathbf{E}} g_n - \limsup \int_{\mathbf{E}} f_n = \int_{\mathbf{E}} g - \limsup \int_{\mathbf{E}} f_n.\end{aligned}$$

Therefore

$$\limsup \int_{\mathbf{E}} f_n \le \int_{\mathbf{E}} f.$$

By the previous inequality and from (2.101), we get

$$\int_{\mathbf{E}} f = \lim_{n\to\infty} \int_{\mathbf{E}} f_n.$$

This completes the proof. □

Theorem 2.79. *Let f be integrable over $\mathbf{E}$ and $\{\mathbf{E}_n\}_{n\in\mathbf{N}}$ be a disjoint countable collection of measurable subsets of $\mathbf{E}$ whose union is $\mathbf{E}$. Then*

$$\int_{\mathbf{E}} f = \sum_{i=1}^{\infty} \int_{\mathbf{E}_i} f.$$

Proof. Let $n \in \mathbf{N}$ be arbitrarily chosen. Let also κ_n be the characteristic function of the measurable set $\bigcup_{k=1}^{n} \mathbf{E}_k$. Define $f_n = f\kappa_n$. Then f_n is measurable on $\mathbf{E}$ and

$$|f_n| \le |f| \quad \text{on} \quad \mathbf{E}.$$

Note that $f_n \to f$, as $n \to \infty$, pointwise on $\mathbf{E}$. Hence, by Theorem 2.77, we get

$$\lim_{n\to\infty} \int_{\mathbf{E}} f_n = \int_{\mathbf{E}} f. \tag{2.102}$$

By Theorem 2.75, we get

$$\int_{\mathbf{E}} f_n = \int_{\bigcup_{k=1}^{n} \mathbf{E}_k} f = \sum_{k=1}^{n} \int_{\mathbf{E}_k} f.$$

Hence, by (2.102), we obtain

$$\int_{\mathbf{E}} f = \lim_{n\to\infty} \sum_{k=1}^{n} \int_{\mathbf{E}_k} f = \sum_{k=1}^{\infty} \int_{\mathbf{E}_k} f.$$

This completes the proof. □

Theorem 2.80. *Let f be integrable over $\mathbf{E}$.*
1. *If $\{\mathbf{E}_n\}_{n\in\mathbf{N}}$ is an ascending countable collection of measurable subsets of $\mathbf{E}$, then*

$$\int_{\bigcup_{k=1}^{\infty} \mathbf{E}_k} f = \lim_{n\to\infty} \int_{\mathbf{E}_n} f.$$

2. *If $\{\mathbf{E}_n\}_{n\in\mathbf{N}}$ is a descending countable collection of measurable subsets of $\mathbf{E}$, then*

$$\int_{\bigcap_{k=1}^{\infty} \mathbf{E}_k} f = \lim_{k\to\infty} \int_{\mathbf{E}_k} f.$$

Proof.
1. Let $\mathbf{A}_0 = \emptyset$, $\mathbf{A}_k = \mathbf{E}_k \setminus \mathbf{E}_{k-1}$, $k \in \mathbf{N}$. Then, using the proof of Theorem 2.25,

$$\bigcup_{k=1}^{\infty} \mathbf{E}_k = \bigcup_{k=1}^{\infty} \mathbf{A}_k, \quad \mathbf{E}_n = \bigcup_{k=1}^{n} \mathbf{A}_k, \quad \mathbf{A}_k \cap \mathbf{A}_l = \emptyset, \quad k \neq l, \quad k, l \in \mathbf{N}.$$

Hence, by the proof of Theorem 2.79, we obtain

$$\int_{\bigcup_{k=1}^{\infty}\mathbf{E}_k} f = \int_{\bigcup_{k=1}^{\infty}\mathbf{A}_k} f = \lim_{n\to\infty}\int_{\bigcup_{k=1}^{n}\mathbf{A}_k} f = \lim_{n\to\infty}\int_{\mathbf{E}_n} f.$$

2. Let $\mathbf{B}_k = \mathbf{E}_1 \setminus \mathbf{E}_k$, $k \in \mathbf{N}$. By the proof of Theorem 2.25, we have

$$\bigcup_{k=1}^{\infty}\mathbf{B}_k = \mathbf{E}_1 \setminus \left(\bigcap_{k=1}^{\infty}\mathbf{E}_k\right), \quad \mathbf{B}_k \cap \mathbf{B}_l = \emptyset, \quad k \neq l, \quad k, l \in \mathbf{N}.$$

Hence, by the proof of Theorem 2.79, we get

$$\int_{\mathbf{E}_1\setminus(\bigcap_{k=1}^{\infty}\mathbf{E}_k)} f = \lim_{n\to\infty}\int_{\mathbf{E}_1\setminus(\bigcap_{k=1}^{n}\mathbf{E}_k)} f,$$

whereupon

$$\int_{\mathbf{E}_1} f - \int_{\bigcap_{k=1}^{\infty}\mathbf{E}_k} f = \int_{\mathbf{E}_1} f - \lim_{n\to\infty}\int_{\bigcap_{k=1}^{n}\mathbf{E}_k} f.$$

Therefore

$$\int_{\bigcap_{k=1}^{\infty}\mathbf{E}_k} f = \lim_{n\to\infty}\int_{\bigcap_{k=1}^{n}\mathbf{E}_k} f.$$

This completes the proof. □

Theorem 2.81. *Let* $\mathbf{E}$ *be a set of finite measure and* $\delta > 0$. *Then* $\mathbf{E}$ *is the disjoint union of a finite collection of sets, each of which has measure less than* δ.

Proof. For any $n \in \mathbf{N}$ we define the sets

$$\mathbf{E}_n = \mathbf{E} \cap [-n, n].$$

Then $\{\mathbf{E}_n\}_{n\in\mathbf{N}}$ is a descending collection of measurable sets. Hence, by Theorem 2.25, it follows that

$$\lim_{n\to\infty} m(\mathbf{E}\setminus[-n,n]) = m\left(\bigcap_{n=1}^{\infty}(\mathbf{E}\setminus[-n,n])\right) = m(\emptyset) = 0.$$

Therefore there exists an $n_0 \in \mathbf{N}$ such that

$$m(\mathbf{E}\setminus[-n_0, n_0]) < \delta.$$

We make a fine enough partition of $[-n_0, n_0]$ so that $\mathbf{E} \cap [-n_0, n_0]$ we represent as the disjoint union of a finite collection $\{\mathbf{E}^1, \ldots, \mathbf{E}^l\}$ of sets, each of which has measure less than δ. We have

$$(\mathbf{E}\setminus[-n_0, n_0]) \cup \mathbf{E}^1 \cup \cdots \cup \mathbf{E}^l = \mathbf{E}.$$

This completes the proof. □

Theorem 2.82. *Let f be a measurable function on $\mathbf{E}$. If f is integrable over $\mathbf{E}$, then for each $\epsilon > 0$, there is a $\delta > 0$ for which*

$$\text{if} \quad \mathbf{A} \subseteq \mathbf{E} \quad \text{is measurable and} \quad m(\mathbf{A}) < \delta, \quad \text{then} \quad \int_{\mathbf{A}} |f| < \epsilon. \tag{2.103}$$

Conversely, in the case $m(\mathbf{E}) < \infty$, if for each $\epsilon > 0$, there is a $\delta > 0$ for which (2.103) *holds, then f is integrable over $\mathbf{E}$.*

Proof.

1. Let f is integrable over $\mathbf{E}$. Let $\epsilon > 0$ be arbitrarily chosen. Consider f^+. We see that f^+ is integrable over $\mathbf{E}$. Then there is a bounded measurable function f_ϵ of finite support for which

$$0 \le f_\epsilon \le f^+ \quad \text{on} \quad \mathbf{E} \quad \text{and} \quad 0 \le \int_{\mathbf{E}} f^+ - \int_{\mathbf{E}} f_\epsilon < \frac{\epsilon}{4}.$$

Since $f^+ - f_\epsilon \ge 0$ on $\mathbf{E}$, if $\mathbf{A} \subseteq \mathbf{E}$ is measurable, then

$$\begin{aligned} \int_{\mathbf{A}} f^+ - \int_{\mathbf{A}} f_\epsilon &= \int_{\mathbf{A}} (f^+ - f_\epsilon) \\ &\le \int_{\mathbf{E}} (f^+ - f_\epsilon) = \int_{\mathbf{E}} f^+ - \int_{\mathbf{E}} f_\epsilon < \frac{\epsilon}{4}. \end{aligned}$$

Since f_ϵ is bounded, there is a positive constant M for which $0 \le f_\epsilon < M$ on $\mathbf{E}$. Therefore, if $\mathbf{A} \subseteq \mathbf{E}$ is measurable, then

$$\int_{\mathbf{A}} f^+ < \int_{\mathbf{A}} f_\epsilon + \frac{\epsilon}{4} \le Mm(\mathbf{A}) + \frac{\epsilon}{4}.$$

We take $\delta = \frac{\epsilon}{4M}$. Then

$$\int_{\mathbf{A}} f^+ < \frac{\epsilon}{4} + \frac{\epsilon}{4} = \frac{\epsilon}{2}.$$

As above,

$$\int_{\mathbf{A}} f^- < \frac{\epsilon}{2}.$$

Therefore

$$\int_{\mathbf{A}} |f| = \int_{\mathbf{A}} (f^+ + f^-) = \int_{\mathbf{A}} f^+ + \int_{\mathbf{A}} f^- < \frac{\epsilon}{2} + \frac{\epsilon}{2} = \epsilon.$$

2. Suppose that $m(\mathbf{A}) < \infty$ and for each $\epsilon > 0$ there is a $\delta > 0$ for which (2.103) holds. By Theorem 2.81, there is a finite disjoint collection of measurable subsets $\{\mathbf{E}_k\}_{k=1}^N$ such that $\mathbf{E} = \bigcup_{k=1}^N \mathbf{E}_k$, $m(\mathbf{E}_k) < \delta$, $k \in \{1, \dots, N\}$. Then

$$\sum_{k=1}^{N} \int_{\mathbf{E}_k} f^+ < \epsilon N.$$

Hence, if h is a bounded measurable function on $\mathbf{E}$ of finite support such that $0 \le h \le f^+$, we get

$$\int_{\mathbf{E}} h = \int_{\bigcup_{k=1}^N \mathbf{E}_k} h = \sum_{k=1}^{N} \int_{\mathbf{E}_k} h \le \sum_{k=1}^{N} \int_{\mathbf{E}_k} f^+ < \epsilon N.$$

Then f^+ is integrable over $\mathbf{E}$. As above, we conclude that f^- is integrable over $\mathbf{E}$. Therefore f is integrable over $\mathbf{E}$. This completes the proof. □

Definition 2.34. A family $\mathcal{F}$ of measurable functions on $\mathbf{E}$ is said to be uniformly integrable over $\mathbf{E}$ provided for each $\epsilon > 0$, there is a $\delta > 0$, such that for each $f \in \mathcal{F}$,

$$\textit{if} \quad \mathbf{A} \subseteq \mathbf{E} \quad \text{is measurable and} \quad m(\mathbf{A}) < \delta, \quad \text{then} \quad \int_{\mathbf{A}} |f| < \epsilon.$$

Theorem 2.83. *Let $\{f_k\}_{k=1}^n$ be a finite collection of functions, each of which is integrable over $\mathbf{E}$. Then $\{f_k\}_{k=1}^n$ is uniformly integrable over $\mathbf{E}$.*

Proof. Let $\epsilon > 0$ be arbitrarily chosen. Then for each $k \in \{1, \dots, n\}$ there is a $\delta_k > 0$ such that

$$\textit{if} \quad \mathbf{A} \subseteq \mathbf{E} \quad \text{is measurable and} \quad m(\mathbf{A}) < \delta_k, \quad \text{then} \quad \int_{\mathbf{A}} |f| < \epsilon.$$

Take $\delta = \max\{\delta_1, \dots, \delta_n\}$. Therefore $\{f_k\}_{k=1}^n$ is uniformly integrable over $\mathbf{E}$. This completes the proof. □

Theorem 2.84. *Assume that $\mathbf{E}$ has finite measure. Let the sequence $\{f_n\}_{n\in\mathbf{N}}$ is uniformly integrable over $\mathbf{E}$. If $f_n \to f$, as $n \to \infty$, pointwise a. e. on $\mathbf{E}$, then f is integrable over $\mathbf{E}$.*

Proof. Let $\epsilon > 0$ be arbitrarily chosen. Let $\delta > 0$ respond to the ϵ challenge in the uniform integrability criteria for the sequence $\{f_n\}_{n\in\mathbf{N}}$. By Theorem 2.81, there is a finite disjoint collection $\{\mathbf{E}_k\}_{k=1}^N$ of measurable sets such that $\mathbf{E} = \bigcup_{k=1}^N \mathbf{E}_k$ and $m(\mathbf{E}_k) < \delta$ for any $k \in \{1, \ldots, N\}$. Hence,

$$\int_{\mathbf{E}} |f_n| = \int_{\bigcup_{k=1}^{N} \mathbf{E}_k} |f_n| = \sum_{k=1}^{N} \int_{\mathbf{E}_k} |f_n| < N\epsilon.$$

By Fatou's lemma, it follows that

$$\int_{\mathbf{E}} |f| \le \liminf \int_{\mathbf{E}} |f_n| < \epsilon N.$$

Therefore f is integrable over $\mathbf{E}$. This completes the proof. □

Theorem 2.85 (The Vitali convergence theorem). *Let $\mathbf{E}$ be of finite measure. Suppose that the sequence $\{f_n\}_{n \in \mathbf{N}}$ is uniformly integrable over $\mathbf{E}$. If $f_n \to f$, as $n \to \infty$, pointwise a. e. on $\mathbf{E}$, then f is integrable over $\mathbf{E}$ and*

$$\lim_{n \to \infty} \int_{\mathbf{E}} f_n = \int_{\mathbf{E}} f.$$

Proof. By Theorem 2.84, it follows that f is integrable over $\mathbf{E}$. Hence, using Theorem 2.76, we see that f is finite a. e. on $\mathbf{E}$. Using Theorem 2.76 once more, by possible excising from $\mathbf{E}$ a set of measure zero, we can assume that $f_n \to f$, as $n \to \infty$, pointwise on all $\mathbf{E}$. Let $\mathbf{A} \subseteq \mathbf{E}$ be arbitrary measurable set and $n \in \mathbf{N}$ be arbitrarily chosen. Then

$$\begin{aligned} \left|\int_{\mathbf{E}} f_n - \int_{\mathbf{E}} f\right| &= \left|\int_{\mathbf{E}} (f_n - f)\right| \le \int_{\mathbf{E}} |f_n - f| = \int_{\mathbf{E}\setminus\mathbf{A}} |f_n - f| + \int_{\mathbf{A}} |f_n - f| \\ &\le \int_{\mathbf{E}\setminus\mathbf{A}} |f_n - f| + \int_{\mathbf{A}} |f_n| + \int_{\mathbf{A}} |f|. \end{aligned} \tag{2.104}$$

Let $\epsilon > 0$ be arbitrarily chosen. Then there is a $\delta > 0$ such that if $\mathbf{A} \subseteq \mathbf{E}$ is a measurable set such that $m(\mathbf{A}) < \delta$ we have

$$\int_{\mathbf{A}} |f_n| < \frac{\epsilon}{3}$$

for any $n \in \mathbf{N}$. Hence, by Fatou's lemma, we get

$$\int_{\mathbf{A}} |f| < \frac{\epsilon}{3}$$

whenever $\mathbf{A} \subseteq \mathbf{E}$ and $m(\mathbf{A}) < \delta$. By Egoroff's theorem, there is a measurable subset $\mathbf{E}_0$ of $\mathbf{E}$ for which $m(\mathbf{E}_0) < \delta$ and $f_n \to f$ uniformly on $\mathbf{E} \setminus \mathbf{E}_0$. Hence, there is an $N \in \mathbf{N}$ such that

$$|f_n - f| < \frac{\epsilon}{3m(\mathbf{E})} \quad \text{on} \quad \mathbf{E} \setminus \mathbf{E}_0$$

for all $n \geq N$. Take $\mathbf{A} = \mathbf{E}_0$ in (2.104) and we get

$$\left|\int_{\mathbf{E}} f_n - \int_{\mathbf{E}} f\right| \leq \int_{\mathbf{E}\setminus\mathbf{E}_0} |f_n - f| + \int_{\mathbf{E}_0} |f_n| + \int_{\mathbf{E}_0} |f|$$

$$< \frac{\epsilon}{3m(\mathbf{E})} m(\mathbf{E} \setminus \mathbf{E}_0) + \frac{\epsilon}{3} + \frac{\epsilon}{3} \leq \frac{\epsilon}{3} + \frac{2\epsilon}{3} = \epsilon.$$

This completes the proof. □

Theorem 2.86. *Let* $\mathbf{E}$ *be of finite measure. Suppose that* $\{f_n\}_{n\in\mathbf{N}}$ *is a sequence of nonnegative integrable functions that converges pointwise a. e. on* $\mathbf{E}$ *to* $f = 0$. *Then*

$$\lim_{n\to\infty} \int_{\mathbf{E}} f_n = 0 \tag{2.105}$$

if and only if $\{f_n\}_{n\in\mathbf{N}}$ *is uniformly integrable over* $\mathbf{E}$.

Proof.
1. Let $\{f_n\}_{n\in\mathbf{N}}$ is uniformly integrable over $\mathbf{E}$. Then by Vitali's theorem, we conclude that (2.105) holds.
2. Let (2.105) holds. Then there is an $N \in \mathbf{N}$ such that

$$\int_{\mathbf{E}} f_n < \epsilon$$

for any $n \geq N$. Since $f_n \geq 0$ on $\mathbf{E}$ for any $n \in \mathbf{N}$, if $\mathbf{A} \subseteq \mathbf{E}$ is measurable and $n \geq N$, then

$$\int_{\mathbf{A}} f_n < \epsilon. \tag{2.106}$$

By Theorem 2.83, it follows that $\{f_n\}_{n=1}^{N-1}$ is uniformly integrable over $\mathbf{E}$. Let $\delta > 0$ respond to the ϵ challenge regarding the criterion for uniform integrability of $\{f_n\}_{n=1}^{N-1}$. Then, using (2.106), we get the criterion for uniform integrability of $\{f_n\}_{n\in\mathbf{N}}$. This completes the proof. □

Theorem 2.87. *Let* f *be integrable over* $\mathbf{E}$. *Then for each* $\epsilon > 0$ *there is a set of finite measure* $\mathbf{E}_0 \subseteq \mathbf{E}$ *such that*

$$\int_{\mathbf{E}\setminus\mathbf{E}_0} |f| < \epsilon.$$

Proof. Let $\epsilon > 0$ be arbitrarily chosen. Since $|f|$ is a nonnegative function on $\mathbf{E}$ that is integrable over $\mathbf{E}$, there exists a bounded nonnegative measurable function g on $\mathbf{E}$ with finite support $\mathbf{E}_0 \subseteq \mathbf{E}$ such that $0 \leq g \leq |f|$ on $\mathbf{E}$ and

$$\int_{\mathbf{E}} |f| - \int_{\mathbf{E}} g < \epsilon.$$

Hence,

$$\int_{\mathbf{E}\setminus\mathbf{E}_0} |f| = \int_{\mathbf{E}\setminus\mathbf{E}_0} (|f| - g)$$
$$\leq \int_{\mathbf{E}} (|f| - g) < \epsilon.$$

This completes the proof. □

Definition 2.35. A family $\mathcal{F}$ of measurable functions on $\mathbf{E}$ is said to be tight over $\mathbf{E}$ provided for each $\epsilon > 0$, there is a subset $\mathbf{E}_0$ of $\mathbf{E}$ of finite measure such that

$$\int_{\mathbf{E}\setminus\mathbf{E}_0} |f| < \epsilon$$

for all $f \in \mathcal{F}$.

Theorem 2.88 (The Vitali convergence theorem). *Let $\{f_n\}_{n\in\mathbf{N}}$ is a sequence of functions on $\mathbf{E}$ that is uniformly integrable and a tight over $\mathbf{E}$. Suppose that $f_n \to f$, as $n \to \infty$, pointwise a. e. on $\mathbf{E}$. Then f is integrable over $\mathbf{E}$ and*

$$\lim_{n\to\infty} \int_{\mathbf{E}} f_n = \int_{\mathbf{E}} f.$$

Proof. Let $\epsilon > 0$ be arbitrarily chosen. Since $\{f_n\}_{n\in\mathbf{N}}$ is a tight, there exists a measurable set $\mathbf{E}_0$ of finite measure such that

$$\int_{\mathbf{E}\setminus\mathbf{E}_0} |f_n| < \frac{\epsilon}{4}$$

for all $n \in \mathbf{N}$. Hence, by Fatou's lemma, we get

$$\int_{\mathbf{E}\setminus\mathbf{E}_0} |f| < \frac{\epsilon}{4}.$$

Therefore $|f|$ is integrable over $\mathbf{E} \setminus \mathbf{E}_0$. Hence,

$$\left|\int_{\mathbf{E}\setminus\mathbf{E}_0} (f_n - f)\right| \leq \int_{\mathbf{E}\setminus\mathbf{E}_0} |f_n - f| \leq \int_{\mathbf{E}\setminus\mathbf{E}_0} (|f_n| + |f|)$$
$$= \int_{\mathbf{E}\setminus\mathbf{E}_0} |f_n| + \int_{\mathbf{E}\setminus\mathbf{E}_0} |f| < \frac{\epsilon}{4} + \frac{\epsilon}{4} = \frac{\epsilon}{2}$$

for all $n \in \mathbf{N}$. Since $\mathbf{E}_0$ has finite measure and $\{f_n\}_{n\in\mathbf{N}}$ is uniformly integrable over $\mathbf{E}_0$, using Theorem 2.85, it follows that f is integrable over $\mathbf{E}_0$ and there is an $N \in \mathbf{N}$ such that

$$\left|\int_{\mathbf{E}_0} (f_n - f)\right| < \frac{\epsilon}{2}$$

for all $n \geq N$. Therefore f is integrable over $\mathbf{E}$ and

$$\begin{aligned}\left|\int_{\mathbf{E}} (f_n - f)\right| &= \left|\int_{\mathbf{E}\setminus\mathbf{E}_0} (f_n - f) + \int_{\mathbf{E}_0} (f_n - f)\right| \\ &\leq \left|\int_{\mathbf{E}\setminus\mathbf{E}_0} (f_n - f)\right| + \left|\int_{\mathbf{E}_0} (f_n - f)\right| < \frac{\epsilon}{2} + \frac{\epsilon}{2} = \epsilon\end{aligned}$$

for all $n \geq N$. This completes the proof. □

Definition 2.36. Let $\{f_n\}_{n\in\mathbf{N}}$ be a sequence of measurable functions on $\mathbf{E}$ and f a measurable function on $\mathbf{E}$ for which f and each f_n are finite a. e. on $\mathbf{E}$. The sequence $\{f_n\}_{n\in\mathbf{N}}$ is said to converge in measure on $\mathbf{E}$ to f provided for each $\epsilon > 0$

$$\lim_{n\to\infty} m(\{x \in \mathbf{E} : |f_n(x) - f(x)| > \epsilon\}) = 0.$$

When we write $f_n \to f$, as $n \to \infty$, in measure, we are assuming that f and each f_n are measurable and finite a. e. on $\mathbf{E}$.

Exercise 2.17. Let $\{f_n\}_{n\in\mathbf{N}}$ be a sequence of measurable functions on $\mathbf{E}$ and f a measurable function on $\mathbf{E}$ for which f and each f_n are finite a. e. on $\mathbf{E}$. If $f_n \to f$, as $n \to \infty$, uniformly on $\mathbf{E}$, prove that $f_n \to f$, as $n \to \infty$, in measure.

Theorem 2.89. *Let* $\mathbf{E}$ *has finite measure. Let also,* $\{f_n\}_{n\in\mathbf{N}}$ *be a sequence of measurable functions on* $\mathbf{E}$ *that converges pointwise a. e. on* $\mathbf{E}$ *to* f *and* f *is finite a. e. on* $\mathbf{E}$*. Then* $f_n \to f$*, as* $n \to \infty$*, in measure on* $\mathbf{E}$*.*

Proof. By Theorem 2.45, we see that f is measurable on $\mathbf{E}$. Let $\epsilon > 0$ be arbitrarily chosen. By Egoroff's theorem, for each $\delta > 0$ there is a measurable subset $\mathbf{E}_0 \subseteq \mathbf{E}$ such that $f_n \to f$, as $n \to \infty$, uniformly on $\mathbf{E}_0$ and $m(\mathbf{E} \setminus \mathbf{E}_0) < \delta$. Then there is an index $N \in \mathbf{N}$ such that

$$|f_n - f| < \epsilon \quad \text{on} \quad \mathbf{E}_0$$

for each $n \geq N$. Thus, for each $n \geq N$, we have

$$m(\{x \in \mathbf{E} : |f_n(x) - f(x)| > \epsilon\}) \leq m(\mathbf{E} \setminus \mathbf{E}_0) < \delta.$$

This completes the proof. □

Theorem 2.90 (Riesz's theorem). *Let* $f_n \to f$*, as* $n \to \infty$*, in measure on* $\mathbf{E}$*. Then there is a subsequence* $\{f_{n_k}\}_{k\in\mathbf{N}}$ *of the sequence* $\{f_n\}_{n\in\mathbf{N}}$ *that converges pointwise a. e. on* $\mathbf{E}$ *to* f*.*

Proof. There is a sequence $\{n_k\}_{k\in\mathbf{N}}$ of natural numbers such that

$$m\left(\left\{x \in \mathbf{E} : |f_j(x) - f(x)| > \frac{1}{k}\right\}\right) < \frac{1}{2^k} \quad \text{for} \quad \text{any} \quad j \geq n_k.$$

Let

$$\mathbf{E}_k = \left\{x \in \mathbf{E} : |f_j(x) - f(x)| > \frac{1}{k}\right\}.$$

Then $m(\mathbf{E}_k) < \frac{1}{2^k}$ and $\sum_{k=1}^{\infty} m(\mathbf{E}_k) < \infty$. By the Borel–Cantelli lemma, it follows that there is an index $K(x)$ such that $x \notin \mathbf{E}_k$ for any $k \geq K(x)$ and

$$|f_{n_k}(x) - f(x)| \leq \frac{1}{k} \quad \text{for} \quad \text{all} \quad k \geq K(x).$$

Therefore

$$\lim_{k\to\infty} f_{n_k}(x) = f(x).$$

This completes the proof. □

Exercise 2.18. Let $\{f_n\}_{n\in\mathbf{N}}$ be a sequence of nonnegative integrable functions on $\mathbf{E}$. Then $\lim_{n\to\infty} \int_{\mathbf{E}} f_n = 0$ if and only if $f_n \to 0$, as $n \to \infty$, in measure on $\mathbf{E}$ and $\{f_n\}_{n\in\mathbf{N}}$ is uniformly integrable and a tight over $\mathbf{E}$.

2.8 Continuity and differentiability of monotone functions. Lebesgue's theorem

Definition 2.37. A real-valued function defined on a set $\mathbf{E}$ of real numbers is said to be increasing if $f(x) \leq f(x')$ whenever $x, x' \in \mathbf{E}$ and $x \leq x'$, and decreasing provided $-f$ is increasing. It is called monotone if it is either decreasing or increasing.

Theorem 2.91. *Let f be a monotone function on the open interval (a, b). Then f is continuous except possibly of a countable number of points in (a, b).*

Proof. Let f is decreasing on (a, b). Assume that (a, b) is bounded and f is decreasing on the closed interval $[a, b]$. Otherwise, we express (a, b) as the union of an ascending sequence of open bounded intervals such that their closures are contained in (a, b) and take the union of discontinuities in each of this countable collection of intervals. For $x_0 \in (a, b)$, we define

$$f(x_0^-) = \lim_{x\to x_0^-} f(x) = \inf\{f(x) : a < x < x_0\},$$

$$f(x_0^+) = \lim_{x\to x_0^+} f(x) = \sup\{f(x) : x_0 < x < b\}.$$

Because f is decreasing on $[a, b]$, we see that $f(x_0^+) \leq f(x_0^-)$. Not that f is not continuous at x_0 if and only if $f(x_0^+) < f(x_0^-)$. If f is discontinuous at x_0, we define the open "jump" intervals

$$J(x_0) = \{y : f(x_0^+) < y < f(x_0^-)\}.$$

Because f is decreasing on $[a, b]$, we have the inclusion

$$J(x_0) \subset [f(b), f(a)].$$

Note that if $x_0 < x_1$, then $f(x_1^-) \leq f(x_0^+)$. Therefore

$$(f(x_1^+), f(x_1^-)) \cap (f(x_0^+), f(x_0^-)) = \emptyset.$$

Then the collection of jump intervals is disjoint. Hence, for each $n \in \mathbb{N}$, there are only a finite number of jump intervals of length greater than $\frac{1}{n}$. Consequently the set of points of discontinuity of f is an union of countable collection of finite sets and then this set is countable. This completes the proof. □

Theorem 2.92. *Let* $\mathbf{A}$ *be a countable subset of the open interval* (a, b)*. Then there is an increasing function on* (a, b) *that is continuous only at points in* $(a, b) \setminus \mathbf{A}$.

Proof. If $\mathbf{A}$ is finite, then the assertion is evident. Let $\mathbf{A}$ be countably infinite. Suppose that $\mathbf{A} = \{a_n\}_{n\in\mathbf{N}}$. Define the function

$$f(x) = \sum_{\{n\in\mathbf{N}_0 : a_n \leq x\}} \frac{1}{2^n}$$

for all $a < x < b$. Since $\sum_{n\in\mathbf{N}_0} \frac{1}{2^n} < \infty$, the function f is well defined. Also, for $x, y \in [a, b]$, $y \geq x$, we have

$$f(y) - f(x) = \sum_{\{n\in\mathbf{N}_0 : a_n \leq y\}} \frac{1}{2^n} - \sum_{\{n\in\mathbf{N}_0 : a_n \leq x\}} \frac{1}{2^n} = \sum_{\{n\in\mathbf{N}_0 : x < a_n \leq y\}} \frac{1}{2^n} > 0. \tag{2.107}$$

Therefore f is an increasing function on (a, b). Let $x_0 = a_k$ for some $k \in \mathbf{N}$. Then, using (2.107), we have

$$f(x_0) - f(x) \geq \frac{1}{2^k}$$

for all $x < x_0$. Hence, f is not continuous at x_0. Let $x_0 \in (a, b) \setminus \mathbf{A}$. Then there is an interval $\mathbf{I}$ such that $x_0 \in \mathbf{I}$ and $a_n \notin \mathbf{I}$ for any $n \in \mathbf{N}$. Then

$$|f(x) - f(x_0)| < \frac{1}{2^n}$$

for all $x, x_0 \in \mathbf{I}$. Therefore f is continuous at x_0. This completes the proof. □

Definition 2.38. A closed bounded interval $[c, d]$ is said to be nondegenerate if $c < d$.

Definition 2.39. A collection $\mathcal{F}$ of closed bounded nondegenerate intervals is said to cover a set $\mathbf{E}$ in the sense of Vitali if for each $x \in \mathbf{E}$ and for each $\epsilon > 0$ there is an interval $\mathbf{I} \in \mathcal{F}$ such that $x \in \mathbf{I}$ and $l(\mathbf{I}) < \epsilon$.

Lemma 2.9 (The Vitali covering lemma). *Let $\mathbf{E}$ be a set of finite measure and $\mathcal{F}$ be a collection of closed bounded intervals that covers $\mathbf{E}$ in the sense of Vitali. Then for each $\epsilon > 0$, there is a finite disjoint subcollection $\{\mathbf{I}_k\}_{k=1}^n$ of $\mathcal{F}$ for which*

$$m^\star\left(\mathbf{E} \setminus \left(\bigcup_{k=1}^n \mathbf{I}_k\right)\right) < \epsilon.$$

Proof. Since $m^\star(\mathbf{E}) < \infty$, there is an open set $\mathbf{O}$, containing $\mathbf{E}$, such that $m(\mathbf{O}) < \infty$. Since $\mathcal{F}$ covers $\mathbf{E}$, we may assume that $\mathcal{F} \subseteq \mathbf{O}$. If $\{\mathbf{I}_k\}_{k\in\mathbf{N}} \subseteq \mathcal{F}$ is a disjoint collection of intervals in $\mathcal{F}$, then

$$\sum_{k=1}^\infty l(\mathbf{I}_k) \le m(\mathbf{O}) \tag{2.108}$$

and for $n \in \mathbf{N}$ we denote

$$\mathcal{F}_n = \left\{\mathbf{I} \in \mathcal{F} : \mathbf{I} \cap \left(\bigcup_{k=1}^n \mathbf{I}_k\right) = \emptyset\right\}.$$

We will prove that

$$\mathbf{E} \setminus \left(\bigcup_{k=1}^\infty \mathbf{I}_k\right) \subseteq \bigcup_{\mathbf{I}\in\mathcal{F}_n} \mathbf{I} \tag{2.109}$$

for any $n \in \mathbf{N}$. Let $x \in \mathbf{E} \setminus (\bigcup_{k=1}^\infty \mathbf{I}_k)$ be arbitrarily chosen. Then $x \in \mathbf{E}$ and $x \notin \bigcup_{k=1}^\infty \mathbf{I}_k$. Hence, $x \notin \mathbf{I}_k$ for any $k \in \mathbf{N}$. Since $\mathcal{F}$ covers the set $\mathbf{E}$ in the sense of Vitali and $x \in \mathbf{E}$, then there is an interval $\mathbf{I} \in \mathcal{F}$ such that $x \in \mathbf{I}$ and $\mathbf{I} \cap (\bigcup_{k=1}^n \mathbf{I}_k) = \emptyset$ for $n \in \mathbf{N}$. Then $x \in \bigcup_{\mathbf{I}\in\mathcal{F}_n} \mathbf{I}$ for $n \in \mathbf{N}$. Because $x \in \mathbf{E} \setminus (\bigcup_{k=1}^\infty \mathbf{I}_k)$ was arbitrarily chosen and we see that it is an element of $\bigcup_{\mathbf{I}\in\mathcal{F}_n} \mathbf{I}$ for $n \in \mathbf{N}$, we obtain equation (2.109). If there is a finite disjoint subcollection $\{\mathbf{I}_k\}_{k=1}^n$ of $\mathcal{F}$ that covers $\mathbf{E}$, the assertion is proved. Otherwise, let $n \in \mathbf{N}$ and $\{\mathbf{I}_k\}_{k=1}^n \in \mathcal{F}$ be a disjoint collection. Since $\mathbf{E} \setminus (\bigcup_{k=1}^n \mathbf{I}_k) \neq \emptyset$, we see that $\mathcal{F}_n \neq \emptyset$. Let

$$s_n = \sup_{\mathbf{I}\in\mathcal{F}_n} l(\mathbf{I}).$$

We have $l(\mathbf{I}) \le m(\mathbf{O})$ for any $\mathbf{I} \in \mathcal{F}_n$. Then $s_n < \infty$. We take $\mathbf{I}_{n+1} \in \mathcal{F}_n$ such that $l(\mathbf{I}_{n+1}) > \frac{s_n}{2}$. This inductively defines $\{\mathbf{I}_k\}_{k\in\mathbf{N}}$, a countable disjoint subcollection of $\mathcal{F}$ such that $l(\mathbf{I}_{n+1}) > \frac{l(\mathbf{I})}{2}$ if $\mathbf{I} \in \mathcal{F}$ and $\mathbf{I} \cap (\bigcup_{k=1}^n \mathbf{I}_k) = \emptyset$. Let $\mathbf{J}_k$ be the closed interval that has the same midpoint as $\mathbf{I}_k$ and 5 times its length. We will prove

$$\mathbf{E} \setminus \left(\bigcup_{k=1}^n \mathbf{I}_k\right) \subseteq \bigcup_{k=n+1}^\infty \mathbf{J}_k \tag{2.110}$$

for any $n \in \mathbf{N}$. Let $n \in \mathbf{N}$ and $x \in \mathbf{E} \setminus (\bigcup_{k=1}^{n} \mathbf{I}_k)$ be arbitrarily chosen. Then there is an $\mathbf{I} \in \mathcal{F}$ such that $x \in \mathbf{I}$ and $\mathbf{I} \cap (\bigcup_{k=1}^{n} \mathbf{I}_k) = \emptyset$. Assume that $\mathbf{I} \cap \mathbf{I}_k = \emptyset$ for any $k \in \mathbf{N}$. Then $l(\mathbf{I}_k) > \frac{l(\mathbf{I})}{2}$ for any $k \in \mathbf{N}$. This is a contradiction. Then there is $m \in \mathbf{N}$ such that $\mathbf{I} \cap \mathbf{I}_m \neq \emptyset$. We take $M > n$ to be the first natural number for which $\mathbf{I} \cap \mathbf{I}_M \neq \emptyset$. Then $\mathbf{I} \cap (\bigcup_{k=1}^{M-1} \mathbf{I}_k) = \emptyset$ and $l(\mathbf{I}_M) > \frac{l(\mathbf{I})}{2}$. Since $x \in \mathbf{I}$ and $\mathbf{I} \cap \mathbf{I}_M \neq \emptyset$, the distance from x to the midpoint of $\mathbf{I}_M$ is at most $l(\mathbf{I}) + \frac{1}{2} l(\mathbf{I}_M)$. Hence, using $l(\mathbf{I}) < 2L(\mathbf{I}_M)$, the distance from x to the midpoint of $\mathbf{I}_M$ is less than $\frac{5}{2} l(\mathbf{I}_M)$. Therefore $x \in \mathbf{J}_M$ and $x \in \bigcup_{k=n+1}^{\infty} \mathbf{J}_k$. Because $x \in \mathbf{E} \setminus (\bigcup_{k=1}^{n} \mathbf{I}_k)$ was arbitrarily chosen and because it is an element of $\bigcup_{k=n+1}^{\infty} \mathbf{J}_k$, we obtain equation (2.110). Let $\epsilon > 0$ be arbitrarily chosen. Then there is an $n \in \mathbf{N}$ such that

$$\sum_{k=n+1}^{\infty} l(\mathbf{I}_k) < \frac{\epsilon}{5}.$$

Hence,

$$\sum_{k=n+1}^{\infty} l(\mathbf{J}_k) < \epsilon$$

and using equation (2.110), we get

$$m^{\star}\left(\mathbf{E} \setminus \left(\bigcup_{k=1}^{n} \mathbf{I}_k\right)\right) < \epsilon.$$

This completes the proof. □

Definition 2.40. Let $f : \mathbf{E} \longmapsto \mathbf{R}$ and $x \in \mathbf{E}$. We define the upper derivative $\overline{D}f(x)$ and the lower derivative $\underline{D}f(x)$ of f at x as follows:

$$\overline{D}f(x) = \lim_{h \to 0}\left(\sup_{0 \le |t| \le h} \frac{f(x+t) - f(x)}{t}\right),$$

$$\underline{D}f(x) = \lim_{h \to 0}\left(\inf_{0 \le |t| \le h} \frac{f(x+t) - f(x)}{t}\right).$$

We have $\underline{D}f(x) \le \overline{D}f(x)$. If $\underline{D}f(x)$ and $\overline{D}f(x)$ are finite and $\underline{D}f(x) = \overline{D}f(x)$, we say that f is differentiable at x and we define $f'(x)$ to be the common value of the upper derivative and the lower derivative.

Theorem 2.93. *Let f be an increasing function on the closed bounded interval $[a, b]$. Then for each $\alpha > 0$, we have*

$$m^{\star}(\{x \in (a,b) : \overline{D}f(x) \ge \alpha\}) \le \frac{1}{\alpha}(f(b) - f(a)) \tag{2.111}$$

and

$$m^{\star}(\{x \in (a,b) : \overline{D}f(x) = \infty\}) = 0. \tag{2.112}$$

Proof. Let $\alpha > 0$ be arbitrarily chosen and

$$\mathbf{E}_\alpha = \{x \in (a,b) : \overline{D}f(x) \geq \alpha\}.$$

We take $\alpha' \in (0, \alpha)$ and $\epsilon > 0$ arbitrarily. With $\mathcal{F}$ we will denote the collection of closed bounded interval $[c,d] \subset (a,b)$ such that

$$f(d) - f(c) \geq \alpha'(d - c).$$

Since $\overline{D}f(x) \geq \alpha$ for $x \in \mathbf{E}_\alpha$, we see that $\mathcal{F}$ is a Vitali covering of $\mathbf{E}_\alpha$. Hence, using the Vitali covering lemma, there is a finite disjoint subcollection $\{[c_k, d_k]\}_{k=1}^n$ of $\mathcal{F}$ such that

$$m^\star\left(\mathbf{E}_\alpha \setminus \left(\bigcup_{k=1}^n [c_k, d_k]\right)\right) < \epsilon.$$

Note that

$$\mathbf{E}_\alpha \subseteq \left(\bigcup_{k=1}^n [c_k, d_k]\right) \cup \left(\mathbf{E}_\alpha \setminus \left(\bigcup_{k=1}^n [c_k, d_k]\right)\right).$$

Therefore

$$\begin{aligned}
m^\star(\mathbf{E}_\alpha) &\leq m^\star\left(\bigcup_{k=1}^n [c_k, d_k]\right) + m^\star\left(\mathbf{E}_\alpha \setminus \left(\bigcup_{k=1}^n [c_k, d_k]\right)\right) \\
&< \sum_{k=1}^n (d_k - c_k) + \epsilon \leq \frac{1}{\alpha'} \sum_{k=1}^n (f(d_k) - f(c_k)) + \epsilon \\
&= \frac{1}{\alpha'}(f(d_1) - f(c_1) + f(d_2) - f(c_2) + \cdots + f(d_n) - f(c_n)) + \epsilon \\
&\leq \frac{1}{\alpha'}(f(c_2) - f(a) + f(c_3) - f(c_2) + \cdots + f(b) - f(c_n)) + \epsilon \\
&= \frac{1}{\alpha'}(f(b) - f(a)) + \epsilon.
\end{aligned}$$

Because $\epsilon > 0$ was arbitrarily chosen, we obtain (2.111). Next, for each natural number n we have

$$\{x \in (a,b) : \overline{D}f(x) = \infty\} \subseteq \mathbb{E}_n.$$

Consequently

$$\begin{aligned}
m^\star(\{x \in (a,b) : \overline{D}f(x) = \infty\}) &\leq m^\star(\mathbb{E}_n) \\
&\leq \frac{1}{n}(f(b) - f(a)) \quad \to 0 \quad \text{as} \quad n \to \infty,
\end{aligned}$$

i. e., we get (2.112). This completes the proof. □

Theorem 2.94 (Lebesgue's theorem). *If the function f is monotone on (a,b), then it is differentiable almost everywhere on (a,b).*

Proof. Let f be increasing on (a,b). Assume that (a,b) is bounded. Otherwise, we express it as the union of an ascending sequence of open bounded intervals contained in (a,b). Let

$$\mathbf{E} = \{x \in (a,b) : \overline{D}f(x) > \underline{D}f(x)\},$$
$$\mathbf{E}_{\alpha,\beta} = \{x \in (a,b) : \overline{D}f(x) > \alpha > \beta > \underline{D}f(x)\}, \quad \alpha,\beta \in \mathbf{Q}, \quad \alpha > \beta.$$

Then

$$\mathbf{E} = \bigcup_{\alpha,\beta\in\mathbf{Q},\alpha>\beta} \mathbf{E}_{\alpha,\beta}.$$

Hence,

$$m^\star(\mathbf{E}) \le \sum_{\alpha,\beta\in\mathbf{Q},\alpha>\beta} m^\star(\mathbf{E}_{\alpha,\beta}). \tag{2.113}$$

We fix $\alpha,\beta \in \mathbf{Q}$, $\alpha > \beta$. Let $\epsilon > 0$ be arbitrarily chosen. We take an open set $\mathbf{O}$ such that

$$\mathbf{E}_{\alpha,\beta} \subseteq \mathbf{O} \subseteq (a,b) \quad \text{and} \quad m^\star(\mathbf{O}) < m^\star(\mathbf{E}_{\alpha,\beta}) + \epsilon. \tag{2.114}$$

Let $\mathcal{F}$ be the collection of closed bounded intervals $[c,d]$ contained in $\mathbf{O}$ for which

$$f(d) - f(c) < \beta(d-c).$$

Since $\underline{D}f(x) < \beta$ for $x \in \mathbf{E}_{\alpha,\beta}$, we see that $\mathcal{F}$ is a Vitali covering of $\mathbf{E}_{\alpha,\beta}$. By the Vitali covering lemma, there is a finite disjoint collection $\{[c_k,d_k]\}_{k=1}^n$ of $\mathcal{F}$ such that

$$m^\star\left(\mathbf{E}_{\alpha,\beta} \setminus \left(\bigcup_{k=1}^n [c_k,d_k]\right)\right) < \epsilon. \tag{2.115}$$

On the other hand, using (2.114), we obtain

$$\begin{aligned}\sum_{k=1}^n (f(d_k) - f(c_k)) &< \beta \sum_{k=1}^n (d_k - c_k)\\ &\le \beta m^\star(\mathbf{O}) < \beta(m^\star(\mathbf{E}_{\alpha,\beta}) + \epsilon).\end{aligned} \tag{2.116}$$

By Theorem 2.93, we get

$$m^\star(\mathbf{E}_{\alpha,\beta} \cap (c_k,d_k)) \le \frac{1}{\alpha}(f(d_k) - f(c_k))$$

for any $k \in \{1,\dots,n\}$. Note that

$$\mathbf{E}_{\alpha,\beta} \subseteq \left(\mathbf{E}_{\alpha,\beta} \cap \left(\bigcup_{k=1}^n (c_k,d_k)\right)\right) \cup \left(\mathbf{E}_{\alpha,\beta} \setminus \left(\bigcup_{k=1}^n [c_k,d_k]\right)\right).$$

Therefore, using (2.115), we get

$$
\begin{aligned}
m^\star(\mathbf{E}_{\alpha,\beta}) &\le m^\star\left(\mathbf{E}_{\alpha,\beta}\cap\left(\bigcup_{k=1}^n (c_k,d_k)\right)\right)+m^\star\left(\mathbf{E}_{\alpha,\beta}\setminus\left(\bigcup_{k=1}^n [c_k,d_k]\right)\right)\\
&< m^\star\left(\mathbf{E}_{\alpha,\beta}\cap\left(\bigcup_{k=1}^n (c_k,d_k)\right)\right)+\epsilon \le \sum_{k=1}^n m^\star(\mathbf{E}_{\alpha,\beta}\cap(c_k,d_k))+\epsilon\\
&\le \frac{1}{\alpha}\sum_{k=1}^n (f(d_k)-f(c_k))+\epsilon.
\end{aligned}
$$

Hence, by (2.116), we obtain

$$m^\star(\mathbf{E}_{\alpha,\beta}) \le \frac{\beta}{\alpha}(m^\star(\mathbf{E}_{\alpha,\beta})+\epsilon)+\epsilon$$

and

$$\frac{\alpha-\beta}{\alpha}m^\star(\mathbf{E}_{\alpha,\beta}) \le \frac{\beta}{\alpha}\epsilon+\epsilon.$$

Because $\epsilon > 0$ was arbitrarily chosen, we get

$$m^\star(\mathbf{E}_{\alpha,\beta}) = 0.$$

Hence, by (2.113), we obtain $m^\star(\mathbf{E}) = 0$. This completes the proof. □

Definition 2.41. Let f be integrable over the closed bounded interval $[a,b]$ and take value $f(b)$ on $(b,b+1]$. For $0 < h \le 1$, we define the divided difference function $\mathrm{Diff}_h f$ and average value function $\mathrm{Av}_h f$ of $[a,b]$ by

$$\mathrm{Diff}_h f(x) = \frac{f(x+h)-f(x)}{h} \quad \text{and} \quad \mathrm{Av}_h f(x) = \frac{1}{h}\int_x^{x+h} f,$$

respectively, for all $x \in [a,b]$.

Let $a \le y < z \le b$ and f be as in Definition 2.41. Then

$$
\begin{aligned}
\int_y^z \mathrm{Diff}_h f(x) &= \int_y^z \frac{f(x+h)-f(x)}{h} = \frac{1}{h}\int_y^z f(x+h) - \frac{1}{h}\int_y^z f(x)\\
&= \frac{1}{h}\int_{y+h}^{z+h} f(x) - \frac{1}{h}\int_y^z f(x) = \frac{1}{h}\int_z^{z+h} f(x) - \frac{1}{h}\int_y^{y+h} f(x)\\
&= \mathrm{Av}_h f(z) - \mathrm{Av}_h f(y).
\end{aligned}
$$

Theorem 2.95. *Let f be an increasing function on the closed bounded interval $[a,b]$. Then f' is integrable over $[a,b]$ and*

$$\int_a^b f' \le f(b)-f(a). \tag{2.117}$$

Proof. Let f take value $f(b)$ on $(b, b+1]$. Then, for $0 < h \le 1$, f is increasing on $[a, b+1]$. Therefore, f and $\mathrm{Diff}_h f$ are measurable on $[a, b]$. By Theorem 2.94, it follows that f is almost everywhere differentiable on $[a, b]$. Therefore $\{\mathrm{Diff}_{\frac{1}{n}} f\}_{n\in\mathbf{N}}$ is a sequence of nonnegative measurable functions that converges pointwise almost everywhere on $[a, b]$ to f'. Hence, by Fatou's lemma, we obtain

$$\int_a^b f' \le \liminf_{n\to\infty}\left(\int_a^b \mathrm{Diff}_{\frac{1}{n}} f\right). \tag{2.118}$$

On the other hand, using the fact that f is increasing on $[a, b+1]$, we have

$$\int_a^b \mathrm{Diff}_{\frac{1}{n}} f = \mathrm{Av}_{\frac{1}{n}} f(b) - \mathrm{Av}_{\frac{1}{n}} f(a) = n \int_b^{b+\frac{1}{n}} f(x) - n \int_a^{a+\frac{1}{n}} f(x) \le f(b) - f(a).$$

Thus

$$\liminf_{n\to\infty} \int_a^b \mathrm{Diff}_{\frac{1}{n}} f \le f(b) - f(a).$$

From this and from (2.118), we obtain (2.117). This completes the proof. □

Definition 2.42. A real-valued function f on a closed bounded interval $[a, b]$ is said to be absolutely continuous on $[a, b]$ if for each $\epsilon > 0$ there is a $\delta > 0$ such that for any finite disjoint collection $\{(a_k, b_k)\}_{k=1}^n$ of open intervals in (a, b) for which $\sum_{k=1}^n (b_k - a_k) < \delta$ we have $\sum_{k=1}^n |f(b_k) - f(a_k)| < \epsilon$.

Theorem 2.96. *Let f be a Lipschitz function on the closed bounded interval $[a, b]$, i. e., there exists a constant $L > 0$ such that*

$$|f(x) - f(y)| \le L|x - y|$$

for any $x, y \in [a, b]$. Then f is absolutely continuous on $[a, b]$.

Proof. Let $\epsilon > 0$ be arbitrarily chosen and $\delta = \frac{\epsilon}{L}$. We take a finite disjoint collection $\{(a_k, b_k)\}_{k=1}^n$ of open intervals in (a, b) such that

$$\sum_{k=1}^n (b_k - a_k) < \frac{\epsilon}{L}.$$

Hence,

$$\sum_{k=1}^n |f(b_k) - f(a_k)| \le L \sum_{k=1}^n |b_k - a_k| < \epsilon.$$

This completes the proof. □

Theorem 2.97. *Let f be an absolutely continuous function on the closed bounded interval $[a,b]$. Then f is difference of two increasing absolutely continuous functions on $[a,b]$.*

Proof. Let $\epsilon = 1$ and $\delta > 0$ responds to the criterion for absolute continuity of f. Let also P be a partition of $[a,b]$ into N closed intervals $\{[c_k,d_k]\}_{k=1}^N$ each of length less than δ. Hence, for the total variation of f on $[c_k,d_k]$ (see the appendix), we have

$$\bigvee_{c_k}^{d_k}(f) \le 1$$

for each $k \in \{1,\dots,N\}$. Therefore

$$\bigvee_a^b(f) = \sum_{k=1}^N \bigvee_{c_k}^{d_k}(f) \le N,$$

i. e., f is a function of bounded variation on $[a,b]$ (see the appendix). Hence, by Jordan's theorem (see the appendix), we have

$$f(x) = \left(f(x) + \bigvee_a^x(f)\right) - \left(\bigvee_a^x(f)\right), \quad x \in [a,b], \tag{2.119}$$

where $f(x) + \bigvee_a^x(f)$ and $\bigvee_a^x(f)$ are increasing functions on $[a,b]$. Let $\epsilon > 0$ be arbitrarily chosen and $\delta > 0$ responds to $\frac{\epsilon}{2}$ regarding the criterion for absolute continuity of f on $[a,b]$. Let also $\{(c_k,d_k)\}_{k=1}^n$ be a disjoint collection of open subintervals of (a,b) for which $\sum_{k=1}^n (d_k - c_k) < \delta$. For $k \in \{1,\dots,n\}$, with P_k we will denote a partition of $[c_k,d_k]$. Then

$$\sum_{k=1}^n \bigvee_{c_k}^{d_k}(f, P_k) \le \frac{\epsilon}{2}.$$

From this,

$$\sum_{k=1}^n \bigvee_{c_k}^{d_k}(f) \le \frac{\epsilon}{2} < \epsilon.$$

Consequently

$$\sum_{k=1}^n \left(\bigvee_a^{d_k}(f) - \bigvee_a^{c_k}(f)\right) < \epsilon.$$

Let

$$f_1(x) = \bigvee_a^x(f), \quad f_2(x) = f(x) + \bigvee_a^x(f), \quad x \in [a,b].$$

Therefore f_1 is absolutely continuous on $[a,b]$. Hence, f_2 is absolutely continuous on $[a,b]$. Using (2.119), we conclude that f is difference of two increasing absolutely continuous functions on $[a,b]$. This completes the proof. □

Exercise 2.19. Let f is absolutely continuous on $[a, b]$. Prove that $\{\mathrm{Diff}_{\frac{1}{n}} f\}_{n\in\mathbf{N}}$ is uniformly integrable over $[a, b]$.

Hint. Use

$$\begin{aligned}
\int_u^v \mathrm{Diff}_h f &= \frac{1}{h}\int_u^v f(t+h) - \frac{1}{h}\int_u^v f(t)\\
&= \frac{1}{h}\int_{u-v+h}^{h} f(t+v). - \frac{1}{h}\int_0^{v-u} f(u+t)\\
&= \frac{1}{h}\int_0^h f(t+v) + \frac{1}{h}\int_{u-v+h}^{0} f(t+v)\\
&\quad - \frac{1}{h}\int_0^h f(u+t) - \frac{1}{h}\int_h^{v-u} f(u+t)\\
&= \frac{1}{h}\int_0^h f(t+v) + \frac{1}{h}\int_{u-v+h}^{0} f(t+v)\\
&\quad - \frac{1}{h}\int_0^h f(u+t) - \frac{1}{h}\int_{u-v+h}^{0} f(t+v)\\
&= \frac{1}{h}\int_0^h (f(t+v) - f(u+t))
\end{aligned}$$

for any $u, v \in [a, b]$.

Theorem 2.98. *Let f be absolutely continuous on the closed bounded interval $[a, b]$. Then f is differentiable almost everywhere on (a, b), its derivative f' is integrable over $[a, b]$ and*

$$\int_a^b f' = f(b) - f(a).$$

Proof. By Theorem 2.97, f is difference of two increasing absolutely continuous functions on $[a, b]$. Hence, by Lebesgue's theorem, it follows that f is differentiable almost everywhere on (a, b). Therefore the sequence $\{\mathrm{Diff}_{\frac{1}{n}} f\}_{n\in\mathbf{N}}$ converges pointwise almost everywhere on (a, b) to f'. Since $\{\mathrm{Diff}_{\frac{1}{n}} f\}_{n\in\mathbf{N}}$ is uniformly integrable over $[a, b]$, using Theorem 2.85, we get

$$f(b) - f(a) = \lim_{n\to\infty}\int_a^b \mathrm{Diff}_{\frac{1}{n}} f = \int_a^b \lim_{n\to\infty} \mathrm{Diff}_{\frac{1}{n}} f = \int_a^b f'.$$

This completes the proof. □

2.9 General measure spaces

Definition 2.43. A collection of subsets of a set $\mathbf{X}$ that contains the empty set and it is closed with respect to the complements in $\mathbf{X}$ and with respect to the formation of countable unions, will be called a σ-algebra of subsets of $\mathbf{X}$.

Definition 2.44. A couple $(\mathbf{X}, \mathcal{M})$ consisting of a set $\mathbf{X}$ and a σ-algebra $\mathcal{M}$ of subsets of $\mathbf{X}$ will be called a measurable space. A subset $\mathbf{E}$ of $\mathbf{X}$ is called measurable (or measurable with respect to $\mathcal{M}$) if $\mathbf{E} \in \mathcal{M}$.

Definition 2.45. Let $(\mathbf{X}, \mathcal{M})$ be a measurable space. An extended real-valued function $\mu : \mathcal{M} \longmapsto [0, \infty]$ for which $\mu(\emptyset) = 0$ and which is countably additive in the sense that for any countable disjoint collection $\{\mathbf{E}_k\}_{k\in\mathbf{N}}$ of measurable sets

$$\mu\left(\bigcup_{k=1}^{\infty} \mathbf{E}_k\right) = \sum_{k=1}^{\infty} \mu(\mathbf{E}_k),$$

will be called a measure.

By a measure space $(\mathbf{X}, \mathcal{M}, \mu)$ we mean a measurable space $(\mathbf{X}, \mathcal{M})$ together with a measure μ defined on $\mathcal{M}$.

Example 2.1. $(\mathbf{R}, \mathcal{L}, m)$, where $\mathcal{L}$ is a collection of Lebesgue measurable sets of real numbers and m is the Lebesgue measure, is a measurable space.

Remark 2.5. Note that the idea for the proof of the assertions in this section is the same as in the previous sections. Therefore we leave the proofs of the assertions in this section.

Theorem 2.99. *Let $(\mathbf{X}, \mathcal{M}, \mu)$ be a measure space.*

1. *For any finite disjoint collection $\{\mathbf{E}_k\}_{k=1}^{n}$ of measurable sets we have*

$$\mu\left(\bigcup_{k=1}^{n} \mathbf{E}_k\right) = \sum_{k=1}^{n} \mu(\mathbf{E}_k).$$

2. *If $\mathbf{A}$ and $\mathbf{B}$ are measurable sets and $\mathbf{A} \subseteq \mathbf{B}$, then*

$$\mu(\mathbf{A}) \le \mu(\mathbf{B}).$$

   Moreover, if $\mathbf{A} \subseteq \mathbf{B}$ and $\mu(\mathbf{A}) < \infty$, then

$$\mu(\mathbf{B} \setminus \mathbf{A}) = \mu(\mathbf{B}) - \mu(\mathbf{A}).$$

3. *For any countable collection $\{\mathbf{E}_k\}_{k\in\mathbf{N}}$ of measurable sets that covers a measurable set $\mathbf{E}$, we have*

$$\mu(\mathbf{E}) \le \sum_{k=1}^{\infty} \mu(\mathbf{E}_k).$$

Theorem 2.100. *Let $(\mathbf{X}, \mathcal{M}, \mu)$ be a measure space.*
1. *If $\{\mathbf{A}_k\}_{k\in\mathbf{N}}$ is an ascending sequence of measurable sets, then*

$$\mu\left(\bigcup_{k=1}^{\infty} \mathbf{A}_k\right) = \lim_{k\to\infty} \mu(\mathbf{A}_k).$$

2. *If $\{\mathbf{A}_k\}_{k\in\mathbf{N}}$ is a descending sequence of measurable sets for which $\mu(\mathbf{A}_1) < \infty$, then*

$$\mu\left(\bigcap_{k=1}^{\infty} \mathbf{A}_k\right) = \lim_{k\to\infty} \mu(\mathbf{A}_k).$$

Definition 2.46. For a measure space $(\mathbf{X}, \mathcal{M}, \mu)$ and a measurable subset $\mathbf{E}$ of $\mathbf{X}$, we say that a property holds almost everywhere on $\mathbf{E}$, or it holds for almost all $x \in \mathbf{E}$, if it holds on $\mathbf{E} \setminus \mathbf{E}_0$ and $\mathbf{E}_0$ is a measurable set for which $\mu(\mathbf{E}_0) = 0$.

Lemma 2.10 (The Borel–Cantelli lemma). *Let $(\mathbf{X}, \mathcal{M}, \mu)$ be a measure space and $\{\mathbf{A}_k\}_{k\in\mathbf{N}}$ is a countable collection of measurable sets for which $\sum_{k=1}^{\infty} \mu(\mathbf{A}_k) < \infty$. Then almost all $x \in \mathbf{X}$ belong to at most a finite number of the $\mathbf{A}_k$.*

Definition 2.47. Let $(\mathbf{X}, \mathcal{M}, \mu)$ be a measure space. The measure μ is called finite if $\mu(\mathbf{X}) < \infty$, and it is called σ-finite if $\mathbf{X}$ is the union of a countable collection of measurable sets, each of which has finite measure. A measurable set $\mathbf{E}$ is said to be of finite measure if $\mu(\mathbf{E}) < \infty$, and it is said to be of σ-finite measure if $\mathbf{E}$ is the union of a countable collection of measurable sets, each of which has finite measure.

2.10 General measurable functions

Definition 2.48. Let $(\mathbf{X}, \mathcal{M})$ be a measurable space. An extended real-valued function f on $\mathbf{X}$ is said to be measurable(or measurable with respect to $\mathcal{M}$) if for each $c \in \mathbf{R}$, the set $\{x \in \mathbf{X} : f(x) > c\}$ is measurable.

Remark 2.6. Note that the idea for the proof of the assertions in this section is the same as the idea for the proofs of the assertions in Section 2.5. Therefore we leave the proofs of the assertions in this section.

Theorem 2.101. *Let $(\mathbf{X}, \mathcal{M})$ be a measurable space and f is an extended real-valued function on $\mathbf{X}$. Then the following assertions are equivalent.*
1. *For each $c \in \mathbf{R}$, the set $\{x \in \mathbf{X} : f(x) > c\}$ is measurable.*
2. *For each $c \in \mathbf{R}$, the set $\{x \in \mathbf{X} : f(x) \geq c\}$ is measurable.*
3. *For each $c \in \mathbf{R}$, the set $\{x \in \mathbf{X} : f(x) < c\}$ is measurable.*
4. *For each $c \in \mathbf{R}$, the set $\{x \in \mathbf{X} : f(x) \leq c\}$ is measurable.*

Theorem 2.102. *Let $(\mathbf{X}, \mathcal{M})$ be a measurable space and f is a real-valued function on $\mathbf{X}$. Then f is measurable if and only if for each open set $\mathbf{O}$ of real numbers, the set $f^{-1}(\mathbf{O})$ is measurable.*

Theorem 2.103. *Let $(\mathbf{X}, \mathcal{M})$ be a measurable space and f and g be measurable real-valued functions. Then*
1. *$\alpha f + \beta g$ is measurable.*
2. *fg is measurable.*
3. *$\max\{f, g\}$ and $\min\{f, g\}$ are measurable.*

Theorem 2.104. *Let $(\mathbf{X}, \mathcal{M})$ be a measurable space, f is a measurable real-valued function on $\mathbf{X}$, $\phi : \mathbf{R} \longmapsto \mathbf{R}$ is continuous. Then $\phi \circ f : \mathbf{X} \longmapsto \mathbf{R}$ is measurable.*

Theorem 2.105. *Let $(\mathbf{X}, \mathcal{M}, \mu)$ be a measure space and $\{f_n\}_{n\in\mathbf{N}}$ be a sequence of measurable functions on $\mathbf{X}$ that converges pointwise to f on all of $\mathbf{X}$. Then f is measurable on $\mathbf{X}$.*

Definition 2.49. Let $(\mathbf{X}, \mathcal{M})$ be a measurable space and $\mathbf{E}$ be a measurable subset of $\mathbf{X}$. We define its characteristic function $\kappa_{\mathbf{E}}$ as follows:

$$\kappa_{\mathbf{E}}(x) = \begin{cases} 1 & \text{if } x \in \mathbf{E}, \\ 0 & \text{if } x \in \mathbf{X} \setminus \mathbf{E}. \end{cases}$$

Definition 2.50. Let $(\mathbf{X}, \mathcal{M})$ be a measurable space. A real-valued function ψ on $\mathbf{X}$ is said to be simple, if there is a finite collection $\{\mathbf{E}_k\}_{k=1}^n$ of measurable sets and a corresponding set of real numbers $\{c_k\}_{k=1}^n$ for which

$$\psi = \sum_{k=1}^{n} c_k \kappa_{\mathbf{E}_k} \quad \text{on} \quad \mathbf{X}.$$

Lemma 2.11 (The simple approximation lemma). *Let $(\mathbf{X}, \mathcal{M})$ be a measurable space and f be a measurable function on $\mathbf{X}$ such that $|f| \le M$ on $\mathbf{X}$ for some positive constant M. Then, for each $\epsilon > 0$, there are simple functions ϕ_ϵ and ψ_ϵ, such that*

$$\phi_\epsilon \le f \le \psi_\epsilon \quad \text{and} \quad 0 \le \psi_\epsilon - \phi_\epsilon < \epsilon \quad \text{on} \quad \mathbf{X}.$$

Theorem 2.106 (The simple approximation theorem). *Let $(\mathbf{X}, \mathcal{M}, \mu)$ be a measure space and f be a measurable function on $\mathbf{X}$. Then there is a sequence $\{\phi_n\}_{n\in\mathbf{N}}$ of simple functions on $\mathbf{X}$ that converges pointwise on $\mathbf{X}$ to f and*

$$|\phi_n| \le |f| \quad \text{on} \quad \mathbf{X}$$

for any $n \in \mathbf{N}$. If $\mathbf{X}$ is σ-finite, then we may choose $\{\phi_n\}_{n\in\mathbf{N}}$ so that ϕ_n vanishes outside a set of finite measure. If f is nonnegative, we may choose $\{\phi_n\}_{n\in\mathbf{N}}$ to be increasing and $\psi_n \ge 0$ on $\mathbf{X}$.

Theorem 2.107 (Egoroff's theorem). *Let $(\mathbf{X}, \mathcal{M}, \mu)$ be a finite measure space and $\{f_n\}_{n\in\mathbf{N}}$ be a sequence of measurable functions on $\mathbf{X}$ that converges pointwise a. e. on $\mathbf{X}$ to a function f that is finite a. e. on $\mathbf{X}$. Then for each $\epsilon > 0$, there is a measurable subset $\mathbf{X}_\epsilon$ of $\mathbf{X}$ for which $f_n \to f$, as $n \to \infty$, uniformly on $\mathbf{X}_\epsilon$ and $\mu(\mathbf{X} \setminus \mathbf{X}_\epsilon) < \epsilon$.*

2.11 Integration over general measure spaces

Definition 2.51. Let $(\mathbf{X}, \mathcal{M}, \mu)$ be a measure space and ψ be a nonnegative simple function on $\mathbf{X}$. If $\psi = 0$ on $\mathbf{X}$, we define

$$\int_{\mathbf{X}} \psi d\mu = 0.$$

Otherwise, let $c_1, \ldots, c_n$ be the positive values taken by ψ on $\mathbf{X}$ and

$$\mathbf{E}_k = \{x \in \mathbf{X} : \psi(x) = c_k\}, \quad k \in \{1, \ldots, n\}.$$

We define

$$\int_{\mathbf{X}} \psi d\mu = \sum_{k=1}^{n} c_k \mu(\mathbf{E}_k),$$

using the convention that the right-hand side is ∞ if $\mu(\mathbf{E}_k) = \infty$ for some $k \in \{1, \ldots, n\}$. For a measurable subset $\mathbf{E}$ of $\mathbf{X}$, we define

$$\int_{\mathbf{E}} \psi d\mu = \int_{\mathbf{X}} \psi \kappa_{\mathbf{E}} d\mu.$$

Remark 2.7. Note that the idea for the proof of the assertions in this section is the same as the idea for the proofs of the assertions in Section 2.7. Therefore we leave the proofs of the assertions in this section.

Theorem 2.108. *Let $(\mathbf{X}, \mathcal{M}, \mu)$ be a measure space and ϕ and ψ be nonnegative simple functions on $\mathbf{X}$.*

1. $\int_{\mathbf{X}} (\alpha\phi + \beta\psi) d\mu = \alpha \int_{\mathbf{X}} \phi d\mu + \beta \int_{\mathbf{X}} \psi d\mu$ *for any* $\alpha, \beta \in \mathbf{R}$.
2. *If $\mathbf{A}$ and $\mathbf{B}$ be disjoint measurable subsets of $\mathbf{X}$, then*

$$\int_{\mathbf{A}\cup\mathbf{B}} \phi d\mu = \int_{\mathbf{A}} \phi d\mu + \int_{\mathbf{B}} \phi d\mu.$$

3. *If $\mathbf{X}_0$ is a measurable subset of $\mathbf{X}$ and $\mu(\mathbf{X} \setminus \mathbf{X}_0) = 0$, then*

$$\int_{\mathbf{X}} \phi d\mu = \int_{\mathbf{X}_0} \phi d\mu.$$

4. *If $\phi \leq \psi$ a. e. on $\mathbf{X}$, then*

$$\int_{\mathbf{X}} \phi d\mu \leq \int_{\mathbf{X}} \psi d\mu.$$

Definition 2.52. Let $(\mathbf{X}, \mathcal{M}, \mu)$ be a measure space and f be a nonnegative extended real-valued measurable function on $\mathbf{X}$. We define

$$\int_{\mathbf{X}} f d\mu = \sup\left\{\int_{\mathbf{X}} \psi d\mu : \psi \quad \text{is} \quad \text{simple}, 0 \leq \psi \leq f\right\}.$$

For a measurable subset $\mathbf{E}$ of $\mathbf{X}$, we define

$$\int_{\mathbf{E}} f d\mu = \int_{\mathbf{X}} f \kappa_{\mathbf{E}} d\mu.$$

Theorem 2.109. *Let $(\mathbf{X}, \mathcal{M}, \mu)$ be a measure space, f and g be nonnegative measurable functions on $\mathbf{X}$.*

1. *If $f \leq g$ a. e. on $\mathbf{X}$, then*

$$\int_{\mathbf{X}} f d\mu \leq \int_{\mathbf{X}} g d\mu.$$

2. *If $\mathbf{X}_0$ is a measurable subset of $\mathbf{X}$ such that $\mu(\mathbf{X} \setminus \mathbf{X}_0) = 0$, then*

$$\int_{\mathbf{X}} f d\mu = \int_{\mathbf{X}_0} f d\mu.$$

Theorem 2.110 (Chebychev's inequality). *Let $(\mathbf{X}, \mathcal{M}, \mu)$ be a measure space, f be a nonnegative measurable function on $\mathbf{X}$ and $\lambda > 0$. Then*

$$\mu(\{x \in \mathbf{X} : f(x) \geq \lambda\}) \leq \frac{1}{\lambda} \int_{\mathbf{X}} f d\mu.$$

Theorem 2.111. *Let $(\mathbf{X}, \mathcal{M}, \mu)$ be a measure space and f be a nonnegative measurable function on $\mathbf{X}$ for which $\int_{\mathbf{X}} f d\mu < \infty$. Then f is finite a. e. on $\mathbf{X}$ and $\{x \in \mathbf{X} : f(x) > 0\}$ is σ-finite.*

Lemma 2.12 (Fatou's lemma). *Let $(\mathbf{X}, \mathcal{M}, \mu)$ be a measure space and $\{f_n\}_{n \in \mathbf{N}}$ be a sequence of nonnegative measurable functions on $\mathbf{X}$ for which $f_n \to f$, as $n \to \infty$, pointwise a. e. on $\mathbf{X}$. Assume that f is measurable. Then*

$$\int_{\mathbf{X}} f d\mu \leq \liminf \int_{\mathbf{X}} f_n d\mu.$$

Theorem 2.112 (The monotone convergence theorem). *Let* $(\mathbf{X}, \mathcal{M}, \mu)$ *be a measure space and* $\{f_n\}_{n\in\mathbf{N}}$ *be an increasing sequence of nonnegative measurable functions on* $\mathbf{X}$. *Define* $f(x) = \lim_{n\to\infty} f_n(x)$ *for each* $x \in \mathbf{X}$. *Then*

$$\lim_{n\to\infty} \int_{\mathbf{X}} f_n d\mu = \int_{\mathbf{X}} f d\mu.$$

Theorem 2.113 (Beppo Levi's lemma). *Let* $(\mathbf{X}, \mathcal{M}, \mu)$ *be a measure space and* $\{f_n\}_{n\in\mathbf{N}}$ *be an increasing sequence of nonnegative measurable functions on* $\mathbf{X}$. *If the sequence*

$$\left\{\int_{\mathbf{X}} f_n d\mu\right\}_{n\in\mathbf{N}}$$

is bounded, then $\{f_n\}_{n\in\mathbf{N}}$ *converges pointwise on* $\mathbf{X}$ *to a measurable function* f *that is finite a. e. on* $\mathbf{X}$ *and*

$$\lim_{n\to\infty} \int_{\mathbf{X}} f_n d\mu = \int_{\mathbf{X}} f d\mu < \infty.$$

Theorem 2.114. *Let* $(\mathbf{X}, \mathcal{M}, \mu)$ *be a measure space and* f *be a nonnegative measurable function on* $\mathbf{X}$. *Then there exists an increasing sequence* $\{\phi_n\}_{n\in\mathbf{N}}$ *of simple functions on* $\mathbf{X}$ *that converges pointwise on* $\mathbf{X}$ *to* f *and*

$$\lim_{n\to\infty} \int_{\mathbf{X}} \phi_n d\mu = \int_{\mathbf{X}} f d\mu.$$

Theorem 2.115. *Let* $(\mathbf{X}, \mathcal{M}, \mu)$ *be a measure space and* f *and* g *be nonnegative measurable functions on* $\mathbf{X}$. *Then*

$$\int_{\mathbf{X}} (\alpha f + \beta g) d\mu = \alpha \int_{\mathbf{X}} f d\mu + \beta \int_{\mathbf{X}} g d\mu$$

for any positive real numbers α *and* β.

Definition 2.53. Let $(\mathbf{X}, \mathcal{M}, \mu)$ be a measure space and f be a nonnegative measurable function on $\mathbf{X}$. Then f is said to be integrable over $\mathbf{X}$ with respect to μ if $\int_{\mathbf{X}} f d\mu < \infty$.

Definition 2.54. Let $(\mathbf{X}, \mathcal{M}, \mu)$ be a measure space. A measurable function f on $\mathbf{X}$ is said to be integrable over $\mathbf{X}$ with respect to μ if $|f|$ is integrable over $\mathbf{X}$ with respect to μ. For such a function, we define the integral of f over $\mathbf{X}$ with respect to μ by

$$\int_{\mathbf{X}} f d\mu = \int_{\mathbf{X}} f^+ d\mu - \int_{\mathbf{X}} f^- d\mu,$$

where $f^+ = \max\{f, 0\}$, $f^- = \max\{-f, 0\}$. For a measurable subset $\mathbf{E}$ of $\mathbf{X}$, we define

$$\int_{\mathbf{E}} f d\mu = \int_{\mathbf{X}} f \kappa_{\mathbf{E}} d\mu.$$

Theorem 2.116. *Let* $(\mathbf{X}, \mathcal{M}, \mu)$ *be a measure space and* f *be a measurable function on* $\mathbf{X}$. *If* g *is integrable over* $\mathbf{X}$ *and* $|f| \leq g$ *a. e. on* $\mathbf{X}$, *then* f *is integrable over* $\mathbf{X}$ *and*

$$\left|\int_{\mathbf{X}} f d\mu\right| \leq \int_{\mathbf{X}} |f| d\mu \leq \int_{\mathbf{X}} g d\mu.$$

Theorem 2.117. *Let* $(\mathbf{X}, \mathcal{M}, \mu)$ *be a measure space and* f *and* g *be integrable over* $\mathbf{X}$.
1. $\int_{\mathbf{X}}(\alpha f + \beta g) d\mu = \alpha \int_{\mathbf{X}} f d\mu + \beta \int_{\mathbf{X}} g d\mu$ *for any real numbers* α *and* β.
2. *If* $f \leq g$ *a. e. on* $\mathbf{X}$, *then*

$$\int_{\mathbf{X}} f d\mu \leq \int_{\mathbf{X}} g d\mu.$$

3. *If* $\mathbf{A}$ *and* $\mathbf{B}$ *be disjoint measurable subsets of* $\mathbf{X}$, *then*

$$\int_{\mathbf{A} \cup \mathbf{B}} f d\mu = \int_{\mathbf{A}} f d\mu + \int_{\mathbf{B}} f d\mu.$$

Theorem 2.118. *Let* $(\mathbf{X}, \mathcal{M}, \mu)$ *be a measure space, the function* f *is integrable over* $\mathbf{X}$, *and* $\{\mathbf{X}_n\}_{n \in \mathbf{N}}$ *be a disjoint countable collection of measurable sets such that* $\mathbf{X} = \bigcup_{n=1}^{\infty} \mathbf{X}_n$. *Then*

$$\int_{\mathbf{X}} f d\mu = \sum_{n=1}^{\infty} \int_{\mathbf{X}_i} f d\mu.$$

Theorem 2.119. *Let* $(\mathbf{X}, \mathcal{M}, \mu)$ *be a measure space and the function* f *be integrable over* $\mathbf{X}$.
1. *If* $\{\mathbf{X}_n\}_{n \in \mathbf{N}}$ *is an ascending countable collection of measurable subsets of* $\mathbf{X}$, $\mathbf{X} = \bigcup_{n=1}^{\infty} \mathbf{X}_n$, *then*

$$\int_{\mathbf{X}} f d\mu = \lim_{n \to \infty} \int_{\mathbf{X}_n} f d\mu.$$

2. *If* $\{\mathbf{X}_n\}_{n \in \mathbf{N}}$ *be a descending countable collection of measurable subsets of* $\mathbf{X}$, *then*

$$\int_{\bigcap_{n=1}^{\infty} \mathbf{X}_n} f d\mu = \lim_{n \to \infty} \int_{\mathbf{X}_n} f d\mu.$$

Theorem 2.120. *Let* $(\mathbf{X}, \mathcal{M}, \mu)$ *be a measure space and* f *be a measurable function on* $\mathbf{X}$. *If* f *is bounded on* $\mathbf{X}$ *and vanishes outside a set of finite measure, then* f *is integrable over* $\mathbf{X}$.

Theorem 2.121 (The Lebesgue dominated convergence theorem). *Let* $(\mathbf{X}, \mathcal{M}, \mu)$ *be a measure space and* $\{f_n\}_{n\in\mathbf{N}}$ *be a sequence of measurable functions on* $\mathbf{X}$ *for which* $f_n \to f$, *as* $n \to \infty$, *pointwise a. e. on* $\mathbf{X}$, *and the function* f *is measurable. Let also there be a nonnegative function* g *that is integrable over* $\mathbf{X}$ *and*

$$|f_n| \leq g \quad a.e. \quad on \quad \mathbf{X}$$

for any $n \in \mathbf{N}$. *Then* f *is integrable over* $\mathbf{X}$ *and*

$$\lim_{n\to\infty} \int_{\mathbf{X}} f_n d\mu = \int_{\mathbf{X}} f d\mu.$$

Definition 2.55. Let $(\mathbf{X}, \mathcal{M}, \mu)$ be a measure space and $\{f_n\}_{n\in\mathbf{N}}$ be a sequence of functions on $\mathbf{X}$, each of which is integrable over $\mathbf{X}$. The sequence $\{f_n\}_{n\in\mathbf{N}}$ is said to be uniformly integrable over $\mathbf{X}$ if for each $\epsilon > 0$, there is a $\delta > 0$ such that, for any $n \in \mathbf{N}$ and for any measurable subset $\mathbf{E}$ of $\mathbf{X}$ for which $\mu(\mathbf{E}) < \delta$, we have

$$\int_{\mathbf{E}} |f_n| d\mu < \epsilon.$$

The sequence $\{f_n\}_{n\in\mathbf{N}}$ is said to be a tight over $\mathbf{X}$ if for each $\epsilon > 0$, there is a subset $\mathbf{X}_0$ of $\mathbf{X}$ that has finite measure and

$$\int_{\mathbf{X}\setminus\mathbf{X}_0} |f_n| d\mu < \epsilon$$

for any $n \in \mathbf{N}$.

Theorem 2.122. *Let* $(\mathbf{X}, \mathcal{M}, \mu)$ *be a measure space and the function* f *is integrable over* $\mathbf{X}$. *Then for each* $\epsilon > 0$, *there is a* $\delta > 0$ *such that, for any measurable subset* $\mathbf{E}$ *of* $\mathbf{X}$ *for which* $\mu(\mathbf{E}) < \delta$, *we have*

$$\int_{\mathbf{E}} |f| d\mu < \epsilon.$$

Moreover, for each $\epsilon > 0$, *there is a subset* $\mathbf{X}_0$ *of* $\mathbf{X}$ *that has finite measure and*

$$\int_{\mathbf{X}\setminus\mathbf{X}_0} |f| d\mu < \epsilon.$$

Theorem 2.123 (The Vitali convergence theorem). *Let* $(\mathbf{X}, \mathcal{M}, \mu)$ *be a measure space and* $\{f_n\}_{n\in\mathbf{N}}$ *be a sequence of functions on* $\mathbf{X}$ *that is uniformly integrable and tight over* $\mathbf{X}$.

Assume that $f_n \to f$, *as* $n \to \infty$, *pointwise a. e. on* $\mathbf{X}$ *and the function* f *is integrable over* $\mathbf{X}$. *Then*

$$\lim_{n\to\infty} \int_{\mathbf{X}} f_n d\mu = \int_{\mathbf{X}} f d\mu.$$

2.12 Advanced practical problems

Problem 2.1. Let $\{\mathbf{E}_k\}_{k\in\mathbf{N}}$ be a countable disjoint subcollection of measurable sets and $\mathbf{A}$ be a measurable set. Prove that

$$m^\star\left(\mathbf{A}\cap\left(\bigcup_{k=1}^{\infty}\mathbf{E}_k\right)\right) = \sum_{k=1}^{\infty} m^\star(\mathbf{A}\cap\mathbf{E}_k).$$

Problem 2.2. Let f be a continuous function and $\mathbf{B}$ be a Borel set. Prove that $f^{-1}(\mathbf{B})$ is a Borel set.

Problem 2.3. Let f be a function with measurable domain $\mathbf{E}$. Prove that f is measurable if and only if the function g, defined on $\mathbf{R}$ by $g(x) = f(x)$ for $x \in \mathbf{E}$ and $g(x) = 0$ for $x \notin \mathbf{E}$, is measurable.

Problem 2.4. Let f be a measurable function on the measurable set $\mathbf{E}$ that is finite a. e. on $\mathbf{E}$ and $m(\mathbf{E}) < \infty$. Prove that for each $\epsilon > 0$ there is a measurable set $\mathbf{E}_0$ contained in $\mathbf{E}$ such that f is bounded on $\mathbf{E}_0$ and $m(\mathbf{E}\setminus\mathbf{E}_0) < \epsilon$.

Problem 2.5. Prove that the sum and product of two simple functions are simple.

Problem 2.6. Let f be a bounded measurable function on a set of finite measure $\mathbf{E}$. Assume that g is bounded and $f = g$ a. e. on $\mathbf{E}$. Prove that

$$\int_{\mathbf{E}} f = \int_{\mathbf{E}} g.$$

Problem 2.7. Let $\{f_n\}_{n\in\mathbf{N}}$ be a sequence of nonnegative measurable functions on $\mathbf{E}$ that converges pointwise on $\mathbf{E}$ to f. Suppose that $f_n \le f$ on $\mathbf{E}$ for each $n \in \mathbf{N}$. Prove that

$$\lim_{n\to\infty} \int_{\mathbf{E}} f_n = \int_{\mathbf{E}} f.$$

Problem 2.8. Let f be a nonnegative measurable function on $\mathbf{R}$. Prove that

$$\lim_{n\to\infty} \int_{-n}^{n} f = \int_{\mathbf{R}} f.$$

Problem 2.9. Let $\{f_n\}_{n\in\mathbf{N}}$ be a sequence of nonnegative integrable functions on $\mathbf{E}$ and $f_n \to 0$, as $n \to \infty$, pointwise a. e. on $\mathbf{E}$. Prove that

$$\lim_{n\to\infty} \int_{\mathbf{E}} f_n = 0$$

if and only if $\{f_n\}_{n\in\mathbf{N}}$ is uniformly integrable and a tight over $\mathbf{E}$.

3 The $\mathbf{L}^p$ spaces

3.1 Definition

Let $\mathbf{E}$ be a measurable set and $\mathcal{F}$ be the collection of all measurable extended real-valued functions on $\mathbf{E}$ that are finite a. e. on $\mathbf{E}$.

Definition 3.1. Define two functions f and g in $\mathcal{F}$ to be equivalent, and we write $f \sim g$, provided

$$f(x) = g(x) \quad \text{for} \quad \text{almost} \quad \text{all} \quad x \in \mathbf{E}.$$

The relation $\sim$ is an equivalent relation, that is, it is reflexive, symmetric and transitive. Therefore it induces a partition of $\mathcal{F}$ into a disjoint collection of equivalence classes, which we denote by $\mathcal{F}/\sim$. For given two functions f and g in $\mathcal{F}$, their equivalence classes $[f]$ and $[g]$ and real numbers α and β, we define $\alpha[f] + \beta[g]$ to be the equivalence class of the functions in $\mathcal{F}$ that take the value $\alpha f(x) + \beta g(x)$ at points $x \in \mathbf{E}$ at which both f and g are finite. Note that these linear combinations are independent of the choice of the representatives of the equivalence classes. The zero element in $\mathcal{F}/\sim$ is the equivalence class of functions that vanish a. e. in $\mathbf{E}$. Thus $\mathcal{F}/\sim$ is a vector space.

Definition 3.2. For $1 \leq p < \infty$, we define $\mathbf{L}^p(\mathbf{E})$ to be the collection of equivalence classes $[f]$ for which

$$\int_{\mathbf{E}} |f|^p < \infty.$$

This is properly defined since, if $f \sim g$, then

$$\int_{\mathbf{E}} |f|^p = \int_{\mathbf{E}} |g|^p.$$

Note that, if $[f], [g] \in \mathbf{L}^p(\mathbf{E})$, then, for any real constants α and β, we have

$$\begin{aligned}\int_{\mathbf{E}} |\alpha f + \beta g|^p &\leq 2^p\left(\int_{\mathbf{E}} |\alpha f|^p + \int_{\mathbf{E}} |\beta g|^p\right) \\ &= |2\alpha|^p \int_{\mathbf{E}} |f|^p + |2\beta|^p \int_{\mathbf{E}} |g|^p < \infty,\end{aligned}$$

i. e., $\alpha[f] + \beta[g] \in \mathbf{L}^p(\mathbf{E})$.

Definition 3.3. We call a function $f \in \mathcal{F}$ essentially bounded provided there is some $M \geq 0$, called an essential upper bound for f, for which

$$|f(x)| \leq M \quad \text{for} \quad \text{almost} \quad \text{all} \quad x \in \mathbf{E}.$$

https://doi.org/10.1515/9783110657722-003

Definition 3.4. We define $\mathbf{L}^\infty(\mathbf{E})$ to be the collection of equivalence classes $[f]$ for which f is essentially bounded.

Note that $\mathbf{L}^\infty(\mathbf{E})$ is properly defined since, if $f \sim g$, then

$$|f(x)| = |g(x)| \leq M \quad \text{for} \quad \text{almost} \quad \text{all} \quad x \in \mathbf{E}.$$

Also, if $[f], [g] \in \mathbf{L}^\infty(\mathbf{E})$ and $\alpha, \beta \in \mathbf{R}$, there are nonnegative constants M_1 and M_2 such that

$$|f(x)| \leq M_1, \quad |g(x)| \leq M_2 \quad \text{for} \quad \text{almost} \quad \text{all} \quad x \in \mathbf{E}.$$

Hence,

$$|\alpha f(x) + \beta g(x)| \leq |\alpha||f(x)| + |\beta||g(x)| \leq |\alpha| M_1 + |\beta| M_2$$

for almost all $x \in \mathbf{E}$. Therefore $\alpha[f] + \beta[g] \in \mathbf{L}^\infty(\mathbf{E})$ and hence, $\mathbf{L}^\infty(\mathbf{E})$ is a vector space. For simplicity and convenience, we refer to the equivalence classes in $\mathcal{F}/\sim$ as functions and denote them by f rather than $[f]$. Thus $f = g$ means $f - g$ vanishes a. e. on $\mathbf{E}$.

Definition 3.5. For $1 \leq p < \infty$, in $\mathbf{L}^p(\mathbf{E})$ we define

$$\|f\|_p = \left(\int_{\mathbf{E}} |f|^p\right)^{\frac{1}{p}}.$$

For $p = \infty$, we define $\|f\|_\infty$ to be the infimum of the essential upper bounds for $|f|$.

3.2 The inequalities of Hölder and Minkowski

Theorem 3.1 (Hölder's inequality). *Let* $\mathbf{E}$ *be a measurable set,* $1 \leq p < \infty$ *and* q *be the conjugate of* p. *If* $f \in \mathbf{L}^p(\mathbf{E})$ *and* $g \in \mathbf{L}^q(\mathbf{E})$, *then the product* fg *is integrable over* $\mathbf{E}$ *and*

$$\int_{\mathbf{E}} |fg| \leq \|f\|_p \|g\|_q. \tag{3.1}$$

Proof.

1. Let $p = 1$. Then $q = \infty$ and

$$\int_{\mathbf{E}} |fg| = \int_{\mathbf{E}} |f||g| \leq \left(\int_{\mathbf{E}} |f|\right)\|g\|_\infty = \|f\|_1 \|g\|_\infty.$$

2. Let $p \in (1, \infty)$. If $\|f\|_p = 0$ or $\|g\|_q = 0$, then the assertion is evident. Let $\|f\|_p \neq 0$ and $\|g\|_q \neq 0$. We set

$$f_1 = \frac{f}{\|f\|_p}, \quad g_1 = \frac{g}{\|g\|_q}.$$

Then

$$\|f_1\|_p = \|g_1\|_q = 1.$$

Now, using Young's inequality, we have

$$|f_1 g_1| \le \frac{1}{p}|f_1|^p + \frac{1}{q}|g_1|^q.$$

Hence,

$$\begin{aligned}\frac{1}{\|f\|_p\|g\|_q}\int_{\mathbf{E}}|fg| &= \int_{\mathbf{E}}|f_1 g_1| \le \int_{\mathbf{E}}\left(\frac{1}{p}|f_1|^p + \frac{1}{q}|g_1|^q\right)\\ &= \int_{\mathbf{E}}\frac{1}{p}|f_1|^p + \int_{\mathbf{E}}\frac{1}{q}|g_1|^q = \frac{1}{p}\int_{\mathbf{E}}|f_1|^p + \frac{1}{q}\int_{\mathbf{E}}|g_1|^q\\ &= \frac{1}{p}\|f_1\|_p^p + \frac{1}{q}\|g_1\|_q^q = \frac{1}{p} + \frac{1}{q} = 1.\end{aligned}$$

From the previous inequality we get the inequality (3.1). This completes the proof. □

Remark 3.1. When $p = q = 2$, the Hölder inequality is known as the Cauchy–Schwartz inequality.

Theorem 3.2 (Minkowski's inequality). *Let* $\mathbf{E}$ *be a measurable set,* $1 \le p \le \infty$. *If* $f, g \in \mathbf{L}^p(\mathbf{E})$, *then*

$$\|f + g\|_p \le \|f\|_p + \|g\|_p.$$

Proof.

1. Let $p = 1$. Then

$$\|f + g\|_1 = \int_{\mathbf{E}}|f + g| \le \int_{\mathbf{E}}(|f| + |g|) = \int_{\mathbf{E}}|f| + \int_{\mathbf{E}}|g| = \|f\|_1 + \|g\|_1.$$

2. Let $p = \infty$. Then

$$\|f + g\|_\infty \le \|f\|_\infty + \|g\|_\infty.$$

3. Let $p \in (1, \infty)$ and q be its conjugate. If $\|f + g\|_p = 0$, then the assertion is evident. Assume that $\|f + g\|_p \ne 0$. Then, applying Hölder's inequality, we get

$$\begin{aligned}\|f + g\|_p^p &= \int_{\mathbf{E}}|f + g|^p = \int_{\mathbf{E}}|f + g||f + g|^{p-1}\\ &\le \int_{\mathbf{E}}(|f||f + g|^{p-1} + |g||f + g|^{p-1}) = \int_{\mathbf{E}}|f||f + g|^{p-1} + \int_{\mathbf{E}}|g||f + g|^{p-1}\end{aligned}$$

$$\leq \left(\int_{\mathbf{E}} |f|^p\right)^{\frac{1}{p}} \left(\int_{\mathbf{E}} |f+g|^{(p-1)q}\right)^{\frac{1}{q}} + \left(\int_{\mathbf{E}} |g|^p\right)^{\frac{1}{p}} \left(\int_{\mathbf{E}} |f+g|^{(p-1)q}\right)^{\frac{1}{q}}$$

$$= \|f+g\|_p^{\frac{p}{q}} \|f\|_p + \|f+g\|_p^{\frac{p}{q}} \|g\|_p.$$

Hence,

$$\|f+g\|_p^{p-\frac{p}{q}} \leq \|f\|_p + \|g\|_p$$

or

$$\|f+g\|_p \leq \|f\|_p + \|g\|_p.$$

This completes the proof. □

3.3 Some properties

Theorem 3.3. *Let* $\mathbf{E}$ *be a measurable set and* $1 < p < \infty$. *Suppose that* $\mathcal{F}$ *is a family of functions in* $\mathbf{L}^p(\mathbf{E})$ *that is bounded in* $\mathbf{L}^p(\mathbf{E})$ *in the sense that there is a constant* $M > 0$ *such that*

$$\|f\|_p \leq M \quad \text{for} \quad \text{all} \quad f \in \mathcal{F}.$$

Then the family $\mathcal{F}$ *is uniformly integrable over* $\mathbf{E}$.

Proof. Let $\epsilon > 0$ be arbitrarily chosen. Suppose that $\mathbf{A}$ is a measurable subset of $\mathbf{E}$ of finite measure. Let also $\frac{1}{p} + \frac{1}{q} = 1$. Define g to be identically equal to 1 on $\mathbf{A}$. Because $m(\mathbf{A}) < \infty$, we have $g \in \mathbf{L}^q(\mathbf{A})$. By Hölder's inequality, for any $f \in \mathcal{F}$, we have

$$\int_{\mathbf{A}} |f| = \int_{\mathbf{A}} |f| g \leq \left(\int_{\mathbf{A}} |f|^p\right)^{\frac{1}{p}} \left(\int_{\mathbf{A}} g^q\right)^{\frac{1}{q}}.$$

On the other hand,

$$\left(\int_{\mathbf{A}} |f|^p\right)^{\frac{1}{p}} \leq M, \quad \left(\int_{\mathbf{A}} |g|^q\right)^{\frac{1}{q}} = (m(\mathbf{A}))^{\frac{1}{q}}$$

for any $f \in \mathcal{F}$. Therefore

$$\int_{\mathbf{A}} |f| \leq (m(\mathbf{A}))^{\frac{1}{q}} M$$

for any $f \in \mathcal{F}$. Let $\delta = (\frac{\epsilon}{M})^q$. Hence, if $m(\mathbf{A}) < \delta$, then

$$\int_{\mathbf{A}} |f| \leq M(m(\mathbf{A}))^{\frac{1}{q}} < M\delta^{\frac{1}{q}} = M\frac{\epsilon}{M} = \epsilon$$

for any $f \in \mathcal{F}$. Therefore $\mathcal{F}$ is uniformly integrable over $\mathbf{E}$. This completes the proof. □

Theorem 3.4. *Let* $\mathbf{E}$ *be a measurable set of finite measure and* $1 \le p_1 < p_2 \le \infty$. *Then*

$$\mathbf{L}^{p_2}(\mathbf{E}) \subseteq \mathbf{L}^{p_1}(\mathbf{E}). \tag{3.2}$$

Proof. Let $p_2 < \infty$ and $f \in \mathbf{L}^{p_2}(\mathbf{E})$ be arbitrarily chosen. Then $p = \frac{p_2}{p_1} > 1$. We take $q > 1$ such that $\frac{1}{p} + \frac{1}{q} = 1$. Then

$$\left(\int_{\mathbf{E}} |f|^{p_1 p}\right)^{\frac{1}{p}} = \left(\int_{\mathbf{E}} |f|^{p_2}\right)^{\frac{p_1}{p_2}} < \infty,$$

i. e., $|f|^{p_1} \in \mathbf{L}^p(\mathbf{E})$. Let $g = \kappa_{\mathbf{E}}$. Since $m(\mathbf{E}) < \infty$, we have $g \in \mathbf{L}^q(\mathbf{E})$. Hence, using Hölder's inequality, we get

$$\begin{aligned} \int_{\mathbf{E}} |f|^{p_1} = \int_{\mathbf{E}} |f|^{p_1} g &\le \left(\int_{\mathbf{E}} |f|^{p_1 p}\right)^{\frac{1}{p}} \left(\int_{\mathbf{E}} g^q\right)^{\frac{1}{q}} \\ &= \|f\|_{p_2}^{p_1} (m(\mathbf{E}))^{\frac{1}{q}} < \infty. \end{aligned}$$

Therefore $f \in \mathbf{L}^{p_1}(\mathbf{E})$. Because $f \in \mathbf{L}^{p_2}(\mathbf{E})$ was arbitrarily chosen and we see that it is an element of $\mathbf{L}^{p_1}(\mathbf{E})$, we obtain equation (3.2). Let $p_2 = \infty$ and $f \in \mathbf{L}^{\infty}(\mathbf{E})$ be arbitrarily chosen. Then there is a positive constant M such that

$$|f(x)| \le M$$

for almost all $x \in \mathbf{E}$. Hence,

$$\int_{\mathbf{E}} |f|^{p_1} \le M^{p_1} m(\mathbf{E}).$$

Then $f \in \mathbf{L}^{p_1}(\mathbf{E})$. Because $f \in \mathbf{L}^{\infty}(\mathbf{E})$ was arbitrarily chosen and we see that it is an element of $\mathbf{L}^{p_1}(\mathbf{E})$, we obtain equation (3.2). This completes the proof. □

Remark 3.2. Let $\{f_n\}_{n\in\mathbb{N}}$ be a sequence of elements of $\mathbf{L}^p(\mathbf{E})$ and $f \in \mathbf{L}^p(\mathbf{E})$. When $\|f_n - f\|_p \to 0$, as $n \to \infty$, we will say that the sequence $\{f_n\}_{n\in\mathbf{N}}$ converges to f in $\mathbf{L}^p(\mathbf{E})$.

3.4 The Riesz–Fischer theorem

Definition 3.6. Let $\mathbf{X}$ be a normed vector space. A sequence $\{f_n\}_{n\in\mathbf{N}}$ in $\mathbf{X}$ is said to be a rapidly Cauchy sequence provided there is a convergent series of positive numbers $\sum_{k=1}^{\infty} \epsilon_k$ for which

$$\|f_{k+1} - f_k\| \le \epsilon_k^2$$

for all $k \in \mathbf{N}$.

Lemma 3.1. *Let $\mathbf{X}$ be a normed vector space and $\{f_n\}_{n\in\mathbf{N}}$ be a sequence in $\mathbf{X}$ such that*

$$\|f_{k+1} - f_k\| \le a_k$$

for all $k \in \mathbf{N}$, where a_k, $k \in \mathbf{N}$, are nonnegative numbers. Then

$$\|f_{n+k} - f_n\| \le \sum_{l=n}^{\infty} a_l \tag{3.3}$$

for all $k, n \in \mathbf{N}$.

Proof. For any $n, k \in \mathbf{N}$, we have

$$f_{n+k} - f_n = f_{n+k} - f_{n+k-1} + f_{n+k-1} - \cdots + f_{n+1} - f_n = \sum_{j=n}^{n+k-1} (f_{j+1} - f_j).$$

Hence,

$$\|f_{n+k} - f_n\| = \left\| \sum_{j=n}^{n+k-1} (f_{j+1} - f_j) \right\| \le \sum_{j=n}^{n+k-1} \|f_{j+1} - f_j\| \le \sum_{j=n}^{n+k-1} a_j \le \sum_{j=n}^{\infty} a_j$$

for any $k, n \in \mathbf{N}$. This completes the proof. □

Theorem 3.5. *Let $\mathbf{X}$ be a normed vector space. Then every rapidly Cauchy sequence in $\mathbf{X}$ is a Cauchy sequence in $\mathbf{X}$. Furthermore, every Cauchy sequence in $\mathbf{X}$ has a rapidly Cauchy subsequence in $\mathbf{X}$.*

Proof.

1. Let $\{f_n\}_{n\in\mathbf{N}}$ be a rapidly Cauchy sequence in $\mathbf{X}$. Then there is a convergent series $\sum_{k=1}^{\infty} \epsilon_k$ of positive numbers such that

$$\|f_{k+1} - f_k\| \le \epsilon_k^2$$

   for any $k \in \mathbf{N}$. Hence, by (3.3), we obtain

$$\|f_{n+k} - f_n\| \le \sum_{l=n}^{\infty} \epsilon_l^2 \to 0 \quad \text{as} \quad n \to \infty,$$

   for any $k \in \mathbf{N}$. Here we have used the fact that the series $\sum_{k=1}^{\infty} \epsilon_k^2$ is a convergent series. Therefore $\{f_n\}_{n\in\mathbf{N}}$ is a Cauchy sequence.
2. Let $\{f_n\}_{n\in\mathbf{N}}$ is a Cauchy sequence in $\mathbf{X}$. Then there are $M_1, M_2, M_3 \in \mathbf{N}$ such that

$$\|f_{m_1^1} - f_{m_2^1}\| < \frac{1}{2} \quad \text{for any} \quad m_1^1, m_2^1 \ge M_1,$$
$$\|f_{m_1^2} - f_{m_2^2}\| < \frac{1}{2^2} \quad \text{for any} \quad m_1^2, m_2^2 \ge M_2,$$

$$\|f_{m_1^3} - f_{m_2^3}\| < \frac{1}{2^3} \quad \text{for} \quad \text{any} \quad m_1^3, m_2^3 \geq M_3.$$

In particular, for $m_1^1, m_2^1, m_2^3 \geq \max\{M_1, M_2, M_3\}$, we have

$$\|f_{m_1^1} - f_{m_2^1}\| < \frac{1}{2}, \quad \|f_{m_2^3} - f_{m_1^1}\| < \frac{1}{2^2}, \quad \|f_{m_1^3} - f_{m_2^3}\| < \frac{1}{2^3}.$$

We set

$$n_1 = m_2^1, \quad n_2 = m_1^1, \quad n_3 = m_2^3,$$

for $m_1^1, m_2^1, m_2^3 \geq \max\{M_1, M_2, M_3\}$. Then

$$\|f_{n_2} - f_{n_1}\| < \frac{1}{2}, \quad \|f_{n_3} - f_{n_2}\| < \frac{1}{2^2}, \quad \|f_{m_1^3} - f_{n_3}\| < \frac{1}{2^3},$$

for $m_1^3 \geq \max\{M_1, M_2, M_3\}$. Continuing this process, we obtain a subsequence $\{f_{n_k}\}_{k\in\mathbf{N}}$ of the sequence $\{f_n\}_{n\in\mathbf{N}}$ such that

$$\|f_{n_{k+1}} - f_{n_k}\| < \frac{1}{2^k}$$

for any $k \in \mathbf{N}$. Since $\sum_{k=0}^{\infty} \frac{1}{(\sqrt{2})^k}$ is convergent, we conclude that $\{f_{n_k}\}_{k\in\mathbf{N}}$ is a rapidly Cauchy sequence in $\mathbf{X}$. This completes the proof. □

Theorem 3.6. *Let* $\mathbf{E}$ *be a measurable set and* $1 \leq p \leq \infty$. *Then every rapidly Cauchy sequence in* $\mathbf{L}^p(\mathbf{E})$ *converges both with respect to the* $\mathbf{L}^p(\mathbf{E})$ *norm and pointwise a. e. in* $\mathbf{E}$ *to a function* f *in* $\mathbf{L}^p(\mathbf{E})$.

Proof. Let $1 \leq p < \infty$. We leave the case $p = \infty$ as an exercise. Let $\{f_n\}_{n\in\mathbf{N}}$ be a rapidly convergent sequence in $\mathbf{L}^p(\mathbf{E})$. We choose $\sum_{k=1}^{\infty} \epsilon_k$ to be a convergent series of positive numbers such that

$$\|f_{k+1} - f_k\|_p \leq \epsilon_k^2 \tag{3.4}$$

for any $k \in \mathbf{N}$. Let

$$\mathbf{E}^k = \{x \in \mathbf{E} : |f_{k+1}(x) - f_k(x)| \geq \epsilon_k\}, \quad k \in \mathbf{N}.$$

Then

$$\mathbf{E}^k = \{x \in \mathbf{E} : |f_{k+1}(x) - f_k(x)|^p \geq \epsilon_k^p\}, \quad k \in \mathbf{N}.$$

Hence,

$$\int_{\mathbf{E}^k} |f_{k+1}(x) - f_k(x)|^p \geq \epsilon_k^p m(\mathbf{E}^k).$$

Now, using (3.4), we get

$$m(\mathbf{E}^k) \le \frac{1}{\epsilon_k^p} \int_{\mathbf{E}} |f_{k+1} - f_k|^p = \frac{1}{\epsilon_k^p} \|f_{k+1} - f_k\|_p^p$$
$$\le \frac{1}{\epsilon_k^p} \epsilon_k^{2p} = \epsilon_k^p.$$

Since $p \ge 1$, the series $\sum_{k=1}^{\infty} \epsilon_k^p$ is convergent and

$$\sum_{k=1}^{\infty} m(\mathbf{E}^k) \le \sum_{k=1}^{\infty} \epsilon_k^p < \infty.$$

Hence, by the Borel–Cantelli lemma, it follows that there is $\mathbf{E}_0 \subset \mathbf{E}$ such that $m(\mathbf{E}_0) = 0$ and for each $x \in \mathbf{E} \setminus \mathbf{E}_0$ there is an index $K(x)$ such that

$$|f_{k+1}(x) - f_k(x)| < \epsilon_k$$

for any $k \ge K(x)$. Let $x \in \mathbf{E} \setminus \mathbf{E}_0$. Then

$$|f_{n+k}(x) - f_n(x)| \le \sum_{j=n}^{n+k-1} |f_{j+1}(x) - f_j(x)| \le \sum_{j=n}^{\infty} \epsilon_j$$

for all $n \ge K(x)$ and for any $k \in \mathbf{N}$. Hence, the sequence of real numbers $\{f_n(x)\}_{n\in\mathbf{N}}$ is a Cauchy sequence in $\mathbf{R}$ for any $x \in \mathbf{E} \setminus \mathbf{E}_0$. Therefore it is convergent for any $x \in \mathbf{E} \setminus \mathbf{E}_0$ and let $\lim_{n\to\infty} f_n(x) = f(x)$ for any $x \in \mathbf{E} \setminus \mathbf{E}_0$. By (3.3) and (3.4), we obtain

$$\int_{\mathbf{E}} |f_{n+k} - f_n|^p \le \left(\sum_{j=n}^{\infty} \epsilon_j^2 \right)^p \tag{3.5}$$

for all $n, k \in \mathbf{N}$. Since $f_n \to f$ pointwise a. e. on $\mathbf{E}$, we take the limit as $k \to \infty$ in (3.5) and using Fatou's lemma, we get

$$\int_{\mathbf{E}} |f - f_n|^p \le \left(\sum_{j=n}^{\infty} \epsilon_j^2 \right)^p$$

for all $n \in \mathbf{N}$. Hence, $f \in \mathbf{L}^p(\mathbf{E})$ and $f_n \to f$, as $n \to \infty$, in $\mathbf{L}^p(\mathbf{E})$. This completes the proof. □

Theorem 3.7 (The Riesz–Fischer theorem). *Let* $\mathbf{E}$ *be a measurable set and* $1 \le p \le \infty$. *Then* $\mathbf{L}^p(\mathbf{E})$ *is a Banach space. Moreover, if* $f_n \to f$, *as* $n \to \infty$, *in* $\mathbf{L}^p(\mathbf{E})$, *a subsequence of* $\{f_n\}_{n\in\mathbf{N}}$ *converges pointwise a. e. on* $\mathbf{E}$ *to* f.

Proof. Let $\{f_n\}_{n\in\mathbf{N}}$ be a Cauchy sequence in $\mathbf{L}^p(\mathbf{E})$. By Theorem 3.5, it follows that there is a subsequence $\{f_{n_k}\}_{k\in\mathbf{N}}$ of the sequence $\{f_n\}_{n\in\mathbf{N}}$ that is a rapidly Cauchy sequence in

$\mathbf{L}^p(\mathbf{E})$. Hence, by Theorem 3.6, it follows that $\{f_{n_k}\}_{k\in\mathbf{N}}$ converges both with respect to the $\mathbf{L}^p(\mathbf{E})$ norm and pointwise a. e. on $\mathbf{E}$. Let $f_{n_k} \to f$, as $k \to \infty$, in $\mathbf{L}^p(\mathbf{E})$. We take $\epsilon > 0$ arbitrarily. Then there exists $K \in \mathbf{N}$ such that

$$\|f_{n_k} - f\|_p < \frac{\epsilon}{2} \quad \text{and} \quad \|f_l - f_m\|_p < \frac{\epsilon}{2}$$

for any $l, m, n_k \geq K$. Hence, for any $n, n_k \geq K$, we have

$$\|f_n - f\|_p = \|f_n - f_{n_k} + f_{n_k} - f\|_p \leq \|f_n - f_{n_k}\|_p + \|f_{n_k} - f\|_p < \frac{\epsilon}{2} + \frac{\epsilon}{2} = \epsilon.$$

Therefore $\{f_n\}_{n\in\mathbf{N}}$ is convergent in $\mathbf{L}^p(\mathbf{E})$ and $\mathbf{L}^p(\mathbf{E})$ is a Banach space. This completes the proof. □

Definition 3.7. A real-valued function f, defined on a set $\mathbf{A}$, is called convex if for each pair of points $x_1, x_2 \in \mathbf{A}$ and for each $\lambda \in [0,1]$, we have

$$f(\lambda x_1 + (1-\lambda)x_2) \leq \lambda f(x_1) + (1-\lambda)f(x_2).$$

Theorem 3.8. *Let $\mathbf{E}$ be a measurable set and $1 \leq p < \infty$. Suppose that $\{f_n\}_{n\in\mathbf{N}}$ is a sequence in $\mathbf{L}^p(\mathbf{E})$ that converges pointwise a. e. on $\mathbf{E}$ to the function $f \in \mathbf{L}^p(\mathbf{E})$. Then $f_n \to f$, as $n \to \infty$, in $\mathbf{L}^p(\mathbf{E})$ if and only if*

$$\lim_{n\to\infty} \int_{\mathbf{E}} |f_n|^p = \int_{\mathbf{E}} |f|^p. \tag{3.6}$$

Proof.

1. Let $f_n \to f$, as $n \to \infty$, in $\mathbf{L}^p(\mathbf{E})$. Then, by Minkowski's inequality, we have

$$\big| \|f_n\|_p - \|f\|_p \big| \leq \|f_n - f\|_p \to 0 \quad \text{as} \quad n \to \infty.$$

   Hence, (3.6) holds.

2. Assume (3.6). Let $\mathbf{E}_0 \subset \mathbf{E}$ be such that $m(\mathbf{E}_0) = 0$ and $f_n(x) \to f(x)$, as $n \to \infty$, for all $x \in \mathbf{E} \setminus \mathbf{E}_0$. Take $\psi(t) = |t|^p$, $t \in \mathbf{R}$. Then ψ is a convex function and

$$\psi\left(\frac{a+b}{2}\right) \leq \frac{\psi(a)}{2} + \frac{\psi(b)}{2}$$

   for any $a, b \in \mathbf{R}$. Hence,

$$0 \leq \frac{|a|^p + |b|^p}{2} - \left|\frac{a-b}{2}\right|^p$$

   for any $a, b \in \mathbf{R}$. For each $n \in \mathbf{N}$, we define the function

$$h_n(x) = \frac{|f_n(x)|^p + |f(x)|^p}{2} - \left|\frac{f_n(x) - f(x)}{2}\right|^p$$

for any $x \in \mathbf{E}$. We have $h_n(x) \to |f(x)|^p$, as $n \to \infty$, for any $x \in \mathbf{E} \setminus \mathbf{E}_0$. Hence, by Fatou's lemma, we obtain

$$\int_{\mathbf{E}} |f|^p \leq \liminf \int_{\mathbf{E}} h_n = \liminf\left(\int_{\mathbf{E}} \frac{|f_n|^p + |f|^p}{2} - \int_{\mathbf{E}} \left|\frac{f_n - f}{2}\right|^p \right)$$
$$= \int_{\mathbf{E}} |f|^p - \limsup \int_{\mathbf{E}} \left|\frac{f_n - f}{2}\right|^p.$$

Therefore

$$\limsup \int_{\mathbf{E}} \left|\frac{f_n - f}{2}\right|^p \leq 0.$$

That is $f_n \to f$, as $n \to \infty$, in $\mathbf{L}^p(\mathbf{E})$. This completes the proof. □

Theorem 3.9. *Let* $\mathbf{E}$ *be a measurable set and* $1 \leq p < \infty$. *Suppose that* $\{f_n\}_{n\in\mathbf{N}}$ *is a sequence in* $\mathbf{L}^p(\mathbf{E})$ *that converges pointwise a. e. on* $\mathbf{E}$ *to the function* f *which belongs to* $\mathbf{L}^p(\mathbf{E})$. *Then* $f_n \to f$, *as* $n \to \infty$, *in* $\mathbf{L}^p(\mathbf{E})$, *if and only if* $\{|f_n|^p\}_{n\in\mathbf{N}}$ *is uniformly integrable and tight over* $\mathbf{E}$.

Proof.
1. Let $f_n \to f$, as $n \to \infty$, in $\mathbf{L}^p(\mathbf{E})$. Hence,

$$\lim_{n\to\infty} \int_{\mathbf{E}} |f_n - f|^p = 0.$$

   Therefore $\{|f_n - f|^p\}_{n\in\mathbf{N}}$ is uniformly integrable and tight over $\mathbf{E}$. Because

$$|f_n|^p \leq 2^p(|f_n - f|^p + |f|^p), \quad n \in \mathbf{N},$$

   we conclude that $\{|f_n|^p\}_{n\in\mathbf{N}}$ is uniformly integrable and tight over $\mathbf{E}$.
2. Let $\{|f_n|^p\}_{n\in\mathbf{N}}$ is uniformly integrable and tight over $\mathbf{E}$. Since $f_n \to f$, as $n \to \infty$, pointwise a. e. on $\mathbf{E}$, we have $|f_n|^p \to |f|^p$, as $n \to \infty$, pointwise a. e. on $\mathbf{E}$. Hence, by the Vitali convergence theorem, we conclude that $|f|^p$ is integrable over $\mathbf{E}$ and

$$\lim_{n\to\infty} \int_{\mathbf{E}} |f_n|^p = \int_{\mathbf{E}} |f|^p.$$

   This completes the proof. □

Theorem 3.10 (The $\mathbf{L}^p$ dominated convergence theorem). *Let* $\{f_n\}_{n\in\mathbf{N}}$ *be a sequence of measurable functions that converges pointwise a. e. on* $\mathbf{E}$ *to* f. *For* $1 \leq p < \infty$, *suppose that there is a function* $g \in \mathbf{L}^p(\mathbf{E})$ *such that* $|f_n| \leq g$ *a. e. on* $\mathbf{E}$ *for all* $n \in \mathbf{N}$. *Then* $f_n \to f$, *as* $n \to \infty$, *in* $\mathbf{L}^p(\mathbf{E})$.

Proof. Since $f_n \to f$, as $n \to \infty$, pointwise a. e. on $\mathbf{E}$, we have $|f_n|^p \to |f|^p$, as $n \to \infty$, pointwise a. e. on $\mathbf{E}$. Also, $|f_n|^p \le g^p$ a. e. on $\mathbf{E}$ and g^p is integrable over $\mathbf{E}$. Hence, by the Lebesgue dominated convergence theorem, we conclude that $|f|^p$ is integrable over $\mathbf{E}$ and

$$\lim_{n\to\infty} \int_{\mathbf{E}} |f_n|^p = \int_{\mathbf{E}} |f|^p.$$

This completes the proof. □

3.5 Separability

Theorem 3.11. *Let $\mathbf{E}$ be a measurable set and $1 \le p \le \infty$. Then the subspace of simple functions in $\mathbf{L}^p(\mathbf{E})$ is dense in $\mathbf{L}^p(\mathbf{E})$.*

Proof. Let $g \in \mathbf{L}^p(\mathbf{E})$.

1. Let $p = \infty$. Then there is $\mathbf{E}_0 \subset \mathbf{E}$ such that $m(\mathbf{E}_0) = 0$ and g is bounded on $\mathbf{E} \setminus \mathbf{E}_0$. By Theorem 2.46, it follows that there is a sequence $\{f_n\}_{n\in\mathbf{N}}$ of simple functions on $\mathbf{E} \setminus \mathbf{E}_0$ that converges uniformly on $\mathbf{E} \setminus \mathbf{E}_0$ to g and therefore with respect to the $\mathbf{L}^\infty(\mathbf{E})$ norm. Consequently the subspace of simple functions in $\mathbf{L}^\infty(\mathbf{E})$ is dense in $\mathbf{L}^\infty(\mathbf{E})$.
2. Let $1 \le p < \infty$. By Theorem 2.47, it follows that there is a sequence of simple functions $\{f_n\}_{n\in\mathbf{N}}$ on $\mathbf{E}$ such that $f_n \to g$, as $n \to \infty$, pointwise on $\mathbf{E}$ and

$$|f_n| \le |g| \quad \text{on} \quad \mathbf{E}$$

for any $n \in \mathbf{N}$. Since $g \in \mathbf{L}^p(\mathbf{E})$, we have $f_n \in \mathbf{L}^p(\mathbf{E})$ for any $n \in \mathbf{N}$. Next,

$$|f_n - g| \le |f_n| + |g| \le 2|g|.$$

Hence, by Theorem 3.10, it follows that $f_n \to g$, as $n \to \infty$, in $\mathbf{L}^p(\mathbf{E})$. This completes the proof. □

Theorem 3.12. *Let $[a, b]$ be a closed bounded interval and $1 \le p < \infty$. Then the subspace of the step functions on $[a, b]$ is dense in $\mathbf{L}^p([a, b])$.*

Proof. Let $\mathbf{A}$ be a measurable subset of $[a, b]$. Let also $g = \kappa_{\mathbf{A}}$. Take $\epsilon > 0$ arbitrarily. By Theorem 2.23, it follows that there is a finite disjoint collection $\{\mathbf{I}_k\}_{k=1}^n$ of open intervals such that if $\mathbf{U} = \bigcup_{k=1}^n \mathbf{I}_k$, then

$$m(\mathbf{A} \setminus \mathbf{U}) + m(\mathbf{U} \setminus \mathbf{A}) < \epsilon^p.$$

Let $f = \kappa_{\mathbf{U}}$. Then

$$\|\kappa_{\mathbf{U}} - \kappa_{\mathbf{A}}\|_p = \left(\int_{\mathbf{E}} |\kappa_{\mathbf{U}} - \kappa_{\mathbf{A}}|^p\right)^{\frac{1}{p}} = (m(\mathbf{A} \setminus \mathbf{U}) + m(\mathbf{U} \setminus \mathbf{A}))^{\frac{1}{p}} < \epsilon.$$

Therefore the step functions are dense in the simple functions with respect to the $\mathbf{L}^p$ norm. Hence, by Theorem 3.11, it follows that the step functions on $[a,b]$ are dense in $\mathbf{L}^p([a,b])$. This completes the proof. □

Theorem 3.13. *Let* $\mathbf{E}$ *be a measurable set and* $1 \le p < \infty$. *Then* $\mathbf{L}^p(\mathbf{E})$ *is separable.*

Proof. Let $[a,b]$ be a closed bounded interval and $\mathbf{S}([a,b])$ be the collection of the step functions on $[a,b]$. Let also $\mathbf{S}^1([a,b])$ be the subcollection of the collection $\mathbf{S}([a,b])$ consisting of the step functions ψ on $[a,b]$ that take rational values and for which there is a partition $\mathbf{P} = \{x_0, \ldots, x_n\}$ of $[a,b]$ so that ψ is a rational constant on (x_{k-1}, x_k), $1 \le k \le n$, and x_k, $1 \le k \le n-1$, are rational numbers. Using the density of the rational numbers in the real numbers, we see that $\mathbf{S}^1([a,b])$ is dense in $\mathbf{S}([a,b])$ with respect to the $\mathbf{L}^p(\mathbf{E})$ norm. Because the set of the rational numbers is countable, we see that $\mathbf{S}^1([a,b])$ is a countable set. We have

$$\mathbf{S}^1([a,b]) \subseteq \mathbf{S}([a,b]) \subseteq \mathbf{L}^p([a,b]).$$

Since $\mathbf{S}^1([a,b])$ is dense in $\mathbf{S}([a,b])$, using Theorem 3.12, it follows that $\mathbf{S}^1([a,b])$ is dense in $\mathbf{L}^p([a,b])$. For each $n \in \mathbf{N}$, we define $\mathcal{F}_n$ to be the collection of the functions that vanishes outside $[-n,n]$ and whose restrictions to $[-n,n]$ belong to $\mathbf{S}^1([-n,n])$. Let

$$\mathcal{F} = \bigcup_{n=1}^{\infty} \mathcal{F}_n.$$

Note that $\mathcal{F}$ is a countable collection of the functions that is dense in $\mathbf{L}^p(\mathbf{R})$. Also, using Theorem 2.80, we have

$$\lim_{n\to\infty} \int_{[-n,n]} |f|^p = \int_{\mathbf{R}} |f|^p$$

for all $f \in \mathbf{L}^p(\mathbf{R})$. Using the definition of $\mathcal{F}$, we conclude that $\mathcal{F}$ is a countable collection of functions that is dense in $\mathbf{L}^p(\mathbf{R})$. Hence, the collection of the restrictions on $\mathbf{E}$ of functions in $\mathcal{F}$ is a countable dense set of $\mathbf{L}^p(\mathbf{E})$. Consequently $\mathbf{L}^p(\mathbf{E})$ is separable. This completes the proof. □

3.6 Duality

Definition 3.8. For a normed vector space $\mathbf{X}$, a linear functional $\mathbb{T}$ on $\mathbf{X}$ is said to be bounded if there is an $M \ge 0$ such that

$$|\mathbb{T}(f)| \le M\|f\| \tag{3.7}$$

for all $f \in \mathbf{X}$. The infimum of all such M will be called the norm of $\mathbb{T}$ and will be denoted by $\|\mathbb{T}\|_\star$.

Let $\mathbb{T}$ be a bounded linear functional on the normed vector space $\mathbf{X}$. Then, for any $f, g \in \mathbf{X}$, we have

$$|\mathbb{T}(f) - \mathbb{T}(g)| \leq M\|f - g\|. \tag{3.8}$$

Hence, if $f_n \to f$, as $n \to \infty$, in $\mathbf{X}$, i. e., $f_n, f \in \mathbf{X}$, $n \in \mathbf{N}$,

$$\|f_n - f\| \to 0 \quad \text{as} \quad n \to \infty,$$

using (3.8), we get

$$\mathbb{T}(f_n) \to \mathbb{T}(f) \quad \text{as} \quad n \to \infty.$$

Proposition 3.1. *Let $\mathbb{T}$ be a bounded linear functional on the normed vector space $\mathbf{X}$. Then*

$$\|\mathbb{T}\|_\star = \sup_{\|f\|\leq 1} |\mathbb{T}(f)|. \tag{3.9}$$

Proof. By the inequality (3.7), we get

$$|\mathbb{T}(f)| \leq \|\mathbb{T}\|_\star \|f\|$$

for any $f \in \mathbf{X}$. Hence,

$$\sup_{\|f\|\leq 1} |\mathbb{T}(f)| \leq \|\mathbb{T}\|_\star. \tag{3.10}$$

Let $\epsilon > 0$ be arbitrarily chosen. Then there exists $g \in \mathbf{X}$, $g \neq 0$, such that

$$|\mathbb{T}(g)| \geq (\|\mathbb{T}\|_\star - \epsilon)\|g\|.$$

Hence,

$$\left|\mathbb{T}\left(\frac{g}{\|g\|}\right)\right| \geq \|\mathbb{T}\|_\star - \epsilon$$

and

$$\sup_{\|f\|\leq 1} |\mathbb{T}(f)| \geq \|\mathbb{T}\|_\star - \epsilon.$$

Because $\epsilon > 0$ was arbitrarily chosen, from the previous inequality, we obtain

$$\sup_{\|f\|\leq 1} |\mathbb{T}(f)| \geq \|\mathbb{T}\|_\star.$$

Hence, by (3.10), we get (3.9). This completes the proof. □

Theorem 3.14. *Let* $\mathbf{E}$ *be a measurable set,* $1 \le p < \infty$, $\frac{1}{p} + \frac{1}{q} = 1$, $g \in \mathbf{L}^q(\mathbf{E})$, $\|g\|_q \neq 0$. *Define the functional* $\mathbb{T}$ *on* $\mathbf{L}^p(\mathbf{E})$ *by*

$$\mathbb{T}(f) = \int_{\mathbf{E}} gf$$

for all $f \in \mathbf{L}^p(\mathbf{E})$. *Then* $\mathbb{T}$ *is a bounded linear functional on* $\mathbf{L}^p(\mathbf{E})$ *and* $\|\mathbb{T}\|_\star = \|g\|_q$.

Proof. Let $f_1, f_2 \in \mathbf{L}^p(\mathbf{E})$ and $\alpha, \beta \in \mathbf{F}$. Then

$$\begin{aligned}\mathbb{T}(\alpha f_1 + \beta f_2) &= \int_{\mathbf{E}} g(\alpha f_1 + \beta f_2) = \int_{\mathbf{E}} (\alpha g f_1 + \beta g f_2) = \int_{\mathbf{E}} \alpha g f_1 + \int_{\mathbf{E}} \beta g f_2 \\ &= \alpha \int_{\mathbf{E}} g f_1 + \beta \int_{\mathbf{E}} g f_2 = \alpha \mathbb{T}(f_1) + \beta \mathbb{T}(f_2).\end{aligned}$$

Therefore $\mathbb{T}$ is a linear functional on $\mathbf{L}^p(\mathbf{E})$. Also, using Hölder's inequality, we have

$$\begin{aligned}|\mathbb{T}(f)| &= \left|\int_{\mathbf{E}} gf\right| \le \int_{\mathbf{E}} |g||f| \\ &\le \left(\int_{\mathbf{E}} |g|^q\right)^{\frac{1}{q}} \left(\int_{\mathbf{E}} |f|^p\right)^{\frac{1}{p}} = \|g\|_q \|f\|_p\end{aligned}$$

for all $f \in \mathbf{L}^p(\mathbf{E})$. Consequently $\mathbb{T}$ is a bounded linear functional on $\mathbf{L}^p(\mathbf{E})$. By the previous inequality, we get

$$\|\mathbb{T}\|_\star \le \|g\|_q. \tag{3.11}$$

Let

$$g_1 = \|g\|_q^{1-q} \operatorname{sign}(g) |g|^{q-1}.$$

We have

$$\int_{\mathbf{E}} |g_1|^p = \int_{\mathbf{E}} \|g\|_q^{p(1-q)} |g|^{p(q-1)} = \|g\|_q^{-q} \int_{\mathbf{E}} |g|^q = 1,$$

i. e., $g_1 \in \mathbf{L}^p(\mathbf{E})$ and $\|g_1\|_p = 1$. Next,

$$\mathbb{T}(g_1) = \int_{\mathbf{E}} g g_1 = \int_{\mathbf{E}} g \|g\|_q^{1-q} \operatorname{sign}(g) |g|^{q-1} = \|g\|_q^{1-q} \int_{\mathbf{E}} |g|^q = \|g\|_q.$$

Therefore

$$\|\mathbb{T}\|_\star \ge \|g\|_q.$$

From the previous inequality and from (3.11), we obtain $\|\mathbb{T}\|_\star = \|g\|_q$. This completes the proof. □

Theorem 3.15. *Let $\mathbb{T}$ and $\mathbb{S}$ be bounded linear functionals on a normed vector space* $\mathbf{X}$. *If $\mathbb{T} = \mathbb{S}$ on a dense subset $\mathbf{X}_0$ of $\mathbf{X}$, then $\mathbb{T} = \mathbb{S}$ on $\mathbf{X}$.*

Proof. Let $g \in \mathbf{X}$ be arbitrarily chosen. Then there exists a sequence $\{g_n\}_{n\in\mathbf{N}}$ of elements of $\mathbf{X}_0$ such that $g_n \to g$, as $n \to \infty$, in $\mathbf{X}$. Hence,

$$\mathbb{T}(g_n) \to \mathbb{T}(g), \quad \mathbb{S}(g_n) \to \mathbb{S}(g), \quad \text{as} \quad n \to \infty,$$

and

$$\mathbb{T}(g_n) = \mathbb{S}(g_n)$$

for any $n \in \mathbf{N}$. Therefore $\mathbb{T}(g) = \mathbb{S}(g)$. Because $g \in \mathbf{X}$ was arbitrarily chosen, we obtain $\mathbb{T} = \mathbb{S}$ on $\mathbf{X}$. This completes the proof. □

Theorem 3.16. *Let $\mathbf{E}$ be a measurable set and $1 \le p < \infty$, $\frac{1}{p} + \frac{1}{q} = 1$, g is integrable over $\mathbf{E}$ and there is an $M \ge 0$ such that*

$$\left|\int_{\mathbf{E}} gf\right| \le M\|f\|_p \tag{3.12}$$

for any simple function f in $\mathbf{L}^p(\mathbf{E})$. Then $g \in \mathbf{L}^q(\mathbf{E})$ and $\|g\|_q \le M$.

Proof. Since g is integrable over $\mathbf{E}$, then it is finite a. e. on $\mathbf{E}$. By excising a set of measure zero from $\mathbf{E}$, we can assume that g is finite on all of $\mathbf{E}$.

1. Let $p > 1$. Because $|g|$ is a nonnegative measurable function on $\mathbf{E}$, by Theorem 2.47, there exists a sequence $\{\phi_n\}_{n\in\mathbf{N}}$ of measurable simple functions on $\mathbf{E}$ that converges pointwise on $\mathbf{E}$ to $|g|$ and $0 \le \phi_n \le |g|$ on $\mathbf{E}$. Hence, $\{\phi_n^q\}_{n\in\mathbf{N}}$ is a sequence of nonnegative simple functions on $\mathbf{E}$ such that

   $$0 \le \phi_n^q \le |g|^q \quad \text{on} \quad \mathbf{E}$$

   and $\phi_n^q \to |g|^q$, as $n \to \infty$, pointwise on $\mathbf{E}$. Hence, by Fatou's lemma,

   $$\int_{\mathbf{E}} |g|^q \le \int_{\mathbf{E}} \phi_n^q \tag{3.13}$$

   for every $n \in \mathbf{N}$. Let $n \in \mathbf{N}$ be arbitrarily chosen. Then

   $$\phi_n^q = \phi_n\phi_n^{q-1} \le |g|\phi_n^{q-1} = g\,\text{sign}(g)\phi_n^{q-1} \quad \text{on} \quad \mathbf{E}.$$

   Let

   $$f_n = \text{sign}(g)\phi_n^{q-1} \quad \text{on} \quad \mathbf{E}.$$

Then f_n is a simple function. Because g is integrable over $\mathbf{E}$, we see that ϕ_n is integrable over $\mathbf{E}$. Then ϕ_n^q is integrable over $\mathbf{E}$ and

$$\int_{\mathbf{E}} |f_n|^p = \int_{\mathbf{E}} \phi_n^q.$$

Therefore $f_n \in \mathbf{L}^p(\mathbf{E})$. Next, by (3.12), we obtain

$$\begin{aligned}\int_{\mathbf{E}} \phi_n^q &= \int_{\mathbf{E}} \phi_n \phi_n^{q-1} \leq \int_{\mathbf{E}} |g| \phi_n^{q-1} = \int_{\mathbf{E}} g \operatorname{sign}(g) \phi_n^{q-1} \\ &= \int_{\mathbf{E}} g f_n \leq M \|f_n\|_p = M\Big(\int_{\mathbf{E}} \phi_n^q\Big)^{\frac{1}{p}}.\end{aligned}$$

From this,

$$\Big(\int_{\mathbf{E}} \phi_n^q\Big)^{\frac{1}{q}} \leq M,$$

i. e.,

$$\|\phi_n\|_q \leq M.$$

Hence, by (3.13), we obtain

$$\|g\|_q \leq M.$$

2. Let $p = 1$. Suppose that M is not an essential upper bound for g. Then there is an $\epsilon > 0$ such that the set

$$\mathbf{E}_\epsilon = \{x \in \mathbf{E} : |g(x)| > M + \epsilon\}$$

has finite positive measure. Let $f = \operatorname{sign}(g)\kappa_{\mathbf{E}_\epsilon}$. Then, by (3.12),

$$\Big|\int_{\mathbf{E}} fg\Big| = \Big|\int_{\mathbf{E}_\epsilon} g \operatorname{sign}(g)\Big| = \int_{\mathbf{E}_\epsilon} |g| > (M + \epsilon) m(\mathbf{E}_\epsilon). \tag{3.14}$$

On the other hand, by (3.12),

$$\Big|\int_{\mathbf{E}} fg\Big| \leq M\|f\|_1 = M \int_{\mathbf{E}} |\operatorname{sign}(g)\kappa_{\mathbf{E}_\epsilon}| = Mm(\mathbf{E}_\epsilon),$$

which contradicts (3.14). Therefore M is an essential upper bound for g. This completes the proof. □

Theorem 3.17. *Let $[a, b]$ be a closed bounded interval, $1 \le p < \infty$, $\frac{1}{p} + \frac{1}{q} = 1$. Suppose that $\mathbb{T}$ is a bounded linear functional on $\mathbf{L}^p([a, b])$. Then there is a function $g \in \mathbf{L}^q([a, b])$ such that*

$$\mathbb{T}(f) = \int_a^b gf$$

for all $f \in \mathbf{L}^p([a, b])$.

Proof. Let $p > 1$. We leave the case $p = \infty$ as an exercise. For $x \in [a, b]$ we define

$$\Phi(x) = \mathbb{T}(\kappa_{[a,x)}).$$

For $[c, d] \subseteq [a, b]$, we have

$$\kappa_{[c,d)} = \kappa_{[a,d)} - \kappa_{[a,c)}.$$

Then

$$\Phi(d) - \Phi(c) = \mathbb{T}(\kappa_{[a,d)}) - \mathbb{T}(\kappa_{[a,c)}) = \mathbb{T}(\kappa_{[a,d)} - \kappa_{[a,c)}) = \mathbb{T}(\kappa_{[c,d)}).$$

If $\{(a_k, b_k)\}_{k=1}^n$ is a finite disjoint collection of intervals in (a, b), then

$$\begin{aligned}\sum_{k=1}^n |\Phi(b_k) - \Phi(a_k)| &= \sum_{k=1}^n \operatorname{sign}(\Phi(b_k) - \Phi(a_k))(\Phi(b_k) - \Phi(a_k)) \\ &= \sum_{k=1}^n \operatorname{sign}(\Phi(b_k) - \Phi(a_k))\mathbb{T}(\kappa_{[a_k,b_k)}) \\ &= \sum_{k=1}^n \mathbb{T}(\operatorname{sign}(\Phi(b_k) - \Phi(a_k))\kappa_{[a_k,b_k)}) \\ &= \mathbb{T}\left(\sum_{k=1}^n \operatorname{sign}(\Phi(b_k) - \Phi(a_k))\kappa_{[a_k,b_k)}\right).\end{aligned}$$

Consider the simple function

$$f = \sum_{k=1}^n \operatorname{sign}(\Phi(b_k) - \Phi(a_k))\kappa_{[a_k,b_k)}.$$

Then

$$|\mathbb{T}(f)| \le \|\mathbb{T}\|_\star \|f\|_p = \|\mathbb{T}\|_\star \left(\sum_{k=1}^n (b_k - a_k)\right)^{\frac{1}{p}}.$$

Consequently

$$\sum_{k=1}^n |\Phi(b_k) - \Phi(a_k)| \le \|\mathbb{T}\|_\star \left(\sum_{k=1}^n (b_k - a_k)\right)^{\frac{1}{p}}$$

and Φ is absolutely continuous on $[a,b]$. Therefore Φ is differentiable a. e. on $[a,b]$, and if $g = \Phi'$, then g is integrable over $[a,b]$ and

$$\Phi(x) = \int_a^x g$$

for all $x \in [a,b]$. Therefore, for each $[c,d] \subseteq (a,b)$,

$$\mathbb{T}(\kappa_{[c,d)}) = \Phi(d) - \Phi(c) = \int_a^b g\kappa_{[c,d)}.$$

Since the functional $\mathbb{T}$ and the functional $f \longmapsto \int_a^b gf$ are linear on the vector space of step functions in $\mathbf{L}^p([a,b])$, it follows that

$$\mathbb{T}(f) = \int_a^b gf$$

for all step functions f in $\mathbf{L}^p([a,b])$. By Theorem 3.12, it follows that there is a sequence $\{\phi_n\}_{n\in\mathbf{N}}$ of step functions that converges to f in $\mathbf{L}^p([a,b])$ and also it is uniformly pointwise bounded on $[a,b]$. Hence, by (3.8), we get

$$\lim_{n\to\infty} \mathbb{T}(\phi_n) = \mathbb{T}(f).$$

On the other hand, by the Lebesgue dominated convergence theorem, we obtain

$$\lim_{n\to\infty} \int_a^b g\phi_n = \int_a^b gf.$$

Therefore

$$\mathbb{T}(f) = \int_a^b gf$$

for all simple functions f in $\mathbf{L}^p([a,b])$. Since $\mathbb{T}$ is bounded,

$$\left|\int_a^b gf\right| = |\mathbb{T}(f)| \le \|\mathbb{T}\|_\star \|f\|_p$$

for all simple functions f in $\mathbf{L}^p([a,b])$. From this and from Theorem 3.16, we have $g \in \mathbf{L}^q([a,b])$. By Theorem 3.14, the functional $f \longmapsto \int_a^b gf$ is bounded on $\mathbf{L}^p([a,b])$. This functional agrees with $\mathbb{T}$ on the simple functions in $\mathbf{L}^p([a,b])$. Because the set of the simple functions in $\mathbf{L}^p([a,b])$ is dense in $\mathbf{L}^p([a,b])$ (see Theorem 3.11), using Theorem 3.15, these two functionals agree on all of $\mathbf{L}^p([a,b])$. This completes the proof. □

Theorem 3.18 (The Riesz representation theorem). *Let* $\mathbf{E}$ *be a measurable set,* $1 \le p < \infty$, $\frac{1}{p} + \frac{1}{q} = 1$. *For each* $g \in \mathbf{L}^p(\mathbf{E})$, *define the bounded linear functional* $\mathbb{T}_g$ *on* $\mathbf{L}^p(\mathbf{E})$ *by*

$$\mathbb{T}_g(f) = \int_{\mathbf{E}} gf$$

for all $f \in \mathbf{L}^p(\mathbf{E})$. *Then, for each bounded linear functional* $\mathbb{T}$ *on* $\mathbf{L}^p(\mathbf{E})$, *there is unique function* $g \in \mathbf{L}^q(\mathbf{E})$ *for which*

$$\mathbb{T}_g = \mathbb{T} \quad \textit{and} \quad \|\mathbb{T}\|_\star = \|g\|_q.$$

Proof. By Theorem 3.14, it follows that $\mathbb{T}_g$ is a bounded linear functional on $\mathbf{L}^p(\mathbf{E})$ and $\|\mathbb{T}_g\|_\star = \|g\|_q$ for each $g \in \mathbf{L}^p(\mathbf{E})$. Also, if $g_1, g_2 \in \mathbf{L}^q(\mathbf{E})$, then

$$\mathbb{T}_{g_1-g_2}(f) = \int_{\mathbf{E}} (g_1 - g_2)f = \int_{\mathbf{E}} (g_1 f - g_2 f) = \int_{\mathbf{E}} g_1 f - \int_{\mathbf{E}} g_2 f = \mathbb{T}_{g_1}(f) - \mathbb{T}_{g_2}(f)$$

for any $f \in \mathbf{L}^p(\mathbf{E})$. Therefore, if $\mathbb{T}_{g_1} = \mathbb{T}_{g_2}$, then $\mathbb{T}_{g_1-g_2} = 0$, and hence, $\|g_1 - g_2\|_q = 0$, so that $g_1 = g_2$. Therefore, for a bounded linear functional $\mathbb{T}$ on $\mathbf{L}^p(\mathbf{E})$ there is at most one $g \in \mathbf{L}^p(\mathbf{E})$ for which $\mathbb{T}_g = \mathbb{T}$.

1. Suppose that $\mathbf{E} = \mathbf{R}$. Let $\mathbb{T}$ be a bounded linear functional on $\mathbf{L}^p(\mathbf{R})$. For any $n \in \mathbf{N}$, we define the linear functional $\mathbb{T}_n$ on $\mathbf{L}^p([-n, n])$ by

   $$\mathbb{T}_n(f) = \mathbb{T}(\hat{f})$$

   for all $f \in \mathbf{L}^p([-n, n])$, where $\hat{f}$ is the extension of f to all of $\mathbf{R}$ that vanishes outside $[-n, n]$. Then

   $$\|f\|_p = \|\hat{f}\|_p$$

   and

   $$|\mathbb{T}_n(f)| = |\mathbb{T}(\hat{f})| \le \|\mathbb{T}\|_\star \|\hat{f}\|_p = \|\mathbb{T}\|_\star \|f\|_p$$

   for any $f \in \mathbf{L}^p([-n, n])$. Hence,

   $$\|\mathbb{T}_n\|_\star \le \|\mathbb{T}\|_\star.$$

   By Theorem 3.17, it follows that there is a function $g_n \in \mathbf{L}^q([-n, n])$ for which

   $$\mathbb{T}_n(f) = \int_{\mathbf{E}} g_n f$$

   for all $f \in \mathbf{L}^p([-n, n])$ and

   $$\|g_n\|_q = \|\mathbb{T}_n\|_\star \le \|\mathbb{T}\|_\star. \tag{3.15}$$

Note that the restriction of g_{n+1} to $[-n,n]$ agrees with g_n a. e. on $[-n,n]$. Define g to be a measurable function on $\mathbf{R}$ that agrees with g_n a. e. on $[-n,n]$ for each $n \in \mathbf{N}$. Hence, for all $f \in \mathbf{L}^p(\mathbf{R})$ that vanish outside a bounded set,

$$\mathbb{T}(f) = \int_{\mathbf{R}} gf.$$

By (3.15),

$$\int_{-n}^{n} |g|^q \le \|\mathbb{T}\|_\star^q.$$

Because the set of all functions of $\mathbf{L}^p(\mathbf{E})$ that vanishes outside a bounded set is dense in $\mathbf{L}^p(\mathbf{R})$, using Theorem 3.15, we conclude that $\mathbb{T}_g$ agrees with $\mathbb{T}$ on all $\mathbf{L}^p(\mathbf{R})$.

2. Let $\mathbf{E}$ be a measurable set and $\mathbb{T}$ be a bounded linear functional on $\mathbf{L}^p(\mathbf{E})$. Define the linear functional $\hat{\mathbb{T}}$ on $\mathbf{L}^p(\mathbf{E})$ by

$$\hat{\mathbb{T}}(f) = \mathbb{T}(f|_{\mathbf{E}}), \quad f \in \mathbf{L}^p(\mathbf{R}).$$

Then $\hat{\mathbb{T}}$ is a bounded linear functional on $\mathbf{L}^p(\mathbf{R})$. Hence, there is a function $\hat{g} \in \mathbf{L}^q(\mathbf{R})$ for which

$$\hat{\mathbb{T}}(f) = \int_{\mathbf{R}} \hat{g}f$$

for any $f \in \mathbf{L}^p(\mathbf{R})$. Define $g = \hat{g}|_{\mathbf{E}}$. Then $\mathbb{T} = \mathbb{T}_g$. This completes the proof. □

Definition 3.9. Let $1 \le p \le \infty$ and q is its conjugate. The space $\mathbf{L}^q(\cdot)$ is called the dual space of the space $\mathbf{L}^p(\cdot)$.

Definition 3.10. Let $\mathbf{X}$ be a normed vector space. A sequence $\{f_n\}_{n\in\mathbf{N}}$ in $\mathbf{X}$ is said to converge weakly in $\mathbf{X}$ to $f \in \mathbf{X}$ if

$$\lim_{n\to\infty} \mathbb{T}(f_n) = \mathbb{T}(f)$$

for any linear functional $\mathbb{T}$ on $\mathbf{X}$. We will write

$$f_n \rightharpoonup f \quad \text{in} \quad \mathbf{X}.$$

Remark 3.3. If $\mathbf{X}$ is a normed vector space, we will write $f_n \to f$ in $\mathbf{X}$ if

$$\|f_n - f\| \to 0 \quad \text{as} \quad n \to \infty.$$

In this case, we will say that the sequence $\{f_n\}_{n\in\mathbf{N}}$ converges strongly to f in $\mathbf{X}$.

Theorem 3.19. *Let* $\mathbf{X}$ *be a normed vector space,* $\{f_n\}_{n\in\mathbf{N}}$ *be a sequence in* $\mathbf{X}$*,* $f \in \mathbf{X}$*. If* $f_n \to f$ *in* $\mathbf{X}$*, then* $f_n \rightharpoonup f$ *in* $\mathbf{X}$*.*

Proof. Since $f_n \to f$ in $\mathbf{X}$, we have

$$\|f_n - f\| \to 0 \quad \text{as} \quad n \to \infty.$$

Let $\mathbb{T}$ be arbitrarily chosen linear functional on $\mathbf{X}$. Then

$$|\mathbb{T}(f_n) - \mathbb{T}(f)| \leq \|\mathbb{T}\|_\star \|f_n - f\| \to 0 \quad \text{as} \quad n \to \infty.$$

Because $\mathbb{T}$ was arbitrarily chosen linear functional on $\mathbf{X}$, we conclude that $f_n \rightharpoonup f$ in $\mathbf{X}$. This completes the proof. □

Theorem 3.20. *Let* $\mathbf{E}$ *be a measurable set,* $1 \leq p < \infty$ *and* q *is its conjugate. Then* $f_n \rightharpoonup f$ *in* $\mathbf{L}^p(\mathbf{E})$ *if and only if*

$$\lim_{n\to\infty} \int_{\mathbf{E}} gf_n = \int_{\mathbf{E}} gf \tag{3.16}$$

for all $g \in \mathbf{L}^q(\mathbf{E})$*.*

Proof.
1. Let $f_n \rightharpoonup f$ in $\mathbf{L}^p(\mathbf{E})$. Then for every linear functional $\mathbb{T}$ we have
$$\lim_{n\to\infty} \mathbb{T}(f_n) = \mathbb{T}(f).$$
Hence, using $h \longmapsto \int_{\mathbf{E}} gh$, $h \in \mathbf{L}^p(\mathbf{E})$, is a linear functional for each $g \in \mathbf{L}^q(\mathbf{E})$, we get (3.16).
2. Assume that (3.16) holds. Let $\mathbb{T}$ be arbitrarily chosen linear functional on $\mathbf{L}^p(\mathbf{E})$. By the Riesz representation theorem, it follows that for any $h \in \mathbf{L}^p(\mathbf{E})$ there is a unique $g \in \mathbf{L}^q(\mathbf{E})$ such that
$$\mathbb{T}(h) = \int_{\mathbf{E}} gh.$$
Hence, by (3.16), we obtain
$$\lim_{n\to\infty} \mathbb{T}(f_n) = \mathbb{T}(f).$$
Because $\mathbb{T}$ was arbitrarily chosen linear functional on $\mathbf{L}^p(\mathbf{E})$, we conclude that $f_n \rightharpoonup f$ in $\mathbf{L}^p(\mathbf{E})$. This completes the proof. □

Theorem 3.21. *Let* $\mathbf{E}$ *be a measurable set,* $1 \leq p < \infty$*. Then a sequence in* $\mathbf{L}^p(\mathbf{E})$ *can converge weakly to at most one function in* $\mathbf{L}^p(\mathbf{E})$*.*

Proof. Let q be the conjugate of p. Let also $\{f_n\}_{n\in\mathbf{N}}$ be a sequence in $\mathbf{L}^p(\mathbf{E})$ that converges weakly to $f_1, f_2 \in \mathbf{L}^p(\mathbf{E})$. Then $f_1 - f_2 \in \mathbf{L}^p(\mathbf{E})$ and

$$\begin{aligned}\int_{\mathbf{E}} \left| \|f_1 - f_2\|_p^{1-p} \operatorname{sign}(f_1 - f_2)|f_1 - f_2|^{p-1}\right|^q &\le \|f_1 - f_2\|_p^{(1-p)q} \int_{\mathbf{E}} |f_1 - f_2|^{q(p-1)} \\ &= \|f_1 - f_2\|_p^{(1-p)q} \int_{\mathbf{E}} |f_1 - f_2|^p = 1,\end{aligned}$$

i. e., $\|f_1 - f_2\|_p^{1-p} \operatorname{sign}(f_1 - f_2)|f_1 - f_2|^{p-1} \in \mathbf{L}^q(\mathbf{E})$. Hence, by Theorem 3.20, we get

$$\begin{aligned}\int_{\mathbf{E}} \|f_1 - f_2\|_p^{1-p} \operatorname{sign}(f_1 - f_2)|f_1 - f_2|^{p-1} f_1 &= \lim_{n\to\infty} \int_{\mathbf{E}} \|f_1 - f_2\|_p^{1-p} \operatorname{sign}(f_1 - f_2)|f_1 - f_2|^{p-1} f_n \\ &= \int_{\mathbf{E}} \|f_1 - f_2\|_p^{1-p} \operatorname{sign}(f_1 - f_2)|f_1 - f_2|^{p-1} f_2.\end{aligned}$$

Therefore

$$\begin{aligned}0 &= \int_{\mathbf{E}} \|f_1 - f_2\|_p^{1-p} \operatorname{sign}(f_1 - f_2)|f_1 - f_2|^{p-1}(f_1 - f_2) \\ &= \|f_1 - f_2\|_p^{1-p} \int_{\mathbf{E}} |f_1 - f_2|^p = \|f_1 - f_2\|_p.\end{aligned}$$

Consequently $f_1 = f_2$. This completes the proof. □

Definition 3.11. Let $\mathbf{E}$ be a measurable set and $1 \le p < \infty$.
1. Let $\{f_n\}_{n\in\mathbf{N}}$ be a sequence in $\mathbf{L}^p(\mathbf{E})$, $f \in \mathbf{L}^p(\mathbf{E})$ and $f_n \rightharpoonup f$ in $\mathbf{L}^p(\mathbf{E})$. The function f will be called the weak sequential limit.
2. Let $f \in \mathbf{L}^p(\mathbf{E})$. The function

$$f^\star = \|f\|_p^{1-p} \operatorname{sign}(f)|f|^{p-1}$$

will be called the conjugate function of f. Note that $f^\star \in \mathbf{L}^q(\mathbf{E})$, where q is the conjugate of p.

Exercise 3.1. Let $\mathbf{E}$ be a measurable set, $1 \le p < \infty$, $f \in \mathbf{L}^p(\mathbf{E})$ and $f^\star$ be the conjugate function of f. Prove that

$$\|f^*\|_q = 1,$$

where q is the conjugate of p.

Remark 3.4. Let $\mathbf{X}$ be a linear normed space and $f_n \rightharpoonup f$ in $\mathbf{X}$. Suppose that the sequence $\{\|f_n\|\}_{n\in\mathbf{N}}$ is unbounded. By taking a subsequence and relabeling, we can assume that $\|f_n\| \ge n3^n$. By taking a further subsequence and relabeling, we can assume that

$$\frac{\|f_n\|}{n3^n} \to \alpha \in [1,\infty), \quad \text{as} \quad n \to \infty.$$

Let $\mathbb{T}$ be arbitrarily chosen linear functional. Then there is a constant $M > 0$ such that

$$|\mathbb{T}(f_n)| \le M, \quad |\mathbb{T}(f)| \le M$$

for any $n \in \mathbf{N}$. Hence,

$$\begin{aligned}
\left|\mathbb{T}\left(\frac{n3^n}{\|f_n\|}f_n\right) - \mathbb{T}\left(\frac{1}{\alpha}f\right)\right| &= \left|\mathbb{T}\left(\frac{n3^n}{\|f_n\|}f_n\right) - \mathbb{T}\left(\frac{1}{\alpha}f_n\right) + \mathbb{T}\left(\frac{1}{\alpha}f_n\right) - \mathbb{T}\left(\frac{1}{\alpha}f\right)\right| \\
&= \left|\mathbb{T}\left(\left(\frac{n3^n}{\|f_n\|} - \frac{1}{\alpha}\right)f_n\right) + \mathbb{T}\left(\frac{1}{\alpha}(f_n - f)\right)\right| \\
&\le \left|\frac{n3^n}{\|f_n\|} - \frac{1}{\alpha}\right| |\mathbb{T}(f_n)| + \frac{1}{\alpha}|\mathbb{T}(f_n) - \mathbb{T}(f)| \\
&\le \left|\frac{n3^n}{\|f_n\|} - \frac{1}{\alpha}\right| M + \frac{1}{\alpha}|\mathbb{T}(f_n) - \mathbb{T}(f)| \to 0 \quad \text{as} \quad n \to \infty.
\end{aligned}$$

Therefore

$$\frac{n3^n}{\|f_n\|}f_n \rightharpoonup \frac{1}{\alpha}f \quad \text{in} \quad \mathbf{X}.$$

Theorem 3.22. *Let* $\mathbf{E}$ *be a measurable set and* $1 \le p < \infty$. *Suppose that* $f_n \rightharpoonup f$ *in* $\mathbf{L}^p(\mathbf{E})$. *Then* $\{f_n\}_{n\in\mathbf{N}}$ *is bounded in* $\mathbf{L}^p(\mathbf{E})$ *and* $\|f\|_p \le \liminf \|f_n\|_p$.

Proof. Let q be the conjugate of p and $f^\star$ be the conjugate function of f. Then $f^\star \in \mathbf{L}^q(\mathbf{E})$ and using Hölder's inequality, we get

$$\int_{\mathbf{E}} f^\star f_n \le \|f^\star\|_q \|f_n\|_p = \|f_n\|_p$$

for any $n \in \mathbf{N}$. Since $f_n \rightharpoonup f$ in $\mathbf{L}^p(\mathbf{E})$, by Theorem 3.20, we get

$$\|f\|_p = \int_{\mathbf{E}} f^\star f = \lim_{n\to\infty} \int_{\mathbf{E}} f^\star f_n \le \liminf \|f_n\|_p.$$

Now we assume that the sequence $\{f_n\}_{n\in\mathbf{N}}$ is unbounded. Using Remark 3.4 and by taking scalar multiples, we can suppose that $\|f_n\|_p = n3^n$, $n \in \mathbf{N}$. Let $f_k^\star$ be the conjugate functions of f_k, $k \in \mathbf{N}$. Let also $\epsilon_1 = \frac{1}{3}$ and we define

$$\epsilon_k = \frac{1}{3^k} \quad \text{if} \quad \int_{\mathbf{E}} \sum_{l=1}^{k-1} \epsilon_l f_l^\star f_k \ge 0$$

and

$$\epsilon_k = -\frac{1}{3^k} \quad \text{if} \quad \int_{\mathbf{E}} \sum_{l=1}^{k-1} \epsilon_l f_l^\star f_k \le 0$$

for $k \in \mathbf{N}$, $k \geq 2$. Then

$$\begin{aligned}\left|\int_{\mathbf{E}}\left(\sum_{k=1}^{n}\epsilon_k f_k^{\star}\right)f_n\right| &= \left|\int_{\mathbf{E}}\left(\sum_{k=1}^{n}\epsilon_k\|f_k\|_p^{1-p}\operatorname{sign}(f_k)|f_k|^{p-1}\right)f_n\right| \\ &= \left|\int_{\mathbf{E}}\sum_{k=1}^{n-1}\epsilon_k\|f_k\|_p^{1-p}\operatorname{sign}(f_k)|f_k|^{p-1}f_n + \int_{\mathbf{E}}\epsilon_n\|f_n\|_p^{1-p}|f_n|^p\right| \\ &= \left|\int_{\mathbf{E}}\sum_{k=1}^{n-1}\epsilon_k\|f_k\|_p^{1-p}\operatorname{sign}(f_k)|f_k|^{p-1}f_n + \epsilon_n\|f_n\|_p\right| \geq \frac{1}{3^n}\|f_n\|_p = n.\end{aligned}$$

Also,

$$\|\epsilon_k f_k^{\star}\|_q = \frac{1}{3^k}, \quad k \in \mathbf{N}.$$

Since the series $\sum_{k=1}^{\infty}\frac{1}{3^k}$ is a convergent series, we see that $\sum_{k=1}^{\infty}\epsilon_k f_k^{\star}$ is a convergent series in $\mathbf{L}^q(\mathbf{E})$ and let

$$g = \sum_{k=1}^{\infty}\epsilon_k f_k^{\star}.$$

For any $n \in \mathbf{N}$, we have

$$\begin{aligned}\left|\int_{\mathbf{E}} g f_n\right| &= \left|\int_{\mathbf{E}}\left(\sum_{k=1}^{\infty}\epsilon_k f_k^{\star}\right)f_n\right| \\ &= \left|\int_{\mathbf{E}}\left(\sum_{k=1}^{n}\epsilon_k f_k^{\star}\right)f_n + \int_{\mathbf{E}}\left(\sum_{k=n+1}^{\infty}\epsilon_k f_k^{\star}\right)f_n\right| \\ &\geq \left|\int_{\mathbf{E}}\left(\sum_{k=1}^{n}\epsilon_k f_k^{\star}\right)f_n\right| - \left|\int_{\mathbf{E}}\left(\sum_{k=n+1}^{\infty}\epsilon_k f_k^{\star}\right)f_n\right| \\ &\geq n - \left\|\sum_{k=n+1}^{\infty}\epsilon_k f_k^{\star}\right\|_q \|f_n\|_p \\ &\geq n - \sum_{k=n+1}^{\infty}\|\epsilon_k f_k^{\star}\|_q\|f_n\|_p = n - \sum_{k=n+1}^{\infty}\frac{1}{3^k}\|f_n\|_p = n - \frac{1}{2(3^n)}\|f_n\|_p = \frac{n}{2},\end{aligned}$$

which is a contradiction, because $f_n \rightharpoonup f$ in $\mathbf{L}^p(\mathbf{E})$ and $g \in \mathbf{L}^q(\mathbf{E})$. This completes the proof. □

Theorem 3.23. *Let* $\mathbf{E}$ *be a measurable set,* $1 \leq p < \infty$ *and* q *is its conjugate. Suppose that* $f_n \rightharpoonup f$ *in* $\mathbf{L}^p(\mathbf{E})$ *and* $g_n \to g$ *strongly in* $\mathbf{L}^q(\mathbf{E})$*. Then*

$$\lim_{n\to\infty}\int_{\mathbf{E}} g_n f_n = \int_{\mathbf{E}} g f. \tag{3.17}$$

Proof. We have

$$\int_{\mathbf{E}} g_n f_n - \int_{\mathbf{E}} gf = \int_{\mathbf{E}} (g_n - g) f_n + \int_{\mathbf{E}} g(f_n - f). \tag{3.18}$$

By Theorem 3.22, it follows that there is a constant $M > 0$ such that $\|f_n\|_p \le M$ for any $n \in \mathbf{N}$. Then, using Hölder's inequality, we obtain

$$\left|\int_{\mathbf{E}} (g_n - g) f_n\right| \le \int_{\mathbf{E}} |g_n - g||f_n| \le \|g_n - g\|_q \|f_n\|_p \le M\|g_n - g\|_q \to 0 \quad \text{as} \quad n \to \infty. \tag{3.19}$$

Since $f_n \rightharpoonup f$ in $\mathbf{L}^p(\mathbf{E})$ and $g \in \mathbf{L}^q(\mathbf{E})$, using Theorem 3.20, we obtain

$$\int_{\mathbf{E}} g(f_n - f) \to 0 \quad \text{as} \quad n \to \infty.$$

Hence, by (3.18) and (3.19), we obtain (3.17). This completes the proof. □

Theorem 3.24. *Let* $\mathbf{E}$ *be a measurable set,* $1 \le p < \infty$ *and* q *be its conjugate. Let also* $\mathcal{F} \subset \mathbf{L}^q(\mathbf{E})$ *and its span is dense in* $\mathbf{L}^q(\mathbf{E})$*. Suppose that* $\{f_n\}_{n \in \mathbf{N}}$ *is a bounded sequence in* $\mathbf{L}^p(\mathbf{E})$ *and* $f \in \mathbf{L}^p(\mathbf{E})$*. Then* $f_n \rightharpoonup f$ *in* $\mathbf{L}^p(\mathbf{E})$ *if and only if*

$$\lim_{n \to \infty} \int_{\mathbf{E}} f_n g = \int_{\mathbf{E}} fg \tag{3.20}$$

for all $g \in \mathcal{F}$.

Proof.
1. Let $f_n \rightharpoonup f$ in $\mathbf{L}^p(\mathbf{E})$. Using Theorem 3.20, we conclude that (3.20) holds.
2. Suppose that (3.20) holds. Let $g_0 \in \mathbf{L}^q(\mathbf{E})$ is arbitrarily chosen. For any $g \in \mathbf{L}^q(\mathbf{E})$ and for any $n \in \mathbf{N}$, we have

$$\int_{\mathbf{E}} (f_n - f) g_0 = \int_{\mathbf{E}} (f_n - f)(g_0 - g) + \int_{\mathbf{E}} (f_n - f) g$$

and hence, using Hölder's inequality, we obtain

$$\begin{aligned} \left|\int_{\mathbf{E}} (f_n - f) g_0\right| &= \left|\int_{\mathbf{E}} (f_n - f)(g_0 - g) + \int_{\mathbf{E}} (f_n - f) g\right| \\ &\le \left|\int_{\mathbf{E}} (f_n - f)(g_0 - g)\right| + \left|\int_{\mathbf{E}} (f_n - f) g\right| \\ &\le \|f_n - f\|_p \|g_0 - g\|_q + \left|\int_{\mathbf{E}} (f_n - f) g\right|. \end{aligned} \tag{3.21}$$

We take $\epsilon > 0$ arbitrarily. Since $\{f_n\}_{n\in\mathbf{N}}$ is bounded in $\mathbf{L}^p(\mathbf{E})$ and the span of $\mathcal{F}$ is dense in $\mathbf{L}^q(\mathbf{E})$, there is $g \in \mathcal{F}$ such that

$$\|f_n - f\|_p \|g - g_0\|_q < \frac{\epsilon}{2}$$

for any $n \in \mathbf{N}$. Hence, by (3.21), we conclude that

$$\int_{\mathbf{E}} f_n g_0 \to \int_{\mathbf{E}} f g_0 \quad \text{as} \quad n \to \infty.$$

Because $g_0 \in \mathbf{L}^q(\mathbf{E})$ was arbitrarily chosen, we obtain (3.16). Therefore $f_n \rightharpoonup f$ in $\mathbf{L}^p(\mathbf{E})$. This completes the proof. □

3.7 General $\mathbf{L}^p$ spaces

Let $(\mathbf{X}, \mathcal{M}, \mu)$ be a measure space. $\mathcal{F}$ be the collection of all measurable extended real-valued functions on $\mathbf{X}$ that are finite a. e. on $\mathbf{X}$.

Definition 3.12. Define two functions f and g in $\mathcal{F}$ to be equivalent, and we write $f \sim g$, provided

$$f(x) = g(x) \quad \text{for} \quad \text{almost} \quad \text{all} \quad x \in \mathbf{X}.$$

The relation $\sim$ is an equivalent relation, that is, it is reflexive, symmetric and transitive. Therefore it induces a partition of $\mathcal{F}$ into a disjoint collection of equivalence classes, which we denote by $\mathcal{F}/\sim$. For given two functions f and g in $\mathcal{F}$, their equivalence classes $[f]$ and $[g]$ and real numbers α and β, we define $\alpha[f] + \beta[g]$ to be the equivalence class of the functions in $\mathcal{F}$ that take the value $\alpha f(x) + \beta g(x)$ at points $x \in \mathbf{X}$ at which both f and g are finite. Note that these linear combinations are independent of the choice of the representatives of the equivalence classes. The zero element in $\mathcal{F}/\sim$ is the equivalence class of functions that vanish a. e. in $\mathbf{X}$. Thus $\mathcal{F}/\sim$ is a vector space.

Definition 3.13. For $1 \le p < \infty$, we define $\mathbf{L}^p(\mathbf{X}, \mu)$ to be the collection of equivalence classes $[f]$ for which

$$\int_{\mathbf{X}} |f|^p d\mu < \infty.$$

This is properly defined since, if $f \sim g$, then

$$\int_{\mathbf{X}} |f|^p d\mu = \int_{\mathbf{X}} |g|^p d\mu.$$

Note that, if $[f], [g] \in \mathbf{L}^p(\mathbf{X}, \mu)$, then, for any real constants α and β, we have

$$\int_{\mathbf{X}} |\alpha f + \beta g|^p \, d\mu \le 2^p \Big(\int_{\mathbf{X}} |\alpha f|^p \, d\mu + \int_{\mathbf{X}} |\beta g|^p \, d\mu \Big)$$
$$= |2\alpha|^p \int_{\mathbf{X}} |f|^p \, d\mu + |2\beta|^p \int_{\mathbf{X}} |g|^p \, d\mu < \infty,$$

i. e., $\alpha[f] + \beta[g] \in \mathbf{L}^p(\mathbf{X}, \mu)$.

Definition 3.14. We call a function $f \in \mathcal{F}$ essentially bounded provided there is some $M \ge 0$, called an essential upper bound for f, for which

$$|f(x)| \le M \quad \text{for} \quad \text{almost} \quad \text{all} \quad x \in \mathbf{X}.$$

Definition 3.15. We define $\mathbf{L}^\infty(\mathbf{X}, \mu)$ to be the collection of equivalence classes $[f]$ for which f is essentially bounded.

Note that $\mathbf{L}^\infty(\mathbf{X}, \mu)$ is properly defined since, if $f \sim g$, then

$$|f(x)| = |g(x)| \le M \quad \text{for} \quad \text{almost} \quad \text{all} \quad x \in \mathbf{X}.$$

Also, if $[f], [g] \in \mathbf{L}^\infty(\mathbf{X}, \mu)$ and $\alpha, \beta \in \mathbf{R}$, there are nonnegative constants M_1 and M_2 such that

$$|f(x)| \le M_1, \quad |g(x)| \le M_2 \quad \text{for} \quad \text{almost} \quad \text{all} \quad x \in \mathbf{X}.$$

Hence,

$$|\alpha f(x) + \beta g(x)| \le |\alpha| |f(x)| + |\beta| |g(x)| \le |\alpha| M_1 + |\beta| M_2$$

for almost all $x \in \mathbf{X}$. Therefore $\alpha[f] + \beta[g] \in \mathbf{L}^\infty(\mathbf{X}, \mu)$ and hence, $\mathbf{L}^\infty(\mathbf{X}, \mu)$ is a vector space. For simplicity and convenience, we refer to the equivalence classes in $\mathcal{F}/\sim$ as functions and denote them by f rather than $[f]$. Thus $f = g$ means $f - g$ vanishes a. e. on $\mathbf{X}$.

Definition 3.16. For $1 \le p < \infty$, in $\mathbf{L}^p(\mathbf{X}, \mu)$ we define

$$\|f\|_p = \Big(\int_{\mathbf{X}} |f|^p \Big)^{\frac{1}{p}}.$$

For $p = \infty$, we define $\|f\|_\infty$ to be the infimum of the essential upper bounds for f.

Remark 3.5. Note that the idea for the proof of next assertions in this section is the same as the idea for the proof of the assertions in the previous sections in this chapter. Therefore we leave the proof of the next assertions in this section.

Theorem 3.25. *Let $(\mathbf{X}, \mathcal{M}, \mu)$ be a measure space, $1 \leq p < \infty$, and q be the conjugate of p. If $f \in \mathbf{L}^p(\mathbf{X}, \mu)$, $g \in \mathbf{L}^q(\mathbf{X}, \mu)$, then $fg \in \mathbf{L}^1(\mathbf{X}, \mu)$ and*

$$\int_{\mathbf{X}} |fg| d\mu \leq \|f\|_p \|g\|_q.$$

Moreover, if $f \neq 0$, the function $f^\star = \|f\|_p^{1-p} \operatorname{sign}(f)|f|^{p-1} \in \mathbf{L}^q(\mathbf{X}, \mu)$,

$$\int_{\mathbf{X}} ff^\star d\mu = \|f\|_p \quad \text{and} \quad \|f^\star\|_q = 1.$$

Theorem 3.26. *Let $(\mathbf{X}, \mathcal{M}, \mu)$ be a finite measure space and $1 \leq p_1 < p_2$. Then $\mathbf{L}^{p_2}(\mathbf{X}, \mu) \subseteq \mathbf{L}^{p_1}(\mathbf{X}, \mu)$ and*

$$\|f\|_{p_1} \leq c\|f\|_{p_2} \quad \text{for} \quad f \in \mathbf{L}^{p_2}(\mathbf{X}, \mu),$$

where

$$c = \begin{cases} (\mu(\mathbf{X}))^{\frac{p_2 - p_1}{p_1 p_2}} & \text{if } p_2 < \infty, \\ (\mu(\mathbf{X}))^{\frac{1}{p_1}} & \text{if } p_2 = \infty. \end{cases}$$

Theorem 3.27. *Let $(\mathbf{X}, \mathcal{M}, \mu)$ be a measure space and $1 < p \leq \infty$. If $\{f_n\}_{n\in\mathbf{N}}$ is a bounded sequence of functions in $\mathbf{L}^p(\mathbf{X}, \mu)$, then $\{f_n\}_{n\in\mathbf{N}}$ is uniformly integrable over $\mathbf{X}$.*

Theorem 3.28. *Let $(\mathbf{X}, \mathcal{M}, \mu)$ be a measure space and $1 \leq p \leq \infty$. Then every rapidly Cauchy sequence in $\mathbf{L}^p(\mathbf{X}, \mu)$ converges to a function in $\mathbf{L}^p(\mathbf{X}, \mu)$, both with respect to the $\mathbf{L}^p(\mathbf{X}, \mu)$ norm and pointwise a. e. in $\mathbf{X}$.*

Theorem 3.29 (The Riesz–Fischer theorem). *Let $(\mathbf{X}, \mathcal{M}, \mu)$ be a measure space and $1 \leq p \leq \infty$. Then $\mathbf{L}^p(\mathbf{X}, \mu)$ is a Banach space. Moreover, if a sequence in $\mathbf{L}^p(\mathbf{X}, \mu)$ converges in $\mathbf{L}^p(\mathbf{X}, \mu)$ to a function $f \in \mathbf{L}^p(\mathbf{X}, \mu)$, then a subsequence converges pointwise a. e. on $\mathbf{X}$ to f.*

Proof. Let $\{f_n\}_{n\in\mathbf{N}}$ be a Cauchy sequence of $\mathbf{L}^p(\mathbf{X}, \mu)$. Then it has a rapidly Cauchy subsequence $\{f_{n_k}\}_{k\in\mathbf{N}}$. By Theorem 3.28, it follows that $\{f_{n_k}\}_{k\in\mathbf{N}}$ converges to a function in $\mathbf{L}^p(\mathbf{X}, \mu)$ both with respect to the $\mathbf{L}^p(\mathbf{X}, \mu)$ norm and pointwise a. e. on $\mathbf{X}$. This completes the proof. □

Theorem 3.30. *Let $(\mathbf{X}, \mathcal{M}, \mu)$ be a measure space and $1 \leq p < \infty$. Then the subspace of simple functions on $\mathbf{X}$ that vanish outside a set of finite measure is dense in $\mathbf{L}^p(\mathbf{X}, \mu)$.*

Theorem 3.31 (The Vitali $\mathbf{L}^p$ convergence theorem). *Let $(\mathbf{X}, \mathcal{M}, \mu)$ be a measure space and $1 \leq p < \infty$. Suppose that $\{f_n\}_{n\in\mathbf{N}}$ is a sequence in $\mathbf{L}^p(\mathbf{X}, \mu)$ that converges pointwise a. e. to $f \in \mathbf{L}^p(\mathbf{X}, \mu)$. Then $f_n \to f$ in $\mathbf{L}^p(\mathbf{X}, \mu)$, as $n \to \infty$, if and only if $\{|f_n|^p\}_{n\in\mathbf{N}}$ is uniformly integrable and tight.*

For $1 \leq p < \infty$, let $f \in \mathbf{L}^q(\mathbf{X}, \mu)$, where q is the conjugate of p. Define the linear functional $\mathbb{T}_f : \mathbf{L}^p(\mathbf{X}, \mu) \longmapsto \mathbf{R}$ by

$$\mathbb{T}_f(g) = \int_{\mathbf{X}} fg d\mu, \quad g \in \mathbf{L}^p(\mathbf{X}, \mu). \tag{3.22}$$

Theorem 3.32 (The Riesz representation theorem for the dual space of the space $\mathbf{L}^p(\mathbf{X}, \mu)$). *Let $(\mathbf{X}, \mathcal{M}, \mu)$ be a σ-finite measure space, $1 \leq p < \infty$, and q be the conjugate to p. For $f \in \mathbf{L}^q(\mathbf{X}, \mu)$, define $\mathbb{T}_f \in (\mathbf{L}^p(\mathbf{X}, \mu))^*$ by (3.22). Then $\mathbb{T}_f$ is an isometric isomorphism of $\mathbf{L}^q(\mathbf{X}, \mu)$ onto the space of the linear functionals on $\mathbf{L}^p(\mathbf{X}, \mu)$.*

Remark 3.6. Let $1 \leq p \leq \infty$. When it is clear from the context what measure is used, μ is omitted and one just writes $\mathbf{L}^p(\mathbf{X})$.

3.8 Advanced practical problems

Problem 3.1. Let $f \in \mathbf{L}^1([a, b])$ and define

$$\|f\| = \int_a^b x^2 |f(x)| dx.$$

Prove that this is a norm in $\mathbf{L}^1([a, b])$.

Problem 3.2. For $f \in \mathbf{L}^\infty([a, b])$, prove that

$$\|f\|_\infty = \min\{M : m\{x \in [a, b] : |f(x)| > M\} = 0\}.$$

Problem 3.3. Let

$$f(x) = \frac{x^{-\frac{1}{2}}}{1 + \log x}, \quad x > 1.$$

Prove that $f \in \mathbf{L}^p((1, \infty))$ if and only if $p = 2$.

Problem 3.4. Let $f(x) = \log(\frac{1}{x})$, $x \in (0, 1]$, $1 \leq p < \infty$. Prove that $f \in \mathbf{L}^p((0, 1])$ and $f \notin \mathbf{L}^\infty((0, 1])$.

Problem 3.5. Let $\mathbf{E}$ be a measurable set, $1 \leq p < \infty$ and q is the conjugate of p, $f \in \mathbf{L}^p(\mathbf{E})$. Prove that $f \equiv 0$ if and only if

$$\int_{\mathbf{E}} fg = 0$$

for any $g \in \mathbf{L}^q(\mathbf{E})$.

Problem 3.6. Let $\mathbf{E}$ be a measurable set of finite measure, $1 \le p_1 < p_2 \le \infty$. Prove that if $f_n \to f$ strongly in $\mathbf{L}^{p_2}(\mathbf{E})$, then $f_n \to f$ strongly in $\mathbf{L}^{p_1}(\mathbf{E})$.

Problem 3.7. Let $\mathbf{E}$ be a measurable set, $1 \le p < \infty$, q is the conjugate of p, $\mathbf{S}$ is dense in $\mathbf{L}^q(\mathbf{E})$. Prove that if $g \in \mathbf{L}^p(\mathbf{E})$ and $\int_{\mathbf{E}} fg = 0$ for any $f \in \mathbf{S}$, then $g = 0$.

Problem 3.8. Let $\mathbf{E}$ be a measurable set, $1 \le p < \infty$. Prove that the functions in $\mathbf{L}^p(\mathbf{E})$ that vanish outside a bounded set are dense in $\mathbf{L}^p(\mathbf{E})$.

Problem 3.9. Let $[a, b]$ be a closed bounded interval and $f_n \rightharpoonup f$ in $\mathbf{C}([a, b])$. Prove that $\{f_n\}_{n\in\mathbf{N}}$ converges pointwise on $[a, b]$ to f.

Problem 3.10. Let $[a, b]$ be a closed bounded interval and $f_n \rightharpoonup f$ in $\mathbf{L}^\infty([a, b])$. Prove that

$$\lim_{n\to\infty} \int_a^x f_n = \int_a^x f$$

for any $x \in [a, b]$.

4 Linear operators

4.1 Definition

An operator is generally a mapping that acts on the elements of a vector space to produce other elements of the same or another vector space. The most common operators which act on vector spaces are linear operators.

Suppose that $\mathbf{X}$ and $\mathbf{Y}$ are vector spaces over $\mathbf{F}$.

Definition 4.1. The operator $\mathbb{A} : \mathbf{X} \longmapsto \mathbf{Y}$ will be called a linear operator, if
1. it is additive, i. e.,

$$\mathbb{A}(x_1 + x_2) = \mathbb{A}x_1 + \mathbb{A}x_2, \quad x_1, x_2 \in \mathbf{X},$$

2. it is homogeneous, i. e.,

$$\mathbb{A}(\lambda x) = \lambda \mathbb{A}x, \quad \lambda \in \mathbf{F}, \quad x \in \mathbf{X}.$$

Example 4.1. Let $K(t, s)$ be a continuous function on the square $0 \le t, s \le 1$. For $x \in \mathbf{C}([0, 1])$ we define the operator

$$y(t) = \int_0^1 K(t, s)x(s)ds, \quad t \in [0, 1], \quad y = \mathbb{A}x.$$

Let $\mathbf{X} = \mathbf{Y} = \mathbf{C}([0, 1])$. It is evident that $\mathbb{A} : \mathbf{X} \longmapsto \mathbf{Y}$. We will prove that it is a linear operator.
1. Let $x_1, x_2 \in \mathbf{X}$ be arbitrarily chosen. Then

$$\mathbb{A}x_1(t) = \int_0^1 K(t, s)x_1(s)ds,$$

$$\mathbb{A}x_2(t) = \int_0^1 K(t, s)x_2(s)ds,$$

$$\begin{aligned}\mathbb{A}(x_1 + x_2)(t) &= \int_0^1 K(t, s)(x_1(s) + x_2(s))ds = \int_0^1 K(t, s)x_1(s)ds + \int_0^1 K(t, s)x_2(s)ds \\ &= \mathbb{A}x_1(t) + \mathbb{A}x_2(t), \quad t \in [0, 1].\end{aligned}$$

2. Let $\lambda \in \mathbf{F}$ and $x \in \mathbf{X}$ be arbitrarily chosen and fixed. Then

$$\mathbb{A}(\lambda x)(t) = \int_0^1 K(t, s)\lambda x(s)ds = \lambda \int_0^1 K(t, s)x(s)ds = \lambda \mathbb{A}x(t), \quad t \in [0, 1].$$

Therefore $\mathbb{A} : \mathbf{X} \longmapsto \mathbf{Y}$ is a linear operator.

https://doi.org/10.1515/9783110657722-004

Example 4.2. For $x \in \mathbf{C}^1([0,1])$ we define the operator

$$y(t) = \frac{d}{dt}x(t), \quad t \in [0,1], \quad y = \mathbb{A}x.$$

Let $\mathbf{X} = \mathbf{C}^1([0,1])$, $\mathbf{Y} = \mathbf{C}([0,1])$. It is evident that $\mathbb{A} : \mathbf{X} \longmapsto \mathbf{Y}$. We will prove that it is a linear operator.

1. Let $x_1, x_2 \in \mathbf{X}$ be arbitrarily chosen and fixed. Then

$$\begin{aligned} \mathbb{A}x_1(t) &= \frac{d}{dt}x_1(t), \quad \mathbb{A}x_2(t) = \frac{d}{dt}x_2(t), \\ \mathbb{A}(x_1 + x_2)(t) &= \frac{d}{dt}(x_1 + x_2)(t) \\ &= \frac{d}{dt}x_1(t) + \frac{d}{dt}x_2(t) = \mathbb{A}x_1(t) + \mathbb{A}x_2(t), \quad t \in [0,1]. \end{aligned}$$

2. Let $\lambda \in \mathbf{F}$ and $x \in \mathbf{X}$ be arbitrarily chosen. Then

$$\mathbb{A}(\lambda x)(t) = \frac{d}{dt}(\lambda x)(t) = \lambda \frac{d}{dt}x(t) = \lambda \mathbb{A}x(t), \quad t \in [0,1].$$

Therefore $\mathbb{A} : \mathbf{X} \longmapsto \mathbf{Y}$ is a linear operator.

Example 4.3. For $x \in \mathbf{C}([0,1])$ we define the operator

$$y = \int_0^1 (x(t))^2 dt, \quad y = \mathbb{A}x.$$

Let $\mathbf{X} = \mathbf{C}([0,1])$, $\mathbf{Y} = \mathbf{F}$. It is evident that $\mathbb{A} : \mathbf{X} \longmapsto \mathbf{Y}$. Let

$$x_1(t) = 1, \quad x_2(t) = t, \quad t \in [0,1].$$

Then

$$\begin{aligned} \mathbb{A}x_1 &= \int_0^1 dt = t\Big|_{t=0}^{t=1} = 1, \\ \mathbb{A}x_2 &= \int_0^1 t^2 dt = \frac{1}{3}t^3\Big|_{t=0}^{t=1} = \frac{1}{3}, \\ \mathbb{A}x_1 + \mathbb{A}x_2 &= 1 + \frac{1}{3} = \frac{4}{3}, \\ \mathbb{A}(x_1 + x_2) &= \int_0^1 (t+1)^2 dt = \int_0^1 (t^2 + 2t + 1)dt = \int_0^1 t^2 dt + 2\int_0^1 t dt + \int_0^1 dt \\ &= \frac{1}{3}t^3\Big|_{t=0}^{t=1} + t^2\Big|_{t=0}^{t=1} + t\Big|_{t=0}^{t=1} = \frac{1}{3} + 1 + 1 = \frac{7}{3}. \end{aligned}$$

Therefore

$$\mathbb{A}(x_1 + x_2) \neq \mathbb{A}x_1 + \mathbb{A}x_2.$$

Consequently $\mathbb{A} : \mathbf{X} \longmapsto \mathbf{Y}$ is not a linear operator.

Exercise 4.1. For $x \in \mathbf{C}([0,1])$ we define the operator

$$y(t) = x(t^2), \quad t \in [0,1], \quad y = \mathbb{A}x.$$

Prove that $\mathbb{A} : \mathbf{C}([0,1]) \longmapsto \mathbf{C}([0,1])$ is a linear operator.

4.2 Linear operators in normed vector spaces

In this section we suppose that $\mathbf{X}$ and $\mathbf{Y}$ are normed vector spaces. The convergence in $\mathbf{X}$ and $\mathbf{Y}$ is a norm convergence.

Definition 4.2. We say that the linear operator $\mathbb{A} : \mathbf{X} \longmapsto \mathbf{Y}$ is continuous at $x \in \mathbf{X}$, if for any $\epsilon > 0$ there is a $\delta = \delta(\epsilon)$ such that

$$\|\mathbb{A}x_1 - \mathbb{A}x\| < \epsilon$$

whenever $\|x_1 - x\| < \delta$, $x_1 \in \mathbf{X}$. In other words, the linear operator $\mathbb{A} : \mathbf{X} \longmapsto \mathbf{Y}$ is said to be continuous at $x \in \mathbf{X}$ if $\mathbb{A}x_n \to \mathbb{A}x$ in $\mathbf{Y}$, as $n \to \infty$, whenever $x_n \to x$ in $\mathbf{X}$, as $n \to \infty$, where $\{x_n\}_{n\in\mathbf{N}}$ is a sequence of elements of $\mathbf{X}$. We say that the linear operator $\mathbb{A} : \mathbf{X} \longmapsto \mathbf{Y}$ is continuous in $\mathbf{X}$, if it is continuous at every point of $\mathbf{X}$.

Example 4.4. Let $\mathbf{X} = \mathbf{Y} = \mathbf{C}([0,1])$. Consider the operator

$$\mathbb{A}x(t) = t^2 \int_0^1 x(s)ds, \quad t \in [0,1], \quad x \in \mathbf{X}.$$

Let $x \in \mathbf{X}$ be arbitrarily chosen and fixed. We take $\epsilon > 0$ arbitrarily and $x_n \in \mathbf{X}$ such that

$$\|x_n - x\| = \max_{t\in[0,1]} |x_n(t) - x(t)| < \epsilon.$$

Hence,

$$\begin{aligned}|\mathbb{A}x_n(t) - \mathbb{A}x(t)| &= \left| t^2 \int_0^1 x_n(s)ds - t^2 \int_0^1 x(s)ds \right| = \left| t^2 \int_0^1 (x_n(s) - x(s))ds \right| \\ &\leq t^2 \int_0^1 |x_n(s) - x(s)|ds \leq \int_0^1 \|x_n - x\|ds < \epsilon, \quad t \in [0,1].\end{aligned}$$

Because $\epsilon > 0$ was arbitrarily chosen, we conclude that $\mathbb{A}$ is continuous at x. Since $x \in \mathbf{X}$ was arbitrarily chosen, we see that the operator $\mathbb{A}$ is continuous in $\mathbf{X}$.

Now we suppose that $\mathbb{A} : \mathbf{X} \longmapsto \mathbf{Y}$ is a linear continuous operator. We take $x = y+z$, $y, z \in \mathbf{X}$. Then

$$\mathbb{A}x = \mathbb{A}(y+z) = \mathbb{A}y + \mathbb{A}z = \mathbb{A}y + \mathbb{A}(x-y).$$

Therefore

$$\mathbb{A}(x-y) = \mathbb{A}x - \mathbb{A}y. \tag{4.1}$$

We set $x = y$ in (4.1) and we get

$$\mathbb{A}0 = \mathbb{A}x - \mathbb{A}x = 0.$$

We set $x = 0$ in (4.1) and we obtain

$$\mathbb{A}(-y) = \mathbb{A}0 - \mathbb{A}y = -\mathbb{A}y. \tag{4.2}$$

Theorem 4.1. *Let $\mathbb{A} : \mathbf{X} \longmapsto \mathbf{Y}$ be a linear operator which is continuous at a single point $x_0 \in \mathbf{X}$. Then it is continuous on the entire space $\mathbf{X}$.*

Proof. Let $\{x_n\}_{n\in\mathbf{N}}$ be a sequence of elements of $\mathbf{X}$ such that $x_n \to x$, as $n \to \infty$, in $\mathbf{X}$, $x \in \mathbf{X}$. Hence,

$$x_n - x + x_0 \to x_0, \quad \text{as} \quad n \to \infty.$$

Therefore

$$\mathbb{A}(x_n - x + x_0) \to \mathbb{A}x_0, \quad \text{as} \quad n \to \infty. \tag{4.3}$$

Since $\mathbb{A} : \mathbf{X} \longmapsto \mathbf{Y}$ is a linear operator, we get

$$\mathbb{A}(x_n - x + x_0) = \mathbb{A}x_n - \mathbb{A}x + \mathbb{A}x_0, \quad n \in \mathbf{N}.$$

Using (4.3), we obtain

$$\mathbb{A}x_n \to \mathbb{A}x, \quad \text{as} \quad n \to \infty.$$

This completes the proof. □

Definition 4.3. Let $\mathbb{A}, \mathbb{B} : \mathbf{X} \longmapsto \mathbf{Y}$ be linear operators. We define the addition of the operators $\mathbb{A}$ and $\mathbb{B}$ by

$$(\mathbb{A} + \mathbb{B})x = \mathbb{A}x + \mathbb{B}x, \quad x \in \mathbf{X},$$

and the scalar multiplication by

$$(\lambda\mathbb{A})x = \lambda\mathbb{A}x, \quad x \in \mathbf{X}, \quad \lambda \in \mathbf{F}.$$

The zero operator $\mathbb{O}$ we define by

$$\mathbb{O}x = 0$$

for any $x \in \mathbf{X}$. The identity operator $\mathbb{I}$ is defined by

$$\mathbb{I}x = x$$

for any $x \in \mathbf{X}$. Let $\mathbb{A}, \mathbb{B} : \mathbf{X} \longmapsto \mathbf{X}$. We define

$$\begin{aligned}(\mathbb{A}\mathbb{B})x &= \mathbb{A}(\mathbb{B}x),\\ \mathbb{A}^2x &= \mathbb{A}(\mathbb{A}x),\\ \mathbb{A}^nx &= \mathbb{A}(\mathbb{A}^{n-1}x), \quad n \geq 3, \quad x \in \mathbf{X}.\end{aligned}$$

Remark 4.1. If $\mathbb{A}, \mathbb{B}, \mathbb{C} : \mathbf{X} \longmapsto \mathbf{Y}$, then

$$\begin{aligned}(\mathbb{A}\mathbb{B})\mathbb{C} &= \mathbb{A}(\mathbb{B}\mathbb{C}),\\ (\mathbb{A} + \mathbb{B})\mathbb{C} &= \mathbb{A}\mathbb{C} + \mathbb{B}\mathbb{C},\\ \mathbb{C}(\mathbb{A} + \mathbb{B}) &= \mathbb{C}\mathbb{A} + \mathbb{C}\mathbb{B}.\end{aligned}$$

In the general case, we have

$$\mathbb{A}\mathbb{B} \neq \mathbb{B}\mathbb{A}.$$

Definition 4.4. Let $\mathbb{A} : \mathbf{X} \longmapsto \mathbf{Y}$ be a linear operator. We say that the linear operator $\mathbb{B} : \mathbf{X} \longmapsto \mathbf{Y}$ is a left inverse of the operator $\mathbb{A}$, if

$$\mathbb{B}\mathbb{A} = \mathbb{I}.$$

We say that the linear operator $\mathbb{C} : \mathbf{X} \longmapsto \mathbf{Y}$ is a right inverse of the operator $\mathbb{A}$, if

$$\mathbb{A}\mathbb{C} = \mathbb{I}.$$

Let $\mathbb{B}, \mathbb{C} : \mathbf{X} \longmapsto \mathbf{Y}$ be left and right inverse, respectively, of the linear operator $\mathbb{A} : \mathbf{X} \longmapsto \mathbf{Y}$. Then

$$\mathbb{B} = \mathbb{B}\mathbb{I} = \mathbb{B}(\mathbb{A}\mathbb{C}) = (\mathbb{B}\mathbb{A})\mathbb{C} = \mathbb{I}\mathbb{C} = \mathbb{C}.$$

In this case it is said that the operator $\mathbb{A}$ has an inverse denoted by $\mathbb{A}^{-1}$. Thus, if $\mathbb{A}^{-1}$ exists, we have

$$\mathbb{A}^{-1}\mathbb{A} = \mathbb{A}\mathbb{A}^{-1} = \mathbb{I}.$$

A linear operator $\mathbb{A} : \mathbf{X} \longmapsto \mathbf{Y}$ can have at most one inverse.

Theorem 4.2. *Let* $\mathbb{A} : \mathbf{X} \longmapsto \mathbf{Y}$ *be a linear operator that is continuous at* 0. *Then* $\mathbb{A}$ *is continuous in* $\mathbf{X}$.

Proof. Let $x \in \mathbf{X}$ be arbitrarily chosen. Since $\mathbb{A}$ is continuous at 0, we get

$$\|\mathbb{A}(x_n - x) - \mathbb{A}0\| \to 0, \quad \text{as} \quad \|x_n - x\| \to 0,$$

or

$$\|\mathbb{A}x_n - \mathbb{A}x\| \to 0, \quad \text{as} \quad \|x_n - x\| \to 0.$$

Therefore $\mathbb{A}$ is continuous at x. Because $x \in \mathbf{X}$ was arbitrarily chosen, we conclude that $\mathbb{A}$ is continuous in $\mathbf{X}$. This completes the proof. □

Definition 4.5. A linear operator $\mathbb{A} : \mathbf{X} \longmapsto \mathbf{Y}$ will be called bounded if there is a constant $M \geq 0$ such that

$$\|\mathbb{A}x\| \leq M\|x\|$$

for any $x \in \mathbf{X}$.

Example 4.5. Let $\mathbf{X} = \mathbf{Y} = \mathbf{C}([0,1])$, $a \in \mathbf{X}$. Consider the operator

$$\mathbb{A}x(t) = \int_0^t a(s)x(s)ds, \quad t \in [0,1], \quad x \in \mathbf{C}([0,1]).$$

We have $\mathbb{A} : \mathbf{X} \longmapsto \mathbf{Y}$. In $\mathbf{X}$ we define a norm as follows:

$$\|x\| = \max_{t\in[0,1]} |x(t)|, \quad x \in \mathbf{X}.$$

Because $a \in \mathbf{C}([0,1])$, there is a positive constant M such that

$$|a(t)| \leq M$$

for any $t \in [0,1]$. Let $x \in \mathbf{X}$ be arbitrarily chosen. Hence, for any $t \in [0,1]$, we have

$$|\mathbb{A}x(t)| = \left|\int_0^t a(s)x(s)ds\right| \leq \int_0^t |a(s)||x(s)|ds \leq M\|x\| \int_0^t ds \leq M\|x\|.$$

From this,

$$\max_{t\in[0,1]} |\mathbb{A}x(t)| \leq M\|x\|,$$

or

$$\|\mathbb{A}x\| \leq M\|x\|.$$

Because $x \in \mathbf{X}$ was arbitrarily chosen, we conclude that $\mathbb{A} : \mathbf{X} \longmapsto \mathbf{Y}$ is a bounded operator.

Theorem 4.3. *A linear operator $\mathbb{A} : \mathbf{X} \longmapsto \mathbf{Y}$ is bounded if and only if it is continuous.*

Proof.

1. Let $\mathbb{A} : \mathbf{X} \longmapsto \mathbf{Y}$ be a continuous operator. Assume that it is not bounded. Then there is a sequence $\{x_n\}_{n\in\mathbf{N}}$ of elements of $\mathbf{X}$ such that

$$\|\mathbb{A}x_n\| > n\|x_n\|, \quad x_n \neq 0,$$

for any $n \in \mathbf{N}$. We set

$$\xi_n = \frac{x_n}{n\|x_n\|}.$$

Then

$$\|\mathbb{A}\xi_n\| > 1, \quad n \in \mathbf{N}. \tag{4.4}$$

On the other hand,

$$\|\xi_n\| = \left\|\frac{x_n}{n\|x_n\|}\right\| = \frac{\|x_n\|}{n\|x_n\|} = \frac{1}{n} \to 0, \quad \text{as} \quad n \to \infty.$$

Because $\mathbb{A} : \mathbf{X} \longmapsto \mathbf{Y}$ is continuous, we get

$$\|\mathbb{A}\xi_n\| \to 0, \quad \text{as} \quad n \to \infty.$$

This contradicts (4.4). Therefore $\mathbb{A} : \mathbf{X} \longmapsto \mathbf{Y}$ is bounded.

2. Let $\mathbb{A} : \mathbf{X} \longmapsto \mathbf{Y}$ be a bounded operator. Then there exists a positive constant M such that

$$\|\mathbb{A}x\| \leq M\|x\|$$

for any $x \in \mathbf{X}$. Let $x_n \to x$, as $n \to \infty$, i. e., $\|x_n - x\| \to 0$, as $n \to \infty$. Then

$$\|\mathbb{A}x_n - \mathbb{A}x\| = \|\mathbb{A}(x_n - x)\| \leq M\|x_n - x\| \to 0, \quad \text{as} \quad n \to \infty.$$

Therefore $\mathbb{A} : \mathbf{X} \longmapsto \mathbf{Y}$ is continuous. This completes the proof. □

The space of all linear bounded operators $\mathbb{A} : \mathbf{X} \longmapsto \mathbf{Y}$ will be denoted with $\mathcal{L}(\mathbf{X}, \mathbf{Y})$. Note that $\mathcal{L}(\mathbf{X}, \mathbf{Y})$ is a vector space.

Theorem 4.4. *Let $\mathbf{X}$ be a Banach space and $\mathbb{A} : \mathbf{X} \longmapsto \mathbf{Y}$ be a linear operator. By $\mathbf{X}_n$ we denote the set of those $x \in \mathbf{X}$ for which $\|\mathbb{A}x\| \leq n\|x\|$. Then $\mathbf{X} = \bigcup_{n=1}^{\infty} \mathbf{X}_n$ and at least one of the sets $\mathbf{X}_n$ is everywhere dense in $\mathbf{X}$.*

Proof. Note that the sets $\mathbf{X}_n$, $n \in \mathbf{N}$, are not empty because $0 \in \mathbf{X}_n$ for any $n \in \mathbf{N}$. Also, any $x \in \mathbf{X}$, $x \neq 0$, occurs in one of the sets $\mathbf{X}_n$ because it is sufficient to take n as the least integer, greater than $\frac{\|\mathbb{A}x\|}{\|x\|}$. Therefore $\mathbf{X} \subseteq \bigcup_{n=1}^{\infty} \mathbf{X}_n$. By the definition of the sets $\mathbf{X}_n$ we have $\bigcup_{n=1}^{\infty} \mathbf{X}_n \subseteq \mathbf{X}$. Consequently $\mathbf{X} = \bigcup_{n=1}^{\infty} \mathbf{X}_n$. By Theorem 1.33, it follows that a complete space $\mathbf{X}$ cannot be a countable sum of nowhere dense sets. Therefore there is an $n_0 \in \mathbf{N}$ and there is an open ball $\mathbf{B}_r(x_0)$ containing $\mathbf{B}_r(x_0) \cap \mathbf{X}_{n_0}$ everywhere dense. We take $x_1 \in \mathbf{B}_r(x_0) \cap \mathbf{X}_{n_0}$ and let $\mathbf{B}_{r_1}[x_1]$ be a closed ball such that

$$\mathbf{B}_{r_1}[x_1] \subset \mathbf{B}_r(x_0).$$

Let $x \in \mathbf{X}$ be an element for which $\|x\| = r_1$. Since

$$\|(x_1 + x) - x_1\| = \|x\| = r_1,$$

we conclude that $x + x_1 \in \mathbf{B}_{r_1}[x_1]$. There is a sequence $\{z_k\}_{k \in \mathbf{N}}$ of elements of $B_{r_1}[x_1] \cap \mathbf{X}_{n_0}$, (this sequence can be stationary if $x_1 + x \in \mathbf{X}_{n_0}$) such that $z_k \to x + x_1$, as $k \to \infty$. Consequently

$$x_k = z_k - x_1 \to x, \quad \text{as} \quad k \to \infty.$$

Because $x_k \to x$, as $k \to \infty$, and $\|x\| = r_1$, we can assume that $\frac{r_1}{2} \leq \|x_k\| \leq r_1$ for any large enough $k \in \mathbf{N}$. Now, using $z_k, x_1 \in \mathbf{X}_{n_0}$, we get

$$\|\mathbb{A}x_k\| = \|\mathbb{A}z_k - \mathbb{A}x_1\| \leq \|\mathbb{A}z_k\| + \|\mathbb{A}x_1\| \leq n_0\big(\|z_k\| + \|x_1\|\big).$$

Also,

$$\|z_k\| = \|x_k + x_1\| \leq \|x_k\| + \|x_1\| \leq r_1 + \|x_1\|.$$

Therefore, for any large enough $k \in \mathbf{N}$,

$$\|\mathbb{A}x_k\| \leq n_0\big(r_1 + 2\|x_1\|\big) \leq \frac{2n_0(r_1 + 2\|x_1\|)}{r_1}\|x_k\|.$$

Let n be the least integer greater than $\frac{2n_0(r_1+2\|x_1\|)}{r_1}$. Then

$$\|\mathbb{A}x_k\| \leq n\|x_k\|$$

for any large enough $k \in \mathbf{N}$. Consequently $x_k \in \mathbf{X}_n$ for any large enough $k \in \mathbf{N}$. Thus every element $x \in \mathbf{X}$ with norm equal to r_1 can be approximated by elements of $\mathbf{X}_n$. Let now $x \in \mathbf{X}$, $x \neq 0$, be arbitrarily chosen. We set $\xi = r_1 \frac{x}{\|x\|}$. Then $\|\xi\| = r_1$. As above, there is a sequence $\{\xi_k\}_{k \in \mathbf{N}} \subseteq \mathbf{X}_n$ convergent to ξ. Then

$$x_k = \xi_k \frac{\|x\|}{r_1} \to x,$$
$$\|\mathbb{A}x_k\| = \frac{\|x\|}{r_1}\|\mathbb{A}\xi_k\| \leq \frac{\|x\|}{r_1} n\|\xi_k\| = n\|x_k\|, \quad k \in \mathbf{N}.$$

Thus, $x_k \in \mathbf{X}_n$, $k \in \mathbf{N}$. Consequently, $\mathbf{X}_n$ is everywhere dense in $\mathbf{X}$. This completes the proof. □

Definition 4.6. Let $\mathbb{A} : \mathbf{X} \longmapsto \mathbf{Y}$ be a bounded linear operator. The smallest number M for which $\|\mathbb{A}x\| \leq M\|x\|$ for any $x \in \mathbf{X}$, will be called the norm of the operator $\mathbb{A}$. It is denoted by $\|\mathbb{A}\|$.

By Definition 4.6, it follows:
1. $\|\mathbb{A}x\| \leq \|\mathbb{A}\|\|x\|$ for any $x \in \mathbf{X}$.
2. For any $\epsilon > 0$ there is an element $x_\epsilon \in \mathbf{X}$, $x_\epsilon \neq 0$, such that

$$\|\mathbb{A}x_\epsilon\| > (\|\mathbb{A}\| - \epsilon)\|x_\epsilon\|.$$

Theorem 4.5. *Let $\mathbb{A} : \mathbf{X} \longmapsto \mathbf{Y}$ be a bounded linear operator. Then*

$$\|\mathbb{A}\| = \sup_{\|x\|\leq 1} \|\mathbb{A}x\|. \tag{4.5}$$

Proof. For any $x \in \mathbf{X}$, $\|x\| \leq 1$, we have

$$\|\mathbb{A}x\| \leq \|\mathbb{A}\|\|x\| \leq \|\mathbb{A}\|. \tag{4.6}$$

Let $\epsilon > 0$ be arbitrarily chosen. Then there exists $x_\epsilon \in \mathbf{X}$, $x_\epsilon \neq 0$, such that

$$\|\mathbb{A}x_\epsilon\| > (\|\mathbb{A}\| - \epsilon)\|x_\epsilon\|.$$

We take $\xi_\epsilon = \frac{x_\epsilon}{\|x_\epsilon\|}$. Then $\|\xi_\epsilon\| = 1$ and

$$\begin{aligned} \|\mathbb{A}\xi_\epsilon\| &= \left\|\mathbb{A}\left(\frac{x_\epsilon}{\|x_\epsilon\|}\right)\right\| = \left\|\frac{1}{\|x_\epsilon\|}\mathbb{A}x_\epsilon\right\| \\ &= \frac{1}{\|x_\epsilon\|}\|\mathbb{A}x_\epsilon\| > \|\mathbb{A}\| - \epsilon. \end{aligned}$$

Because $\|\xi_\epsilon\| = 1$, from the previous inequality, we get

$$\sup_{\|x\|\leq 1} \|\mathbb{A}x\| \geq \|\mathbb{A}\| - \epsilon.$$

Hence, using the fact that $\epsilon > 0$ was arbitrarily chosen, we get

$$\sup_{\|x\|\leq 1} \|\mathbb{A}x\| \geq \|\mathbb{A}\|.$$

From the previous inequality and from (4.6), we obtain the inequality (4.5). This completes the proof. □

Remark 4.2. By Theorem 4.5, it follows that

$$\|\mathbb{A}\| = \sup_{x\in\mathbf{X}, x\neq 0} \frac{\|\mathbb{A}x\|}{\|x\|}.$$

Example 4.6. Let $\mathbf{X} = \mathbf{Y} = \mathbf{C}([0,1])$. Consider the operator

$$\mathbb{A}x(t) = \int_0^1 x(s)ds, \quad x \in \mathbf{X}, \quad t \in [0,1].$$

In $\mathbf{X}$ we take the norm $\|x\| = \max_{t\in[0,1]} |x(t)|$. We have $\mathbb{A} : \mathbf{X} \longmapsto \mathbf{X}$. We will find $\|\mathbb{A}\|$. For any $x \in \mathbf{X}$, $\|x\| \leq 1$, we have

$$\begin{aligned}|\mathbb{A}x(t)| &= \left|\int_0^1 x(s)ds\right| \leq \int_0^1 |x(s)|ds \\ &\leq \int_0^1 \max_{s\in[0,1]} |x(s)|ds = \|x\| \leq 1, \quad t \in [0,1].\end{aligned}$$

Hence,

$$\max_{t\in[0,1]} |\mathbb{A}x(t)| \leq 1$$

or

$$\|\mathbb{A}x\| \leq 1.$$

Therefore

$$\|\mathbb{A}\| \leq 1. \tag{4.7}$$

Now we take $y(t) = 1$, $t \in [0,1]$. Then $\|y\| = 1$ and

$$\mathbb{A}y(t) = \int_0^1 ds = 1, \quad t \in [0,1].$$

Hence, using $\|y\| = 1$, we get

$$\sup_{\|x\|\leq 1} \|\mathbb{A}x\| \geq 1,$$

i. e., $\|\mathbb{A}\| \geq 1$. From this and from (4.7), we obtain $\|\mathbb{A}\| = 1$.

Example 4.7. Let $K \in \mathbf{C}([0,1] \times [0,1])$, $\mathbf{X} = \mathbf{Y} = \mathbf{C}([0,1])$. Consider the operator

$$\mathbb{A}x(t) = \int_0^1 K(t,s)x(s)ds, \quad x \in \mathbf{X}, \quad t \in [0,1].$$

In $\mathbf{X}$ we take the norm $\|x\| = \max_{t\in[0,1]} |x(t)|$. We have $\mathbb{A} : \mathbf{X} \longmapsto \mathbf{Y}$. We will find $\|\mathbb{A}\|$. Let $x \in \mathbf{X}$ be arbitrarily chosen. Then

$$\begin{aligned}|\mathbb{A}x(t)| &= \left|\int_0^1 K(t,s)x(s)ds\right| \le \int_0^1 |K(t,s)||x(s)|ds \\ &\le \|x\| \max_{t\in[0,1]} \int_0^1 |K(t,s)|ds, \quad t \in [0,1].\end{aligned}$$

Hence,

$$\max_{t\in[0,1]} |\mathbb{A}x(t)| \le \|x\| \max_{t\in[0,1]} \int_0^1 |K(t,s)|ds$$

or

$$\|\mathbb{A}x\| \le \|x\| \max_{t\in[0,1]} \int_0^1 |K(t,s)|ds.$$

Then

$$\begin{aligned}\sup_{\|x\|\le 1} \|\mathbb{A}x\| &= \sup_{\|x\|\le 1}\left(\max_{t\in[0,1]} |\mathbb{A}x(t)|\right) \le \sup_{\|x\|\le 1}\left(\|x\| \max_{t\in[0,1]} \int_0^1 |K(t,s)|ds\right) \\ &\le \max_{t\in[0,1]} \int_0^1 |K(t,s)|ds.\end{aligned}$$

Therefore

$$\|\mathbb{A}\| \le \max_{t\in[0,1]} \int_0^1 |K(t,s)|ds. \tag{4.8}$$

For each $n \in \mathbf{N}$ we take a set $\mathbf{E}_n \subseteq [0,1]$ such that

$$m(\mathbf{E}_n) \le \frac{1}{2Mn},$$

where $M = \max_{(t,s)\in[0,1]\times[0,1]} |K(t,s)|$. Since $t \longmapsto \int_0^1 |K(t,s)|ds$ is a continuous function on $[0,1]$, there exists $t_0 \in [0,1]$ such that

$$\int_0^1 |K(t_0,s)|ds = \max_{t\in[0,1]} \int_0^1 |K(t,s)|ds.$$

Let

$$z_0(s) = \operatorname{sign} K(t_0, s), \quad s \in [0,1].$$

For each $n \in \mathbf{N}$ we take a continuous function x_n on $[0,1]$ such that $|x_n(s)| \leq 1$ for any $s \in [0,1]$ and $x_n(s) = z_0(s)$ for $s \in [0,1] \setminus \mathbf{E}_n$. We have

$$\left|x_n(s) - z_0(s)\right| \leq \left|x_n(s)\right| + \left|z_0(s)\right| \leq 1 + 1 = 2$$

for any $s \in \mathbf{E}_n$. Then

$$\begin{aligned}\left|\int_0^1 K(t,s)z_0(s)ds - \int_0^1 K(t,s)x_n(s)ds\right| &= \left|\int_0^1 K(t,s)(z_0(s) - x_n(s))ds\right| \\ &\leq \int_0^1 \left|K(t,s)\right|\left|z_0(s) - x_n(s)\right|ds \\ &= \int_{\mathbf{E}_n} \left|K(t,s)\right|\left|z_0(s) - x_n(s)\right|ds \leq 2Mm(\mathbf{E}_n) \leq \frac{1}{n}\end{aligned}$$

for any $t \in [0,1]$ and for any $n \in \mathbf{N}$. Therefore

$$\int_0^1 K(t,s)z_0(s)ds \leq \int_0^1 K(t,s)x_n(s)ds + \frac{1}{n} = \mathbb{A}x_n(t) + \frac{1}{n} \leq \|\mathbb{A}\|\|x_n\| + \frac{1}{n} \leq \|\mathbb{A}\| + \frac{1}{n}$$

for any $t \in [0,1]$ and for any $n \in \mathbf{N}$. In particular,

$$\int_0^1 K(t_0,s)z_0(s)ds \leq \|\mathbb{A}\| + \frac{1}{n}$$

for any $n \in \mathbf{N}$. Therefore

$$\int_0^1 K(t_0,s)z_0(s)ds \leq \|\mathbb{A}\|,$$

i. e.,

$$\max_{t\in[0,1]} \int_0^1 \left|K(t,s)\right|ds \leq \|\mathbb{A}\|.$$

From the previous inequality and from (4.8), we get

$$\|\mathbb{A}\| = \max_{t\in[0,1]} \int_0^1 \left|K(t,s)\right|ds.$$

Example 4.8. Let $\mathbf{X} = \mathbf{Y} = \mathbf{C}([0,1])$. Consider the operator

$$\mathbb{A}x(t) = \frac{t^2}{2}\int_0^t x(s)ds, \quad t \in [0,1], \quad x \in \mathbf{X}.$$

In $\mathbf{X}$ we take the norm $\|x\| = \max_{t\in[0,1]} |x(t)|$. We have $\mathbb{A} : \mathbf{X} \longmapsto \mathbf{Y}$. We will find $\|\mathbb{A}\|$. Let $x \in \mathbf{X}$ be arbitrarily chosen. Then

$$\left|\mathbb{A}x(t)\right| = \left|\frac{t^2}{2}\int_0^t x(s)ds\right| \le \frac{t^2}{2}\int_0^t |x(s)|ds \le \frac{1}{2}\|x\|$$

for any $t \in [0,1]$. Hence,

$$\|\mathbb{A}x\| = \max_{t\in[0,1]}\left|\mathbb{A}x(t)\right| \le \frac{1}{2}\|x\|.$$

Then

$$\|\mathbb{A}\| = \sup_{|x|\le 1}\|\mathbb{A}x\| \le \frac{1}{2}. \tag{4.9}$$

For each $n \in \mathbf{N}$ we take a set $\mathbf{E}_n \subset [0,1]$ such that $m(\mathbf{E}_n) \le \frac{1}{n}$. Also, for each $n \in \mathbf{N}$ we take a continuous function x_n on $[0,1]$ such that $|x_n(s)| \le 1$ for any $s \in [0,1]$ and $x_n(s) = 1$ for $s \in [0,1] \setminus \mathbf{E}_n$. Then

$$\left|x_n(s) - 1\right| \le 2$$

for any $s \in \mathbf{E}_n$. Hence,

$$\begin{aligned}\left|\frac{t^3}{2} - \frac{t^2}{2}\int_0^t x_n(s)ds\right| &= \left|\frac{t^2}{2}\int_0^t (1 - x_n(s))ds\right| \le \frac{t^2}{2}\int_0^t \left|1 - x_n(s)\right|ds \le \frac{t^2}{2}\int_0^1 \left|1 - x_n(s)\right|ds \\ &= \frac{t^2}{2}\int_{\mathbf{E}_n} \left|1 - x_n(s)\right|ds \le m(\mathbf{E}_n) \le \frac{1}{n}\end{aligned}$$

for any $t \in [0,1]$ and for any $n \in \mathbf{N}$. Hence,

$$\frac{t^3}{2} \le \frac{1}{n} + \frac{t^2}{2}\int_0^t x_n(s)ds = \frac{1}{n} + \mathbb{A}x_n(t) \le \frac{1}{n} + \|\mathbb{A}\|\|x_n\| \le \frac{1}{n} + \|\mathbb{A}\|$$

for any $t \in [0,1]$ and for any $n \in \mathbf{N}$. In particular, for $t = 1$, we get

$$\frac{1}{2} \le \frac{1}{n} + \|\mathbb{A}\|$$

for any $n \in \mathbf{N}$. Consequently

$$\frac{1}{2} \le \|\mathbb{A}\|.$$

From the previous inequality and from (4.9), we get $\|\mathbb{A}\| = \frac{1}{2}$.

Exercise 4.2. Let $\mathbf{X} = \mathbf{Y} = \mathbf{C}([0,1])$, $\|x\| = \max_{t\in[0,1]} |x(t)|$, $x \in \mathbf{X}$. Find $\|\mathbb{A}\|$, where

$$\mathbb{A}x(t) = \frac{t^2}{3}\int_0^t s^2 x(s)ds, \quad x \in \mathbf{X}, \quad t \in [0,1].$$

Answer. $\frac{1}{9}$.

Theorem 4.6. *Let $\mathbb{A} : \mathbf{X} \longmapsto \mathbf{Y}$ be a linear operator. Then $\mathbb{A}$ is a bounded operator if and only if there is a constant $M \geq 0$ such that $\|\mathbb{A}\| \leq M$.*

Proof.
1. Let $\mathbb{A}$ be a bounded operator. Then there is a constant $M \geq 0$ such that

$$\|\mathbb{A}x\| \leq M\|x\|$$

for any $x \in \mathbf{X}$. Hence,

$$\|\mathbb{A}\| = \sup_{\|x\|\leq 1} \|\mathbb{A}x\| \leq \sup_{\|x\|\leq 1}(M\|x\|) = M.$$

2. Let there be a constant $M \geq 0$ such that $\|\mathbb{A}\| \leq M$. Then

$$\|\mathbb{A}x\| \leq \|\mathbb{A}\|\|x\| \leq M\|x\|$$

for any $x \in \mathbf{X}$. Therefore $\mathbb{A}$ is a bounded operator. This completes the proof. □

Theorem 4.7. *Let $\mathbb{A}, \mathbb{B} : \mathbf{X} \longmapsto \mathbf{Y}$ be linear operators. Then*

$$\|\mathbb{A} + \mathbb{B}\| \leq \|\mathbb{A}\| + \|\mathbb{B}\| \quad \textit{and} \quad \|\lambda\mathbb{A}\| = |\lambda|\|\mathbb{A}\|$$

for any $\lambda \in \mathbf{F}$.

Proof. We have

$$\begin{aligned}\|\mathbb{A} + \mathbb{B}\| &= \sup_{x\in\mathbf{X},\|x\|\leq 1} \|(\mathbb{A} + \mathbb{B})x\| = \sup_{x\in\mathbf{X},\|x\|\leq 1} \|\mathbb{A}x + \mathbb{B}x\| \\ &\leq \sup_{x\in\mathbf{X},\|x\|\leq 1} (\|\mathbb{A}x\| + \|\mathbb{B}x\|) \leq \sup_{x\in\mathbf{X},\|x\|\leq 1} \|\mathbb{A}x\| + \sup_{x\in\mathbf{X},\|x\|\leq 1} \|\mathbb{B}x\| = \|\mathbb{A}\| + \|\mathbb{B}\|\end{aligned}$$

and

$$\begin{aligned}\|\lambda\mathbb{A}\| &= \sup_{x\in\mathbf{X},\|x\|\leq 1} \|\lambda\mathbb{A}x\| = \sup_{x\in\mathbf{X},\|x\|\leq 1} (|\lambda|\|\mathbb{A}x\|) \\ &= |\lambda| \sup_{x\in\mathbf{X},\|x\|\leq 1} \|\mathbb{A}x\| = |\lambda|\|\mathbb{A}\|\end{aligned}$$

for any $\lambda \in \mathbf{F}$. This completes the proof. □

Definition 4.7. We say that a sequence $\{\mathbb{A}_n\}_{n\in\mathbf{N}}$ of elements of $\mathcal{L}(\mathbf{X},\mathbf{Y})$ is uniformly convergent to $\mathbb{A} \in \mathcal{L}(\mathbf{X},\mathbf{Y})$ if

$$\|\mathbb{A}_n - \mathbb{A}\| \to 0, \quad \text{as} \quad n \to \infty.$$

We will write $\mathbb{A}_n \to \mathbb{A}$, as $n \to \infty$, or $\lim_{n\to\infty} \mathbb{A}_n = \mathbb{A}$.

Theorem 4.8. *Let $\{\mathbb{A}_n\}_{n\in\mathbf{N}} \subseteq \mathcal{L}(\mathbf{X},\mathbf{Y})$ and $\mathbb{A} \in \mathcal{L}(\mathbf{X},\mathbf{Y})$. Then $\mathbb{A}_n \to \mathbb{A}$, as $n \to \infty$, if and only if*

$$\|\mathbb{A}_n x - \mathbb{A}x\| \to 0, \quad \text{as} \quad n \to \infty,$$

for any $x \in \mathbf{X}$ such that $\|x\| \le 1$.

Proof.
1. Let $\mathbb{A}_n \to \mathbb{A}$, as $n \to \infty$, uniformly. Then

$$\|\mathbb{A}_n - \mathbb{A}\| \to 0, \quad \text{as} \quad n \to \infty.$$

Hence, for $x \in \mathbf{X}$, $\|x\| \le 1$, we get

$$\|\mathbb{A}_n x - \mathbb{A}x\| \le \|\mathbb{A}_n - \mathbb{A}\|\|x\| \le \|\mathbb{A}_n - \mathbb{A}\| \to 0, \quad \text{as} \quad n \to \infty.$$

2. Let $\|\mathbb{A}_n x - \mathbb{A}x\| \to 0$, as $n \to \infty$, for any $x \in \mathbf{X}$ such that $\|x\| \le 1$. Hence,

$$\sup_{x\in\mathbf{X},\|x\|\le 1} \|\mathbb{A}_n x - \mathbb{A}x\| \le \frac{\epsilon}{2}$$

for any $n \ge N$. Therefore

$$\|\mathbb{A}_n - \mathbb{A}\| \to 0, \quad \text{as} \quad n \to \infty.$$

This completes the proof. □

Corollary 4.1. *Let $\{\mathbb{A}_n\}_{n\in\mathbf{N}} \subseteq \mathcal{L}(\mathbf{X},\mathbf{Y})$ be such that $\mathbb{A}_n \to \mathbb{A}$, as $n \to \infty$, for $\mathbb{A} \in \mathcal{L}(\mathbf{X},\mathbf{Y})$. Let also, $\mathbf{U}$ be an arbitrary bounded set in $\mathbf{X}$. Then*

$$\|\mathbb{A}_n x - \mathbb{A}x\| \to 0, \quad \text{as} \quad n \to \infty,$$

for any $x \in \mathbf{U}$.

Proof. Because $\mathbf{U}$ is a bounded set in $\mathbf{X}$, there is an $R > 0$ such that $\|x\| \le R$ for any $x \in \mathbf{U}$. Let $\epsilon > 0$ be arbitrarily chosen. Since $\mathbb{A}_n \to \mathbb{A}$, as $n \to \infty$, we have

$$\|\mathbb{A}_n - \mathbb{A}\| \to 0, \quad \text{as} \quad n \to \infty.$$

Hence, for $x \in \mathbf{U}$, we have

$$\|\mathbb{A}_n x - \mathbb{A}x\| \le \|\mathbb{A}_n - \mathbb{A}\|\|x\| \le R\|\mathbb{A}_n - \mathbb{A}\| \to 0, \quad \text{as} \quad n \to \infty.$$

This completes the proof. □

Definition 4.8. We say that a sequence $\{\mathbb{A}_n\}_{n\in\mathbf{N}}$ of elements of $\mathcal{L}(\mathbf{X},\mathbf{Y})$ is a Cauchy sequence, if for any $\epsilon > 0$ there is an $N \in \mathbf{N}$ such that

$$\|\mathbb{A}_{n+p} - \mathbb{A}_n\| < \epsilon$$

for any $n \geq N$, $p \in \mathbf{N}$.

Theorem 4.9. *Let* $\mathbf{Y}$ *be a Banach space. Then* $\mathcal{L}(\mathbf{X},\mathbf{Y})$ *is a Banach space.*

Proof. Let $\{\mathbb{A}_n\}_{n\in\mathbf{N}}$ be a Cauchy sequence of elements of $\mathcal{L}(\mathbf{X},\mathbf{Y})$. We fix $\epsilon > 0$. Then there is an $N \in \mathbf{N}$ such that

$$\|\mathbb{A}_{n+p} - \mathbb{A}_n\| < \epsilon$$

for any $n \geq N$, $p \in \mathbf{N}$. Let $x \in \mathbf{X}$ be arbitrarily chosen. We have

$$\|\mathbb{A}_{n+p}x - \mathbb{A}x\| = \big\|(\mathbb{A}_n - \mathbb{A})x\big\| \leq \|\mathbb{A}_n - \mathbb{A}\|\|x\|$$

for any $n \geq N$, $p \in \mathbf{N}$. Therefore $\{\mathbb{A}_n x\}_{n\in\mathbb{N}}$ is a Cauchy sequence in $\mathbf{Y}$. Since $\mathbf{Y}$ is a Banach space, we see that there exists $\lim_{n\to\infty} \mathbb{A}_n x$. We define the operator $\mathbb{A} : \mathbf{X} \to \mathbf{Y}$ as follows:

$$\mathbb{A}x = \lim_{n\to\infty} \mathbb{A}_n x, \quad x \in \mathbf{X}.$$

For any $\alpha, \beta \in \mathbf{F}$ and $x_1, x_2 \in \mathbf{X}$, we have

$$\begin{aligned}\mathbb{A}(\alpha x_1 + \beta x_2) &= \lim_{n\to\infty} \mathbb{A}_n(\alpha x_1 + \beta x_2) = \lim_{n\to\infty}(\alpha\mathbb{A}_n x_1 + \beta\mathbb{A}_n x_2)\\ &= \alpha \lim_{n\to\infty} \mathbb{A}_n x_1 + \beta \lim_{n\to\infty} \mathbb{A}_n x_2 = \alpha\mathbb{A}x_1 + \beta\mathbb{A}x_2.\end{aligned}$$

Therefore $\mathbb{A} : \mathbf{X} \longmapsto \mathbf{Y}$ is a linear operator. Note that

$$\big|\|\mathbb{A}_{n+p}\| - \|\mathbb{A}_n\|\big| \leq \|\mathbb{A}_{n+p} - \mathbb{A}_n\|$$

for any $n, p \in \mathbf{N}$. Therefore $\{\|\mathbb{A}_n\|\}_{n\in\mathbf{N}}$ is a Cauchy sequence in $\mathbf{R}$. Hence, it is a bounded sequence in $\mathbf{R}$ and there exists a constant $c > 0$ such that

$$\|\mathbb{A}_n\| \leq c$$

for any $n \in \mathbb{N}$. Hence,

$$\|\mathbb{A}_n x\| \leq c\|x\|$$

for any $x \in \mathbf{X}$ and for any $n \in \mathbf{N}$. Consequently

$$\|\mathbb{A}x\| \leq c\|x\|$$

for any $x \in \mathbf{X}$. From this, $\mathbb{A} \in \mathcal{L}(\mathbf{X},\mathbf{Y})$. This completes the proof. □

Definition 4.9. Let $\mathbb{A}_n \in \mathcal{L}(\mathbf{X}, \mathbf{Y})$, $n \in \mathbf{N}$.

1. The series $\sum_{n=1}^{\infty} \mathbb{A}_n$ is said to be uniformly convergent, if the sequence $\{S_n = \sum_{k=1}^{n} \mathbb{A}_k\}_{n\in\mathbf{N}}$ is uniformly convergent.
2. The series $\sum_{n=1}^{\infty} \mathbb{A}_n$ is said to be absolutely convergent if the series $\sum_{n=1}^{\infty} \|\mathbb{A}_n\|$ is convergent.

Theorem 4.10. *Let $\mathcal{L}(\mathbf{X}, \mathbf{Y})$ be a Banach space. If the series $\sum_{n=1}^{\infty} \mathbb{A}_n$ is an absolutely convergent series, then it is uniformly convergent.*

Proof. Since the series $\sum_{n=1}^{\infty} \mathbb{A}_n$ is absolutely convergent, we see that the series $\sum_{n=1}^{\infty} \|\mathbb{A}_n\|$ is convergent. We fix $\epsilon > 0$. Then there is an $N \in \mathbf{N}$ such that

$$\sum_{n=N+1}^{N+p} \|\mathbb{A}_n\| < \epsilon$$

for any $p \in \mathbf{N}$. Hence,

$$\|S_{N+p} - S_N\| = \left\| \sum_{n=N+1}^{N+p} \mathbb{A}_n \right\| \leq \sum_{n=N+1}^{N+p} \|\mathbb{A}_n\| < \epsilon$$

for any $p \in \mathbf{N}$. Therefore $\{S_n\}_{n\in\mathbf{N}}$ is a Cauchy sequence in $\mathcal{L}(\mathbf{X}, \mathbf{Y})$. Because $\mathcal{L}(\mathbf{X}, \mathbf{Y})$ is a Banach space, we see that $\{S_n\}_{n\in\mathbf{N}}$ is uniformly convergent. This completes the proof. □

Below by $\mathcal{L}(\mathbf{X})$ we will denote the space $\mathcal{L}(\mathbf{X}, \mathbf{X})$.

Exercise 4.3. Let $\mathbb{A}, \mathbb{B} \in \mathcal{L}(\mathbf{X})$. Prove that $\mathbb{A}\mathbb{B} \in \mathcal{L}(\mathbf{X})$ and $\mathbb{A}^k \in \mathcal{L}(\mathbf{X})$ for any $k \in \mathbf{N}$.

Exercise 4.4. For $\mathbb{A} \in \mathcal{L}(\mathbf{X})$, we define

$$e^{\mathbb{A}} = \sum_{k=1}^{\infty} \frac{\mathbb{A}^k}{k!}.$$

Prove that $e^{\mathbb{A}} \in \mathcal{L}(\mathbf{X})$ and $\|e^{\mathbb{A}}\| \leq e^{\|\mathbb{A}\|}$.

Theorem 4.11. *Let $\{\mathbb{A}_n\}_{n\in\mathbf{N}}, \{\mathbb{B}_n\}_{n\in\mathbf{N}} \subset \mathcal{L}(\mathbf{X})$, $\mathbb{A}, \mathbb{B} \in \mathcal{L}(\mathbf{X})$. If $\mathbb{A}_n \to \mathbb{A}$, $\mathbb{B}_n \to \mathbb{B}$, as $n \to \infty$, then $\mathbb{A}_n\mathbb{B}_n \to \mathbb{A}\mathbb{B}$, as $n \to \infty$.*

Proof. Because $\mathbb{B}_n \in \mathcal{L}(\mathbf{X})$, $n \in \mathbf{N}$, $\mathbb{B}_n \to \mathbb{B}$, as $n \to \infty$, we see that the sequence $\{\|\mathbb{B}_n\|\}_{n\in\mathbf{N}}$ is a bounded sequence. Therefore there is a constant $M_1 > 0$ such that $\|\mathbb{B}_n\| \leq M_1$ for any $n \in \mathbf{N}$. Since $\mathbb{A} \in \mathcal{L}(\mathbf{X})$, then there is a constant $M_2 > 0$ such that $\|\mathbb{A}\| \leq M_2$. We take $\epsilon > 0$ arbitrarily. Since $\mathbb{A}_n \to \mathbb{A}$, $\mathbb{B}_n \to \mathbb{B}$, as $n \to \infty$, there is an $N \in \mathbf{N}$ such that

$$\|\mathbb{A}_n - \mathbb{A}\| < \frac{\epsilon}{2M_1} \quad \text{and} \quad \|\mathbb{B}_n - \mathbb{B}\| < \frac{\epsilon}{2M_2}$$

for any $n \geq N$. Hence,

$$\begin{aligned}\|\mathbb{A}_n\mathbb{B}_n - \mathbb{A}\mathbb{B}\| &= \|\mathbb{A}_n\mathbb{B}_n - \mathbb{A}\mathbb{B}_n + \mathbb{A}\mathbb{B}_n - \mathbb{A}\mathbb{B}\| \leq \|(\mathbb{A}_n - \mathbb{A})\mathbb{B}_n\| + \|\mathbb{A}(\mathbb{B}_n - \mathbb{B})\| \\ &\leq \|\mathbb{A}_n - \mathbb{A}\|\|\mathbb{B}_n\| + \|\mathbb{B}_n - \mathbb{B}\|\|\mathbb{A}\| < \frac{\epsilon}{2M_1}M_1 + \frac{\epsilon}{2M_2}M_2 = \frac{\epsilon}{2} + \frac{\epsilon}{2} = \epsilon\end{aligned}$$

for any $n \geq N$. This completes the proof. □

Definition 4.10. We will say that the sequence $\{\mathbb{A}_n\}_{n\in\mathbf{N}} \subset \mathcal{L}(\mathbf{X}, \mathbf{Y})$ is strongly convergent to the operator $\mathbb{A} \in \mathcal{L}(\mathbf{X}, \mathbf{Y})$, if for any $x \in \mathbf{X}$ we have

$$\|\mathbb{A}_n x - \mathbb{A}x\| \to 0,$$

as $n \to \infty$.

Theorem 4.12. *If $\{\mathbb{A}_n\}_{n\in\mathbf{N}} \subseteq \mathcal{L}(\mathbf{X}, \mathbf{Y})$ is uniformly convergent to $\mathbb{A} \in \mathcal{L}(\mathbf{X}, \mathbf{Y})$, then it is strongly convergent to $\mathbb{A}$.*

Proof. Take $\epsilon > 0$ arbitrarily. Then there is an $N \in \mathbf{N}$ such that

$$\|\mathbb{A}_n - \mathbb{A}\| < \epsilon$$

for any $n \geq N$. Hence,

$$\|\mathbb{A}_n x - \mathbb{A}x\| \leq \|\mathbb{A}_n - \mathbb{A}\|\|x\| < \epsilon\|x\|$$

for any $n \geq N$ and for any $x \in \mathbf{X}$. This completes the proof. □

Theorem 4.13. *Let $\{\mathbb{A}_n\}_{n\in\mathbf{N}} \subset \mathcal{L}(\mathbf{X}, \mathbf{Y})$ and there exist $c > 0$ and a closed ball $\mathbf{B}_r[x_0]$ such that $\|\mathbb{A}_n x\| \leq c$ for any $x \in \mathbf{B}_r[x_0]$. Then the sequence $\{\|\mathbb{A}_n\|\}_{n\in\mathbf{N}}$ is bounded.*

Proof. Let $x \in \mathbf{X}$, $x \neq 0$. Then $x_0 + \frac{x}{\|x\|}r \in \mathbf{B}_r[x_0]$. Hence,

$$c \geq \left\|\mathbb{A}_n\left(r\frac{x}{\|x\|} + x_0\right)\right\| = \left\|\frac{r}{\|x\|}\mathbb{A}_n x + \mathbb{A}_n x_0\right\| \geq \left\|\frac{r}{\|x\|}\mathbb{A}_n x\right\| - \|\mathbb{A}_n x_0\| \geq \frac{r}{\|x\|}\|\mathbb{A}_n x\| - c,$$

whereupon

$$\frac{r}{\|x\|}\|\mathbb{A}_n x\| \leq 2c \quad \text{or} \quad \frac{\|\mathbb{A}_n x\|}{\|x\|} \leq \frac{2c}{r}.$$

Therefore

$$\|\mathbb{A}_n\| \leq \frac{2c}{r}.$$

This completes the proof. □

Theorem 4.14 (Uniform boundedness principle). *Let $\mathbf{X}$ be a Banach space and $\{\mathbb{A}_n\}_{n\in\mathbf{N}} \subset \mathcal{L}(\mathbf{X}, \mathbf{Y})$. If $\{\mathbb{A}_n x\}_{n\in\mathbf{N}}$ is bounded for any fixed $x \in \mathbf{X}$, then the sequence $\{\|\mathbb{A}_n\|\}_{n\in\mathbf{N}}$ is bounded.*

Proof. Assume the contrary. Suppose that there is a closed ball $\mathbf{B}_r[x_0]$ such that the sequences $\{\|\mathbb{A}_n x\|\}_{n\in\mathbf{N}}$ is unbounded for some $x \in \mathbf{B}_r[x_0]$. Then there is $x_1 \in \mathbf{B}_r[x_0]$ and $n_1 \in \mathbf{N}$ such that $\|\mathbb{A}_{n_1} x_1\| > 1$. Since $\mathbb{A}_{n_1}$ is continuous, there is a closed ball $\mathbf{B}_{r_1}[x_1] \subset \mathbf{B}_r[x_0]$ such that $\|\mathbb{A}_{n_1} x\| > 1$ for any $x \in \mathbf{B}_{r_1}[x_1]$. By Theorem 4.13, it follows that the sequences $\{\|\mathbb{A}_n x\|\}_{n\in\mathbf{N}}$ is unbounded for any $x \in \mathbf{B}_{r_1}[x_1]$. Then there is $x_2 \in \mathbf{B}_{r_1}[x_1]$ and $n_2 > n_1$, $n_2 \in \mathbf{N}$ such that $\|\mathbb{A}_{n_2} x_2\| > 2$. Since $\mathbb{A}_{n_2}$ is continuous, there is a closed ball $\mathbf{B}_{r_2}[x_2] \subset \mathbf{B}_{r_1}[x_1]$ such that $\|\mathbb{A}_{n_2} x\| > 2$ for any $x \in \mathbf{B}_{r_2}[x_2]$ and so on. In this way get the sequences $\{x_k\}_{k\in\mathbf{N}}$ and $\{\mathbf{B}_{r_k}[x_k]\}_{k\in\mathbf{N}}$ such that

$$\cdots \subset \mathbf{B}_{r_k}[x_k] \subset \mathbf{B}_{r_{k-1}}[x_{k-1}] \subset \cdots \subset \mathbf{B}_{r_1}[x_1]$$

and $\|\mathbb{A}_{n_k} x\| \geq k$ for any $k \in \mathbf{N}$ and for any $x \in \mathbf{B}_{r_k}[x_k]$. Hence, by Theorem 1.30, it follows that there is an $\overline{x} \in \mathbf{X}$ such that $\overline{x} \in \mathbf{B}_{r_k}[x_k]$ for any $k \in \mathbf{N}$. Then $\|\mathbb{A}_{n_k}\overline{x}\| \geq k$ for any $k \in \mathbf{N}$, i. e., the sequence $\{\|\mathbb{A}_{n_k}\overline{x}\|\}_{k\in\mathbf{N}}$ is unbounded. This is a contradiction. Consequently $\{\|\mathbb{A}_n\|\}_{n\in\mathbf{N}}$ is bounded. This completes the proof. □

Theorem 4.15 (Banach–Steinhaus' theorem). *Let $\{\mathbb{A}_n\}_{n\in\mathbf{N}} \subseteq \mathcal{L}(\mathbf{X},\mathbf{Y})$ and $\mathbb{A} \in \mathcal{L}(\mathbf{X},\mathbf{Y})$. Then $\mathbb{A}_n \to \mathbb{A}$, as $n \to \infty$, strongly if and only if*
(H1) *the sequence $\{\|\mathbb{A}_n\|\}_{n\in\mathbf{N}}$ is bounded and*

$$\|\mathbb{A}_n x - \mathbb{A}x\| \to 0, \quad \text{as} \quad n \to \infty,$$

for any $x \in \mathbf{X}'$, where $\mathbf{X}'$ is dense in $\mathbf{X}$.

Proof.
1. Let $\mathbb{A}_n \to \mathbb{A}$, as $n \to \infty$, strongly. Then $\mathbb{A}_n x \to \mathbb{A}x$, as $n \to \infty$, for any $x \in \mathbf{X}$. Therefore $\{\|\mathbb{A}_n x\|\}_{n\in\mathbf{N}}$ is a bounded sequence for any $x \in \mathbf{X}$. Hence, by Theorem 4.14, it follows that the sequence $\{\|\mathbb{A}_n\|\}_{n\in\mathbf{N}}$ is a bounded sequence. As $\mathbf{X}'$ we can take $\mathbf{X}$.
2. Suppose (H1) holds. We set

$$c = \sup_{n\in\mathbf{N}_0} \|\mathbb{A}_n\|,$$

where $\mathbb{A}_0 = \mathbb{A}$. We take $x \in \mathbf{X}$, $x \notin \mathbf{X}'$. Let $\epsilon > 0$ be arbitrarily chosen. Because $\mathbf{X}'$ is dense in $\mathbf{X}$, there is an element $x_1 \in \mathbf{X}'$ such that $\|x - x_1\| < \frac{\epsilon}{3c}$. Also, there is an $N \in \mathbf{N}$ such that

$$\|\mathbb{A}_n x_1 - \mathbb{A}x_1\| < \frac{\epsilon}{3}$$

for any $n \geq N$. From this,

$$\begin{aligned}\|\mathbb{A}_n x - \mathbb{A}x\| &= \big\|\mathbb{A}_n(x - x_1) + (\mathbb{A}_n x_1 - \mathbb{A}x_1) + \mathbb{A}(x_1 - x)\big\| \\ &\leq \big\|\mathbb{A}_n(x - x_1)\big\| + \|\mathbb{A}_n x_1 - \mathbb{A}x_1\| + \big\|\mathbb{A}(x - x_1)\big\| \\ &\leq \|\mathbb{A}_n\|\|x - x_1\| + \|\mathbb{A}_n x_1 - \mathbb{A}x_1\| + \|\mathbb{A}\|\|x - x_1\|\end{aligned}$$

$$\le 2c\|x - x_1\| + \|\mathbb{A}_n x_1 - \mathbb{A}x_1\| < 2c\frac{\epsilon}{3c} + \frac{\epsilon}{3} = \epsilon$$

for any $n \ge N$. This completes the proof. □

Now we assume that $\mathbb{A} : \mathbf{D}(\mathbf{A}) \longmapsto \mathbf{Y}$ is a linear operator, where $\mathbf{D}(\mathbf{A}) \subseteq \mathbf{X}$. For $\mathbb{A}$ we can define its norm in the following way:

$$\|\mathbb{A}\| = \sup_{\substack{x \in \mathbf{D}(\mathbf{A}) \\ \|x\| \le 1}} \|\mathbb{A}x\|$$

and we will say that $\mathbb{A}$ is bounded if $\|\mathbb{A}\| < \infty$.

Theorem 4.16. *Let $\mathbf{Y}$ be a Banach space and $\mathbb{A} : \mathbf{D}(\mathbf{A}) \longmapsto \mathbf{Y}$ be a linear bounded operator, where $\mathbf{D}(\mathbf{A}) \subseteq \mathbf{X}$ and $\mathbf{D}(\mathbb{A})$ is dense in $\mathbf{X}$. Then there exists a linear bounded operator $\widetilde{\mathbb{A}} : \mathbf{X} \longmapsto \mathbf{Y}$ such that $\widetilde{\mathbb{A}}x = \mathbb{A}x$ for any $x \in \mathbf{D}(\mathbf{A})$ and $\|\mathbb{A}\| = \|\widetilde{\mathbb{A}}\|$.*

Proof. For $x \in \mathbf{D}(\mathbf{A})$ we define $\widetilde{\mathbb{A}}x = \mathbb{A}x$. Let $x \notin \mathbf{D}(\mathbf{A})$ and $x \in \mathbf{X}$. Since $\mathbf{D}(\mathbf{A})$ is dense in $\mathbf{X}$, there exists a sequence $\{x_n\}_{n \in \mathbf{N}}$ of elements of $\mathbf{D}(\mathbf{A})$ such that $x_n \to x$, as $n \to \infty$. Then we set

$$\widetilde{\mathbb{A}}x = \lim_{n \to \infty} \mathbb{A}x_n. \tag{4.10}$$

We will prove that $\widetilde{\mathbb{A}}$ is well defined, i. e., we will prove that the limit (4.10) exists and it does not depend on the choice of the sequence $\{x_n\}_{n \in \mathbf{N}}$. Note that

$$\|\mathbb{A}x_n - \mathbb{A}x_m\| \le \|\mathbb{A}\|\|x_n - x_m\| \to 0, \quad \text{as} \quad m, n \to \infty.$$

Therefore $\{\mathbb{A}x_n\}_{n \in \mathbf{N}}$ is a Cauchy sequence in $\mathbf{Y}$. Since $\mathbf{Y}$ is a Banach space, it is convergent. Consequently (4.10) exists. Suppose that $\{x_n\}_{n \in \mathbf{N}}$ and $\{x'_n\}_{n \in \mathbf{N}}$ are two sequences of elements of $\mathbf{X}$ such that $x_n \to x$, $x'_n \to x$, as $n \to \infty$. Let

$$\alpha = \lim_{n \to \infty} \mathbb{A}x_n, \quad \beta = \lim_{n \to \infty} \mathbb{A}x'_n.$$

Then

$$\begin{aligned} \|\alpha - \beta\| &= \|\alpha - \mathbb{A}x_n + \mathbb{A}x_n - \mathbb{A}x'_n + \mathbb{A}x'_n - \beta\| \le \|\alpha - \mathbb{A}x_n\| + \|\mathbb{A}x_n - \mathbb{A}x'_n\| + \|\beta - \mathbb{A}x'_n\| \\ &\le \|\mathbb{A}x_n - \alpha\| + \|\mathbb{A}\|\|x_n - x'_n\| + \|\mathbb{A}x'_n - \beta\| \to 0, \quad \text{as} \quad n \to \infty, \end{aligned}$$

i. e., $\alpha = \beta$. Next,

$$\|\mathbb{A}x_n\| \le \|\mathbb{A}\|\|x_n\|.$$

Hence,

$$\lim_{n \to \infty} \|\mathbb{A}x_n\| \le \lim_{n \to \infty} \|\mathbb{A}\|\|x_n\|,$$

whereupon

$$\|\widetilde{\mathbb{A}}x\| \le \|\mathbb{A}\|\|x\|.$$

Consequently

$$\|\widetilde{\mathbb{A}}\| \le \|\mathbb{A}\|. \tag{4.11}$$

On the other hand,

$$\begin{aligned}\|\widetilde{\mathbb{A}}\| &= \sup_{\substack{\|x\|\le 1,\\ x\in\mathbf{X}}} \|\widetilde{\mathbb{A}}x\| \\ &\ge \sup_{\substack{\|x\|\le 1,\\ x\in\mathbf{D}(\mathbf{A})}} \|\widetilde{\mathbb{A}}x\| = \sup_{\substack{\|x\|\le 1,\\ x\in\mathbf{D}(\mathbf{A})}} \|\mathbb{A}x\| = \|\mathbb{A}\|.\end{aligned}$$

From the previous inequality and from (4.11), we get $\|\mathbb{A}\| = \|\widetilde{\mathbb{A}}\|$. This completes the proof. □

4.3 Inverse operators

Suppose that $\mathbf{X}$ and $\mathbf{Y}$ are normed vector spaces.

Theorem 4.17. *Let $\mathbb{A} : \mathbf{X} \longmapsto \mathbf{Y}$ be a linear operator,*

$$\mathbf{R}(\mathbf{A}) = \{\mathbb{A}(x) : x \in \mathbf{X}\}.$$

Suppose that $\mathbb{A}^{-1}$ exists. Then $\mathbb{A}^{-1} : \mathbf{R}(\mathbf{A}) \longmapsto \mathbf{X}$ is a linear operator.

Proof. Let $y_1, y_2 \in \mathbf{R}(\mathbb{A})$. Then there exist $x_1, x_2 \in \mathbf{X}$ such that

$$\mathbb{A}x_1 = y_1, \quad \mathbb{A}x_2 = y_2.$$

For $\alpha, \beta \in \mathbf{F}$, we have

$$\alpha\mathbb{A}x_1 = \alpha y_1, \quad \beta\mathbb{A}x_2 = \beta y_2,$$

or

$$\mathbb{A}(\alpha x_1) = \alpha y_1, \quad \mathbb{A}(\beta x_2) = \beta y_2, \quad \mathbb{A}(\alpha x_1 + \beta x_2) = \alpha y_1 + \beta y_2.$$

Hence,

$$\begin{aligned}x_1 = \mathbb{A}^{-1}y_1, \quad x_2 = \mathbb{A}^{-1}y_2, \quad \alpha x_1 = \mathbb{A}^{-1}(\alpha y_1), \quad \beta x_2 = \mathbb{A}^{-1}(\beta y_2),\\ \alpha x_1 + \beta x_2 = \mathbb{A}^{-1}(\alpha y_1 + \beta y_2),\end{aligned}$$

and

$$\mathbb{A}^{-1}(\alpha y_1 + \beta y_2) = \alpha x_1 + \beta x_2 = \alpha\mathbb{A}^{-1}y_1 + \beta\mathbb{A}^{-1}y_2.$$

This completes the proof. □

Theorem 4.18. *Let $\mathbb{A} : \mathbf{X} \longmapsto \mathbf{Y}$ be a linear operator such that*

$$\|\mathbb{A}x\| \geq m\|x\| \tag{4.12}$$

for any $x \in \mathbf{X}$ and for some constant $m > 0$. Then $\mathbb{A}$ has an inverse bounded operator $\mathbb{A}^{-1} : \mathbf{R}(\mathbf{A}) \longmapsto \mathbf{X}$.

Proof. Let $x_1, x_2 \in \mathbf{X}$ be such that $x_1 \neq x_2$ and $\mathbb{A}x_1 = \mathbb{A}x_2$. Then, using (4.12), we get

$$0 = \|\mathbb{A}x_1 - \mathbb{A}x_2\| = \|\mathbb{A}(x_1 - x_2)\| \geq m\|x_1 - x_2\|,$$

which is a contradiction. Therefore $\mathbb{A}$ has an inverse operator $\mathbb{A}^{-1} : \mathbf{R}(\mathbf{A}) \longmapsto \mathbf{X}$. We have

$$\|\mathbb{A}^{-1}y\| \leq \frac{1}{m}\|\mathbb{A}\mathbb{A}^{-1}y\| = \frac{1}{m}\|y\|$$

for any $y \in \mathbf{R}(\mathbf{A})$. This completes the proof. □

Theorem 4.19. *Let $\mathbf{X}$ be a Banach space and a bounded linear operator $\mathbb{A}$ maps $\mathbf{X}$ onto $\mathbf{X}$ and $\|\mathbb{A}\| \leq q < 1$. Then the operator $\mathbb{I} + \mathbb{A}$ has an inverse which is a bounded linear operator.*

Proof. Consider the series

$$\mathbb{I} - \mathbb{A} + \mathbb{A}^2 - \cdots + (-1)^n \mathbb{A}^n + \cdots. \tag{4.13}$$

Let

$$S_n = \mathbb{I} - \mathbb{A} + \mathbb{A}^2 - \cdots + (-1)^n \mathbb{A}^n, \quad n \in \mathbf{N}.$$

Since $\|\mathbb{A}^n\| \leq \|\mathbb{A}\|^n$, it follows that

$$\begin{aligned}\|S_{n+p} - S_n\| &= \|(-1)^{n+1}\mathbb{A}^{n+1} + \cdots + (-1)^{n+p}\mathbb{A}^{n+p}\| \\ &\leq \|\mathbb{A}\|^{n+1} + \cdots + \|\mathbb{A}\|^{n+p} \leq q^{n+1} + \cdots + q^{n+p} \to 0, \quad \text{as} \quad n \to \infty\end{aligned}$$

for any $p \in \mathbf{N}$. Therefore the sequence $\{S_n\}_{n\in\mathbf{N}}$ is a Cauchy sequence in $\mathcal{L}(\mathbf{X})$. Since $\mathbf{X}$ is a Banach space, using Theorem 4.9, we see that $\mathcal{L}(\mathbf{X})$ is a Banach space. Consequently the sequence $\{S_n\}_{n\in\mathbf{N}}$ is convergent. Let

$$S = \lim_{n\to\infty} S_n.$$

Hence,

$$\begin{aligned}S(\mathbb{I} + \mathbb{A}) &= \lim_{n\to\infty} S_n(\mathbb{I} + \mathbb{A}) \\ &= \lim_{n\to\infty} (\mathbb{I} - \mathbb{A} + \mathbb{A}^2 - \cdots + (-1)^n\mathbb{A}^n)(\mathbb{I} + \mathbb{A})\end{aligned}$$

$$
\begin{aligned}
&= \lim_{n\to\infty} \left(\mathbb{I} - \mathbb{A} + \mathbb{A}^2 - \cdots + (-1)^n \mathbb{A}^n + \mathbb{A} - \mathbb{A}^2 + \cdots + (-1)^{n+1}\mathbb{A}^{n+1}\right) \\
&= \lim_{n\to\infty} \left(\mathbb{I} - \mathbb{A}^{n+1}\right) = \mathbb{I}.
\end{aligned}
$$

As above,

$$(\mathbb{I} + \mathbb{A})S = \mathbb{I}.$$

Therefore $(\mathbb{I} + \mathbb{A})^{-1} : \mathbf{X} \longmapsto \mathbf{X}$ exists and

$$(\mathbb{I} + \mathbb{A})^{-1} = S.$$

Since $\mathbb{I} + \mathbb{A}$ is a linear operator, by Theorem 4.17, it follows that $(\mathbb{I} + \mathbb{A})^{-1} : \mathbf{X} \longmapsto \mathbf{X}$ is a linear operator. Besides,

$$\|S\| = \left\| \sum_{n=0}^{\infty} \mathbb{A}^n \right\| \le \sum_{n=0}^{\infty} \|\mathbb{A}\|^n \le \sum_{n=0}^{\infty} q^n = \frac{1}{1-q}.$$

Therefore $(\mathbb{I} + \mathbb{A})^{-1} : \mathbf{X} \longmapsto \mathbf{X}$ is bounded. This completes the proof. □

Theorem 4.20. *Let* $\mathbb{A} : \mathbf{X} \longmapsto \mathbf{Y}$ *has an inverse* $\mathbb{A}^{-1} : \mathbf{R}(\mathbf{A}) \longmapsto \mathbf{X}$ *and there is an operator* $\mathbb{B} : \mathbf{X} \longmapsto \mathbf{Y}$ *such that*

$$\|\mathbb{B}\| < \left\|\mathbb{A}^{-1}\right\|^{-1}.$$

Then $\mathbb{C} = \mathbb{A} + \mathbb{B}$ *has an inverse* $\mathbb{C}^{-1} : \mathbf{R}(\mathbf{A} + \mathbf{B}) \longmapsto \mathbf{X}$ *and*

$$\left\|\mathbb{C}^{-1} - \mathbb{A}^{-1}\right\| \le \frac{\|\mathbb{B}\|}{1 - \|\mathbb{A}^{-1}\|\|\mathbb{B}\|} \left\|\mathbb{A}^{-1}\right\|^2.$$

Proof. We have

$$\mathbb{A} + \mathbb{B} = \mathbb{A}\left(\mathbb{I} + \mathbb{A}^{-1}\mathbb{B}\right).$$

Since

$$\left\|\mathbb{A}^{-1}\mathbb{B}\right\| \le \left\|\mathbb{A}^{-1}\right\|\|\mathbb{B}\| < 1,$$

by Theorem 4.19, it follows that $\mathbb{I} + \mathbb{A}^{-1}\mathbb{B}$ has an inverse and

$$\left(\mathbb{I} + \mathbb{A}^{-1}\mathbb{B}\right)^{-1} = \sum_{n=0}^{\infty} \left(-\mathbb{A}^{-1}\mathbb{B}\right)^n.$$

Note that

$$
\begin{aligned}
\mathbb{I} &= \mathbb{A}\left(\mathbb{I} + \mathbb{A}^{-1}\mathbb{B}\right)\left(\mathbb{I} + \mathbb{A}^{-1}\mathbb{B}\right)^{-1}\mathbb{A}^{-1}, \\
\mathbb{I} &= \left(\mathbb{I} + \mathbb{A}^{-1}\mathbb{B}\right)^{-1}\mathbb{A}^{-1}\mathbb{A}\left(\mathbb{I} + \mathbb{A}^{-1}\mathbb{B}\right).
\end{aligned}
$$

Therefore

$$(\mathbb{A}(\mathbb{I}+\mathbb{A}^{-1}\mathbb{B}))^{-1} = (\mathbb{I}+\mathbb{A}^{-1}\mathbb{B})^{-1}\mathbb{A}^{-1},$$

i. e.,

$$\mathbb{C}^{-1} = (\mathbb{I}+\mathbb{A}^{-1}\mathbb{B})^{-1}\mathbb{A}^{-1}.$$

Besides,

$$\begin{aligned}
\|(\mathbb{A}+\mathbb{B})^{-1}-\mathbb{A}^{-1}\| &= \|(\mathbb{A}(\mathbb{I}+\mathbb{A}^{-1}\mathbb{B}))^{-1}-\mathbb{A}^{-1}\| = \|(\mathbb{I}+\mathbb{A}^{-1}\mathbb{B})^{-1}\mathbb{A}^{-1}-\mathbb{A}^{-1}\| \\
&= \|((\mathbb{I}+\mathbb{A}^{-1}\mathbb{B})^{-1}-\mathbb{I})\mathbb{A}^{-1}\| \le \|\mathbb{I}-(\mathbb{I}+\mathbb{A}^{-1}\mathbb{B})^{-1}\|\|\mathbb{A}^{-1}\| \\
&= \left\|\sum_{n=1}^{\infty}(-\mathbb{A}^{-1}\mathbb{B})^{n}\right\|\|\mathbb{A}^{-1}\| \le \sum_{n=1}^{\infty}\|\mathbb{A}^{-1}\mathbb{B}\|^{n}\|\mathbb{A}^{-1}\| \\
&= \frac{\|\mathbb{A}^{-1}\|\|\mathbb{A}^{-1}\mathbb{B}\|}{1-\|\mathbb{A}^{-1}\mathbb{B}\|} \le \frac{\|\mathbb{A}^{-1}\|^{2}\|\mathbb{B}\|}{1-\|\mathbb{A}^{-1}\mathbb{B}\|}.
\end{aligned}$$

This completes the proof. □

Theorem 4.21. *Let $\mathbb{A}$ be a bounded linear operator that maps the Banach space $\mathbf{X}$ onto the whole of the Banach space $\mathbf{Y}$ in a one-one fashion. Then there exists a bounded linear operator $\mathbb{A}^{-1}$, inverse to the operator $\mathbb{A}$, which maps $\mathbf{Y}$ onto $\mathbf{X}$.*

Proof. Since $\mathbb{A} : \mathbf{X} \longmapsto \mathbf{Y}$ is onto and one-one, there exists $\mathbb{A}^{-1}$ that maps $\mathbf{Y}$ onto $\mathbf{X}$. By Theorem 4.17, it follows that $\mathbb{A}^{-1}$ is a linear operator. By Theorem 4.4, we see that the Banach space $\mathbf{Y}$ can be represented in the form $\mathbf{Y} = \bigcup_{k=1}^{\infty}\mathbf{Y}_k$, where $\mathbf{Y}_k \subset \mathbf{Y}$, $\|\mathbb{A}^{-1}y\| \le k\|y\|$ for any $y \in \mathbf{Y}_k$ and at least one $\mathbf{Y}_l$ is everywhere dense in $\mathbf{Y}$. We denote it by $\mathbf{Y}_n$. Let $y \in \mathbf{Y}$ be arbitrarily chosen and $\|y\| = a$. Because $\mathbf{B}_a[0] \cap \mathbf{Y}_n$ is everywhere dense in $\mathbf{B}_1[0]$, there exists $y_1 \in \mathbf{Y}_n$ such that

$$\|y-y_1\| \le \frac{a}{2} \quad \text{and} \quad \|y_1\| \le a.$$

Since $\mathbf{B}_{\frac{a}{2}}[0] \cap \mathbf{Y}_n$ is everywhere dense in $\mathbf{B}_{\frac{a}{2}}[0]$, there is $y_2 \in \mathbf{Y}_n$ such that

$$\|(y-y_1)-y_2\| \le \frac{a}{2^2}, \quad \|y_2\| \le \frac{a}{2},$$

and so on, there exists $y_m \in \mathbf{Y}_n$ such that

$$\|(y-y_1-\cdots-y_{m-1})-y_m\| \le \frac{a}{2^m}, \quad \|y_m\| \le \frac{a}{2^{m-1}}.$$

Thus,

$$y = \lim_{m\to\infty}\sum_{k=1}^{m} y_k.$$

We set

$$x_k = \mathbb{A}^{-1} y_k.$$

Then

$$\|x_k\| \le n\|y_k\| \le \frac{na}{2^{k-1}}.$$

Let

$$s_k = \sum_{m=1}^{k} x_m.$$

Then

$$\|s_{r+p} - s_r\| = \left\| \sum_{l=r+1}^{r+p} x_l \right\| \le \sum_{l=r+1}^{r+p} \|x_l\| \le na \sum_{l=r+1}^{r+p} \frac{1}{2^{l-1}} < \frac{na}{2^{r-1}}.$$

Since $\mathbf{X}$ is a complete normed space, the sequence $\{s_k\}_{k\in\mathbf{N}}$ is convergent to some element $x \in \mathbf{X}$. Consequently

$$x = \lim_{k\to\infty} \sum_{i=1}^{k} x_i = \sum_{i=1}^{\infty} x_i.$$

We have

$$\mathbb{A}x = \mathbb{A}\left(\lim_{k\to\infty} \sum_{i=1}^{k} x_i \right) = \lim_{k\to\infty} \mathbb{A}\left(\sum_{i=1}^{k} x_i \right) = \lim_{k\to\infty} \sum_{i=1}^{k} \mathbb{A}x_i = \lim_{k\to\infty} \sum_{i=1}^{k} y_i = y.$$

Hence,

$$\|\mathbb{A}^{-1}y\| = \|x\| = \lim_{k\to\infty} \left\| \sum_{i=1}^{k} x_i \right\| \le \lim_{k\to\infty} \sum_{i=1}^{k} \|x_i\| \le \lim_{k\to\infty} \sum_{i=1}^{k} \frac{na}{2^{k-1}} = 2na = 2n\|y\|.$$

Because $y \in \mathbf{Y}$ was arbitrarily chosen, we conclude that $\mathbb{A}^{-1}$ is bounded. This completes the proof. □

4.4 Advanced practical problems

Problem 4.1. Let $\mathbf{X} = \mathbf{Y} = \mathbf{C}([0,1])$. For $x \in \mathbf{X}$ we define the operator

$$y(t) = t \int_0^1 x(s)ds, \quad t \in [0,1], \quad y = \mathbb{A}x.$$

Prove that $\mathbb{A} : \mathbf{X} \longmapsto \mathbf{X}$ is a linear operator.

Problem 4.2. Let $\mathbf{X} = \mathbf{Y} = \mathbf{C}([0,1])$, $\|x\| = \max_{t\in[0,1]} |x(t)|$, $x \in \mathbf{C}([0,1])$. Find $\|\mathbb{A}\|$, where

$$\mathbb{A}x(t) = t^7 \int_0^t s^4 x(s)ds, \quad x \in \mathbf{X}, \quad t \in [0,1].$$

Answer. $\frac{1}{5}$.

Problem 4.3. Let $\mathbf{X}$ be a normed vector space, and $\mathbf{Y}$ and $\mathbf{Z}$ be Banach spaces. Suppose that $\mathbf{B} \in \mathcal{L}(\mathbf{X}, \mathbf{Y})$, $\mathbb{A} \in \mathcal{L}(\mathbf{Y}, \mathbf{Z})$. Prove that

$$\|\mathbb{A}\mathbb{B}\| \leq \|\mathbb{A}\|\|\mathbb{B}\|.$$

Problem 4.4. Let $\mathbf{X}$, $\mathbf{Y}$ and $\mathbf{Z}$ be normed vector spaces. Suppose that $\{\mathbb{B}_n\}_{n\in\mathbf{N}} \subset \mathcal{L}(\mathbf{X}, \mathbf{Y})$, $\mathbb{B} \in \mathcal{L}(\mathbf{X}, \mathbf{Y})$, $\{\mathbb{A}_n\}_{n\in\mathbf{N}} \subset \mathcal{L}(\mathbf{Y}, \mathbf{Z})$, $\mathbb{A} \in \mathcal{L}(\mathbf{Y}, \mathbf{Z})$. Suppose that $\mathbb{A}_n \to \mathbb{A}$, $\mathbb{B}_n \to \mathbb{B}$, as $n \to \infty$. Prove that $\mathbb{A}_n\mathbb{B}_n \to \mathbb{A}\mathbb{B}$, as $n \to \infty$.

Problem 4.5. Let $\mathbf{X}$ be a normed vector space and $\mathbb{A} \in \mathcal{L}(\mathbf{X})$. Define

$$\sin(\mathbb{A}) = \sum_{k=0}^{\infty} \frac{(-1)^k \mathbb{A}^{2k+1}}{(2k+1)!}, \quad \cos(\mathbb{A}) = \sum_{k=0}^{\infty} \frac{(-1)^k \mathbb{A}^{2k}}{(2k)!},$$
$$\sinh(\mathbb{A}) = \sum_{k=0}^{\infty} \frac{\mathbb{A}^{2k+1}}{(2k+1)!}, \quad \cosh(\mathbb{A}) = \sum_{k=0}^{\infty} \frac{\mathbb{A}^{2k}}{(2k)!}.$$

Prove that

$$e^{\mathbb{A}} = \sinh(\mathbb{A}) + \cosh(\mathbb{A}), \quad \|\sinh(\mathbb{A})\| \leq \sinh(\|\mathbb{A}\|),$$
$$\|\cosh(\mathbb{A})\| \leq \cosh(\|\mathbb{A}\|), \quad \|\sin(\mathbb{A})\| \leq \sinh(\|\mathbb{A}\|),$$
$$\|\cos(\mathbb{A})\| \leq \cosh(\|\mathbb{A}\|).$$

5 Linear functionals

5.1 The Hahn–Banach extension theorem

Let $\mathbf{X}$ be a normed vector space.

Definition 5.1. A linear operator $f : \mathbf{X} \longmapsto \mathbf{R}$ will be called a linear functional on $\mathbf{X}$.

Since the set $\mathbf{R}$ of real numbers is a Banach space, all previous definitions and theorems derived for linear continuous operators are preserved for linear continuous functionals.

Theorem 5.1 (The Hahn–Banach extension theorem). *Every linear bounded functional f defined on a linear subspace $\mathbf{L}$ of $\mathbf{X}$ can be extended to the entire space with preservation of the norm, i. e., for every linear bounded functional f on $\mathbf{L}$ there exists a linear functional F, defined on $\mathbf{X}$, such that*
1. $F(x) = f(x)$, $x \in \mathbf{L}$,
2. $\|F\|_{\mathbf{X}} = \|f\|_{\mathbf{L}}$.

Proof. If $\mathbf{L} = \mathbf{X}$, then the assertion is evident. Let $\mathbf{L} \subset \mathbf{X}$. Let also, $x_0 \notin \mathbf{L}$ and

$$\mathbf{L}_1 = \{x + tx_0 : x \in \mathbf{L}, \quad t \in \mathbf{R}\}.$$

Note that $\mathbf{L}_1$ is a linear subspace of $\mathbf{X}$. We will prove that each element of $\mathbf{L}_1$ has a unique representation of the form $x + tx_0$. Assume the contrary. Let $y \in \mathbf{L}_1$ be such that

$$y = x_1 + t_1 x_0 \quad \text{and} \quad y = x_2 + t_2 x_0,$$

where $x_1, x_2 \in \mathbf{L}$, $t_1, t_2 \in \mathbf{R}$, and $t_1 \neq t_2$. If $t_1 = t_2$, then $x_1 = x_2$. Let $t_1 \neq t_2$. We have

$$x_1 + t_1 x_0 = x_2 + t_2 x_0,$$

whereupon

$$x_1 - x_2 = (t_2 - t_1)x_0,$$

or

$$x_0 = \frac{x_1 - x_2}{t_2 - t_1}.$$

Since $\mathbf{L}$ is a linear subspace of $\mathbf{X}$ and $x_1, x_2 \in \mathbf{L}$, $t_1, t_2 \in \mathbf{R}$, $t_1 \neq t_2$, we get $x_0 \in \mathbf{L}$. This is a contradiction. Now we take two elements y_1 and y_2 of $\mathbf{L}$. We have

$$\begin{aligned} f(y_1) - f(y_2) = f(y_1 - y_2) &\leq \|f\|\|y_1 - y_2\| = \|f\|\|y_1 + x_0 - (y_2 + x_0)\| \\ &\leq \|f\|(\|y_1 + x_0\| + \|y_2 + x_0\|). \end{aligned}$$

https://doi.org/10.1515/9783110657722-005

Hence,

$$f(y_1) - \|f\|\|y_1 + x_0\| \le f(y_2) + \|f\|\|y_2 + x_0\|.$$

Since $y_1, y_2 \in \mathbf{L}$ were arbitrarily chosen, independently of each other, we get

$$\sup_{x\in\mathbf{L}}\{f(x) - \|f\|\|x + x_0\|\} \le \inf_{x\in\mathbf{L}}\{f(x) + \|f\|\|x + x_0\|\}.$$

Therefore there exists a real constant c such that

$$\sup_{x\in\mathbf{L}}\{f(x) - \|f\|\|x + x_0\|\} \le c \le \inf_{x\in\mathbf{L}}\{f(x) + \|f\|\|x + x_0\|\}. \tag{5.1}$$

We fix such a constant c. Let now $y \in \mathbf{L}_1$ be arbitrarily chosen. Then it has the representation

$$y = x + tx_0,$$

where $x \in \mathbf{L}$ and t is a uniquely defined real number. Define the functional $\phi : \mathbf{L}_1 \longmapsto \mathbf{R}$ as follows:

$$\phi(y) = f(x) - tc, \quad y \in \mathbf{L}_1.$$

For $y \in \mathbf{L}$ we have

$$\phi(y) = f(y).$$

Let $y_1, y_2 \in \mathbf{L}_1$ and

$$y_1 = x_1 + t_1x_0, \quad y_2 = x_2 + t_2x_0, \quad x_1, x_2 \in \mathbf{L}, \quad t_1, t_2 \in \mathbf{R}.$$

Then

$$y_1 + y_2 = x_1 + x_2 + (t_1 + t_2)x_0.$$

Hence,

$$\begin{aligned}\phi(y_1 + y_2) &= f(x_1 + x_2) - (t_1 + t_2)c = f(x_1) + f(x_2) - t_1c - t_2c\\ &= (f(x_1) - t_1c) + (f(x_2) - t_2c) = \phi(y_1) + \phi(y_2),\end{aligned}$$

i. e., $\phi : \mathbf{L}_1 \longmapsto \mathbf{R}$ is additive. If $\alpha \in \mathbf{R}$, then

$$\alpha y_1 = \alpha x_1 + \alpha t_1 c$$

and

$$\phi(\alpha y_1) = f(\alpha x_1) - \alpha t_1 c = \alpha f(x_1) - \alpha t_1 c = \alpha(f(x_1) - t_1 c) = \alpha\phi(y_1).$$

Therefore $\phi : \mathbf{L}_1 \longmapsto \mathbf{R}$ is a linear functional. Now we will prove that $\phi : \mathbf{L}_1 \longmapsto \mathbf{R}$ is bounded and $\|\phi\|_{\mathbf{L}_1} = \|f\|_{\mathbf{L}}$. Let $y \in \mathbf{L}_1$ be such that

$$y = x + tx_0, \quad x \in \mathbf{L}, \quad t \in \mathbf{R}.$$

1. Let $t > 0$. Then

$$\begin{aligned}\phi(y) &= f(x) - tc \leq |f(x) - tc| = t\left|f\left(\frac{x}{t}\right) - c\right| \leq t\|f\|_{\mathbf{L}}\left\|\frac{x}{t} + x_0\right\| \\ &= \|f\|\|_{\mathbf{L}}x + tx_0\| = \|f\|_{\mathbf{L}}\|y\|.\end{aligned} \tag{5.2}$$

2. Let $t < 0$. Then

$$\begin{aligned}f\left(\frac{x}{t}\right) - c &\geq -\|f\|_{\mathbf{L}}\left\|\frac{x}{t} + x_0\right\| = -\frac{1}{|t|}\|f\|_{\mathbf{L}}\|x + tx_0\| \\ &= \frac{1}{t}\|f\|\|_{\mathbf{L}}x + tx_0\| = \frac{1}{t}\|f\|_{\mathbf{L}}\|y\|\end{aligned}$$

   and

$$\phi(y) = t\left(f\left(\frac{x}{t}\right) - c\right) \leq \|f\|_{\mathbf{L}}\|y\|. \tag{5.3}$$

   Substituting $-y$ for y in (5.2) and (5.3), we get

$$-\phi(y) \leq \|f\|\|y\|$$

   or

$$\phi(y) \geq -\|f\|_{\mathbf{L}}\|y\|.$$

Therefore

$$|\phi(y)| \leq \|f\|_{\mathbf{L}}\|y\|$$

and

$$\|\phi\|_{\mathbf{L}_1} \leq \|f\|_{\mathbf{L}}.$$

Since the functional ϕ is an extension of f from $\mathbf{L}$ to $\mathbf{L}_1$, we get

$$\|\phi\|_{\mathbf{L}_1} \geq \|f\|_{\mathbf{L}}.$$

Therefore

$$\|\phi\|_{\mathbf{L}_1} = \|f\|_{\mathbf{L}}.$$

If there is a $x_1 \notin \mathbf{L}_1$, $x_1 \in \mathbf{X}$, then we extend the functional f to the functional ϕ_1 on

$$\mathbf{L}_2 = \{x + tx_1 : x \in \mathbf{L}_1, \quad t \in \mathbf{R}\}$$

such that

$$\phi_1(x) = f(x), \quad x \in \mathbf{L}, \quad \|\phi_1\|_{\mathbf{L}_2} = \|f\|_{\mathbf{L}}.$$

Thus proceeding, we construct a linear functional ϕ_w, defined on the linear subspace $\mathbf{L}_w$, which is everywhere dense in $\mathbf{X}$ and it is equal to the union of all $\mathbf{L}_n$. Moreover, $\|\phi_w\|_{\mathbf{L}_w} = \|f\|_{\mathbf{L}}$. Hence, by Theorem 4.16, we extend ϕ_w to a linear functional $\tilde{\phi}_w$ on $\mathbf{X}$ such that

$$\tilde{\phi}_w(x) = f(x), \quad x \in \mathbf{L},$$

and

$$\|\tilde{\phi}_w\|_{\mathbf{X}} = \|f\|_{\mathbf{L}}.$$

This completes the proof. □

Corollary 5.1. *Let $x_0 \neq 0$ be any fixed element of $\mathbf{X}$. Then there exists a linear functional f, defined on the entire space $\mathbf{X}$, such that*
1. $\|f\| = 1$,
2. $f(x_0) = \|x_0\|$.

Proof. Let

$$\mathbf{L} = \{tx_0 : t \in \mathbf{R}\}.$$

Note that L is a linear subspace of $\mathbf{X}$, spanned by x_0. Define $\phi : \mathbf{L} \longmapsto \mathbf{R}$ as follows:

$$\phi(x) = t\|x_0\|, \quad x = tx_0.$$

Then, if $x_1, x_2 \in \mathbf{L}$,

$$x_1 = t_1x_0, \quad x_2 = t_2x_0,$$

and $\alpha, \beta \in \mathbf{R}$, we have

$$\begin{aligned} \alpha x_1 + \beta x_2 &= \alpha t_1 x_0 + \beta t_2 x_0 = (\alpha t_1 + \beta t_2)x_0, \\ \phi(\alpha x_1 + \beta x_2) &= (\alpha t_1 + \beta t_2)\|x_0\| = \alpha(t_1\|x_0\|) + \beta(t_2\|x_0\|) = \alpha\phi(x_1) + \beta\phi(x_2), \end{aligned}$$

i. e., $\phi : \mathbf{L} \longmapsto \mathbf{R}$ is a linear functional. Also,

$$\phi(x_0) = \|x_0\|,$$

and, if $x = tx_0$, we get

$$|\phi(x)| = |t|\|x_0\| = \|tx_0\| = \|x\|.$$

Hence,

$$\|\phi\| = 1.$$

Now we apply the Hahn–Banach extension theorem and we see that there exists a linear functional f, defined on $\mathbf{X}$, such that

1. $f(x) = \phi(x)$, $x \in \mathbf{L}$. In particular,

$$f(x_0) = \|x_0\|.$$

2.

$$\|f\|_X = \|\phi\|_{\mathbf{L}} = 1.$$

This completes the proof. □

Corollary 5.2. *Let* $\mathbf{L}$ *be a linear subspace of* $\mathbf{X}$ *and* $x_0 \notin \mathbf{L}$, $x_0 \in \mathbf{X}$. *Let also*

$$d = \inf_{x\in\mathbf{L}} \|x_0 - x\|.$$

Then there exists a linear functional f, *defined on* $\mathbf{X}$, *such that*

1. $f(x) = 0$, $x \in \mathbf{L}$,
2. $f(x_0) = 1$,
3. $\|f\| = \frac{1}{d}$.

Proof. Let

$$\mathbf{L}_1 = \{x + tx_0 : x \in \mathbf{L}, \quad t \in \mathbf{R}\}.$$

Note that every element y of $\mathbf{L}_1$ is uniquely representable in the form

$$y = x + tx_0, \quad x \in \mathbf{L}, \quad t \in \mathbf{R}.$$

Define $\phi : \mathbf{L}_1 \longmapsto \mathbf{R}$ as follows:

$$\phi(y) = t, \quad y = x + tx_0.$$

Then, if $y_1, y_2 \in \mathbf{L}_1$,

$$y_1 = x_1 + t_1x_0, \quad y_2 = x_2 + t_2x_0, \quad x_1, x_2 \in \mathbf{L}, \quad t_1, t_2 \in \mathbf{R},$$

$\alpha, \beta \in \mathbf{R}$, then

$$\alpha y_1 + \beta y_2 = \alpha(x_1 + t_1x_0) + \beta(x_2 + t_2x_0) = (\alpha x_1 + \beta x_2) + (\alpha t_1 + \beta t_2)x_0,$$
$$\phi(\alpha y_1 + \beta y_2) = \alpha t_1 + \beta t_2 = \alpha\phi(y_1) + \beta\phi(y_2),$$

i. e., $\phi : \mathbf{L}_1 \to \mathbf{R}$ is a linear functional. Next, if $x \in \mathbf{L}$, then

$$\phi(x) = 0$$

and

$$\phi(x_0) = 1.$$

Moreover, for $y = x + tx_0 \in \mathbf{L}_1$, we obtain

$$\begin{aligned}|\phi(y)| = |t| = |t|\frac{\|y\|}{\|y\|} = |t|\frac{\|y\|}{\|x + tx_0\|} = \frac{\|y\|}{\|\frac{x}{t} + x_0\|}\\ = \frac{\|y\|}{\|x_0 - (-\frac{x}{t})\|} \le \frac{\|y\|}{d}.\end{aligned}$$

Hence,

$$\|\phi\|_{\mathbf{L}_1} \le \frac{1}{d}.$$

Let $\{x_n\}_{n\in\mathbf{N}}$ be a sequence in $\mathbf{L}$ such that

$$\lim_{n\to\infty} \|x_n - x_0\| = d.$$

Then

$$|\phi(x_n - x_0)| \le \|\phi\|_{\mathbf{L}_1} \|x_n - x_0\|.$$

Since

$$|\phi(x_n - x_0)| = |\phi(x_n) - \phi(x_0)| = \phi(x_0) = 1,$$

we get

$$1 \le \|\phi\|_{\mathbf{L}_1} \|x_n - x_0\|.$$

Hence,

$$1 \le \|\phi\|_{\mathbf{L}_1} \lim_{n\to\infty} \|x_n - x_0\| = d\|\phi\|_{\mathbf{L}_1},$$

or

$$\|\phi\|_{\mathbf{L}_1} \ge \frac{1}{d}.$$

Consequently

$$\|\phi\|_{\mathbf{L}_1} = \frac{1}{d}.$$

By the Hahn–Banach extension theorem, it follows that there exists a linear functional f, defined on $\mathbf{X}$, such that

1. $f(x) = \phi(x)$, $x \in \mathbf{L}_1$. In particular,

   $$f(x) = \phi(x) = 0, \quad x \in \mathbf{L},$$

   and

   $$f(x_0) = \phi(x_0) = 1.$$

2.

$$\|f\|_{\mathbf{X}} = \|\phi\|_{\mathbf{L}_1} = \frac{1}{d}.$$

This completes the proof. □

Theorem 5.2. *Let $\{x_n\}_{n\in\mathbf{N}} \subset \mathbf{X}$. Then, in order for $x_0 \in \mathbf{X}$ to be the limit of some sequence of linear combinations of the form $\sum_{j=1}^n c_j x_j$ it is necessary and sufficient that $f(x_0) = 0$ for all linear continuous functionals f, defined on $\mathbf{X}$ for which $f(x_n) = 0$, $n \in \mathbf{N}$.*

Proof. Let $\mathbf{L}$ be the vector space spanned by $\{x_n\}_{n\in\mathbf{N}}$.

1. Let f be a linear functional on $\mathbf{X}$ such that $f(x_0) = 0$ and $f(x_n) = 0$, $n \in \mathbf{N}$. Assume that

$$d = \inf_{x\in\mathbf{L}} \|x - x_0\| > 0.$$

Hence, by Corollary 5.2, it follows that there exists a linear functional f_0, defined on $\mathbf{X}$, such that

$$f_0(x_n) = 0, \quad n \in \mathbf{N}, \quad \text{and} \quad f_0(x_0) = 1.$$

This is a contradiction. Therefore $d = 0$. Hence, $x_0 \in \mathbf{L}$ or x_0 can be approximated by elements of the form $\sum_{j=1}^n c_j x_j$.

2. Let x_0 be the limit of a sequence $\{\sum_{j=1}^n c_j x_j\}$ and let $f(x_l) = 0$, $l \in \mathbf{N}$, for some linear functional f, defined on $\mathbf{X}$. Then

$$\begin{aligned} f(x_0) &= f\left(\lim_{n\to\infty} \sum_{j=1}^n c_j x_j\right) = \lim_{n\to\infty} f\left(\sum_{j=1}^n c_j x_j\right) \\ &= \lim_{n\to\infty}\left(\sum_{j=1}^n f(c_j x_j)\right) = \lim_{n\to\infty}\left(\sum_{j=1}^n c_j f(x_j)\right) = 0. \end{aligned}$$

This completes the proof. □

5.2 The general form of the linear functionals on $\mathbf{E}_n$ in the case $\mathbf{F} = \mathbf{R}$

Consider the vector space $\mathbf{E}_n$ on $\mathbf{R}$. Let $e_1, \ldots, e_n$ be a basis of $\mathbf{E}_n$. Let also, $f : \mathbf{E}_n \longmapsto \mathbf{R}$ be an arbitrary linear functional. For

$$x = \sum_{i=1}^n \xi_i e_i, \quad \xi_i \in \mathbf{R}, \quad i \in \{1, \ldots, n\},$$

we get

$$f(x) = f\left(\sum_{i=1}^{n} \xi_i e_i\right) = \sum_{i=1}^{n} f(\xi_i e_i) = \sum_{i=1}^{n} \xi_i f(e_i) = \sum_{i=1}^{n} \xi_i f_i,$$

where $f_i = f(e_i)$, $i \in \{1, \dots, n\}$. If $f : \mathbf{E}_n \longmapsto \mathbf{R}$ is expressed of the form

$$f(x) = \sum_{i=1}^{n} \xi_i f_i, \quad x \in \mathbf{E}_n, \quad x = \sum_{i=1}^{n} \xi_i e_i, \quad \xi_i \in \mathbf{R}, \quad i \in \{1, \dots, n\},$$

where $f_i \in \mathbf{R}$, $i \in \{1, \dots, n\}$, are arbitrarily chosen, then it is a linear functional on $\mathbf{E}_n$.

5.3 The general form of the linear functionals on Hilbert spaces

Let $\mathbf{H}$ be a Hilbert space.

Theorem 5.3. *Every linear bounded functional f on the Hilbert space $\mathbf{H}$ can be represented of the form*

$$f(x) = (x, u),$$

where the element $u \in \mathbf{H}$ is defined uniquely by the functional f. Moreover,

$$\|f\| = \|u\|.$$

Proof. Let f be an arbitrary linear functional on $\mathbf{H}$. Define

$$\mathbf{L} = \{x \in \mathbf{H} : f(x) = 0\}.$$

Note that $\mathbf{L}$ is a closed linear subspace of $\mathbf{H}$. If $\mathbf{L} = \mathbf{H}$, then $f = 0$ and we take $u = 0$. Let $\mathbf{L} \neq \mathbf{H}$. Then there exists $x_0 \in \mathbf{H} \setminus \mathbf{L}$ such that $x_0 \perp \mathbf{L}$. Denote

$$\alpha = f(x_0).$$

By the definition of $\mathbf{L}$, it follows that $\alpha \neq 0$. We set

$$x_1 = \frac{x_0}{\alpha}.$$

Then

$$f(x_1) = f\left(\frac{x_0}{\alpha}\right) = \frac{1}{\alpha} f(x_0) = 1.$$

Take $x \in \mathbf{H}$ arbitrarily and set

$$\beta = f(x).$$

Then

$$0 = f(x) - \beta f(x_1) = f(x) - f(\beta x_1) = f(x - \beta x_1).$$

Therefore $z = x - \beta x_1 \in \mathbf{L}$ and

$$x = z + \beta x_1.$$

Also, $x_1 \perp z$ and

$$(x, x_1) = (z + \beta x_1, x_1) = (z, x_1) + (\beta x_1, x_1) = \beta(x_1, x_1) = \beta \|x_1\|^2,$$

whereupon

$$f(x) = \frac{1}{\|x_1\|^2}(x, x_1) = \left(x, \frac{x_1}{\|x_1\|^2}\right).$$

Let

$$u = \frac{x_1}{\|x_1\|^2}.$$

Then

$$f(x) = (x, u).$$

Assume that there is a $v \in \mathbf{H}$ such that

$$f(x) = (x, v) \quad \text{for} \quad \text{any} \quad x \in \mathbf{H}.$$

Then

$$(x, u - v) = 0 \quad \text{for} \quad \text{any} \quad x \in \mathbf{H}.$$

In particular, for $x = u - v$, we get

$$(u - v, u - v) = 0.$$

Therefore $u = v$. Next,

$$|f(x)| = |(x, u)| \leq \|x\| \|u\|, \quad x \in \mathbf{H}.$$

Therefore

$$\|f\| \leq \|u\|. \tag{5.4}$$

Also,

$$|f(u)| = |(u, u)| = \|u\|^2.$$

Consequently

$$\|f\| \geq \|u\|.$$

Hence, by (5.4), we conclude that

$$\|f\| = \|u\|.$$

This completes the proof. □

5.4 Weak convergence of sequences of functionals

Let $\mathbf{X}$ and $\mathbf{Y}$ be normed vector spaces.

Definition 5.2. A sequence $\{f_n\}_{n\in\mathbf{N}}$ of linear functionals defined on $\mathbf{X}$ is said to be weakly convergent to the linear functional f_0 defined on $\mathbf{X}$, if

$$f_n(x) \to f(x) \quad \text{as} \quad n \to \infty,$$

for any $x \in \mathbf{X}$.

Theorem 5.4. *Let $\{x_n\}_{n\in\mathbf{N}}$ be a sequence of elements of $\mathbf{X}$ that converges weakly to an element $x_0 \in \mathbf{X}$. Then there exists a sequence of linear combinations $\{\sum_{k=1}^{k_n} c_k^{(n)} x_k\}_{k_n\in\mathbf{N}}$, converging strongly to x_0.*

Proof. Assume the contrary. By Corollary 5.2, it follows that there exists a linear functional f, defined on $\mathbf{X}$, such that $f(x_0) = 1$ and $f(x_n) = 0$ for any $n \in \mathbf{N}$. Hence, the sequence $\{f(x_n)\}_{n\in\mathbf{N}}$ does not converge to $f(x_0)$. This is a contradiction. This completes the proof. □

Theorem 5.5. *Let $\mathbb{A} : \mathbf{X} \longmapsto \mathbf{Y}$ be a linear bounded operator and $\{x_n\}_{n\in\mathbf{N}} \subset \mathbf{X}$ be a sequence that converges weakly to $x_0 \in \mathbf{X}$. Then $\{\mathbb{A}x_n\}_{n\in\mathbf{N}}$ converges weakly to $\mathbb{A}x_0$.*

Proof. Let ϕ be arbitrarily chosen linear functional defined on $\mathbf{Y}$. Define f on $\mathbf{X}$ as follows:

$$f(x) = \phi(\mathbb{A}x), \quad x \in \mathbf{X}.$$

Note that f is a linear functional on $\mathbf{X}$. Since

$$x_n \rightharpoonup x_0, \quad \text{as} \quad n \to \infty,$$

we get

$$\phi(\mathbb{A}x_n) = f(x_n) \to f(x_0) = \phi(\mathbb{A}x_0), \quad \text{as} \quad n \to \infty.$$

Therefore

$$\mathbb{A}x_n \rightharpoonup \mathbb{A}x_0, \quad \text{as} \quad n \to \infty.$$

This completes the proof. □

5.5 Advanced practical problems

Problem 5.1. Let $\mathbf{X} = C([-1,1])$. Prove that

1. $f(x) = \frac{1}{7}(x(-1) + x(1))$, $x \in \mathbf{X}$,

2. $f(x) = 4(x(1) + x(0))$, $x \in \mathbf{X}$,
3. $f(x) = \int_{-1}^{1} x(t)dt + x(0)$, $x \in \mathbf{X}$,
4. $f(x) = \int_{-1}^{0} x(t)dt - 2\int_{0}^{1} x(t)dt$, $x \in \mathbf{X}$,

are linear functionals on $\mathbf{X}$.

Problem 5.2. Let $\mathbb{A} : \mathbf{L}^2([0,1]) \longmapsto \mathbf{L}^2([0,1])$ and
1. $\mathbb{A}x(t) = \int_{0}^{t} x(s)ds$, $x \in \mathbf{L}^2([0,1])$,
2. $\mathbb{A}x(t) = tx(t)$, $t \in [0,1]$, $x \in \mathbf{L}^2([0,1])$.

Find $\|\mathbb{A}\|$.

6 Relatively compact sets in metric and normed spaces. Compact operators

6.1 Definitions. General theorems

Let $\mathbf{X}$ be a metric space.

Definition 6.1. A set $\mathbf{K} \subset \mathbf{X}$ is called relatively compact if every sequence of elements of this set contains a convergent subsequence. If the limits of the sequences belong to $\mathbf{K}$, then $\mathbf{K}$ is called sequentially compact. If the limits belong to $\mathbf{X}$ and possibly not to the set $\mathbf{K}$, then $\mathbf{K}$ is called relatively compact in $\mathbf{X}$ or relatively compact with respect to $\mathbf{X}$.

Exercise 6.1. Let $\mathbf{K} \subset \mathbf{X}$. Prove that $\mathbf{K}$ is sequentially compact if and only if $\mathbf{K}$ is closed and relatively compact in $\mathbf{X}$.

Definition 6.2. If every infinite subset of the metric space $\mathbf{X}$ contains a sequence which converges to an element of $\mathbf{X}$, then $\mathbf{X}$ is called (sequentially)compact.

Theorem 6.1 (Cantor's theorem). *Given a nested sequence*

$$\mathbf{K}_1 \supset \mathbf{K}_2 \supset \cdots \supset \mathbf{K}_n \supset \cdots$$

of nonempty sequentially compact sets of $\mathbf{X}$. *Then the intersection*

$$\mathbf{K} = \bigcap_{i=1}^{\infty} \mathbf{K}_i$$

is nonempty.

Proof. Let $x_i \in \mathbf{K}_i$ be arbitrarily chosen for any $i \in \mathbf{N}$. Consider the sequence $\{x_n\}_{n\in\mathbf{N}}$. We have $\{x_n\}_{n\geq i} \subset \mathbf{K}_i$ for any $i \in \mathbf{N}$. Let $j \in \mathbf{N}$ be arbitrarily chosen and fixed. Since $\mathbf{K}_j$ is sequentially compact set in $\mathbf{X}$, there is a subsequence $\{x_{j_n}\}_{n\in\mathbf{N}}$ that converges to x_0 and $x_0 \in \mathbf{K}_j$. For arbitrary $m \in \mathbf{N}$, $m \geq j$, since all terms x_{j_n}, $j_n \geq m$, of this sequence belong to $\mathbf{K}_m$ and $\mathbf{K}_m$ is sequentially compact, we have $x_0 \in \mathbf{K}_m$. Therefore

$$x_0 \in \bigcap_{i=1}^{\infty} \mathbf{K}_i.$$

This completes the proof. □

Theorem 6.2. *Let* $\mathbf{K}$ *be a sequentially compact set of* $\mathbf{X}$ *and* f *be a linear continuous functional on* $\mathbf{K}$. *Then:*
1. f *is bounded on* $\mathbf{K}$.
2. f *assumes its least upper (supremum) and greatest lower (infimum) bounds on* $\mathbf{K}$.

https://doi.org/10.1515/9783110657722-006

Proof.

1. Assume that f is not bounded above on $\mathbf{K}$. Then there is a sequence $\{x_n\}_{n\in\mathbf{N}}$ of elements of $\mathbf{K}$ such that

$$f(x_n) > n \quad \text{for} \quad \text{any} \quad n \in \mathbf{N}.$$

Since $\mathbf{K}$ is sequentially compact, there is a subsequence $\{x_{n_k}\}_{k\in\mathbf{N}}$ of the sequence $\{x_n\}_{n\in\mathbf{N}}$ such that

$$x_{n_k} \longrightarrow x_0, \quad \text{as} \quad k \longrightarrow \infty, \quad x_0 \in \mathbf{K}.$$

Then

$$f(x_{n_k}) > n_k,$$
$$f(x_0) = \lim_{k\to\infty} f(x_{n_k}) \geq \lim_{k\to\infty} n_k = \infty,$$

which is a contradiction. Therefore the functional f is bounded above on $\mathbf{K}$. Now we assume that the functional f is not bounded below on $\mathbf{K}$. Then there exists a sequence $\{y_n\}_{n\in\mathbf{N}}$ of elements of $\mathbf{K}$ such that

$$f(y_n) < -n \quad \text{for} \quad \text{any} \quad n \in \mathbf{N}.$$

Because $\mathbf{K}$ is a sequentially compact set, there is a subsequence $\{y_{n_k}\}_{k\in\mathbf{N}}$ of the sequence $\{y_n\}_{n\in\mathbf{N}}$ such that

$$y_{n_k} \longrightarrow y_0, \quad \text{as} \quad k \longrightarrow \infty, \quad y_0 \in \mathbf{K}.$$

We have

$$f(y_0) = \lim_{k\to\infty} f(y_{n_k}) \leq -\lim_{k\to\infty} n_k = -\infty.$$

This is a contradiction. Therefore f is a bounded below functional on $\mathbf{K}$. Consequently f is a bounded functional on $\mathbf{K}$.

2. Let

$$\alpha = \inf_{x\in\mathbf{K}} f(x).$$

Then

$$f(x) \geq \alpha \quad \text{for} \quad \text{any} \quad x \in \mathbf{K}$$

and there exists a sequence $\{x_n\}_{n\in\mathbf{N}}$ of elements of $\mathbf{K}$ such that

$$f(x_n) < \alpha + \frac{1}{n} \quad \text{for} \quad \text{any} \quad n \in \mathbf{N}.$$

Since **K** is sequentially compact, there is a subsequence $\{x_{n_k}\}_{k\in\mathbf{N}}$ of the sequence $\{x_n\}_{n\in\mathbf{N}}$ such that

$$x_{n_k} \to x_0, \quad \text{as} \quad k \to \infty, \quad x_0 \in \mathbf{K}.$$

Then

$$\alpha \leq f(x_{n_k}) < \alpha + \frac{1}{n_k}, \quad k \in \mathbf{N}.$$

Therefore

$$\alpha \leq \lim_{k\to\infty} f(x_{n_k}) = f(x_0) \leq \alpha,$$

i. e.,

$$\alpha = f(x_0).$$

Let

$$\beta = \sup_{x\in\mathbf{K}} f(x).$$

Then

$$f(x) \leq \beta \quad \text{for} \quad \text{any} \quad x \in \mathbf{K}$$

and there exists a sequence $\{y_n\}_{n\in\mathbf{N}}$ of elements of **K** such that

$$f(y_n) > \beta - \frac{1}{n} \quad \text{for} \quad \text{any} \quad n \in \mathbf{N}.$$

Since **K** is sequentially compact, there is a subsequence $\{y_{n_k}\}_{k\in\mathbf{N}}$ of the sequence $\{y_n\}_{n\in\mathbf{N}}$ so that

$$y_{n_k} \to y_0, \quad \text{as} \quad k \to \infty, \quad y_0 \in \mathbf{K}.$$

Then

$$\beta - \frac{1}{n_k} < f(y_{n_k}) \leq \beta, \quad k \in \mathbf{N}.$$

Hence,

$$\beta = \lim_{k\to\infty}\left(\beta - \frac{1}{n_k}\right) \leq f(y_0) = \lim_{k\to\infty} f(y_{n_k}) \leq \beta.$$

This completes the proof. □

6.2 Criteria for compactness of sets in metric spaces

Let $\mathbf{X}$ be a metric space with a metric d.

Definition 6.3. A set $\mathbf{A}$ in the metric space $\mathbf{X}$ is called an ϵ-net for the set $\mathbf{B}$ in the metric space $\mathbf{X}$ if for every $x \in \mathbf{B}$ there is an element $x_\epsilon \in \mathbf{A}$ such that

$$d(x, x_\epsilon) < \epsilon.$$

Theorem 6.3 (Hausdorff's theorem). *For a set $\mathbf{K}$ in the metric space $\mathbf{X}$ to be relatively compact, it is necessary, and in the case of completeness of $\mathbf{X}$, sufficient that there is a finite ϵ-net for the set $\mathbf{K}$ for any $\epsilon > 0$.*

Proof.

1. Necessity. Let $\mathbf{K}$ be a relatively compact set in the metric space $\mathbf{X}$ and $x_1 \in \mathbf{K}$ be arbitrarily chosen. Take $\epsilon > 0$ arbitrarily. If

   $$d(x, x_1) < \epsilon \quad \text{for} \quad \text{any} \quad x \in \mathbf{K},$$

   then a finite ϵ-net is already constructed. Otherwise, there is an element $x_2 \in \mathbf{K}$ such that

   $$d(x_2, x_1) \geq \epsilon.$$

   If either

   $$d(x, x_1) < \epsilon \quad \text{or} \quad d(x, x_2) < \epsilon \quad \text{for} \quad \text{any} \quad x \in \mathbf{K},$$

   then a finite ϵ-net is found. If this does not hold, then there is an element $x_3 \in \mathbf{K}$ so that

   $$d(x_1, x_3) \geq \epsilon \quad \text{or} \quad d(x_2, x_3) \geq \epsilon.$$

   We continue this process. If this process continues indefinitely, we get a sequence $\{x_n\}_{n\in\mathbf{N}}$ of elements of $\mathbf{K}$ such that

   $$d(x_i, x_j) \geq \epsilon, \quad i, j \in \mathbf{N}, \quad i \neq j.$$

   Hence, there is not any subsequence of the sequence $\{x_n\}_{n\in\mathbf{N}}$ that is convergent. This is a contradiction because $\mathbf{K}$ is compact set of $\mathbf{X}$.
2. Sufficiency. Assume that $\mathbf{X}$ is a complete metric space and there is a finite ϵ-net for the set $\mathbf{K}$ for any $\epsilon > 0$. We take a sequence $\{\epsilon_n\}_{n\in\mathbf{N}}$ of positive numbers such that $\epsilon_n \to 0$, as $n \to \infty$. For any ϵ_n we construct a finite ϵ_n-net

   $$\{x_1^{(n)}, x_2^{(n)}, \ldots, x_{k_n}^{(n)}\}$$

for the set **K**. Let $\mathbf{L} \subset \mathbf{K}$ be an infinite set. We describe a closed ball of radius ϵ_1 around each of the points

$$\{x_1^{(1)}, x_2^{(1)}, \dots, x_{k_1}^{(1)}\}$$

in the ϵ_1-net. Then each of the elements of **L** belongs to one of these balls. Since the number of the balls is finite, there is at least one of these balls that contains an infinite set of elements of **L**. Denote such a subset by $\mathbf{L}_1$. Now we describe a closed ball of radius ϵ_2 around each of the elements

$$\{x_1^{(2)}, x_2^{(2)}, \dots, x_{k_2}^{(2)}\}$$

in the ϵ_2-net. Then there is an infinite subset $\mathbf{L}_2$ of $\mathbf{L}_1$ lying in a ball of radius ϵ_2. Continuing this process, we get the sequence

$$\mathbf{L}_1 \supset \mathbf{L}_2 \supset \cdots \supset \mathbf{L}_n \supset \cdots$$

and each $\mathbf{L}_j$ is contained in a closed ball of radius ϵ_j and if $x_1, x_2 \in \mathbf{L}_j$, we have

$$d(x_1, x_2) \le 2\epsilon_j.$$

Let

$$\xi_1 \in \mathbf{L}_1, \quad \xi_2 \in \mathbf{L}_2, \quad \xi_2 \neq \xi_1, \quad \dots, \quad \xi_n \in \mathbf{L}_n, \quad \xi_n \neq \xi_j, \quad j \in \{1, \dots, n-1\},$$

and so on. In this way, we get a sequence $\{\xi_n\}_{n\in\mathbf{N}}$ for which

$$\xi_{n+p} \in \mathbf{L}_{n+p} \subset \mathbf{L}_n$$

for any $p \in \mathbf{N}$. Consequently

$$d(\xi_{n+p}, \xi_n) \le 2\epsilon_{n+p} \to 0, \quad \text{as} \quad n \to \infty, \quad p > 0.$$

Because **X** is complete, the sequence $\{\xi_n\}_{n\in\mathbf{N}}$ is convergent to an element $\xi \in \mathbf{X}$. Therefore **K** is a relatively compact set of **X**. This completes the proof. □

Theorem 6.4. *For a set* **K** *of a complete metric space* **X** *to be a relatively compact set in* **X**, *it is necessary and sufficient that there is a relatively compact ϵ-net for the set* **K** *for every $\epsilon > 0$.*

Proof.

1. Let **L** be a relatively compact $\frac{\epsilon}{2}$-net for the set **K**. Since **L** is a relatively compact set in **X**, by Theorem 6.3, it follows that there is a finite $\frac{\epsilon}{2}$-net for the set **L**, which we denote by $\mathbf{L}_1$. Let $x \in \mathbf{K}$ be arbitrarily chosen. Then there are $x_1 \in \mathbf{L}$ and $x_2 \in \mathbf{L}_1$ such that

$$d(x, x_1) < \frac{\epsilon}{2} \quad \text{and} \quad d(x_1, x_2) < \frac{\epsilon}{2}.$$

Hence,

$$d(x, x_2) \le d(x, x_1) + d(x_1, x_2) < \frac{\epsilon}{2} + \frac{\epsilon}{2} = \epsilon. \tag{6.1}$$

Because $x \in \mathbf{K}$ was arbitrarily chosen and for it there is an element $x_2 \in \mathbf{L}_1$ so that (6.1) holds and since $\mathbf{X}$ is complete, using Theorem 6.3, we conclude that $\mathbf{K}$ is a relatively compact set in $\mathbf{X}$.

2. Let $\mathbf{K}$ is a relatively compact set in $\mathbf{X}$. Then, by Theorem 6.3, it follows that there is a finite ϵ-net for $\mathbf{K}$, which is a relatively compact ϵ-net for $\mathbf{K}$. This completes the proof. □

Theorem 6.5. *A relatively compact space is separable.*

Proof. Let $\mathbf{Y}$ be a relatively compact metric space. We take a sequence $\{\epsilon_n\}_{n\in\mathbf{N}}$ of positive numbers so that

$$\epsilon_n \to 0, \quad \text{as} \quad n \to \infty$$

For every $n \in \mathbf{N}$ we construct a finite ϵ_n-net

$$\mathbf{N}_n = \{x_1^{(n)}, \dots, x_{k_n}^{(n)}\}.$$

We set

$$\mathbf{N} = \bigcup_{n=1}^{\infty} \mathbf{N}_n.$$

We see that $\mathbf{N}$ is a countable everywhere dense set in $\mathbf{Y}$. Therefore the metric space $\mathbf{Y}$ is separable. This completes the proof. □

Theorem 6.6. *A relatively compact set* $\mathbf{K}$ *of the metric space* $\mathbf{X}$ *is bounded.*

Proof. Let $\mathbf{L}$ be a finite 1-net for $\mathbf{K}$,

$$\mathbf{L} = \{x_1, \dots, x_p\}.$$

Let also, $a \in \mathbf{K}$ be arbitrarily chosen and fixed. We set

$$r = \max_{j\in\{1,\dots,p\}} d(a, x_j).$$

We take $x \in \mathbf{K}$ arbitrarily. Then there is a $j \in \{1, \dots, p\}$ such that

$$d(x, x_j) < 1.$$

Hence,

$$d(a, x) \le d(a, x_j) + d(x, x_j) < 1 + r. \tag{6.2}$$

Since $x \in \mathbf{K}$ was arbitrarily chosen, we conclude that the inequality (6.2) holds for any $x \in \mathbf{K}$. Therefore $\mathbf{K}$ is bounded. This completes the proof. □

Theorem 6.7. *A closed set* $\mathbf{K}$ *of the metric space* $\mathbf{X}$ *is sequentially compact if and only if every covering of* $\mathbf{K}$ *of open sets of* $\mathbf{X}$ *contains a covering consisting of a finite number of these open sets.*

Proof.

1. Necessity. Let $\{\mathbf{G}_\alpha\}$ be a system of open sets that covers the set $\mathbf{K}$ and suppose that it is not possible to be extracted from it a finite covering. Take a sequence $\{\epsilon_n\}_{n\in\mathbf{N}}$ of positive numbers such that

$$\epsilon_n \to 0, \quad \text{as} \quad n \to \infty.$$

Let

$$\{x_1^{(1)}, \dots, x_{k_1}^{(1)}\}$$

be an ϵ_1-net for the set $\mathbf{K}$. Let

$$\mathbf{K}_j = \mathbf{B}_{\epsilon_{x_j}}[x_j^{(1)}] \cap \mathbf{K}, \quad j \in \{1, \dots, k_1\}.$$

Then

$$\mathbf{K} = \bigcup_{j=1}^{k_1} \mathbf{K}_j.$$

Note that $\mathbf{K}_j$, $j \in \{1, \dots, k_1\}$ is a sequentially compact set. If $\mathbf{K}$ cannot be covered by any finite subsystem in $\{\mathbf{G}_\alpha\}$, then the same is true for some of these sets $\mathbf{K}_j$, $j \in \{1, \dots, k_1\}$, which we denote by $\mathbf{K}_{j_1}$. Continuing this process, we extract $\mathbf{K}_{j_1 j_2}$ from the sequentially compact set $\mathbf{K}_{j_1}$ if it is not possible to extract from $\{\mathbf{G}_\alpha\}$ any finite covering of $\mathbf{K}_{j_1 j_2}$. And so on. In this way, we get the system

$$\mathbf{K}_{j_1} \supset \mathbf{K}_{j_1 j_2} \supset \cdots \supset \mathbf{K}_{j_1 j_2 \dots j_n} \supset \cdots.$$

Let x_0 be an element that belongs to all of these sets. Since $x_0 \in \mathbf{K}$ and $\{\mathbf{G}_\alpha\}$ covers $\mathbf{K}$, there is a set $\mathbf{G}_{\alpha_0}$ so that $x_0 \in \mathbf{G}_{\alpha_0}$. Because $\mathbf{G}_{\alpha_0}$ is an open set, there is a neighborhood $\mathbf{B}_\epsilon(x_0)$ of the element x_0 such that

$$\mathbf{B}_\epsilon(x_0) \subset \mathbf{G}_{\alpha_0}.$$

Now we take $n \in \mathbf{N}$ large enough so that

$$d(\tilde{x}, \tilde{\tilde{x}}) < \epsilon$$

for any $\tilde{x}, \tilde{\tilde{x}} \in K_{j_1 j_2 \dots j_n}$. Then

$$\mathbf{K}_{j_1 j_2 \dots j_n} \subset \mathbf{B}_\epsilon(x_0) \subset \mathbf{G}_{\alpha_0}.$$

This is a contradiction because it is not possible to extract from the system $\{\mathbf{G}_\alpha\}$ a finite covering of $\mathbf{K}_{j_1 j_2 \dots j_n}$.

2. Sufficiency. Suppose that $\mathbf{M}$ is a subset of $\mathbf{K}$ which has not any limit point. Then for every $x \in \mathbf{K}$ there is an $\epsilon_x > 0$ so that

$$\mathbf{B}_{\epsilon_x}(x) \cap \mathbf{M} = \{x\} \quad \text{or} \quad \mathbf{B}_{\epsilon_x}(x) \cap \mathbf{M} = \emptyset.$$

Note that the neighborhood $\mathbf{B}_{\epsilon_x}(x)$ forms a covering of $\mathbf{M}$. We extract a finite coverings

$$\mathbf{B}_{\epsilon_{x_1}}(x_1), \quad \mathbf{B}_{\epsilon_{x_2}}(x_2), \quad \ldots, \mathbf{B}_{\epsilon_{x_n}}(x_1).$$

Since the entire set $\mathbf{M}$ is located in these neighborhoods and each of these neighborhoods cannot contain more than one element of $\mathbf{M}$, then $\mathbf{M}$ is finite. Therefore every infinite subset of $\mathbf{K}$ must have a limit point, i. e., $\mathbf{K}$ is a relatively compact set in $\mathbf{X}$. This completes the proof. □

Theorem 6.8. *Every continuous image of a relatively compact set* $\mathbf{K}$ *in the metric space* $\mathbf{X}$ *is a relatively compact set in* $\mathbf{X}$.

Proof. Let $\mathbf{Y}$ be a metric space and $f : \mathbf{X} \longmapsto \mathbf{Y}$ be a continuous map such that $f(\mathbf{K}) \subset \mathbf{Y}$. Let also, $\{y_n\}_{n\in\mathbf{N}}$ be an arbitrary sequence of elements of $f(\mathbf{K})$. With x_n, $n \in \mathbf{N}$, we denote the elements of $\mathbf{K}$ such that

$$y_n = f(x_n), \quad n \in \mathbf{N}.$$

Since $\mathbf{K}$ is a relatively compact set in the metric space $\mathbf{X}$, there exists a subsequence $\{x_{n_k}\}_{k\in\mathbf{N}}$ and $x_0 \in \mathbf{X}$ such that

$$x_{n_k} \to x_0, \quad \text{as} \quad k \to \infty.$$

Because $f : \mathbf{K} \longmapsto \mathbf{Y}$ is continuous, we get

$$y_{n_k} = f(x_{n_k}) \to f(x_0) \in \mathbf{Y}, \quad \text{as} \quad k \to \infty,$$

i. e., the sequence $\{y_n\}_{n\in\mathbf{N}}$ contains a convergent subsequence to an element of $\mathbf{Y}$. Consequently $f(\mathbf{K})$ is compact. This completes the proof. □

6.3 A Criteria for relative compactness in the space C([*a*, *b*])

Let $\mathbf{M}$ be a set of elements of $\mathbf{C}([a,b])$. We provide $\mathbf{C}([a,b])$ with the metric

$$d(x_1, x_2) = \max_{t\in[a,b]} |x_1(t) - x_2(t)|, \quad x_1, x_2 \in \mathbf{C}([a,b]).$$

Definition 6.4. We say that the set $\mathbf{M}$ is uniformly bounded if there exists a constant $c > 0$ such that

$$|x(t)| \le c \quad \text{for} \quad \text{all} \quad t \in [a,b]$$

and for all $x \in \mathbf{M}$.

Definition 6.5. We say that the set $\mathbf{M}$ is equi-continuous if for every $\epsilon > 0$ there is a $\delta = \delta(\epsilon) > 0$ such that the inequality $|t_1 - t_2| < \delta$, $t_1, t_2 \in [a, b]$, implies

$$|x(t_1) - x(t_2)| < \epsilon$$

for any $x \in \mathbf{M}$.

Theorem 6.9 (Arzela theorem). *The set* $\mathbf{K} \subset \mathbf{C}([a, b])$ *is a relatively compact set in* $\mathbf{C}([a, b])$ *if and only if it is uniformly bounded and equi-continuous.*

Proof.

1. Let $\mathbf{K}$ be a relatively compact set in $\mathbf{C}([a, b])$. By Theorem 6.6, it follows that it is uniformly bounded. Let $\epsilon > 0$ be arbitrarily chosen. We construct a finite $\frac{\epsilon}{3}$-net

$$\{x_1, \ldots, x_k\}$$

for the set $\mathbf{K}$. Because x_j, $j \in \{1, \ldots, k\}$, are continuous on $[a, b]$, they are uniformly continuous on $[a, b]$. For any x_j, $j \in \{1, \ldots, k\}$, we take a positive number δ_j such that

$$|x_j(t_1) - x_j(t_2)| < \frac{\epsilon}{3},$$

whenever $|t_1 - t_2| < \delta_j$, $t_1, t_2 \in [a, b]$. Let

$$\delta = \min\{\delta_1, \ldots, \delta_k\}.$$

Then, if $|t_1 - t_2| < \delta$, $t_1, t_2 \in [a, b]$, and $x \in \mathbf{K}$, we see that there is a $j \in \{1, \ldots, k\}$, such that

$$|x_j(t_1) - x(t_1)| < \frac{\epsilon}{3}, \quad |x_j(t_2) - x(t_2)| < \frac{\epsilon}{3},$$

whereupon

$$\begin{aligned} |x(t_1) - x(t_2)| &= |x(t_1) - x_j(t_1) + x_j(t_1) + x_j(t_2) - x_j(t_2) + x(t_2)| \\ &\leq |x(t_1) - x_j(t_1)| + |x_j(t_1) - x_j(t_2)| \\ &\quad + |x_j(t_2) - x(t_2)| < \frac{\epsilon}{3} + \frac{\epsilon}{3} + \frac{\epsilon}{3} = \epsilon. \end{aligned}$$

Consequently $\mathbf{K}$ is equi-continuous.

2. Let $\mathbf{K} \subset \mathbf{C}([a, b])$ be uniformly bounded and equi-continuous. Let $c > 0$ be chosen so that

$$|x(t)| \leq c, \quad t \in [a, b],$$

for any $x \in \mathbf{K}$. We take $\epsilon > 0$ arbitrarily. Then there exists a $\delta = \delta(\epsilon) > 0$ such that if $|t_1 - t_2| < \delta$, $t_1, t_2 \in [a, b]$, we have

$$|x(t_1) - x(t_2)| < \epsilon$$

for any $x \in \mathbf{K}$. Let $n \in \mathbf{N}$ be arbitrarily chosen so that

$$(b-a)\frac{1}{n} < \delta.$$

We divide the interval $[a,b]$ into n equal parts

$$\left[a+(b-a)\frac{k}{n}, a+(b-a)\frac{k+1}{n}\right], \quad k \in \{0,1,\ldots,n-1\}.$$

Then

$$|x(t_1)-x(t_2)| < \epsilon$$

for every $x \in \mathbf{K}$ and $t_1, t_2 \in [a,b]$, $|t_1-t_2| < (b-a)\frac{1}{n}$, in particular for

$$t_1, t_2 \in \left[a+(b-a)\frac{k}{n}, a+(b-a)\frac{k+1}{n}\right], \quad k \in \{0,1,\ldots,n-1\}.$$

We construct the function x_n to the function x such that

(a) $x_n(a+(b-a)\frac{k}{n}) = x(a+(b-a)\frac{k}{n})$, $k \in \{0,\ldots,n-1\}$,

(b) x_n is a linear function on $[a+(b-a)\frac{k}{n}, a+(b-a)\frac{k+1}{n}]$, $k \in \{0,1,\ldots,n-1\}$.

Let $x \in \mathbf{K}$ be such that

$$x\left(a+(b-a)\frac{k}{n}\right) \le x\left(a+(b-a)\frac{k+1}{n}\right), \quad k \in \{0,\ldots,n-1\}.$$

Hence,

$$x_n\left(a+(b-a)\frac{k}{n}\right) \le x_n(t) \le x_n\left(a+(b-a)\frac{k+1}{n}\right), \quad k \in \{0,1,\ldots,n-1\},$$

for $t \in [a+(b-a)\frac{k}{n}, a+(b-a)\frac{k+1}{n}]$, and

$$-\epsilon < x(t) - x\left(a+(b-a)\frac{k+1}{n}\right) \le x(t)-x_n(t) \le x(t) - x\left(a+(b-a)\frac{k}{n}\right) < \epsilon$$

for any $t \in [a+(b-a)\frac{k}{n}, a+(b-a)\frac{k+1}{n}]$, $k \in \{0,\ldots,n-1\}$, i. e.,

$$|x(t)-x_n(t)| < \epsilon$$

for all $t \in [a,b]$. Therefore

$$d(x_n, x) < \epsilon$$

and the set $\mathbf{L}$ of functions x_n is an ϵ-net for $\mathbf{K}$. Also,

$$|x_n(t)| \le |x(t)| + |x(t)-x_n(t)| < c+\epsilon, \quad t \in [a,b].$$

Therefore **L** is uniformly bounded. Now we associate to every function x_n of **L** the points of an $(n+1)$-dimensional space $\widetilde{\mathbf{X}}$, having as coordinates the ordinates of the vertices of a polygon, the graph of x_n. This correspondence is one-to-one and continuous. Therefore if the sequence of functions $\{x_n^{(k)}\}$ converges to $x_n^{(0)}$ in the sense of the metric space $\mathbf{C}([a,b])$, the sequence of points $\{\tilde{x}^{(k)}\}$ converges to the points $\tilde{x}^{(0)}$ in the sense of the metric space $\mathbf{E}_{n+1}$. The set $\tilde{\mathbf{K}} = \{\tilde{x}\}$ is bounded and consequently it is relatively compact in $\mathbf{E}_{n+1}$. Therefore **L** is relatively compact in $\mathbf{C}([a,b])$ and for any $\epsilon > 0$ we construct a relatively compact ϵ-net for **K**. Since $\mathbf{C}([a,b])$ is complete, using Theorem 6.4, we conclude that **K** is a relatively compact set in $\mathbf{C}([a,b])$. This completes the proof. □

6.4 A Criteria for compactness in the space $L^p([a,b])$, $p > 1$

Consider the space $\mathbf{L}^p([a,b])$. We extend all functions $x \in \mathbf{L}^p([a,b])$ beyond the interval $[a,b]$ and put $x(t) = 0$ if t lies outside the interval $[a,b]$.

Theorem 6.10 (Riesz's theorem). *A set **K** is the space $\mathbf{L}^p([a,b])$ is a relatively compact set in $\mathbf{L}^p([a,b])$ if and only if there exists a constant $c > 0$ such that*

$$\int_a^b |x(t)|^p dt \le c^p \tag{6.3}$$

for every $x \in \mathbf{L}^p([a,b])$, and for every $\epsilon > 0$ there exists a $\delta = \delta(\epsilon) > 0$ such that

$$\int_a^b |x(t+h) - x(t)|^p dt < \epsilon^p \quad \text{for} \quad 0 < h < \delta, \tag{6.4}$$

*simultaneously for all functions of the set **K**.*

Proof.

1. Let **K** is a relatively compact set in $\mathbf{L}^p([a,b])$. Then **K** is bounded and from this, the condition (6.3) holds. Now we will check the condition (6.4). Let $\epsilon > 0$ be arbitrarily chosen. Since **K** is a relatively compact set in $\mathbf{L}^p([a,b])$, there exists a finite $\frac{\epsilon}{3}$-net for **K**

   $$\{x_1, \ldots, x_n\}.$$

   Note that, for any $i \in \{1, \ldots, n\}$, there exists $\delta_i = \delta_i(\epsilon) > 0$ such that

   $$\int_a^b |x_i(t+h) - x_i(t)|^p dt < \left(\frac{\epsilon}{3}\right)^p, \quad 0 < h < \delta_i.$$

Let

$$\delta = \min_{i\in\{1,\dots,n\}} \delta_i.$$

Then

$$\int_a^b |x_i(t+h) - x_i(t)|^p dt < \left(\frac{\epsilon}{3}\right)^p, \quad 0 < h < \delta,$$

for any $i \in \{1, \dots, n\}$. Let $x \in \mathbf{K}$ be arbitrarily chosen. Then there exists $j \in \{1, \dots, n\}$, such that

$$\int_a^b |x(t) - x_j(t)|^p dt < \left(\frac{\epsilon}{3}\right)^p,$$

hence, for $0 < h < \delta$, we have

$$\begin{aligned}\int_a^b |x(t+h) - x_j(t+h)|^p dt &= \int_{a+h}^{b+h} |x(t) - x_j(t)|^p dt \le \int_a^b |x(t) - x_j(t)|^p dt \\ &< \left(\frac{\epsilon}{3}\right)^p, \quad 0 < h < \delta.\end{aligned}$$

Then

$$\begin{aligned}&\left(\int_a^b |x(t+h) - x(t)|^p dt\right)^{\frac{1}{p}} \\ &\quad = \left(\int_a^b |x(t+h) - x_j(t+h) + x_j(t+h) - x_j(t) + x_j(t) - x(t)|^p dt\right)^{\frac{1}{p}} \\ &\quad \le \left(\int_a^b |x(t+h) - x_j(t+h)|^p dt\right)^{\frac{1}{p}} + \left(\int_a^b |x_j(t+h) - x_j(t)|^p dt\right)^{\frac{1}{p}} \\ &\quad + \left(\int_a^b |x_j(t) - x(t)|^p dt\right)^{\frac{1}{p}} < \frac{\epsilon}{3} + \frac{\epsilon}{3} + \frac{\epsilon}{3} = \epsilon, \quad 0 < h < \delta.\end{aligned}$$

2. Suppose that all functions of the set $\mathbf{K}$ satisfy the conditions (6.3) and (6.4). For $h > 0$ and $x \in \mathbf{K}$ we define

$$x_h(t) = \frac{1}{2h} \int_{t-h}^{t+h} x(\tau) d\tau, \quad t \in [a, b].$$

Let q be the conjugate of p. Then

$$|x_h(t)| = \frac{1}{2h}\left|\int_{t-h}^{t+h} x(\tau)d\tau\right| \le \frac{1}{2h}\int_{t-h}^{t+h}|x(\tau)|d\tau \le \frac{1}{2h}\left(\int_{t-h}^{t+h}d\tau\right)^{\frac{1}{q}}\left(\int_{t-h}^{t+h}|x(\tau)|^p d\tau\right)^{\frac{1}{p}}$$
$$= \frac{1}{2h}(2h)^{\frac{1}{q}}\left(\int_{t-h}^{t+h}|x(\tau)|^p d\tau\right)^{\frac{1}{p}} = \left(\frac{1}{2h}\right)^{\frac{1}{p}}\left(\int_{t-h}^{t+h}|x(\tau)|^p d\tau\right)^{\frac{1}{p}}$$

and

$$\begin{aligned}|x_h(t+y)-x_h(t)| &= \frac{1}{2h}\left|\int_{t+y-h}^{t+y+h} x(\tau)d\tau - \int_{t-h}^{t+h} x(\tau)d\tau\right| \\ &= \frac{1}{2h}\left|\int_{t-h}^{t+h} x(\tau+y)d\tau - \int_{t-h}^{t+h} x(\tau)d\tau\right| \\ &= \frac{1}{2h}\left|\int_{t-h}^{t+h}(x(\tau+y)-x(\tau))d\tau\right| \\ &\le \frac{1}{2h}\int_{t-h}^{t+h}|x(\tau+y)-x(\tau)|d\tau \\ &\le \left(\frac{1}{2h}\right)^{\frac{1}{p}}\left(\int_{t-h}^{t+h}|x(\tau+y)-x(\tau)|^p d\tau\right)^{\frac{1}{p}} \\ &\le \left(\frac{1}{2h}\right)^{\frac{1}{p}}\left(\int_a^b|x(\tau+y)-x(\tau)|^p d\tau\right)^{\frac{1}{p}}.\end{aligned}$$

Therefore, for fixed $h>0$, the family $\{x_h\}$ for $x \in \mathbf{K}$ is uniformly bounded and equi-continuous. Next,

$$\begin{aligned}|x(t)-x_h(t)| &= \frac{1}{2h}\left|\int_{t-h}^{t+h} x(t)d\tau - \int_{t-h}^{t+h} x(\tau)d\tau\right| = \frac{1}{2h}\left|\int_{t-h}^{t+h}(x(t)-x(\tau))d\tau\right| \\ &\le \frac{1}{2h}\int_{t-h}^{t+h}|x(t)-x(\tau)|d\tau = \frac{1}{2h}\int_{-h}^{h}|x(t)-x(t+\tau)|d\tau \\ &\le \left(\frac{1}{2h}\right)^{\frac{1}{p}}\left(\int_{-h}^{h}|x(t)-x(t+\tau)|^p d\tau\right)^{\frac{1}{p}}.\end{aligned}$$

Hence,

$$\begin{aligned}\int_a^b |x(t)-x_h(t)|^p dt &\le \frac{1}{2h}\int_a^b\int_{-h}^h |x(t)-x(t+\tau)|^p d\tau dt\\ &= \frac{1}{2h}\int_{-h}^h\int_a^b |x(t)-x(t+\tau)|^p dt d\tau\\ &< \frac{\epsilon^p}{2h}\int_{-h}^h d\tau = \epsilon^p \quad \text{if} \quad 0<h<\delta.\end{aligned}$$

Therefore $\{x_h\}$ is an ϵ-net for $\mathbf{K}$ and since it is a relatively compact set in $\mathbf{L}^p([a,b])$, we conclude that $\mathbf{K}$ is a relatively compact set in $\mathbf{L}^p([a,b])$. This completes the proof. □

6.5 Compact operators

Let $\mathbf{X}$ and $\mathbf{Y}$ be normed vector spaces.

Definition 6.6. A linear operator $\mathbb{A} : \mathbf{X} \longmapsto \mathbf{Y}$ is called compact if it maps each bounded set of $\mathbf{X}$ into a relatively compact set of $\mathbf{Y}$.

Example 6.1. Let $\mathbf{X} = \mathbf{Y} = \mathbf{C}([0,1])$. Suppose that $K \in \mathbf{C}([0,1]\times[0,1])$ and consider the operator

$$\mathbb{A}x(t) = \int_0^1 K(t,s)x(s)ds, \quad t\in[0,1], \quad x\in\mathbf{X}.$$

Evidently, $\mathbb{A} : \mathbf{X} \longmapsto \mathbf{Y}$. We will prove that it is a compact operator. Let $\mathbf{K}$ be an arbitrary bounded set of $\mathbf{X}$. Then there exists a constant $r>0$ such that

$$\|x\| \le r$$

for any $x \in \mathbf{K}$. Let $\epsilon > 0$ be arbitrarily chosen. Because $K \in \mathbf{C}([0,1]\times[0,1])$, there exists a constant $l>0$ such that

$$|K(t,s)| \le l, \quad t,s\in[0,1],$$

and there exists a $\delta = \delta(\epsilon) > 0$ so that

$$|K(t_1,s)-K(t_2,s)| < \frac{\epsilon}{r}, \quad t_1,t_2,s\in[0,1], \quad |t_1-t_2|<\delta.$$

Hence, for $x \in \mathbf{K}$,

$$|\mathbb{A}x(t)| = \left|\int_0^1 K(t,s)x(s)ds\right| \le \int_0^1 |K(t,s)||x(s)|ds \le lr, \quad t \in [0,1].$$

Therefore $\mathbb{A}(\mathbf{K})$ is uniformly bounded. Next,

$$\begin{aligned}|\mathbb{A}x(t_1) - \mathbb{A}x(t_2)| &= \left|\int_0^1 K(t_1,s)x(s)ds - \int_0^1 K(t_2,s)x(s)ds\right| \\ &= \left|\int_0^1 (K(t_1,s) - K(t_2,s))x(s)ds\right| \le \int_0^1 |K(t_1,s) - K(t_2,s)||x(s)|ds \\ &< \frac{\epsilon}{r}r = \epsilon \quad \text{for} \quad |t_1 - t_2| < \delta, \quad t_1, t_2 \in [0,1],\end{aligned}$$

and for any $x \in \mathbf{K}$. Consequently $\mathbb{A}(\mathbf{K})$ is equi-continuous. Because $\mathbb{A}(\mathbf{K})$ is bounded and equi-continuous, we conclude that it is compact. Consequently $\mathbb{A} : \mathbf{X} \longmapsto \mathbf{Y}$ is a compact operator.

Lemma 6.1. *If a sequence $\{x_n\}_{n\in\mathbf{N}} \subset \mathbf{X}$ is weakly convergent to $x_0 \in \mathbf{X}$ and a relatively compact in $\mathbf{X}$, then it is strongly convergent to x_0.*

Proof. Assume the contrary. Then there exist an $\epsilon > 0$ and a subsequence $\{x_{n_k}\}_{k\in\mathbf{N}}$ of the sequence $\{x_n\}_{n\in\mathbf{N}}$ such that

$$\|x_{n_k} - x_0\| \ge \epsilon, \quad k \in \mathbf{N}.$$

Because $\{x_{n_k}\}_{k\in\mathbf{N}}$ is a relatively compact set, there is a subsequence $\{x_{n_{k_l}}\}_{l\in\mathbf{N}}$ strongly convergent to $y_0 \in \mathbf{X}$. We have

$$\|x_{n_{k_l}} - x_0\| \ge \epsilon. \tag{6.5}$$

Now, using $x_{n_{k_l}} \rightharpoonup x_0$, we conclude that $x_0 = y_0$. Therefore

$$\|x_{n_{k_l}} - x_0\| < \epsilon,$$

which contradicts (6.5). This completes the proof. □

Lemma 6.2. *Let $\mathbb{A} : \mathbf{X} \longmapsto \mathbf{Y}$ be a linear bounded operator. If $x_n \rightharpoonup x_0$, as $n \to \infty$, then $\mathbb{A}x_n \rightharpoonup \mathbb{A}x_0$, as $n \to \infty$.*

Proof. Let ϕ be arbitrarily chosen linear functional on $\mathbf{Y}$. Let also,

$$f(x_0) = \phi(\mathbb{A}x_0), \quad f(x_n) = \phi(\mathbb{A}x_n), \quad n \in \mathbf{N}.$$

Since $\mathbb{A} : \mathbf{X} \longmapsto \mathbf{Y}$ is a linear operator, f is a linear functional on $\mathbf{X}$. Because $x_n \rightharpoonup x_0$, as $n \to \infty$, we get

$$f(x_n) \to f(x_0), \quad \text{as} \quad n \to \infty.$$

Hence,

$$\phi(\mathbb{A}x_n) \to \phi(\mathbb{A}x_0), \quad \text{as} \quad n \to \infty.$$

Since ϕ was arbitrarily chosen linear functional on $\mathbf{Y}$, we conclude that

$$\mathbb{A}x_n \rightharpoonup \mathbb{A}x_0, \quad \text{as} \quad n \to \infty.$$

This completes the proof. □

Theorem 6.11. *A compact operator $\mathbb{A} : \mathbf{X} \longmapsto \mathbf{Y}$ maps a weakly convergent sequence in $\mathbf{X}$ into a strongly convergent sequence in $\mathbf{Y}$.*

Proof. Let $\{x_n\}_{n\in\mathbf{N}} \subset \mathbf{X}$ is a weakly convergent sequence to $x_0 \in \mathbf{X}$. Then $\{\|x_n\|\}_{n\in\mathbf{N}}$ is a bounded sequence. Because the operator $\mathbb{A}$ is compact, we see that $\{\mathbb{A}x_n\}_{n\in\mathbf{N}}$ is compact. By Lemma 6.2, we see that $\{\mathbb{A}x_n\}_{n\in\mathbf{N}}$ is weakly convergent to $\mathbb{A}x_0$. From this and from Lemma 6.1, it follows that $\{\mathbb{A}x_n\}_{n\in\mathbf{N}}$ is strongly convergent to $\mathbb{A}x_0$. This completes the proof. □

Theorem 6.12. *Let $\mathbb{A}_n : \mathbf{X} \longmapsto \mathbf{Y}$, $n \in \mathbf{N}$, be compact operators, and*

$$\|\mathbb{A}_n - \mathbb{A}\| \to 0, \quad \textit{as} \quad n \to \infty.$$

Then $\mathbb{A} : \mathbf{X} \longmapsto \mathbf{Y}$ is a compact operator.

Proof. Let $\mathbf{M} \subset \mathbf{X}$ be a bounded set. Then there exists a constant $r > 0$ such that

$$\|x\| \le r \quad \text{for} \quad x \in \mathbf{M}.$$

We take $\epsilon > 0$ arbitrarily. Then there exists an $n_0 \in \mathbf{N}$ such that

$$\|\mathbb{A}_n - \mathbb{A}\| < \frac{\epsilon}{r}$$

for any $n \in \mathbf{N}$, $n \ge n_0$. Set

$$\mathbb{A}(\mathbf{M}) = \mathbf{K} \quad \text{and} \quad \mathbb{A}_{n_0}(\mathbf{M}) = \mathbf{L}.$$

Now, we take $y \in \mathbf{K}$ arbitrarily. Let $x \in \mathbf{M}$ be such that $y = \mathbb{A}x$ and set

$$y_0 = \mathbb{A}_{n_0}x.$$

Then

$$\begin{aligned}\|y - y_0\| &= \|\mathbb{A}x - \mathbb{A}_{n_0}x\| = \|(\mathbb{A} - \mathbb{A}_{n_0})x\| \\ &\le \|\mathbb{A} - \mathbb{A}_n\|\|x\| < \frac{\epsilon}{r}r = \epsilon.\end{aligned}$$

Therefore **L** is an ϵ-net of **K**. Since $\mathbb{A}_{n_0} : \mathbf{X} \longmapsto \mathbf{Y}$ is a compact operator and **M** is a bounded set, we see that **L** is relatively compact. Hence, $\mathbb{A} : \mathbf{X} \longmapsto \mathbf{Y}$ is compact. This completes the proof. □

6.6 Advanced practical problems

Problem 6.1. Prove that the set of all functions $x \in \mathbf{C}^1([a,b])$ for which

$$|x(0)| \le k_1, \quad \int_1^b |x'(t)|dt \le k_2,$$

where k_1 and k_2 are positive constants, is relatively compact in the space $\mathbf{C}([a,b])$.

Problem 6.2. Prove that the set of all functions $x \in \mathbf{C}^1([a,b])$ for which

$$\int_a^b (|x(t)|^2 + |x'(t)|^2)dt \le k,$$

where k is a positive constant, is relatively compact in $\mathbf{C}([a,b])$.

Problem 6.3. Prove that every compact set in $\mathbf{C}^1([a,b])$ is a relatively compact set in $\mathbf{C}([a,b])$.

Problem 6.4. Let **X**, **Y** and **Z** are normed vector spaces, $\mathbb{A} \in \mathcal{L}(\mathbf{X},\mathbf{Y})$ is a compact operator and $\mathbb{B} \in \mathcal{L}(\mathbf{Y},\mathbf{Z})$. Prove that $\mathbb{B}\mathbb{A}$ is a compact operator.

Problem 6.5. Prove that the following operators $\mathbb{A} : \mathbf{C}([0,1]) \longmapsto \mathbf{C}([0,1])$ are compact:
1. $\mathbb{A}x(t) = \int_0^t x(s)ds$,
2. $\mathbb{A}x(t) = x(0) + tx(1)$,
3. $\mathbb{A}x(t) = \int_0^1 e^{ts}x(s)ds$.

7 Self-adjoint operators in Hilbert spaces

7.1 Adjoint operators. Self-adjoint operators

Let $\mathbf{H}$ be a Hilbert space and $\mathbb{A} \in \mathcal{L}(\mathbf{H})$. For $x, y \in \mathbf{H}$ consider a linear functional f_y defined as follows:

$$f_y(x) = (\mathbb{A}x, y).$$

As a linear functional on $\mathbf{H}$, f_y has the form

$$f_y(x) = (x, y^*),$$

where $y^* \in \mathbf{H}$ is uniquely defined by f_y for any $x \in \mathbf{H}$. Note that f_y varies with a change of y and so does y^*. Therefore we get the operator

$$\mathbb{A}^* y = y^*, \quad y \in \mathbf{H},$$

and $\mathbb{A}^* : \mathbf{H} \longmapsto \mathbf{H}$. This operator $\mathbb{A}^*$ is associated with the operator $\mathbb{A}$ in the following manner:

$$(\mathbb{A}x, y) = (x, \mathbb{A}^* y). \tag{7.1}$$

Definition 7.1. The operator $\mathbb{A}^*$ is called the adjoint operator of the operator $\mathbb{A}$.

Assume that for $x, y \in \mathbf{H}$ we have

$$(\mathbb{A}x, y) = (x, \mathbb{A}^* y) = (x, \mathbb{A}_1^* y),$$

where $\mathbb{A}^*, \mathbb{A}_1^* : \mathbf{H} \longmapsto \mathbf{H}$ are adjoint operators of the operator $\mathbb{A}$. Therefore

$$\mathbb{A}^* y = \mathbb{A}_1^* y \quad \text{for} \quad \text{all} \quad y \in \mathbf{H}.$$

Thus

$$\mathbb{A}^* = \mathbb{A}_1^*,$$

i. e., the operator $\mathbb{A}^*$ is uniquely determined by (7.1). From Theorem 5.3, it follows that

$$\|\mathbb{A}\| = \|\mathbb{A}^*\|.$$

By (7.1), we get

$$(\mathbb{A}^* x, y) = \overline{(y, \mathbb{A}^* x)} = \overline{(\mathbb{A}y, x)} = (x, \mathbb{A}y), \quad x, y \in \mathbf{H}.$$

https://doi.org/10.1515/9783110657722-007

Thus

$$\mathbb{A}^{**} = \mathbb{A}.$$

Similarly

$$\mathbb{A}^{***} = \mathbb{A}^{*},$$

and so on.

Exercise 7.1. Let $\mathbb{A}, \mathbb{B} \in \mathcal{L}(\mathbf{H})$. Prove that
1. $(\mathbb{A} + \mathbb{B})^{*} = \mathbb{A}^{*} + \mathbb{B}^{*}$,
2. $(\lambda\mathbb{A})^{*} = \bar{\lambda}\mathbb{A}^{*}, \lambda \in \mathbf{F}$,
3. $(\mathbb{A}\mathbb{B})^{*} = \mathbb{B}^{*}\mathbb{A}^{*}$,
4. $(\mathbb{A}^{*})^{-1} = (\mathbb{A}^{-1})^{*}$ if $\mathbb{A}^{-1}$ exists.

Definition 7.2. An operator $\mathbb{A} \in \mathcal{L}(\mathbf{H})$ is called self-adjoint, if $\mathbb{A} = \mathbb{A}^{*}$.

7.2 Unitary operators

Definition 7.3. A linear operator $\mathbb{U} : \mathbf{H} \longmapsto \mathbf{H}$ is called unitary if it maps the Hilbert space $\mathbf{H}$ onto all of $\mathbf{H}$ with preservation of the norm, i. e.,

$$\|\mathbb{U}x\| = \|x\|, \quad x \in \mathbf{H}.$$

Remark 7.1. If $\mathbb{U} : \mathbf{H} \longmapsto \mathbf{H}$ is unitary, then it is one-one. In fact, if $x_1, x_2 \in \mathbf{H}$, then

$$\mathbb{U}x_1 = \mathbb{U}x_2$$

or

$$\mathbb{U}(x_1 - x_2) = 0,$$

and hence

$$\|\mathbb{U}(x_1 - x_2)\| = \|x_1 - x_2\| = 0,$$

i. e., $x_1 = x_2$. Furthermore,

$$(\mathbb{U}x, \mathbb{U}x) = \|\mathbb{U}x\|^2 = \|x\|^2 = (x, x) = (\mathbb{U}^{*}\mathbb{U}x, x), \quad x \in \mathbf{H}.$$

Therefore

$$\mathbb{U}^{*}\mathbb{U} = \mathbb{I} \tag{7.2}$$

and

$$\mathbb{U}^*\mathbb{U}\mathbb{U}^{-1} = \mathbb{U}^* = \mathbb{U}^{-1}.$$

Hence,

$$\mathbb{U}\mathbb{U}^* = \mathbb{U}\mathbb{U}^{-1} = \mathbb{I}. \tag{7.3}$$

Consequently

$$\mathbb{U}^* = \mathbb{U}^{-1}.$$

By (7.2), it also follows that

$$(\mathbb{U}x, \mathbb{U}y) = (x, y), \quad x, y \in \mathbf{H}.$$

Conversely, by (7.2) and (7.3), it follows that $\mathbb{U}$ is an unitary operator, since these imply $\mathbb{U}^{-1} = \mathbb{U}^*$ exists. Then $\mathbf{H}$ is mapped one-one onto $\mathbf{H}$ and

$$\|\mathbb{U}x\|^2 = (\mathbb{U}x, \mathbb{U}x) = \left(\mathbb{U}^*\mathbb{U}x, x\right) = (x, x) = \|x\|^2,$$

whereupon

$$\|\mathbb{U}x\| = \|x\|, \quad x \in \mathbf{H}.$$

Definition 7.4. Let $\mathbb{A} : \mathbf{H} \longmapsto \mathbf{H}$ be a linear operator, $\mathbb{U} : \mathbf{H} \longmapsto \mathbf{H}$ is an unitary operator. The operator

$$\mathbb{B} = \mathbb{U}\mathbb{A}\mathbb{U}^{-1}$$

is called an operator unitarily equivalent to $\mathbb{A}$.

7.3 Projection operators

Let $\mathbf{L}$ be a closed linear subspace of $\mathbf{H}$. Then every element $x \in \mathbf{H}$ is uniquely representable in the form

$$x = y + z,$$

where $y \in \mathbf{L}$ and $z \perp \mathbf{L}$. We set

$$\mathbb{P}x = y.$$

Then

$$\mathbb{P} : \mathbf{H} \longmapsto \mathbf{L}.$$

Definition 7.5. The operator $\mathbb{P} : \mathbf{H} \longmapsto \mathbf{L}$ is called the operator of orthogonal projection upon $\mathbf{L}$, or a projection operator, or a projector. It is denoted by $\mathbb{P}_{\mathbf{L}}$. The space $\mathbf{L}$ will be called the corresponding space of the projector $\mathbb{P}$.

Theorem 7.1. *The operator* $\mathbb{P}_{\mathbf{L}} : \mathbf{H} \longmapsto \mathbf{L}$ *is a linear self-adjoint operator and*

$$\|\mathbb{P}_{\mathbf{L}}\| = 1, \quad \mathbb{P}_{\mathbf{L}}^2 = \mathbb{P}_{\mathbf{L}}.$$

Proof. Firstly, we will prove that $\mathbb{P}_{\mathbf{L}} : \mathbf{H} \longmapsto \mathbf{L}$ is a linear operator. Let $x_1, x_2 \in \mathbf{H}$ and

$$x_1 = y_1 + z_1, \quad x_2 = y_2 + z_2, \quad y_1, y_2 \in \mathbf{L}, \quad z_1, z_2 \perp \mathbf{L}.$$

Let also $\alpha \in \mathbf{F}$. Then

$$\begin{aligned} \mathbb{P}_{\mathbf{L}} x_1 &= y_1, \\ \alpha x_1 &= \alpha y_1 + \alpha z_1, \\ \mathbb{P}_{\mathbf{L}}(\alpha x_1) &= \alpha y_1 = \alpha \mathbb{P}_{\mathbf{L}} x_1, \end{aligned}$$

and

$$\begin{aligned} x_1 + x_2 &= (y_1 + y_2) + (z_1 + z_2), \\ \mathbb{P}_{\mathbf{L}} x_2 &= y_2, \\ \mathbb{P}_{\mathbf{L}}(x_1 + x_2) &= y_1 + y_2 = \mathbb{P}_{\mathbf{L}} x_1 + \mathbb{P}_{\mathbf{L}} x_2. \end{aligned}$$

Consequently the operator $\mathbb{P}_{\mathbf{L}} : \mathbf{H} \longmapsto \mathbf{L}$ is a linear operator. Furthermore,

$$\begin{aligned} (\mathbb{P}_{\mathbf{L}} x_1, x_2) &= (y_1, x_2) = (y_1, y_2 + z_2) = (y_1, y_2) + (y_1, z_2) = (y_1, y_2) \\ &= (z_1, y_2) + (y_1, y_2) = (z_1 + y_1, y_2) = (x_1, y_2) = (x_1, \mathbb{P}_{\mathbf{L}} x_2), \end{aligned}$$

i. e., $\mathbb{P}_{\mathbf{L}} : \mathbf{H} \longmapsto \mathbf{L}$ is a self-adjoint operator. Note that, for $x \in \mathbf{H}$, we have $\mathbb{P}_{\mathbf{L}} x \in \mathbf{L}$ and

$$\mathbb{P}_{\mathbf{L}}^2 x = \mathbb{P}_{\mathbf{L}}(\mathbb{P}_{\mathbf{L}} x) = \mathbb{P}_{\mathbf{L}} x,$$

i. e.,

$$\mathbb{P}_{\mathbf{L}}^2 = \mathbb{P}_{\mathbf{L}}.$$

Next, for $x \in \mathbf{H}$, $x = y + z$, $y \in \mathbf{L}$, $z \perp \mathbf{L}$, we have

$$\begin{aligned} \|x\|^2 &= \|y + z\|^2 = (y + z, y + z) = (y, y) + (y, z) + (z, y) + (z, z) \\ &= (y, y) + (z, z) = \|y\|^2 + \|z\|^2, \end{aligned}$$

and then

$$\|y\| \le \|x\|,$$

or

$$\|\mathbb{P}_{\mathbf{L}}x\| \le \|x\|.$$

Therefore

$$\|\mathbb{P}_{\mathbf{L}}\| \le 1. \tag{7.4}$$

For $x \in \mathbf{L}$, we have $\mathbb{P}_{\mathbf{L}}x = x$ and

$$\|\mathbb{P}_{\mathbf{L}}x\| = \|x\|,$$

whereupon

$$\|\mathbb{P}_{\mathbf{L}}\| \ge 1.$$

Hence, by (7.4), we conclude that

$$\|\mathbb{P}_{\mathbf{L}}\| = 1.$$

This completes the proof. □

Theorem 7.2. *Let $\mathbb{P} \in \mathcal{L}(\mathbf{H})$ be a self-adjoint operator such that $\mathbb{P}^2 = \mathbb{P}$. Then $\mathbb{P}$ is an orthogonal projection on some linear subspace $\mathbf{L} \subseteq \mathbf{H}$.*

Proof. Let

$$\mathbf{L} = \{y = \mathbb{P}x : x \in \mathbf{H}\}.$$

Since $\mathbb{P} \in \mathcal{L}(\mathbf{H})$, we see that $\mathbf{L}$ is a closed linear subspace of $\mathbf{H}$. Note that

$$(x - \mathbb{P}x, \mathbb{P}x) = (\mathbb{P}(x - \mathbb{P}x), x) = (\mathbb{P}x - \mathbb{P}^2x, x) = (\mathbb{P}x - \mathbb{P}x, x) = 0, \quad x \in \mathbf{H}.$$

This completes the proof. □

Remark 7.2. By the proof of Theorem 7.2, it follows that $\mathbb{I} - \mathbb{P}$ is a projection operator.

Definition 7.6. Two projection operators $\mathbb{P}_1$ and $\mathbb{P}_2$ are called orthogonal if $\mathbb{P}_1\mathbb{P}_2 = \mathbb{O}$. Hence, using $(\mathbb{P}_1\mathbb{P}_2)^* = \mathbb{P}_2\mathbb{P}_1$, we get $\mathbb{P}_2\mathbb{P}_1 = \mathbb{O}$.

Definition 7.7. Two linear subspaces $\mathbf{L}_1$ and $\mathbf{L}_2$ of the Hilbert space $\mathbf{H}$ are called orthogonal if

$$(x_1, x_2) = 0$$

for any $x_1 \in \mathbf{L}_1$ and for any $x_2 \in \mathbf{L}_2$. We will write $\mathbf{L}_1 \perp \mathbf{L}_2$.

Theorem 7.3. *Let $\mathbb{P}_1$ and $\mathbb{P}_2$ be two projection operators with corresponding spaces $\mathbf{L}_1$ and $\mathbf{L}_2$. Then $\mathbb{P}_1$ and $\mathbb{P}_2$ are orthogonal if and only if $\mathbf{L}_1 \perp \mathbf{L}_2$.*

Proof.

1. Let $\mathbb{P}_1$ and $\mathbb{P}_2$ be orthogonal. Then, for $x_1 \in \mathbf{L}_1$ and $x_2 \in \mathbf{L}_2$, we have

$$(x_1, x_2) = (\mathbb{P}_1 x_1, \mathbb{P}_2 x_2) = (\mathbb{P}_2 \mathbb{P}_1 x_1, x_2) = 0.$$

   Therefore $\mathbf{L}_1 \perp \mathbf{L}_2$.

2. Let $\mathbf{L}_1 \perp \mathbf{L}_2$. Then, for $x \in \mathbf{H}$, we have $\mathbb{P}_2 x \in \mathbf{L}_2$. Hence, $\mathbb{P}_1 \mathbb{P}_2 x = 0$, $x \in \mathbf{H}$. Therefore $\mathbb{P}_1 \mathbb{P}_2 = \mathbb{O}$. This completes the proof. □

Theorem 7.4. *Let $\mathbb{P}_1$ and $\mathbb{P}_2$ be projection operators with corresponding spaces $\mathbf{L}_1$ and $\mathbf{L}_2$. Then $\mathbb{P}_1 + \mathbb{P}_2$ is a projection operator if and only if $\mathbb{P}_1$ and $\mathbb{P}_2$ are orthogonal. In this case $\mathbb{P}_1 + \mathbb{P}_2 = \mathbb{P}_{\mathbf{L}_1 + \mathbf{L}_2}$.*

Proof.

1. Let $\mathbb{P} = \mathbb{P}_1 + \mathbb{P}_2$ be a projection operator. Then

$$\begin{aligned} \mathbb{P}_1 + \mathbb{P}_2 = (\mathbb{P}_1 + \mathbb{P}_2)^2 &= \mathbb{P}_1^2 + \mathbb{P}_1 \mathbb{P}_2 + \mathbb{P}_2 \mathbb{P}_1 + \mathbb{P}_2^2 \\ &= \mathbb{P}_1 + \mathbb{P}_1 \mathbb{P}_2 + \mathbb{P}_2 \mathbb{P}_1 + \mathbb{P}_2, \end{aligned}$$

   whereupon

$$\mathbb{P}_1 \mathbb{P}_2 + \mathbb{P}_2 \mathbb{P}_1 = \mathbb{O}. \tag{7.5}$$

   Hence,

$$\mathbb{P}_1^2 \mathbb{P}_2 + \mathbb{P}_1 \mathbb{P}_2 \mathbb{P}_1 = \mathbb{O},$$

   or

$$\mathbb{P}_1 \mathbb{P}_2 + \mathbb{P}_1 \mathbb{P}_2 \mathbb{P}_1 = \mathbb{O},$$

   and

$$\mathbb{P}_1 \mathbb{P}_2 \mathbb{P}_1 + \mathbb{P}_1 \mathbb{P}_2 \mathbb{P}_1^2 = \mathbb{O},$$

   or

$$\mathbb{P}_1 \mathbb{P}_2 \mathbb{P}_1 = \mathbb{O}.$$

   From the previous equality and from (7.5), we obtain

$$\mathbb{P}_1^2 \mathbb{P}_2 + \mathbb{P}_1 \mathbb{P}_2 \mathbb{P}_1 = \mathbb{O} \quad \text{and} \quad \mathbb{P}_1 \mathbb{P}_2 \mathbb{P}_1 + \mathbb{P}_2 \mathbb{P}_1^2 = \mathbb{O},$$

   or

$$\mathbb{P}_1 \mathbb{P}_2 = \mathbb{O} \quad \text{and} \quad \mathbb{P}_2 \mathbb{P}_1 = \mathbb{O}.$$

2. Let

$$\mathbb{P}_1\mathbb{P}_2 = \mathbb{P}_2\mathbb{P}_1 = \mathbb{O}.$$

Then

$$\begin{aligned}(\mathbb{P}_1 + \mathbb{P}_2)^2 &= \mathbb{P}_1^2 + \mathbb{P}_1\mathbb{P}_2 + \mathbb{P}_2\mathbb{P}_1 + \mathbb{P}_2^2 = \mathbb{P}_1 + \mathbb{P}_2,\\ (\mathbb{P}_1 + \mathbb{P}_2)^* &= \mathbb{P}_1^* + \mathbb{P}_2^* = \mathbb{P}_1 + \mathbb{P}_2.\end{aligned}$$

Therefore $\mathbb{P}_1 + \mathbb{P}_2$ is a projection operator. This completes the proof. □

Theorem 7.5. *Let $\mathbb{P}_1$ and $\mathbb{P}_2$ be projection operators with corresponding spaces $\mathbf{L}_1$ and $\mathbf{L}_2$. Then $\mathbb{P}_1\mathbb{P}_2$ is a projection operator if and only if $\mathbb{P}_1\mathbb{P}_2 = \mathbb{P}_2\mathbb{P}_1$.*

Proof.
1. Let $\mathbb{P}_1\mathbb{P}_2$ be a projection operator. Then

$$\mathbb{P}_1\mathbb{P}_2 = (\mathbb{P}_1\mathbb{P}_2)^* = \mathbb{P}_2^*\mathbb{P}_1^* = \mathbb{P}_2\mathbb{P}_1. \tag{7.6}$$

2. Let

$$\mathbb{P}_1\mathbb{P}_2 = \mathbb{P}_2\mathbb{P}_1.$$

Then, by (7.6), we get

$$\mathbb{P}_1\mathbb{P}_2 = (\mathbb{P}_1\mathbb{P}_2)^*$$

and

$$(\mathbb{P}_1\mathbb{P}_2)^2 = \mathbb{P}_1\mathbb{P}_2\mathbb{P}_1\mathbb{P}_2 = \mathbb{P}_1^2\mathbb{P}_2^2 = \mathbb{P}_1\mathbb{P}_2.$$

Therefore $\mathbb{P}_1\mathbb{P}_2$ is a projection operator. This completes the proof. □

8 The method of the small parameter

8.1 Abstract functions of a real variable

Let $\mathbf{A} \subseteq \mathbf{R}$, $\mathbf{X}$ be a normed vector space.

Definition 8.1. Any function $f : \mathbf{A} \longmapsto \mathbf{X}$ will be called an abstract function of a real variable.

Definition 8.2. Suppose that $\lambda_0 \in \mathbf{A}$ and the function f is defined in a neighborhood $\mathbf{U}$ of λ_0 and $f : \mathbf{U} \longmapsto \mathbf{X}$. We will say that $a \in \mathbf{X}$ is a limit of the function f when $\lambda \to \lambda_0$ and we will write

$$a = \lim_{\lambda \to \lambda_0} f(\lambda) \quad \text{or} \quad f(\lambda) \to a, \quad \text{as} \quad \lambda \to \lambda_0,$$

if

$$\|f(\lambda) - a\| \to 0, \quad \text{as} \quad \lambda \to \lambda_0.$$

Below we suppose that $\lambda_0 \in \mathbf{A}$ and $\mathbf{U}$ is a neighborhood of λ_0.

Theorem 8.1. *Let $\phi : \mathbf{U} \longmapsto \mathbf{R}$, $f, g : \mathbf{U} \longmapsto \mathbf{X}$ and $\phi(\lambda) \to \alpha$, $f(\lambda) \to a$, $g(\lambda) \to b$, as $\lambda \to \lambda_0$, where $\alpha \in \mathbf{R}$, $a, b \in \mathbf{X}$. Then:*
1. *$\phi(\lambda)f(\lambda) \to \alpha a$, as $\lambda \to \lambda_0$,*
2. *$f(\lambda) + g(\lambda) \to a + b$, as $\lambda \to \lambda_0$.*

Proof.
1. Because $f(\lambda) \to a$, as $\lambda \to \lambda_0$, there is a constant $r > 0$ such that

$$\|f(\lambda)\| \le r$$

for any $\lambda \in \mathbf{U}$. We have

$$\begin{aligned}
\|\phi(\lambda)f(\lambda) - \alpha a\| &= \|\phi(\lambda)f(\lambda) - \alpha f(\lambda) + \alpha f(\lambda) - \alpha a\| \\
&= \|(\phi(\lambda) - \alpha)f(\lambda) + \alpha(f(\lambda) - a)\| \\
&\le \|(\phi(\lambda) - \alpha)f(\lambda)\| + \|\alpha(f(\lambda) - a)\| \\
&= |\phi(\lambda) - \alpha|\|f(\lambda)\| + |\alpha|\|f(\lambda) - a\| \\
&\le r|\phi(\lambda) - \alpha| + |\alpha|\|f(\lambda) - a\| \to 0, \quad \text{as} \quad \lambda \to \lambda_0.
\end{aligned}$$

2. We have

$$\begin{aligned}
\|f(\lambda) + g(\lambda) - (a + b)\| &= \|(f(\lambda) - a) + (g(\lambda) - b)\| \\
&\le \|f(\lambda) - a\| + \|g(\lambda) - b\| \to 0, \quad \text{as} \quad \lambda \to \lambda_0.
\end{aligned}$$

This completes the proof. □

https://doi.org/10.1515/9783110657722-008

Definition 8.3. Let $f : \mathbf{U} \longmapsto \mathbf{X}$ and $f : \mathbf{A} \longmapsto \mathbf{X}$.

1. We will say that f is continuous at λ_0 if
$$\|f(\lambda) - f(\lambda_0)\| \to 0, \quad \text{as} \quad \lambda \to \lambda_0.$$
2. We will say that the abstract function f is continuous on $\mathbf{A}$ if it is continuous at every point of $\mathbf{A}$.
3. We will say that the abstract function f is continuous on the right of λ_0 if $f(\lambda) \to f(\lambda_0)$, as $\lambda \to \lambda_0+$.
4. We will say that the abstract function f is continuous on the left of λ_0 if $f(\lambda) \to f(\lambda_0)$, as $\lambda \to \lambda_0-$.

Theorem 8.2. *Let* $\mathbf{A}$ *be a bounded subset of* $\mathbf{F}$, *the abstract function* $f : \mathbf{A} \longmapsto \mathbf{X}$ *be a continuous function on the set* $\mathbf{A}$. *Then there is a constant* $M > 0$ *such that*
$$\|f(\lambda)\| \leq M$$
for any $\lambda \in \mathbf{A}$.

Proof. Suppose that for any $n \in \mathbf{N}$ there is a $\lambda_n \in \mathbf{A}$ such that
$$\|f(\lambda_n)\| \geq n.$$
There is a subsequence $\{\lambda_{n_k}\}_{k\in\mathbf{N}}$ of the sequence $\{\lambda_n\}_{n\in\mathbf{N}}$ such that $\lambda_{n_k} \to \lambda_0$, as $k \to \infty$, $\lambda_0 \in \mathbf{A}$, and
$$\|f(\lambda_{n_k})\| \geq n_k. \tag{8.1}$$
Since f is continuous on $\mathbf{A}$, we have $f(\lambda_{n_k}) \to f(\lambda_0)$, as $k \to \infty$. Hence, by (8.1), we obtain
$$\|f(\lambda_0)\| \geq n_k$$
for any $k \in \mathbf{N}$, which is a contradiction. This completes the proof. □

Theorem 8.3. *Let* $\phi : \mathbf{U} \longmapsto \mathbf{R}, f, g : \mathbf{U} \longmapsto \mathbf{X}$ *be continuous at* λ_0. *Then:*

1. ϕf *is continuous at* λ_0,
2. $f + g$ *is continuous at* λ_0.

Proof.

1. Since $f : \mathbf{U} \longmapsto \mathbf{X}$ is continuous at λ_0, there is an $r > 0$ such that
$$\|f(\lambda)\| \leq r \quad \text{for} \quad \text{any} \quad \lambda \in \mathbf{U}.$$
We have
$$\begin{aligned}\|\phi(\lambda)f(\lambda) - f(\lambda_0)\phi(\lambda_0)\| &= \|\phi(\lambda)f(\lambda) - \phi(\lambda_0)f(\lambda) + \phi(\lambda_0)f(\lambda) - \phi(\lambda_0)f(\lambda_0)\| \\ &= \|(\phi(\lambda) - \phi(\lambda_0))f(\lambda) + \phi(\lambda_0)(f(\lambda) - f(\lambda_0))\|\end{aligned}$$

$$\begin{aligned}&\le \|(\phi(\lambda)-\phi(\lambda_0))f(\lambda)\| + \|\phi(\lambda_0)(f(\lambda)-f(\lambda_0))\|\\ &= |\phi(\lambda)-\phi(\lambda_0)|\|f(\lambda)\| + |\phi(\lambda_0)|\|f(\lambda)-f(\lambda_0)\|\\ &\le r|\phi(\lambda)-\phi(\lambda_0)| + |\phi(\lambda_0)|\|f(\lambda)-f(\lambda_0)\| \to 0,\end{aligned}$$

as $\lambda \to \lambda_0$.

2. We have

$$\begin{aligned}\|f(\lambda)+g(\lambda)-(f(\lambda_0)+g(\lambda_0))\| &= \|(f(\lambda)-f(\lambda_0))+(g(\lambda)-g(\lambda_0))\|\\ &\le \|f(\lambda)-f(\lambda_0)\| + \|g(\lambda)-g(\lambda_0)\|\\ &\to 0, \quad \text{as} \quad \lambda \to \lambda_0.\end{aligned}$$

This completes the proof. □

Suppose that $\mathbf{Y}$ is a normed vector space. For $\lambda \in \mathbf{A}$ we consider the operator $\mathbb{A}(\lambda) \in \mathcal{L}(\mathbf{X}, \mathbf{Y})$.

Definition 8.4.

1. A linear operator $\mathbb{B} : \mathbf{X} \longmapsto \mathbf{Y}$ is said to be a limit of the operator $\mathbb{A}(\lambda)$ when $\lambda \to \lambda_0$, if

$$\|\mathbb{A}(\lambda) - B\| \to 0, \quad \text{as} \quad \lambda \to \lambda_0.$$

   We write $\mathbb{A}(\lambda) \to \mathbb{B}$, as $\lambda \to \lambda_0$.
2. We say that the operator $\mathbb{A}(\lambda)$ is continuous at λ_0, if $\mathbb{A}(\lambda) \to \mathbb{A}(\lambda_0)$, as $\lambda \to \lambda_0$.
3. We say that the operator $\mathbb{A}(\lambda)$ is continuous on $\mathbf{A}$, if it is continuous at every point of $\lambda_0 \in \mathbf{A}$.
4. We say that the operator $\mathbb{A}(\lambda)$ is continuous on the right of λ_0, if $\mathbb{A}(\lambda) \to \mathbb{A}(\lambda_0)$, as $\lambda \to \lambda_0+$.
5. We say that the operator $\mathbb{A}(\lambda)$ is continuous on the left of λ_0, if $\mathbb{A}(\lambda) \to \mathbb{A}(\lambda_0)$, as $\lambda \to \lambda_0-$.

Below we suppose that $\mathbb{A}(\lambda), \mathbb{B} \in \mathcal{L}(\mathbf{X}, \mathbf{Y})$, $\lambda \in \mathbf{A}$, $f : \mathbf{A} \longmapsto \mathbf{X}$, $a \in \mathbf{X}$.

Theorem 8.4. *Let $\mathbb{A}(\lambda) \to \mathbb{B}$, $f(\lambda) \to a$, as $\lambda \to \lambda_0$. Then $\mathbb{A}(\lambda)f(\lambda) \to \mathbb{B}a$, as $\lambda \to \lambda_0$.*

Proof. We have

$$\begin{aligned}\|\mathbb{A}(\lambda)f(\lambda) - \mathbb{B}a\| &= \|\mathbb{A}(\lambda)f(\lambda) - \mathbb{B}f(\lambda) + \mathbb{B}f(\lambda) - \mathbb{B}a\|\\ &= \|(\mathbb{A}(\lambda) - \mathbb{B})f(\lambda) + \mathbb{B}(f(\lambda) - a)\|\\ &\le \|(\mathbb{A}(\lambda) - \mathbb{B})f(\lambda)\| + \|\mathbb{B}(f(\lambda) - a)\|\\ &\le \|\mathbb{A}(\lambda) - \mathbb{B}\|\|f(\lambda)\| + \|\mathbb{B}\|\|f(\lambda) - a\| \to 0, \quad \text{as} \quad \lambda \to \lambda_0.\end{aligned}$$

This completes the proof. □

Theorem 8.5. *Let $\mathbb{A}(\lambda)$ and f be continuous at λ_0. Then $\mathbb{A}(\lambda)f$ is continuous at λ_0.*

Proof. We have

$$\begin{aligned}\|\mathbb{A}(\lambda)f(\lambda)-\mathbb{A}(\lambda_0)f(\lambda_0)\| &= \|\mathbb{A}(\lambda)f(\lambda)-\mathbb{A}(\lambda_0)f(\lambda)+\mathbb{A}(\lambda_0)f(\lambda)-\mathbb{A}(\lambda_0)f(\lambda_0)\|\\ &= \|(\mathbb{A}(\lambda)-\mathbb{A}(\lambda_0))f(\lambda)+\mathbb{A}(\lambda_0)(f(\lambda)-f(\lambda_0))\|\\ &\le \|(\mathbb{A}(\lambda)-\mathbb{A}(\lambda_0))f(\lambda)\|+\|\mathbb{A}(\lambda_0)(f(\lambda)-f(\lambda_0))\|\\ &\le \|\mathbb{A}(\lambda)-\mathbb{A}(\lambda_0)\|\|f(\lambda)\|+\|\mathbb{A}(\lambda_0)\|\|f(\lambda)-f(\lambda_0)\|\to 0,\end{aligned}$$

as $\lambda \to \lambda_0$. This completes the proof. □

Exercise 8.1. Let $\mathbb{A}(\lambda)$ and f be continuous at λ_0. Prove that $\alpha\mathbb{A}(\lambda)+\beta f(\lambda)$ is continuous at λ_0 for any $\alpha, \beta \in \mathbf{F}$.

Definition 8.5. We will say that the abstract function $f : \mathbf{A} \longmapsto \mathbf{X}$ is differentiable at λ_0 if the limit

$$\lim_{\lambda\to\lambda_0}\left\|\frac{f(\lambda)-f(\lambda_0)}{\lambda-\lambda_0}\right\| \tag{8.2}$$

exists. In this case, we will say that the limit (8.2) is the derivative of the abstract function $f(\lambda)$ at λ_0 and we will write $\frac{d}{d\lambda}f(\lambda_0)$. We have

$$\left\|\frac{f(\lambda)-f(\lambda_0)}{\lambda-\lambda_0}-\frac{d}{d\lambda}f(\lambda_0)\right\|\to 0, \quad \text{as} \quad \lambda\to\lambda_0.$$

Theorem 8.6. *Suppose that the abstract functions $f, g : \mathbf{A} \longmapsto \mathbf{X}$ are differentiable at $\lambda_0 \in \mathbf{A}$. Then:*

1. *$\alpha f : \mathbf{A} \longmapsto \mathbf{X}$ is differentiable at λ_0 and*

$$\frac{d}{d\lambda}(\alpha f)(\lambda_0) = \alpha\frac{d}{d\lambda}f(\lambda_0),$$

for any $\alpha \in \mathbf{F}$.

2. *$f + g : \mathbf{A} \longmapsto \mathbf{X}$ is differentiable at λ_0 and*

$$\frac{d}{d\lambda}(f+g)(\lambda_0) = \frac{d}{d\lambda}f(\lambda_0)+\frac{d}{d\lambda}g(\lambda_0).$$

Proof.

1. Let $\alpha \in \mathbf{F}$ be arbitrarily chosen. Then

$$\begin{aligned}\left\|\frac{(\alpha f)(\lambda)-(\alpha f)(\lambda_0)}{\lambda-\lambda_0}-\alpha\frac{d}{d\lambda}f(\lambda_0)\right\| &= \left\|\alpha\frac{f(\lambda)-f(\lambda_0)}{\lambda-\lambda_0}-\alpha\frac{d}{d\lambda}f(\lambda_0)\right\|\\ &= |\alpha|\left\|\frac{f(\lambda)-f(\lambda_0)}{\lambda-\lambda_0}-\frac{d}{d\lambda}f(\lambda_0)\right\|\to 0,\end{aligned}$$

as $\lambda \to \lambda_0$.

2. We have

$$\begin{aligned}
&\left\| \frac{(f+g)(\lambda) - (f+g)(\lambda_0)}{\lambda - \lambda_0} - \frac{d}{d\lambda} f(\lambda_0) - \frac{d}{d\lambda} g(\lambda_0) \right\| \\
&\quad = \left\| \frac{f(\lambda) + g(\lambda) - f(\lambda_0) - g(\lambda_0)}{\lambda - \lambda_0} - \frac{d}{d\lambda} f(\lambda_0) - \frac{d}{d\lambda} g(\lambda_0) \right\| \\
&\quad = \left\| \frac{f(\lambda) - f(\lambda_0)}{\lambda - \lambda_0} - \frac{d}{d\lambda} f(\lambda_0) + \frac{g(\lambda) - g(\lambda_0)}{\lambda - \lambda_0} - \frac{d}{d\lambda} g(\lambda_0) \right\| \\
&\quad \le \left\| \frac{f(\lambda) - f(\lambda_0)}{\lambda - \lambda_0} - \frac{d}{d\lambda} f(\lambda_0) \right\| + \left\| \frac{g(\lambda) - g(\lambda_0)}{\lambda - \lambda_0} - \frac{d}{d\lambda} g(\lambda_0) \right\| \to 0, \quad \text{as} \quad \lambda \to \lambda_0,
\end{aligned}$$

as $\lambda \to \lambda_0$. This completes the proof. □

Theorem 8.7. *Let the abstract function $f : \mathbf{A} \longmapsto \mathbf{X}$ and the scalar function $\phi : \mathbf{A} \longmapsto \mathbf{R}$ be differentiable at $\lambda_0 \in \mathbf{A}$. Then ϕf is differentiable at λ_0 and*

$$\frac{d}{d\lambda}(\phi f)(\lambda_0) = \phi(\lambda_0)\frac{d}{d\lambda} f(\lambda_0) + \frac{d}{d\lambda}\phi(\lambda_0) f(\lambda_0).$$

Proof. We have

$$\begin{aligned}
&\left\| \frac{(\phi f)(\lambda) - (\phi f)(\lambda_0)}{\lambda - \lambda_0} - \phi(\lambda_0)\frac{d}{d\lambda} f(\lambda_0) - \frac{d}{d\lambda}\phi(\lambda_0) f(\lambda_0) \right\| \\
&\quad = \left\| \frac{\phi(\lambda) f(\lambda) - \phi(\lambda_0) f(\lambda_0)}{\lambda - \lambda_0} - \phi(\lambda_0)\frac{d}{d\lambda} f(\lambda_0) - \frac{d}{d\lambda}\phi(\lambda_0) f(\lambda_0) \right\| \\
&\quad = \left\| (\phi(\lambda) - \phi(\lambda_0)) \frac{f(\lambda) - f(\lambda_0)}{\lambda - \lambda_0} + \left(\frac{f(\lambda) - f(\lambda_0)}{\lambda - \lambda_0}\phi(\lambda_0) - \frac{d}{d\lambda} f(\lambda_0)\phi(\lambda_0) \right) \right. \\
&\qquad \left. + \left(\frac{\phi(\lambda) - \phi(\lambda_0)}{\lambda - \lambda_0} - \frac{d}{d\lambda}\phi(\lambda_0) \right) f(\lambda_0) \right\| \\
&\quad \le \left\| (\phi(\lambda) - \phi(\lambda_0)) \frac{f(\lambda) - f(\lambda_0)}{\lambda - \lambda_0} \right\| + \left\| \phi(\lambda_0) \left(\frac{f(\lambda) - f(\lambda_0)}{\lambda - \lambda_0} - \frac{d}{d\lambda} f(\lambda_0) \right) \right\| \\
&\qquad + \left\| \left(\frac{\phi(\lambda) - \phi(\lambda_0)}{\lambda - \lambda_0} - \frac{d}{d\lambda}\phi(\lambda_0) \right) f(\lambda_0) \right\| \\
&\quad = |\phi(\lambda) - \phi(\lambda_0)| \left\| \frac{f(\lambda) - f(\lambda_0)}{\lambda - \lambda_0} \right\| + |\phi(\lambda_0)| \left\| \frac{f(\lambda) - f(\lambda_0)}{\lambda - \lambda_0} - \frac{d}{d\lambda} f(\lambda_0) \right\| \\
&\qquad + \|f(\lambda_0)\| \left| \frac{\phi(\lambda) - \phi(\lambda_0)}{\lambda - \lambda_0} - \frac{d}{d\lambda}\phi(\lambda_0) \right| \to 0, \quad \text{as} \quad \lambda \to \lambda_0.
\end{aligned}$$

This completes the proof. □

Exercise 8.2. Let the abstract function $f : \mathbf{A} \longmapsto \mathbf{X}$ be differentiable at $\lambda_0 \in \mathbf{A}$. Prove that it is continuous at λ_0.

Definition 8.6. We say that the operator $\mathbb{A}(\lambda) \in \mathcal{L}(\mathbf{X}, \mathbf{Y})$ is differentiable at $\lambda_0 \in \mathbf{A}$, if the limit

$$\lim_{\lambda \to \lambda_0} \left\| \frac{1}{\lambda - \lambda_0} (\mathbb{A}(\lambda) - \mathbb{A}(\lambda_0)) \right\|$$

exists. In this case we call this limit the derivative of the operator $\mathbb{A}(\lambda)$ at λ_0. We will denote $\frac{d}{d\lambda}\mathbb{A}(\lambda_0)$. We have

$$\left\| \frac{1}{\lambda - \lambda_0} (\mathbb{A}(\lambda) - \mathbb{A}(\lambda_0)) - \frac{d}{d\lambda}\mathbb{A}(\lambda_0) \right\| \to 0, \quad \text{as} \quad \lambda \to \lambda_0.$$

Theorem 8.8. *Suppose that the abstract function $f : \mathbf{A} \longmapsto \mathbf{X}$ and the operator $\mathbb{A}(\lambda) \in \mathcal{L}(\mathbf{X}, \mathbf{Y})$ are differentiable at $\lambda_0 \in \mathbf{A}$. Then $\mathbb{A}(\lambda)f$ is differentiable at λ_0 and*

$$\frac{d}{d\lambda}(\mathbb{A}f)(\lambda_0) = \frac{d}{d\lambda}\mathbb{A}(\lambda_0)f(\lambda_0) + \mathbb{A}(\lambda_0)\frac{d}{d\lambda}f(\lambda_0).$$

Proof. We have

$$\begin{aligned}
&\left\| \frac{\mathbb{A}(\lambda)f(\lambda) - \mathbb{A}(\lambda_0)f(\lambda_0)}{\lambda - \lambda_0} - \frac{d}{d\lambda}\mathbb{A}(\lambda_0)f(\lambda_0) - \mathbb{A}(\lambda_0)\frac{d}{d\lambda}f(\lambda_0) \right\| \\
&= \left\| \frac{\mathbb{A}(\lambda) - \mathbb{A}(\lambda_0)}{\lambda - \lambda_0}(f(\lambda) - f(\lambda_0)) + \frac{\mathbb{A}(\lambda) - \mathbb{A}(\lambda_0)}{\lambda - \lambda_0} f(\lambda_0) \right. \\
&\quad \left. + \frac{\mathbb{A}(\lambda_0)f(\lambda) - \mathbb{A}(\lambda_0)f(\lambda_0)}{\lambda - \lambda_0} - \frac{d}{d\lambda}\mathbb{A}(\lambda_0)f(\lambda_0) - \mathbb{A}(\lambda_0)\frac{d}{d\lambda}f(\lambda_0) \right\| \\
&= \left\| \frac{\mathbb{A}(\lambda) - \mathbb{A}(\lambda_0)}{\lambda - \lambda_0}(f(\lambda) - f(\lambda_0)) + \left(\frac{\mathbb{A}(\lambda) - \mathbb{A}(\lambda_0)}{\lambda - \lambda_0} - \frac{d}{d\lambda}\mathbb{A}(\lambda_0) \right) f(\lambda_0) \right. \\
&\quad \left. + \mathbb{A}(\lambda_0)\left(\frac{f(\lambda) - f(\lambda_0)}{\lambda - \lambda_0} - \frac{d}{d\lambda}f(\lambda_0) \right) \right\| \\
&\le \left\| \frac{\mathbb{A}(\lambda) - \mathbb{A}(\lambda_0)}{\lambda - \lambda_0}(f(\lambda) - f(\lambda_0)) \right\| + \left\| \left(\frac{\mathbb{A}(\lambda) - \mathbb{A}(\lambda_0)}{\lambda - \lambda_0} - \frac{d}{d\lambda}\mathbb{A}(\lambda_0) \right) f(\lambda_0) \right\| \\
&\quad + \left\| \mathbb{A}(\lambda_0)\left(\frac{f(\lambda) - f(\lambda_0)}{\lambda - \lambda_0} - \frac{d}{d\lambda}f(\lambda_0) \right) \right\| \\
&\le \left\| \frac{\mathbb{A}(\lambda) - \mathbb{A}(\lambda_0)}{\lambda - \lambda_0} \right\| \|f(\lambda) - f(\lambda_0)\| + \left\| \frac{\mathbb{A}(\lambda) - \mathbb{A}(\lambda_0)}{\lambda - \lambda_0} - \frac{d}{d\lambda}\mathbb{A}(\lambda_0) \right\| \|f(\lambda_0)\| \\
&\quad + \|\mathbb{A}(\lambda_0)\| \left\| \frac{f(\lambda) - f(\lambda_0)}{\lambda - \lambda_0} - \frac{d}{d\lambda}f(\lambda_0) \right\| \to 0, \quad \text{as} \quad \lambda \to \lambda_0.
\end{aligned}$$

This completes the proof. □

Exercise 8.3. Let the operator $\mathbb{A}(\lambda) \in \mathcal{L}(\mathbf{X}, \mathbf{Y})$ is differentiable at λ_0. Prove that it is continuous at λ_0.

Remark 8.1. Below, for convenience, we will write $\mathbb{A}^{-1}(\lambda)$ instead of $(\mathbb{A}(\lambda))^{-1}$ when $(\mathbb{A}(\lambda))^{-1}$ exists.

Theorem 8.9. *Let the operator $\mathbb{A}(\lambda) \in \mathcal{L}(\mathbf{X}, \mathbf{Y})$ be differentiable at $\lambda_0 \in \mathbf{A}$ and $\mathbb{A}^{-1}(\lambda)$ exists and let it be continuous in a neighborhood of λ_0. Then $\mathbb{A}^{-1}(\lambda)$ is differentiable at λ_0 and*

$$\frac{d}{d\lambda}\mathbb{A}^{-1}(\lambda_0) = -\mathbb{A}^{-1}(\lambda_0)\frac{d}{d\lambda}\mathbb{A}(\lambda_0)\mathbb{A}^{-1}(\lambda_0).$$

Proof. We have

$$\begin{aligned}
&\left\|\frac{\mathbb{A}^{-1}(\lambda) - \mathbb{A}^{-1}(\lambda_0)}{\lambda - \lambda_0} + \mathbb{A}^{-1}(\lambda_0)\frac{d}{d\lambda}\mathbb{A}(\lambda_0)\mathbb{A}^{-1}(\lambda_0)\right\| \\
&= \left\|\mathbb{A}^{-1}(\lambda_0)\frac{\mathbb{A}(\lambda_0) - \mathbb{A}(\lambda)}{\lambda - \lambda_0}\mathbb{A}^{-1}(\lambda) + \mathbb{A}^{-1}(\lambda_0)\frac{d}{d\lambda}\mathbb{A}(\lambda_0)\mathbb{A}^{-1}(\lambda)\right. \\
&\quad \left. - \mathbb{A}^{-1}(\lambda_0)\frac{d}{d\lambda}\mathbb{A}(\lambda_0)\mathbb{A}^{-1}(\lambda) + \mathbb{A}^{-1}(\lambda_0)\frac{d}{d\lambda}\mathbb{A}(\lambda_0)\mathbb{A}^{-1}(\lambda_0)\right\| \\
&= \left\|\mathbb{A}^{-1}(\lambda_0)\left(\frac{d}{d\lambda}\mathbb{A}(\lambda_0) - \frac{\mathbb{A}(\lambda) - \mathbb{A}(\lambda_0)}{\lambda - \lambda_0}\right)\mathbb{A}^{-1}(\lambda)\right. \\
&\quad \left. + \mathbb{A}^{-1}(\lambda_0)\frac{d}{d\lambda}\mathbb{A}(\lambda_0)(\mathbb{A}^{-1}(\lambda_0) - \mathbb{A}^{-1}(\lambda))\right\| \\
&\le \left\|\mathbb{A}^{-1}(\lambda_0)\left(\frac{d}{d\lambda}\mathbb{A}(\lambda_0) - \frac{\mathbb{A}(\lambda) - \mathbb{A}(\lambda_0)}{\lambda - \lambda_0}\right)\mathbb{A}^{-1}(\lambda)\right\| \\
&\quad + \left\|\mathbb{A}^{-1}(\lambda_0)\frac{d}{d\lambda}\mathbb{A}(\lambda_0)(\mathbb{A}^{-1}(\lambda_0) - \mathbb{A}^{-1}(\lambda))\right\| \\
&\le \|\mathbb{A}^{-1}(\lambda_0)\|\left\|\frac{d}{d\lambda}\mathbb{A}(\lambda_0) - \frac{\mathbb{A}(\lambda) - \mathbb{A}(\lambda_0)}{\lambda - \lambda_0}\right\|\|\mathbb{A}^{-1}(\lambda)\| \\
&\quad + \|\mathbb{A}^{-1}(\lambda_0)\|\left\|\frac{d}{d\lambda}\mathbb{A}(\lambda_0)\right\|\|\mathbb{A}^{-1}(\lambda_0) - \mathbb{A}^{-1}(\lambda)\| \to 0, \quad \text{as} \quad \lambda \to \lambda_0.
\end{aligned}$$

This completes the proof. □

Definition 8.7. For an abstract function $f : \mathbf{A} \longmapsto \mathbf{X}$ and an operator $\mathbb{A}(\lambda) \in \mathcal{L}(\mathbf{X}, \mathbf{Y})$, defined in a neighborhood of $\lambda_0 \in \mathbf{A}$, we define

$$\begin{aligned}
\frac{d^k}{d\lambda^k}f(\lambda_0) &= \frac{d}{d\lambda}\left(\frac{d^{k-1}}{d\lambda^{k-1}}f\right)(\lambda_0), \\
\frac{d^k}{d\lambda^k}A(\lambda_0) &= \frac{d}{d\lambda}\left(\frac{d^{k-1}}{d\lambda^{k-1}}A\right)(\lambda_0), \quad k = 2, 3, \ldots,
\end{aligned}$$

whenever $\frac{d^{k-1}}{d\lambda^{k-1}}f(\lambda)$ and $\frac{d^{k-1}}{d\lambda^{k-1}}\mathbb{A}(\lambda)$ exist in a neighborhood of λ_0.

8.2 Power series

Suppose that $\mathbf{X}$ is a normed vector space. Consider the power series

$$\sum_{k=0}^{\infty} x_k \lambda^k, \tag{8.3}$$

where $x_k \in \mathbf{X}$, $k \in \mathbf{N}_0$, $\lambda \in \mathbf{F}$.

Definition 8.8. Let $\mathbf{A}$ be the set of all points $\lambda \in \mathbf{F}$ for which (8.3) is convergent. The set $\mathbf{A}$ will be called the convergence domain of the power series (8.3).

Note that $0 \in \mathbf{A}$. The sum of (8.3) for $\lambda \in \mathbf{A}$ will be denoted by $S(\lambda)$. We will write

$$S(\lambda) = \sum_{k=0}^{\infty} x_k \lambda^k, \quad \lambda \in \mathbf{A}.$$

The partial sums of (8.3) will be denoted by

$$S_n(\lambda) = \sum_{k=0}^{n} x_k \lambda^k, \quad n \in \mathbf{N}.$$

Theorem 8.10. *Let $\lambda_0 \neq 0$, $\lambda_0 \in \mathbf{A}$. Then $\mathbf{B}_{|\lambda_0|}(0) \subseteq \mathbf{A}$ and for any $r < |\lambda_0|$, the series* (8.3) *is convergent for $\lambda \in \mathbf{B}_r[0]$.*

Proof. Since $\lambda_0 \in \mathbf{A}$, we have $x_n \lambda_0^n \to 0$, as $n \to \infty$. Therefore the sequence $\{x_n \lambda_0^n\}_{n\in\mathbf{N}}$ is bounded. There exists a positive constant M such that

$$\|x_n \lambda_0^n\| \leq M$$

for any $n \in \mathbf{N}$. Suppose that $|\lambda| < |\lambda_0|$. Then

$$\begin{aligned} \|x_n \lambda^n\| &= \left\| x_n \lambda_0^n \frac{\lambda^n}{\lambda_0^n} \right\| \\ &= \left| \frac{\lambda}{\lambda_0} \right|^n \|x_n \lambda_0^n\| \leq M \left| \frac{\lambda}{\lambda_0^n} \right|^n. \end{aligned}$$

Since $|\lambda| < |\lambda_0|$, we see that the series $\sum_{n=0}^{\infty} |\frac{\lambda}{\lambda_0}|^n$ is convergent. Therefore the series (8.3) is convergent for $\lambda \in \mathbf{B}_{|\lambda_0|}(0)$. Let $|\lambda| \leq r < |\lambda_0|$. Then

$$\|x_n \lambda^n\| \leq M \left| \frac{\lambda}{\lambda_0} \right|^n \leq M \left(\frac{r}{|\lambda_0|} \right)^n.$$

Hence, (8.3) is convergent for $\lambda \in \mathbf{B}_r[0]$. This completes the proof. □

Definition 8.9. The number

$$R = \sup_{\lambda \in \mathbf{A}} |\lambda|$$

will be called the convergence radius of (8.3).

By Definition 8.9, it follows that:
1. if $R = 0$, then $\mathbf{A} = \{0\}$;
2. if $R = \infty$, then $\mathbf{A} = \mathbf{F}$.

Theorem 8.11 (Cauchy–Hadamard's formula). *We have*

$$R = \frac{1}{\limsup_{n\to\infty} \sqrt[n]{\|x_n\|}}.$$

Proof. Let

$$\rho = \frac{1}{\limsup_{n\to\infty} \sqrt[n]{\|x_n\|}}.$$

We will prove that $\rho = R$. By the definition of ρ, for any $\epsilon \in (0,1)$, there is an $N \in \mathbf{N}$ such that

$$(1-\epsilon)^n \le \|x_n\|\rho^n \le (1+\epsilon)^n$$

for any $n \ge N$. If $|\lambda| > \rho$, we have

$$\|x_n\lambda^n\| = \|x_n\||\lambda|^n > \|x_n\|\rho^n \ge (1-\epsilon)^n$$

for any $n \ge N$. Hence, $\|x_n\lambda^n\| \ge 1$ for any $n \ge N$ and the power series (8.3) diverges. Therefore

$$\rho \ge R. \tag{8.4}$$

On the other hand, if $|\lambda| < \rho$, we have

$$\|x_n\lambda^n\| = \|x_n\||\lambda|^n = \|x_n\|\rho^n \frac{|\lambda|^n}{\rho^n} \le (1+\epsilon)^n \frac{|\lambda|^n}{\rho^n} \tag{8.5}$$

for any $n \ge N$. Let

$$k = (1+\epsilon)\frac{|\lambda|}{\rho}.$$

We can choose $\epsilon \in (0,1)$ small enough so that $k < 1$. Hence, by (8.5), we obtain

$$\|x_n\lambda^n\| \le k^n$$

for any $n \ge N$. Therefore (8.3) is convergent. From this, $\rho \le R$. From the last inequality and from (8.4), we get $\rho = R$. This completes the proof. □

Theorem 8.12. *Suppose that there are constants $M > 0$, $k > 0$ and $N \in \mathbf{N}$ such that $\|x_n\| \le Mk^n$ for any $n \ge N$. Then $R \ge \frac{1}{k}$.*

Proof. Let $|\lambda| < \frac{1}{k}$. We set $q = k|\lambda|$. Then $q < 1$ and

$$\|x_n\lambda^n\| = \|x_n\||\lambda|^n = \frac{\|x_n\|}{k^n}q^n \le Mq^n$$

for any $n \ge N$. Therefore (8.3) is convergent. Thus, $R \ge \frac{1}{k}$. This completes the proof. □

Theorem 8.13. *Suppose that the power series $\sum_{k=0}^{\infty} x_k\lambda^k$ and $\sum_{k=0}^{\infty} y_k\lambda^k$, $x_k, y_k \in \mathbf{X}$, $k \in \mathbf{N}_0$, $\lambda \in \mathbf{F}$, have the same convergence radius R. If*

$$\sum_{k=0}^{\infty} x_k\lambda^k = \sum_{k=0}^{\infty} y_k\lambda^k \tag{8.6}$$

for $\lambda \in \mathbf{B}_R(0)$, then $x_k = y_k$ for any $k \in \mathbf{N}$.

Proof. Since $\lambda = 0 \in \mathbf{B}_R(0)$, we put $\lambda = 0$ in (8.6) and we obtain $x_0 = y_0$. Hence, by (8.6), we get

$$\sum_{k=1}^{\infty} x_k\lambda^k = \sum_{k=1}^{\infty} y_k\lambda^k,$$

whereupon

$$\sum_{k=1}^{\infty} x_k\lambda^{k-1} = \sum_{k=1}^{\infty} y_k\lambda^{k-1}. \tag{8.7}$$

We put $\lambda = 0$ in (8.7) and we obtain $x_1 = y_1$, and so on. This completes the proof. □

8.3 Analytic abstract functions and Taylor's series

Suppose that $\mathbf{X}$ is a normed vector space and $\mathbf{A} \subseteq \mathbf{F}$, $0 \in \mathbf{A}$, $\mathbf{A} \neq \{0\}$.

Definition 8.10. The abstract function $f : \mathbf{A} \longmapsto \mathbf{X}$ will be called analytic at $\lambda = 0$, if in a neighborhood $\mathbf{U}$ of $\lambda = 0$ it can be represented as an uniformly convergent power series

$$f(\lambda) = \sum_{k=0}^{\infty} x_k\lambda^k, \tag{8.8}$$

$x_k \in \mathbf{X}$, $k \in \mathbf{N}_0$, $\lambda \in \mathbf{U}$.

Theorem 8.14. *Let the abstract function $f : \mathbf{A} \longmapsto \mathbf{X}$ be analytic at $\lambda = 0$. Then it is continuous in $\mathbf{B}_R(0)$, where R is the convergence radius of* (8.8).

Proof. Let $\rho \in (0, R)$. Consider the series

$$\sum_{k=1}^{\infty} k\|x_k\|\rho^{k-1}. \tag{8.9}$$

For any $\rho_1 \in (\rho, R)$ we see that the series $\sum_{k=0}^{\infty} x_k\rho_1^k$ is convergent. Therefore $\|x_n\rho_1^n\| \to 0$, as $n \to \infty$. Hence, the sequence $\{\rho_1^n\|x_n\|\}_{n\in\mathbf{N}}$ is bounded. Consequently there is a constant $M > 0$ such that $\|x_n\|\rho_1^n \leq M$ for any $n \in \mathbf{N}$. Furthermore,

$$n\|x_n\|\rho^{n-1} = \|x_n\|\rho_1^n \frac{n}{\rho_1}\left(\frac{\rho}{\rho_1}\right)^{n-1} \leq \frac{M}{\rho_1} n\left(\frac{\rho}{\rho_1}\right)^{n-1}. \tag{8.10}$$

Let $q = \frac{\rho}{\rho_1}$. Since $q < 1$ and the series $\sum_{n=0}^{\infty} nq^{n-1}$ is convergent, using (8.10), we see that the series (8.9) is convergent. Let

$$S(\rho) = \sum_{k=1}^{\infty} k\|x_k\|\rho^{k-1}.$$

Then, for any $\lambda_1, \lambda_2 \in \mathbf{B}_\rho[0]$, we have

$$\begin{aligned} f(\lambda_1) - f(\lambda_2) &= \sum_{k=1}^{\infty} x_k\lambda_1^k - \sum_{k=1}^{\infty} x_k\lambda_2^k \\ &= \sum_{k=1}^{\infty} x_k(\lambda_1^k - \lambda_2^k) = (\lambda_1 - \lambda_2)\sum_{k=1}^{\infty} x_k(\lambda_1^{k-1} + \lambda_1^{k-2}\lambda_2 + \cdots + \lambda_2^{k-1}). \end{aligned}$$

Hence,

$$\begin{aligned} \|f(\lambda_1) - f(\lambda_2)\| &= \left\|\sum_{k=1}^{\infty} x_k(\lambda_1^{k-1} + \lambda_1^{k-2}\lambda_2 + \cdots + \lambda_2^{k-1})(\lambda_1 - \lambda_2)\right\| \\ &\leq \sum_{k=1}^{\infty} \|x_k(\lambda_1^{k-1} + \lambda_1^{k-2}\lambda_2 + \cdots + \lambda_2^{k-1})(\lambda_1 - \lambda_2)\| \\ &\leq |\lambda_1 - \lambda_2| \sum_{k=1}^{\infty} \|x_k\|(|\lambda_1|^{k-1} + |\lambda_1|^{k-2}|\lambda_2| + \cdots + |\lambda_2|^{k-2}) \\ &\leq |\lambda_1 - \lambda_2| \sum_{k=1}^{\infty} k\rho^{k-1}\|x_k\| = |\lambda_1 - \lambda_2|S(\rho) \to 0, \quad \text{as} \quad \lambda_1 \to \lambda_2, \end{aligned}$$

because (8.9) is convergent. This completes the proof. □

Corollary 8.1. *The series $\sum_{k=1}^{\infty} kx_k\lambda^{k-1}$ is convergent for $\lambda \in \mathbf{B}_R(0)$.*

Proof. Let $|\lambda| < R$ and $\rho \in (|\lambda|, R)$. Hence, using the fact that (8.9) is convergent, we get

$$\begin{aligned} \left\|\sum_{k=1}^{\infty} kx_k\lambda^{k-1}\right\| &\leq \sum_{k=1}^{\infty} \|kx_k\lambda^{k-1}\| = \sum_{k=1}^{\infty} k\|x_k\||\lambda|^{k-1} \\ &\leq \sum_{k=1}^{\infty} k\|x_k\|\rho^{k-1} < \infty. \end{aligned}$$

This completes the proof. □

Theorem 8.15. *Let the abstract function $f : \mathbf{A} \longmapsto \mathbf{X}$ be analytic at $\lambda = 0$. Then f is differentiable for any $\lambda \in \mathbf{B}_R(0)$, where R is its convergence radius.*

Proof. Let $\rho \in (0, R)$. Consider the series

$$\sum_{k=2}^{\infty} k(k-1)\|x_k\|\rho^{k-2}. \tag{8.11}$$

Take $\rho_1 \in (\rho, R)$ arbitrarily. Then $\sum_{k=0}^{\infty} x_k\rho_1^k$ is convergent and hence $\|x_k\rho_1^k\| \to 0$, as $k \to \infty$. Therefore there exists a constant $M > 0$ such that $\|x_n\rho_1^n\| \le M$ for any $n \in \mathbf{N}$. Note that

$$k(k-1)\|x_k\|\rho^{k-2} = k(k-1)\|x_k\|\rho_1^k \frac{1}{\rho_1^2}\left(\frac{\rho}{\rho_1}\right)^{k-2} \le \frac{Mk(k-1)}{\rho_1^2}\left(\frac{\rho}{\rho_1}\right)^{k-2}. \tag{8.12}$$

We set $q = \frac{\rho}{\rho_1}$. Since $q < 1$ and $\sum_{k=2}^{\infty} q^{k-2}$ is convergent, using (8.12), we conclude that the series (8.11) is convergent. Let

$$S(\rho) = \sum_{k=2}^{\infty} k(k-1)\|x_k\|\rho^{k-2}.$$

Now we take $\lambda \neq \mu$. Then

$$\begin{aligned}
&(\mu - \lambda)\int_0^1 (1-z)((1-z)\lambda + z\mu)^{k-2}dz \\
&= (\mu - \lambda)\int_0^1 (1-z)(\lambda + z(\mu - \lambda))^{k-2}dz \\
&= \int_0^1 (1-z)(\lambda + z(\mu - \lambda))^{k-2}d((\mu - \lambda)z) \\
&= \frac{1}{k-1}\int_0^1 (1-z)d((\lambda + (\mu - \lambda)z)^{k-1}) \\
&= \frac{1}{k-1}(1-z)(\lambda + (\mu - \lambda)z)^{k-1}\Big|_{z=0}^{z=1} + \frac{1}{k-1}\int_0^1 (\lambda + (\mu - \lambda)z)^{k-1}dz \\
&= -\frac{\lambda^{k-1}}{k-1} + \frac{1}{k(k-1)(\mu - \lambda)}(\lambda + (\mu - \lambda)z)^k\Big|_{z=0}^{z=1} \\
&= -\frac{\lambda^{k-1}}{k-1} + \frac{1}{k(k-1)(\mu - \lambda)}(\mu^k - \lambda^k),
\end{aligned}$$

whereupon

$$\frac{\mu^k - \lambda^k}{\mu - \lambda} - k\lambda^{k-1} = k(k-1)(\mu - \lambda)\int_0^1 (1-z)((1-z)\lambda + z\mu)^{k-2}dz.$$

Hence, for $\mu, \lambda \in \mathbf{B}_\rho(0)$, we get

$$|(1-z)\lambda + z\mu| \le (1-z)|\lambda| + z|\mu| \le (1-z)\rho + z\rho = \rho, \quad z \in [0,1],$$

and

$$\left|\frac{\mu^k - \lambda^k}{\mu - \lambda} - k\lambda^{k-1}\right| \le k(k-1)|\mu - \lambda|\rho^{k-2}.$$

Let

$$g(\lambda) = \sum_{k=1}^{\infty} k x_k \lambda^{k-1}, \quad \lambda \in \mathbf{B}_\rho(0).$$

Then

$$\begin{aligned}
\left\|\frac{f(\mu) - f(\lambda)}{\mu - \lambda} - g(\lambda)\right\| &= \left\|\frac{1}{\mu - \lambda}\left(\sum_{k=0}^{\infty} x_k \mu^k - \sum_{k=0}^{\infty} x_k \lambda^k\right) - \sum_{k=1}^{\infty} k x_k \lambda^{k-1}\right\| \\
&= \left\|\frac{1}{\mu - \lambda}\sum_{k=1}^{\infty} x_k(\mu^k - \lambda^k) - \sum_{k=1}^{\infty} k x_k \lambda^{k-1}\right\| \\
&= \left\|\sum_{k=1}^{\infty} x_k\left(\frac{\mu^k - \lambda^k}{\mu - \lambda} - k\lambda^{k-1}\right)\right\| \\
&\le \sum_{k=1}^{\infty} \|x_k\| \left|\frac{\mu^k - \lambda^k}{\mu - \lambda} - k\lambda^{k-1}\right| \\
&\le |\mu - \lambda| \sum_{k=1}^{\infty} k(k-1)\|x_k\|\rho^{k-2} \\
&= |\mu - \lambda| S(\rho) \to 0, \quad \text{as} \quad \mu \to \lambda, \quad \mu, \lambda \in \mathbf{B}_\rho(0).
\end{aligned}$$

This completes the proof. □

Corollary 8.2. *Let the abstract function $f : \mathbf{A} \longmapsto \mathbf{X}$ be analytic at $\lambda = 0$. Then f is infinitely many times differentiable in $\mathbf{B}_R(0)$, where R is its convergence radius. Moreover,*

$$\frac{d^n}{d\lambda^n} f(\lambda) = \sum_{k=n}^{\infty} k(k-1)\cdots(k-n+1)x_k \lambda^{k-n}, \quad \lambda \in \mathbf{B}_\rho(0).$$

Proof. The proof many times invokes Theorem 8.15. □

Definition 8.11. Let the abstract function $f : \mathbf{A} \longmapsto \mathbf{X}$ be infinitely many times differentiable at $\lambda = 0$. The series

$$\sum_{k=0}^{\infty} \frac{1}{k!} \frac{d^k}{d\lambda^k} f(0) \lambda^k$$

is called the Taylor series of the abstract function f.

If the abstract function $f : \mathbf{A} \longmapsto \mathbf{X}$ is analytic at $\lambda = 0$, using Theorem 8.13, its Taylor series coincides with (8.8), and hence, it is convergent for $\lambda \in \mathbf{B}_R(0)$.

8.4 The method of the smaller parameter

Suppose that $\mathbf{X}$ and $\mathbf{Y}$ are normed vector spaces and $\mathbb{A}, \mathbb{B} \in \mathcal{L}(\mathbf{X}, \mathbf{Y})$. Consider the equation

$$\mathbb{A}x - \lambda\mathbb{B}x = y, \tag{8.13}$$

where $y \in \mathbf{Y}$ is given, λ is a scalar parameter, $|\lambda| < \rho$, $x \in \mathbf{X}$ is unknown. Suppose that $\mathbb{A}^{-1}$ exists and $\mathbb{A}^{-1} \in \mathcal{L}(\mathbf{X}, \mathbf{Y})$ and

$$|\lambda|\|\mathbb{B}\|\|\mathbb{A}^{-1}\| < 1. \tag{8.14}$$

By Theorem 4.20, we see that the operator $(\mathbb{A} - \lambda\mathbb{B})^{-1}$ exists and that it belongs to $\mathcal{L}(\mathbf{X}, \mathbf{Y})$. Hence, equation (8.13) has a unique solution,

$$x(\lambda) = (\mathbb{A} - \lambda\mathbb{B})^{-1}y, \quad |\lambda| < \rho,$$

λ satisfies (8.14). Assume that the solution x is an analytic function of λ. Suppose that its representation is

$$x(\lambda) = \sum_{k=0}^{\infty} x_k\lambda^k. \tag{8.15}$$

We put (8.15) in (8.13) and we get

$$\mathbb{A}\left(\sum_{k=0}^{\infty} x_k\lambda^k\right) - \lambda\mathbb{B}\left(\sum_{k=0}^{\infty} x_k\lambda^k\right) = y,$$

or

$$\sum_{k=0}^{\infty} \lambda^k\mathbb{A}x_k - \sum_{k=0}^{\infty} \lambda^{k+1}\mathbb{B}x_k = y,$$

or

$$\sum_{k=0}^{\infty} \lambda^k\mathbb{A}x_k = y + \sum_{k=0}^{\infty} \lambda^{k+1}\mathbb{B}x_k,$$

or

$$\mathbb{A}x_0 + \lambda\mathbb{A}x_1 + \lambda^2\mathbb{A}x_2 + \cdots + \lambda^k\mathbb{A}x_k + \cdots = y + \lambda\mathbb{B}x_0 + \lambda^2\mathbb{B}x_1 + \cdots + \lambda^{k+1}\mathbb{B}x_k + \cdots.$$

Hence,

$$\mathbb{A}x_0 = y, \quad \mathbb{A}x_{k+1} = \mathbb{B}x_k, \quad k \in \mathbf{N}_0.$$

Then

$$\begin{aligned} x_0 &= \mathbb{A}^{-1}y, \\ x_1 &= \mathbb{A}^{-1}\mathbb{B}x_0 = \mathbb{A}^{-1}\mathbb{B}\mathbb{A}^{-1}y, \\ x_2 &= \mathbb{A}^{-1}\mathbb{B}x_1 = \mathbb{A}^{-1}\mathbb{B}(\mathbb{A}^{-1}\mathbb{B}\mathbb{A}^{-1}y) = \mathbb{A}^{-1}(\mathbb{B}\mathbb{A}^{-1})^2 y. \end{aligned}$$

Suppose that

$$x_k = \mathbb{A}^{-1}(\mathbb{B}\mathbb{A}^{-1})^k y.$$

Then

$$x_{k+1} = \mathbb{A}^{-1}\mathbb{B}x_k = \mathbb{A}^{-1}\mathbb{B}(\mathbb{A}^{-1}(\mathbb{B}\mathbb{A}^{-1})^k y) = \mathbb{A}^{-1}(\mathbb{B}\mathbb{A}^{-1})^{k+1} y.$$

Therefore

$$x(\lambda) = \sum_{k=0}^{\infty} \mathbb{A}^{-1}(\mathbb{B}\mathbb{A}^{-1})^k y\lambda^k.$$

Let

$$x_n(\lambda) = \sum_{k=0}^{n} \mathbb{A}^{-1}(\mathbb{B}\mathbb{A}^{-1})^k y\lambda^k.$$

Then

$$\begin{aligned} \|x(\lambda) - x_n(\lambda)\| &= \left\| \sum_{k=n+1}^{\infty} \mathbb{A}^{-1}(\mathbb{B}\mathbb{A}^{-1})^k y\lambda^k \right\| \\ &\le \sum_{k=n+1}^{\infty} \|\mathbb{A}^{-1}(\mathbb{B}\mathbb{A}^{-1})^k y\lambda^k\| \\ &\le \sum_{k=n+1}^{\infty} \|\mathbb{A}^{-1}\|(\|\mathbb{B}\mathbb{A}^{-1}\||\lambda|)^k \|y\| \\ &= \|\mathbb{A}^{-1}\| \frac{(|\lambda|\|\mathbb{B}\mathbb{A}^{-1}\|)^{n+1}}{1 - |\lambda|\|\mathbb{B}\mathbb{A}^{-1}\|} \|y\|. \end{aligned}$$

Now we consider the equation

$$\mathbb{A}(\lambda)x = y(\lambda), \tag{8.16}$$

where $x \in \mathbf{X}$ is unknown, $y(\lambda) \in \mathbf{Y}$ is given, $\mathbb{A}(\lambda) \in \mathcal{L}(\mathbf{X}, \mathbf{Y})$ is given, λ is a scalar parameter, $|\lambda| < \rho$, $\rho > 0$. Suppose that $\mathbb{A}(\lambda)$ and $y(\lambda)$ are analytic at $\lambda = 0$ and $\mathbb{A}^{-1}(0)$ exists and it belongs to $\mathcal{L}(\mathbf{Y}, \mathbf{X})$. Since $\mathbb{A}(\lambda)$ and $y(\lambda)$ are analytic at $\lambda = 0$, they can be represented in the form of a power series with convergence radiuses ρ_1 and ρ_2, respectively,

$$\mathbb{A}(\lambda) = \sum_{k=0}^{\infty} \mathbb{A}_k \lambda^k, \quad y(\lambda) = \sum_{k=0}^{\infty} y_k \lambda^k. \tag{8.17}$$

Since $\mathbb{A}(\lambda)$ is analytic at $\lambda = 0$, there exists an $r > 0$ such that

$$\|(\mathbb{A}(\lambda) - \mathbb{A}(0))\mathbb{A}^{-1}(0)\| < 1$$

for $|\lambda| < r$. Hence, equation (8.16) has an unique solution

$$x(\lambda) = \mathbb{A}^{-1}(\lambda)y(\lambda), \quad |\lambda| < \min\{\rho_2, r\},$$

and x is analytic at $\lambda = 0$ with convergence radius $R = \min\{\rho_2, r\}$. We will represent the solution x in the form

$$x(\lambda) = \sum_{k=0}^{\infty} x_k \lambda^k \quad \text{for} \quad |\lambda| < R. \tag{8.18}$$

We put (8.17) and (8.18) in (8.16) and we get

$$\left(\sum_{k=0}^{\infty} \mathbb{A}_k \lambda^k\right)\left(\sum_{k=0}^{\infty} x_k \lambda^k\right) = \sum_{k=0}^{\infty} y_k \lambda^k, \quad |\lambda| < R,$$

or

$$\sum_{m=0}^{\infty}\left(\sum_{l=0}^{m} \mathbb{A}_l x_{m-l}\right)\lambda^m = \sum_{k=0}^{\infty} y_k \lambda^k, \quad |\lambda| < R.$$

Therefore

$$\mathbb{A}_0 x_0 = y_0, \quad \sum_{l=0}^{m} \mathbb{A}_l x_{m-l} = y_m.$$

Let

$$\|y_n\| \le M_1 \alpha^n, \quad \|\mathbb{A}_n \mathbb{A}_0^{-1}\| \le M\beta^{n-1},$$

for any $n \in \mathbf{N}$ and for some $\alpha > 0$, $\beta > 0$, $M_1 > 0$ and $M > 0$. By Theorem 8.12, it follows that $\rho_2 \ge \frac{1}{\alpha}$. Moreover, for $|\lambda| < \frac{1}{M+\beta}$, we have

$$\begin{aligned}\|(\mathbb{A}(\lambda) - \mathbb{A}(0))\mathbb{A}^{-1}(0)\| &= \left\|\sum_{n=1}^{\infty} \mathbb{A}_n \mathbb{A}_0^{-1} \lambda^n\right\| \\ &\le \sum_{n=1}^{\infty} \|\mathbb{A}_n \mathbb{A}_0^{-1}\| |\lambda|^n\end{aligned}$$

$$\leq M|\lambda| \sum_{n=1}^{\infty} (\beta|\lambda|)^{n-1}$$
$$= \frac{M|\lambda|}{1 - |\lambda|\beta} < 1.$$

Therefore

$$R \geq \min\{\alpha^{-1}, (M+\beta)^{-1}\}.$$

8.5 An application to integral equations

Consider the Fredholm integral equation of the second kind,

$$\phi(x) = u(x) + \frac{1}{2\pi} \int_{-\pi}^{\pi} \cos(x - y + \lambda xy)\phi(y)dy, \quad x \in [-\pi, \pi], \tag{8.19}$$

where $u \in \mathbf{C}([-\pi,\pi])$ is a given function, $\phi \in \mathbf{C}([-\pi,\pi])$ is unknown function, $\lambda \in \mathbf{F}$ is a scalar parameter. Suppose that $\mathbf{C}([-\pi,\pi])$ is endowed with the maximum norm. Define the operator $\mathbb{A}(\lambda) : \mathbf{C}([-\pi,\pi]) \longmapsto \mathbf{C}([-\pi,\pi])$ as follows:

$$(\mathbb{A}(\lambda)\phi)(x) = \phi(x) - \frac{1}{2\pi} \int_{-\pi}^{\pi} \cos(x - y + \lambda xy)\phi(y)dy, \quad x \in [-\pi,\pi], \quad \lambda \in \mathbf{F}.$$

For $\phi_1, \phi_2 \in \mathbf{C}([-\pi,\pi])$ and for $\alpha \in \mathbf{F}$, we have

$$\begin{aligned}
(\mathbb{A}(\lambda)(\phi_1 + \phi_2))(x) &= (\phi_1 + \phi_2)(x) - \frac{1}{2\pi} \int_{-\pi}^{\pi} \cos(x - y + \lambda xy)(\phi_1 + \phi_2)(y)dy \\
&= \phi_1(x) + \phi_2(x) - \frac{1}{2\pi} \int_{-\pi}^{\pi} \cos(x - y + \lambda xy)(\phi_1(y) + \phi_2(y))dy \\
&= \phi_1(x) - \frac{1}{2\pi} \int_{-\pi}^{\pi} \cos(x - y + \lambda xy)\phi_1(y)dy \\
&\quad + \phi_2(x) - \frac{1}{2\pi} \int_{-\pi}^{\pi} \cos(x - y + \lambda xy)\phi_2(y)dy \\
&= (\mathbb{A}(\lambda)\phi_1)(x) + (\mathbb{A}(\lambda)\phi_2)(x), \\
(\mathbb{A}(\lambda)(\alpha\phi))(x) &= (\alpha\phi)(x) - \frac{1}{2\pi} \int_{-\pi}^{\pi} \cos(x - y + \lambda xy)(\alpha\phi)(y)dy \\
&= \alpha\phi(x) - \frac{1}{2\pi} \int_{-\pi}^{\pi} \cos(x - y + \lambda xy)(\alpha\phi(y))dy
\end{aligned}$$

$$= \alpha\phi(x) - \frac{\alpha}{2\pi}\int_{-\pi}^{\pi}\cos(x - y + \lambda xy)\phi(y)dy$$
$$= \alpha(\mathbb{A}(\lambda)\phi)(x), \quad x \in [-\pi, \pi], \quad \lambda \in \mathbf{F}.$$

Therefore $\mathbb{A}(\lambda) : \mathbf{C}([-\pi,\pi]) \longmapsto \mathbf{C}([-\pi,\pi])$ is a linear operator. Also, for any $\phi \in \mathbf{C}([-\pi,\pi])$, we have

$$\begin{aligned}
\left|(\mathbb{A}(\lambda)\phi)(x)\right| &= \left|\phi(x) - \frac{1}{2\pi}\int_{-\pi}^{\pi}\cos(x - y + \lambda xy)\phi(y)dy\right| \\
&\le |\phi(x)| + \frac{1}{2\pi}\int_{-\pi}^{\pi}|\cos(x - y + \lambda xy)||\phi(y)|dy \\
&\le \|\phi\| + \frac{1}{2\pi}\int_{-\pi}^{\pi}\|\phi\|dy \\
&= \|\phi\| + \|\phi\| = 2\|\phi\|, \quad \lambda \in \mathbf{F},
\end{aligned}$$

whereupon

$$\|\mathbb{A}(\lambda)\phi\| \le 2\|\phi\| \quad \text{and} \quad \|\mathbb{A}(\lambda)\| \le 2, \quad \lambda \in \mathbf{F}.$$

Consequently $\mathbb{A}(\lambda) \in \mathcal{L}(\mathbf{C}([-\pi,\pi]))$ for $\lambda \in \mathbf{F}$. We observe that

$$\begin{aligned}
\frac{d}{d\lambda}\cos(x - y + \lambda xy) &= -(xy)\sin(x - y + \lambda xy) \\
&= xy\cos\left(x - y + \lambda xy + \frac{\pi}{2}\right), \\
\frac{d^2}{d\lambda^2}\cos(x - y + \lambda xy) &= -(xy)^2\sin\left(x - y + \lambda xy + \frac{\pi}{2}\right) \\
&= (xy)^2\cos(x - y + \lambda xy + \pi), \quad x, y \in [-\pi,\pi], \quad \lambda \in \mathbf{F}.
\end{aligned}$$

Assume that

$$\frac{d^k}{d\lambda^k}\cos(x - y + \lambda xy) = (xy)^k\cos\left(x - y + \lambda xy + k\frac{\pi}{2}\right), \quad x, y \in [-\pi,\pi], \quad \lambda \in \mathbf{F}, \tag{8.20}$$

for some $k \in \mathbf{N}$. We will prove that

$$\frac{d^{k+1}}{d\lambda^{k+1}}\cos(x - y + \lambda xy) = (xy)^{k+1}\cos\left(x - y + \lambda xy + (k+1)\frac{\pi}{2}\right), \quad x, y \in [-\pi,\pi], \quad \lambda \in \mathbf{F}.$$

Indeed, we have

$$\begin{aligned}
\frac{d^{k1}}{d\lambda^{k+1}}\cos(x - y + \lambda xy) &= \frac{d}{d\lambda}\left(\frac{d^k}{d\lambda^k}\cos(x - y + \lambda xy)\right) \\
&= \frac{d}{d\lambda}\left((xy)^k\cos\left(x - y + \lambda xy + k\frac{\pi}{2}\right)\right)
\end{aligned}$$

$$= -(xy)^{k+1}\sin\left(x - y + \lambda xy + k\frac{\pi}{2}\right)$$
$$= (xy)^{k+1}\cos\left(x - y + \lambda xy + k\frac{\pi}{2} + \frac{\pi}{2}\right)$$
$$= (xy)^{k+1}\cos\left(x - y + \lambda xy + (k+1)\frac{\pi}{2}\right), \quad x, y \in [-\pi, \pi], \quad \lambda \in \mathbf{F}.$$

Consequently (8.20) holds for any $k \in \mathbf{N}$ and $\lambda \in \mathbf{F}$. Therefore

$$\cos(x - y + \lambda xy) = \sum_{k=0}^{\infty} (xy)^k \cos\left(x - y + k\frac{\pi}{2}\right)\lambda^k, \quad x, y \in [-\pi, \pi],$$

which is convergent for $|\lambda| < \frac{1}{\pi^2}$. Note that $\mathbb{A}(\lambda)$ is analytic at $\lambda = 0$. Equation (8.19) can be rewritten in the form

$$\mathbb{A}(\lambda)\phi = u. \tag{8.21}$$

We have

$$(\mathbb{A}(0)\phi)(x) = \phi(x) - \frac{1}{2\pi}\int_{-\pi}^{\pi} \cos(x - y)\phi(y)dy$$
$$= \phi(x) - \frac{1}{2\pi}\int_{-\pi}^{\pi} (\cos x \cos y + \sin x \sin y)\phi(y)dy$$
$$= \phi(x) - \frac{\cos x}{2\pi}\int_{-\pi}^{\pi} \cos y\phi(y)dy - \frac{\sin x}{2\pi}\int_{-\pi}^{\pi} \sin y\phi(y)dy.$$

Let

$$A = \frac{1}{2\pi}\int_{-\pi}^{\pi} \cos y\phi(y)dy, \quad B = \frac{1}{2\pi}\int_{-\pi}^{\pi} \sin y\phi(y)dy.$$

Then

$$(\mathbb{A}(0)\phi)(x) = \phi(x) - A\cos x - B\sin x, \quad x \in [-\pi, \pi].$$

Let

$$\psi(x) = \phi(x) - A\cos x - B\sin x, \quad x \in [-\pi, \pi].$$

Hence,

$$\phi(x) = \psi(x) + A\cos x + B\sin x$$

and

$$A = \frac{1}{2\pi}\int_{-\pi}^{\pi} \cos y(\psi(y) + A\cos y + B\sin y)dy$$

$$= \frac{1}{2\pi}\int_{-\pi}^{\pi} \cos y\psi(y)dy + \frac{A}{2\pi}\int_{-\pi}^{\pi} (\cos y)^2 dy + \frac{B}{2\pi}\int_{-\pi}^{\pi} \cos y \sin y dy$$

$$= \frac{1}{2\pi}\int_{-\pi}^{\pi} \cos y\psi(y)dy + \frac{A}{2\pi}\int_{-\pi}^{\pi} \frac{1+\cos(2y)}{2}dy$$

$$+ \frac{B}{4\pi}(\sin y)^2\Big|_{y=-\pi}^{y=\pi}$$

$$= \frac{1}{2\pi}\int_{-\pi}^{\pi} \cos y\psi(y)dy + \frac{A}{4\pi}\int_{-\pi}^{\pi} dy + \frac{A}{4\pi}\int_{-\pi}^{\pi} \cos(2y)dy$$

$$= \frac{A}{2} + \frac{1}{2\pi}\int_{-\pi}^{\pi} \cos y\psi(y)dy,$$

or

$$\frac{A}{2} = \frac{1}{2\pi}\int_{-\pi}^{\pi} \cos y\psi(y)dy,$$

or

$$A = \frac{1}{\pi}\int_{-\pi}^{\pi} \cos y\psi(y)dy.$$

Next,

$$B = \frac{1}{2\pi}\int_{-\pi}^{\pi} \sin y(\psi(y) + A\cos y + B\sin y)dy$$

$$= \frac{1}{2\pi}\int_{-\pi}^{\pi} \sin y\psi(y)dy + \frac{A}{2\pi}\int_{-\pi}^{\pi} \cos y \sin y dy$$

$$+ \frac{B}{2\pi}\int_{-\pi}^{\pi} (\sin y)^2 dy$$

$$= \frac{1}{2\pi}\int_{-\pi}^{\pi} \sin y\psi(y)dy + \frac{B}{4\pi}\int_{-\pi}^{\pi} (1-\cos(2y))dy$$

$$= \frac{1}{2\pi}\int_{-\pi}^{\pi} \sin y\psi(y)dy + \frac{B}{2},$$

or

$$\frac{B}{2} = \frac{1}{2\pi} \int_{-\pi}^{\pi} \sin y \psi(y) dy,$$

or

$$B = \frac{1}{\pi} \int_{-\pi}^{\pi} \sin y \psi(y) dy.$$

Consequently

$$\begin{aligned}\phi(x) &= \psi(x) + \frac{\cos x}{\pi} \int_{-\pi}^{\pi} \cos y \psi(y) dy + \frac{\sin x}{\pi} \int_{-\pi}^{\pi} \sin y \psi(y) dy \\ &= \psi(x) + \frac{1}{\pi} \int_{-\pi}^{\pi} \cos(x-y)\psi(y) dy,\end{aligned}$$

whereupon

$$(\mathbb{A}^{-1}(0)\psi)(x) = \psi(x) + \frac{1}{\pi} \int_{-\pi}^{\pi} \cos(x-y)\psi(y) dy, \quad x \in [-\pi, \pi].$$

Also, for $\psi \in \mathbf{C}([-\pi, \pi])$, we have

$$\begin{aligned}|(\mathbb{A}^{-1}(0)\psi)(x)| &= \left| \psi(x) + \frac{1}{\pi} \int_{-\pi}^{\pi} \cos(x-y)\psi(y) dy \right| \\ &\le |\psi(x)| + \frac{1}{\pi} \int_{-\pi}^{\pi} |\cos(x-y)||\psi(y)| dy \\ &\le \|\psi\| + \frac{4}{\pi}\|\psi\| = \left(1 + \frac{4}{\pi}\right)\|\psi\|.\end{aligned}$$

Consequently

$$\|\mathbb{A}^{-1}(0)\psi\| \le \left(1 + \frac{4}{\pi}\right)\|\psi\| \quad \text{and} \quad \|\mathbb{A}^{-1}(0)\| \le 1 + \frac{4}{\pi}. \tag{8.22}$$

We conclude that $\mathbb{A}^{-1}(0)$ exists and it belongs to $\mathcal{L}(\mathbf{C}([-\pi, \pi]))$. Since $\mathbb{A}(\lambda)$ is analytic at $\lambda = 0$, there exists $r > 0$ such that

$$\|(\mathbb{A}(\lambda) - \mathbb{A}(0))\mathbb{A}^{-1}(0)\| < 1$$

for $|\lambda| < r$. Hence, equation (8.21) has an unique solution

$$\phi(\lambda) = \mathbb{A}^{-1}(\lambda)u \quad \text{for} \quad |\lambda| < r,$$

and $\phi(\lambda)$ is analytic at $\lambda = 0$ with convergence radius $R = r$. We have

$$\begin{aligned}
&\phi(x) - \frac{1}{2\pi}\int_{-\pi}^{\pi}\sum_{k=0}^{\infty}(xy)^k\cos\left(x - y + k\frac{\pi}{2}\right)\lambda^k\phi(y)dy \\
&= \phi(x) - \frac{1}{2\pi}\sum_{k=0}^{\infty}\left(\int_{-\pi}^{\pi}(xy)^k\cos\left(x - y + k\frac{\pi}{2}\right)\phi(y)dy\right)\lambda^k.
\end{aligned}$$

Let

$$\begin{aligned}
(\mathbb{A}_0\phi)(x) &= \phi(x) - \frac{1}{2\pi}\int_{-\pi}^{\pi}\cos(x - y)\phi(y)dy, \\
(\mathbb{A}_k\phi)(x) &= -\frac{1}{2\pi}\int_{-\pi}^{\pi}(xy)^k\cos\left(x - y + k\frac{\pi}{2}\right)\phi(y)dy, \quad k \in \mathbf{N}.
\end{aligned}$$

Then

$$(\mathbb{A}(\lambda)\phi)(x) = \sum_{k=0}^{\infty}(\mathbb{A}_k\phi)(x)\lambda^k, \quad x \in [-\pi, \pi], \quad |\lambda| < r.$$

We represent the solution $\phi(\lambda)$ of (8.21) in the form

$$\phi(\lambda) = \sum_{k=0}^{\infty}\phi_k\lambda^k.$$

Then

$$\begin{aligned}
u(x) &= \sum_{k=0}^{\infty}\phi_k(x)\lambda^k \\
&\quad - \frac{1}{2\pi}\int_{-\pi}^{\pi}\left(\sum_{k=0}^{\infty}(xy)^k\cos\left(x - y + k\frac{\pi}{2}\right)\lambda^k\right)\left(\sum_{k=0}^{\infty}\phi_k(y)\lambda^k\right)dy \\
&= \sum_{k=0}^{\infty}\phi_k(x)\lambda^k \\
&\quad - \frac{1}{2\pi}\sum_{k=0}^{\infty}\left(\sum_{l=0}^{k}\int_{-\pi}^{\pi}(xy)^l\cos\left(x - y + l\frac{\pi}{2}\right)\phi_{k-l}(y)dy\right)\lambda^k, \quad x \in [-\pi, \pi].
\end{aligned}$$

Hence,

$$u(x) = \phi_0(x) - \frac{1}{2\pi}\int_{-\pi}^{\pi}\cos(x - y)\phi_0(y)dy,$$

$$0=\phi_k(x)-\frac{1}{2\pi}\sum_{l=0}^{k}\int_{-\pi}^{\pi}(xy)^l\cos\left(x-y+l\frac{\pi}{2}\right)\phi_{k-l}(y)dy,\quad k\in\mathbf{N},\quad x\in[-\pi,\pi],$$

and

$$\phi_0(x)=u(x)+\frac{1}{\pi}\int_{-\pi}^{\pi}\cos(x-y)u(y)dy=\mathbb{A}(0)u(x),\quad x\in[-\pi,\pi].$$

Next,

$$\begin{aligned}0&=\phi_1(x)-\frac{1}{2\pi}\sum_{l=0}^{1}\int_{-\pi}^{\pi}(xy)^l\cos\left(x-y+l\frac{\pi}{2}\right)\phi_{1-l}(y)dy\\&=\phi_1(x)-\frac{1}{2\pi}\int_{-\pi}^{\pi}\cos(x-y)\phi_1(y)dy\\&\quad-\frac{1}{2\pi}\int_{-\pi}^{\pi}xy\cos\left(x-y+\frac{\pi}{2}\right)\phi_0(y)dy.\end{aligned}$$

Therefore

$$\begin{aligned}\phi_1(x)&=\frac{1}{2\pi}\int_{-\pi}^{\pi}xy\cos\left(x-y+\frac{\pi}{2}\right)\phi_0(y)dy\\&\quad+\frac{1}{2\pi^2}\int_{-\pi}^{\pi}\cos(x-y)\int_{-\pi}^{\pi}yz\cos\left(y-z+\frac{\pi}{2}\right)\phi_0(z)dzdy,\quad x\in[-\pi,\pi],\end{aligned}$$

and so on for $\phi_2,\phi_3,\ldots$. Now we will make an estimate for R. For $\phi\in\mathbf{C}([-\pi,\pi])$, we have

$$\begin{aligned}|(\mathbb{A}_1\phi)(x)|&=\frac{1}{2\pi}\left|\int_{-\pi}^{\pi}xy\cos\left(x-y+\frac{\pi}{2}\right)\phi(y)dy\right|\\&\le\frac{1}{2\pi}\int_{-\pi}^{\pi}|x||y|\left|\cos\left(x-y+\frac{\pi}{2}\right)\right||\phi(y)|dy\le\frac{2}{\pi}(\pi^2)\|\phi\|,\end{aligned}$$

whereupon

$$\|\mathbb{A}_1\phi\|\le\frac{2}{\pi}(\pi^2)\|\phi\|$$

and

$$\|\mathbb{A}_1\|\le\frac{2}{\pi}(\pi^2).$$

Also,

$$\begin{aligned}|(\mathbb{A}_2\phi)(x)| &= \frac{1}{2\pi}\left|\int_{-\pi}^{\pi} x^2y^2\cos(x-y+\pi)\phi(y)dy\right|\\ &\le \frac{1}{2\pi}\int_{-\pi}^{\pi} x^2y^2|\cos(x-y+\pi)||\phi(y)|dy \le \frac{2}{\pi}(\pi^4)\|\phi\|,\end{aligned}$$

whereupon

$$\|\mathbb{A}_2\phi\| \le \frac{2}{\pi}(\pi^4)\|\phi\|$$

and

$$\|\mathbb{A}_2\| \le \frac{2}{\pi}(\pi^4).$$

Assume that

$$\|\mathbb{A}_k\| \le \frac{2}{\pi}(\pi^{2k}) \tag{8.23}$$

for some $k \in \mathbb{N}$. We will prove that

$$\|\mathbb{A}_{k+1}\| \le \frac{2}{\pi}(\pi^{2k+2}).$$

Really, for $\phi \in \mathbf{C}([-\pi,\pi])$, we have

$$\begin{aligned}|(\mathbb{A}_{k+1}\phi)(x)| &= \left|\frac{1}{2\pi}\int_{-\pi}^{\pi}(xy)^{k+1}\cos\left(x-y+(k+1)\frac{\pi}{2}\right)\phi(y)dy\right|\\ &\le \frac{1}{2\pi}\int_{-\pi}^{\pi}|xy|^{k+1}\left|\cos\left(x-y+(k+1)\frac{\pi}{2}\right)\right||\phi(y)|dy\\ &\le \frac{1}{2\pi}(\pi^{2k+2})\left(\int_{-\pi}^{\pi}\left|\cos\left(x-y+(k+1)\frac{\pi}{2}\right)\right|dy\right)\|\phi\|\\ &\le \frac{2}{\pi}(\pi^{2k+2})\|\phi\|,\end{aligned}$$

whereupon

$$\|\mathbb{A}_{k+1}\phi\| \le \frac{2}{\pi}(\pi^{2k+2})\|\phi\|$$

and

$$\|\mathbb{A}_{k+1}\| \le \frac{2}{\pi}(\pi^{2k+2}).$$

Consequently (8.23) holds for any $k \in \mathbf{N}$. Hence, using (8.22), we get

$$\begin{aligned}\|\mathbb{A}_k \mathbb{A}_0^{-1}\| &\leq \|\mathbb{A}_k\| \|\mathbb{A}_0^{-1}\| \\ &\leq \frac{2}{\pi}(\pi^2)^k \left(1 + \frac{4}{\pi}\right) \\ &= (2\pi + 8)\pi^{2k-2} = (2\pi + 8)(\pi^2)^{k-1}.\end{aligned}$$

Hence,

$$\alpha = \beta = \pi^2, \quad M = 2\pi + 8,$$

and

$$R > \min\left\{\frac{1}{\pi^2}, \frac{1}{\pi^2 + 2\pi + 8}\right\} = \frac{1}{\pi^2 + 2\pi + 8}.$$

9 The parameter continuation method

9.1 Statement of the basic result

We will start with the following useful lemma which we will use for the proof of the main result in this section.

Lemma 9.1. *Let* $\mathbf{A}$ *be an open and a closed subset of* $[0,1]$. *Then* $\mathbf{A} = [0,1]$.

Proof. Suppose that $\mathbf{A} \neq [0,1]$. Let

$$\mathbf{B} = [0,1] \setminus \mathbf{A}.$$

We see that $\mathbf{B}$ is an open and a closed subset of $[0,1]$ and $\mathbf{A} \cap \mathbf{B} = \emptyset$. Because $\mathbf{A}$ is a bounded set, there exist $a = \sup \mathbf{A}$ and $b = \inf \mathbf{A}$. Since $\mathbf{A}$ is closed, we have $a, b \in \mathbf{A}$. If $a < 1$, using the fact that $\mathbf{A}$ is an open set, we conclude that there exists $x > a$, $x \in \mathbf{A}$, which is a contradiction. Then $a = 1$. If $b > 0$, using the fact that $\mathbf{A}$ is an open set, there is an $y \in \mathbf{A}$, $y < b$, which is a contradiction. Consequently $b = 0$. As above, we have $0, 1 \in \mathbf{B}$. Consequently $0, 1 \in \mathbf{A} \cap \mathbf{B}$. This is a contradiction. Therefore $\mathbf{A} = [0,1]$. □

Remark 9.1. Below, for convenience, we will write $\mathbb{A}^{-1}(\lambda)$ instead of $(\mathbb{A}(\lambda))^{-1}$ when it exists.

Theorem 9.1. *Let* $\mathbf{X}$ *and* $\mathbf{Y}$ *be Banach spaces,* $\mathbb{A}(\lambda) \in \mathcal{L}(\mathbf{X}, \mathbf{Y})$ *and it is continuous with respect to* λ, $\lambda \in [0,1]$. *Suppose that there exists* $\gamma > 0$ *such that*

$$\|\mathbb{A}(\lambda)x\| \geq \gamma \|x\| \tag{9.1}$$

for any $x \in \mathbf{X}$ *and for any* $\lambda \in [0,1]$. *Let also* $\mathbb{A}^{-1}(0)$ *exist and belong to* $\mathcal{L}(\mathbf{Y}, \mathbf{X})$. *Then* $\mathbb{A}^{-1}(1)$ *exists and it belongs to* $\mathcal{L}(\mathbf{Y}, \mathbf{X})$ *and*

$$\|\mathbb{A}^{-1}(1)\| \leq \gamma^{-1}.$$

Proof. Let $\mathbf{A}$ be the set of all $\lambda \in [0,1]$ for which $\mathbb{A}^{-1}(\lambda)$ exists and belongs to $\mathcal{L}(\mathbf{Y}, \mathbf{X})$. Then, using (9.1), we have

$$\|\mathbb{A}^{-1}(\lambda)\| \leq \gamma^{-1}$$

for any $\lambda \in \mathbf{A}$. Note that $0 \in \mathbf{A}$ and $\mathbf{A} \neq \emptyset$. Also, for $\lambda_0 \in \mathbf{A}$,

$$\begin{aligned} \|(\mathbb{A}(\lambda) - \mathbb{A}(\lambda_0))\mathbb{A}^{-1}(\lambda_0)\| &\leq \|\mathbb{A}(\lambda) - \mathbb{A}(\lambda_0)\| \|\mathbb{A}^{-1}(\lambda_0)\| \\ &\leq \gamma^{-1} \|\mathbb{A}(\lambda) - \mathbb{A}(\lambda_0)\| \end{aligned} \tag{9.2}$$

for any $\lambda \in [0,1]$. Let $\lambda_0 \in \mathbf{A}$ be arbitrarily chosen. Since $\mathbb{A}(\lambda)$ is continuous for any $\lambda \in [0,1]$, there is a $\delta = \delta(\gamma) > 0$ such that

$$\|\mathbb{A}(\lambda) - \mathbb{A}(\lambda_0)\| < \gamma$$

https://doi.org/10.1515/9783110657722-009

whenever $|\lambda - \lambda_0| < \delta$. Hence, using (9.2), we find

$$\|(\mathbb{A}(\lambda) - \mathbb{A}(\lambda_0))\mathbb{A}^{-1}(\lambda_0)\| < 1$$

for $|\lambda - \lambda_0| < \delta$. Therefore $\mathbb{A}^{-1}(\lambda)$ exists and it belongs to $\mathcal{L}(\mathbf{Y}, \mathbf{X})$ for any λ for which $|\lambda - \lambda_0| < \delta$. Because $\lambda_0 \in \mathbf{A}$ was arbitrarily chosen, we conclude that $\mathbf{A}$ is open. Let $\{\lambda_n\}_{n\in\mathbf{N}}$ be a sequence of elements of $\mathbf{A}$ such that $\lambda_n \to \mu_0$, as $n \to \infty$. Since

$$\|\mathbb{A}^{-1}(\lambda_n)\| \leq \gamma^{-1},$$

we get

$$\begin{aligned}\|(\mathbb{A}(\lambda_n) - \mathbb{A}(\mu_0))\mathbb{A}^{-1}(\lambda_n)\| &\leq \|\mathbb{A}(\lambda_n) - \mathbb{A}(\mu_0)\|\|\mathbb{A}^{-1}(\lambda_n)\| \\ &\leq \gamma^{-1}\|\mathbb{A}(\lambda_n) - \mathbb{A}(\mu_0)\|.\end{aligned} \tag{9.3}$$

Because $\mathbb{A}(\lambda)$ is continuous in λ, there exists an $N \in \mathbf{N}$ such that

$$\|\mathbb{A}(\lambda_n) - \mathbb{A}(\mu_0)\| < \gamma$$

for any $n > N$. Hence, by (9.3), we find

$$\|(\mathbb{A}(\lambda_n) - \mathbb{A}(\mu_0))\mathbb{A}^{-1}(\mu_0)\| < 1$$

for any $n > N$. Therefore $\mathbb{A}^{-1}(\mu_0)$ exists and it belongs to $\mathcal{L}(\mathbf{Y}, \mathbf{X})$, i. e., $\mu_0 \in \mathbf{A}$. Consequently $\mathbf{A}$ is closed. Since $\mathbf{A}$ is open and closed, using Lemma 9.1, we have $\mathbf{A} = [0, 1]$. Consequently $\mathbb{A}^{-1}(1)$ exists and it belongs to $\mathcal{L}(\mathbf{Y}, \mathbf{X})$. This completes the proof. □

9.2 An application to a boundary value problem for a class of second order ordinary differential equations

Consider the boundary value problem

$$-y'' + b(t)y' + c(t)y = a(t), \quad 0 < t < 1, \tag{9.4}$$

$$y(0) = y(1) = 0, \tag{9.5}$$

where $a, c \in \mathbf{C}([0, 1])$, $b \in \mathbf{C}^1([0, 1])$, and

$$c(t) - \frac{1}{2}b'(t) \geq \alpha > -\frac{8}{\pi}, \quad t \in [0, 1],$$

α is a given negative constant. Let $\mathbf{X} = \mathbf{C}^2([0, 1])$, $\mathbf{Y} = \mathbf{C}([0, 1])$. We endow $\mathbf{Y}$ with the maximum norm and $\mathbf{X}$ with the norm

$$\|y\|_{\mathbf{X}} = \max_{t\in[0,1]}|y(t)| + \max_{t\in[0,1]}|y'(t)| + \max_{t\in[0,1]}|y''(t)|,$$

$y \in \mathbf{X}$. Define the operators $\mathbb{A}, \mathbb{B} : \mathbf{X} \longmapsto \mathbf{Y}$ as follows:

$$\begin{aligned}\mathbb{A}y &= -\frac{d^2}{dt^2}y,\\ \mathbb{B}y &= -\frac{d^2}{dt^2}y + b(t)\frac{d}{dt}y + c(t), \quad t \in [0,1], \quad y \in \mathbf{X}.\end{aligned}$$

Consider the boundary value problem

$$-y'' = z, \quad 0 < t < 1, \tag{9.6}$$

$$y(0) = y(1) = 0, \tag{9.7}$$

where $z \in \mathbf{Y}$. We integrate (9.6) from 0 to t and we get

$$-y'(t) = c_1 + \int_0^t z(s)ds, \quad t \in [0,1],$$

where c_1 is a constant. Again we integrate from 0 to t and we find

$$-y(t) = c_2 + c_1 t + \int_0^t \int_0^\tau z(s)dsd\tau = c_2 + c_1 t + \int_0^t (t-s)z(s)ds$$

or

$$y(t) = c_3 + c_4 t - \int_0^t (t-s)z(s)ds,$$

where c_2 is a constant, $c_3 = -c_2$, $c_4 = -c_1$. We will find the constants c_3 and c_4 using (9.7). We have

$$\begin{aligned}y(0) &= c_3 = 0,\\ y(1) &= c_3 + c_4 - \int_0^1 (1-s)z(s)ds = c_4 - \int_0^1 (1-s)z(s)ds = 0,\end{aligned}$$

whereupon

$$c_4 = \int_0^1 (1-s)z(s)ds.$$

Therefore

$$y(t) = \int_0^1 t(1-s)z(s)ds - \int_0^t (t-s)z(s)ds, \quad t \in [0,1].$$

We have $y \in \mathbf{X}$. Therefore the problem (9.6) and (9.7) has a solution in $\mathbf{X}$. Assume that the problem (9.6) and (9.7) has two solutions $y_1, y_2 \in \mathbf{X}$. Let

$$v = y_1 - y_2.$$

Then

$$v'' = 0, \quad 0 < t < 1, \quad v(0) = v(1) = 0.$$

Hence,

$$v'(t) = c_5, \quad t \in [0, 1],$$

where c_5 is a constant, and

$$v(t) = c_5 t + c_6, \quad t \in [0, 1],$$

where c_6 is a constant. Also,

$$v(0) = c_6 = 0, \quad v(1) = c_5 = 0,$$

i. e.,

$$v(t) = 0, \quad t \in [0, 1].$$

Consequently the problem (9.6) and (9.7) has a unique solution in $\mathbf{X}$ for any $z \in \mathbf{Y}$ and there exists $\mathbb{A}^{-1} : \mathbf{Y} \longmapsto \mathbf{X}$ for which we have

$$(\mathbb{A}^{-1}(z))(t) = \int_0^1 (t - ts)z(s)ds - \int_0^t (t - s)z(s)ds, \quad t \in [0, 1], \quad z \in \mathbf{Y}.$$

Let $z_1, z_2 \in \mathbf{Y}$ and $\alpha \in \mathbf{F}$. Then

$$\begin{aligned}
(\mathbb{A}^{-1}(z_1 + z_2))(t) &= \int_0^1 (t - ts)(z_1 + z_2)(s)ds - \int_0^t (t - s)(z_1 + z_2)(s)ds \\
&= \int_0^1 (t - ts)(z_1(s) + z_2(s))ds \\
&\quad - \int_0^t (t - s)(z_1(s) + z_2(s))ds \\
&= \int_0^1 (t - ts)z_1(s)ds - \int_0^t (t - s)z_1(s)ds
\end{aligned}$$

$$+\int_0^t (t-ts)z_2(s)ds - \int_0^t (t-s)z_2(s)ds$$
$$= (\mathbb{A}^{-1}(z_1))(t) + (\mathbb{A}^{-1}(z_2))(t)$$

and

$$\begin{aligned}(\mathbb{A}^{-1}(\alpha z_1))(t) &= \int_0^1 (t-ts)(\alpha z_1)(s)ds - \int_0^t (t-s)(\alpha z_1)(s)ds \\ &= \alpha\int_0^1 (t-ts)z_1(s)ds - \alpha\int_0^t (t-s)z_1(s)ds = \alpha(\mathbb{A}^{-1}(z_1))(t), \quad t \in [0,1].\end{aligned}$$

Therefore $\mathbb{A}^{-1} : \mathbf{Y} \longmapsto \mathbf{X}$ is a linear operator. Next, for $z \in \mathbf{Y}$, we have

$$\begin{aligned}|(\mathbb{A}^{-1}(z))(t)| &= \left|\int_0^1 t(1-s)z(s)ds - \int_0^t (t-s)z(s)ds\right| \\ &\le \int_0^1 (t-ts)|z(s)|ds + \int_0^t (t-s)|z(s)|ds \\ &\le \left(t\int_0^1 (1-s)ds + \int_0^t (t-s)ds\right)\|z\| \\ &= \left(-t\frac{(1-s)^2}{2}\Big|_{s=0}^{s=1} - \frac{(t-s)^2}{2}\Big|_{s=0}^{s=t}\right)\|z\| \\ &= \left(\frac{t}{2}+\frac{t}{2}\right)\|z\| \\ &= t\|z\| \le \|z\|, \quad t \in [0,1],\end{aligned}$$

whereupon

$$\|\mathbb{A}^{-1}(z)\| \le \|z\| \quad \text{and} \quad \|\mathbb{A}^{-1}\| \le 1.$$

Consequently $\mathbb{A}^{-1} \in \mathcal{L}(\mathbf{Y}, \mathbf{X})$. Now we define

$$\mathbb{A}(\lambda) = -\frac{d^2}{dt^2} + \lambda b(t)\frac{d}{dt} + \lambda c(t), \quad \lambda \in [0,1].$$

Note that $\mathbb{A}(\lambda)$ is continuous with respect to λ, $\lambda \in [0,1]$. Now we will find an a priori estimate of the solutions of the following BVP:

$$-x'' + \lambda b(t)x' + \lambda c(t)x = y(t), \quad 0 < t < 1, \tag{9.8}$$
$$x(0) = x(1) = 0. \tag{9.9}$$

We multiply both sides of (9.8) by x and we get

$$-x''(t)x(t) + \lambda b(t)x'(t)x(t) + \lambda c(t)(x(t))^2 = x(t)y(t), \quad 0 < t < 1,$$

which we integrate from 0 to 1 and we get

$$\begin{aligned}\int_0^1 x(t)y(t)dt &= -\int_0^1 x''(t)x(t)dt + \lambda\int_0^1 b(t)x'(t)x(t)dt \\ &\quad + \lambda\int_0^1 c(t)(x(t))^2 dt \\ &= -x'(t)x(t)\Big|_{t=0}^{t=1} + \int_0^1 (x'(t))^2 dt \\ &\quad + \frac{\lambda}{2}(x(t))^2 b(t)\Big|_{t=0}^{t=1} - \frac{\lambda}{2}\int_0^1 b'(t)(x(t))^2 dt \\ &\quad + \lambda\int_0^1 c(t)(x(t))^2 dt \\ &= \int_0^1 (x'(t))^2 dt - \frac{\lambda}{2}\int_0^1 b'(t)(x(t))^2 dt \\ &\quad + \lambda\int_0^1 c(t)(x(t))^2 dt \\ &= \int_0^1 (x'(t))^2 dt + \lambda\int_0^1 \left(c(t) - \frac{1}{2}b'(t)\right)(x(t))^2 dt.\end{aligned} \tag{9.10}$$

We have

$$\begin{aligned}|x(s)| &= \left|\int_0^s x'(t)dt\right| \\ &\le \int_0^s |x'(t)|dt \\ &\le \left(\int_0^s dt\right)^{\frac{1}{2}}\left(\int_0^s (x'(t))^2 dt\right)^{\frac{1}{2}} \\ &\le \sqrt{s}\left(\int_0^1 (x'(t))^2 dt\right)^{\frac{1}{2}}, \quad s \in [0,1],\end{aligned}$$

and

$$\begin{aligned}
|x(s)| &= \left|\int_s^1 x'(t)dt\right| \\
&\le \int_s^1 |x'(t)|dt \\
&\le \left(\int_s^1 dt\right)^{\frac{1}{2}} \left(\int_s^1 |x'(t)|^2 dt\right)^{\frac{1}{2}} \\
&\le \sqrt{1-s}\left(\int_0^1 |x'(t)|^2 dt\right)^{\frac{1}{2}}, \quad s \in [0,1].
\end{aligned}$$

Therefore

$$|x(s)|^2 \le \sqrt{s(1-s)}\left(\int_0^1 (x'(t))^2 dt\right), \quad s \in [0,1], \tag{9.11}$$

and

$$\begin{aligned}
\int_0^1 (x(s))^2 ds &\le \left(\int_0^1 (x'(t))^2 dt\right) \int_0^1 \sqrt{s(1-s)}ds \\
&= \frac{\pi}{8}\left(\int_0^1 (x'(t))^2 dt\right), \quad s \in [0,1],
\end{aligned} \tag{9.12}$$

where we have used

$$\begin{aligned}
\int_0^1 \sqrt{s(1-s)}ds &= 2\int_0^{\frac{\pi}{2}} \sin^2 t \cos^2 t dt \\
&= \frac{1}{2}\int_0^{\frac{\pi}{2}} (\sin(2t))^2 dt \\
&= \frac{1}{2}\int_0^{\frac{\pi}{2}} \frac{1-\cos(4t)}{2} dt \\
&= \frac{1}{4}\int_0^{\frac{\pi}{2}} dt - \frac{1}{8}\sin(4t)\Big|_{t=0}^{t=\frac{\pi}{2}} = \frac{\pi}{8}.
\end{aligned}$$

Also,

$$\int_0^1 \left(c(t) - \frac{1}{2}b'(t)\right)(x(t))^2 dt \geq \alpha \int_0^1 (x(t))^2 dt \tag{9.13}$$

and

$$\int_0^1 x(t)y(t)dt \leq \epsilon \int_0^1 (x(t))^2 dt + \frac{1}{4\epsilon}\int_0^1 (y(t))^2 dt$$

for some $\epsilon > 0$, which will be determined below. Now we apply the previous inequality and (9.12) and (9.13), and we get

$$\begin{aligned}\epsilon \int_0^1 (x(t))^2 dt + \frac{1}{4\epsilon}\int_0^1 (y(t))^2 dt &\geq \frac{8}{\pi}\int_0^1 |x(s)|^2 ds \\ &\quad + \lambda\alpha \int_0^1 (x(t))^2 dt\end{aligned}$$

or

$$\frac{1}{4\epsilon}\int_0^1 (y(t))^2 dt \geq \left(\frac{8}{\pi} - \epsilon + \alpha\lambda\right)\int_0^1 (x(t))^2 dt,$$

or

$$\frac{1}{4\epsilon(\frac{8}{\pi} - \epsilon + \alpha\lambda)}\int_0^1 (y(t))^2 dt \geq \int_0^1 (x(t))^2 dt.$$

Let

$$\epsilon = \frac{4}{\pi} + \frac{\alpha\lambda}{2}.$$

Then

$$\begin{aligned}\frac{1}{4\epsilon(\frac{8}{\pi} - \epsilon + \lambda\alpha)} &= \frac{1}{4(\frac{4}{\pi} + \frac{\alpha\lambda}{2})(\frac{8}{\pi} - \frac{4}{\pi} + \frac{\alpha\lambda}{2})} \\ &= \frac{1}{(\frac{8}{\pi} + \alpha\lambda)^2}.\end{aligned}$$

We set

$$c_1 = \frac{1}{(\frac{8}{\pi} + \alpha\lambda)^2}.$$

Therefore

$$\int_0^1 (x(t))^2 dt \le c_1 \int_0^1 (y(t))^2 dt.$$

Also,

$$\epsilon \int_0^1 (x(t))^2 dt + \frac{1}{4\epsilon} \int_0^1 (y(t))^2 dt \ge \int_0^1 (x'(t))^2 dt + \lambda\alpha \int_0^1 (x(t))^2 dt,$$

whereupon

$$\begin{aligned}
\int_0^1 (x'(t))^2 dt &\le (\epsilon - \lambda\alpha) \int_0^1 (x(t))^2 dt + \frac{1}{4\epsilon} \int_0^1 (y(t))^2 dt \\
&\le c_1(\epsilon - \lambda\alpha) \int_0^1 (y(t))^2 dt + \frac{1}{4\epsilon} \int_0^1 (y(t))^2 dt \\
&= (c_1 c_2 + c_3) \int_0^1 (y(t))^2 dt,
\end{aligned}$$

where

$$c_2 = \epsilon - \lambda\alpha, \quad c_3 = \frac{1}{4\epsilon}.$$

By (9.11), we obtain

$$|x(t)| \le \frac{1}{\sqrt{2}} \left(\int_0^1 (x'(t))^2 dt \right)^{\frac{1}{2}}$$

and by the inequality

$$\left(\int_0^1 (y(t))^2 dt \right)^{\frac{1}{2}} \le \|y\|_{\mathbf{Y}},$$

we get

$$\begin{aligned}
|x(t)| &\le \frac{1}{\sqrt{2}} (c_1 c_2 + c_3)^{\frac{1}{2}} \left(\int_0^1 (y(t))^2 dt \right)^{\frac{1}{2}} \\
&\le \sqrt{\frac{c_1 c_2 + c_3}{2}} \|y\|_{\mathbf{Y}}.
\end{aligned}$$

Consequently

$$\max_{t\in[0,1]}|x(t)| \le \sqrt{\frac{c_1c_2+c_3}{2}}\|y\|_{\mathbf{Y}}. \tag{9.14}$$

Since $x(0) = x(1) = 0$, there is a $\xi \in (0,1)$ such that $x'(\xi) = 0$. Note that we can rewrite equation (9.8) in the form

$$\frac{d}{dt}\left(x'(t)e^{-\lambda\int_0^t b(s)ds}\right) = (\lambda c(t)x(t) - y(t))e^{-\lambda\int_0^t b(s)ds}$$

and integrating from ξ to t, we obtain

$$x'(t)e^{-\lambda\int_0^t b(s)ds} = \int_\xi^t (\lambda c(\tau)x(\tau) - y(\tau))e^{-\lambda\int_0^\tau b(s)ds}d\tau$$

or

$$x'(t) = \int_\xi^t (\lambda c(\tau)x(\tau) - y(\tau))e^{-\lambda\int_t^\tau b(s)ds}d\tau, \quad t \in [0,1].$$

Therefore there exists a constant $m > 0$ such that

$$\begin{aligned}\max_{t\in[0,1]}|x'(t)| &\le m\Big(\max_{t\in[0,1]}|x(t)| + \|y\|_{\mathbf{Y}}\Big)\\ &\le m\left(\sqrt{\frac{c_1c_2+c_3}{2}}+1\right)\|y\|_{\mathbf{Y}}.\end{aligned} \tag{9.15}$$

From equation (9.8), it follows that there exists a constant $m_1 > 0$ such that

$$\begin{aligned}\max_{t\in[0,1]}|x''(t)| &\le m_1\Big(\max_{t\in[0,1]}|x'(t)| + \max_{t\in[0,1]}|x(t)|\\ &\quad + \max_{t\in[0,1]}|y(t)|\Big)\\ &\le m_1\left(m\left(\sqrt{\frac{c_1c_2+c_3}{2}}+1\right) + \sqrt{\frac{c_1c_2+c_3}{2}}+1\right)\|y\|_{\mathbf{Y}}.\end{aligned}$$

By the previous inequality and from (9.14) and (9.15), we conclude that there exists a positive constant M such that

$$\|x\|_X \le M\|y\|_{\mathbf{Y}}.$$

Consequently the problem (9.8) and (9.9) has a unique solution that belongs to **X**. Hence, by Theorem 9.1, we conclude that the problem (9.4) and (9.5) has a unique solution in **X**.

10 Fixed-point theorems and applications

The aim of this chapter is the study of some fixed-point theorems. We start with the simplest and best-known of them: Banach's fixed-point theorem for contraction maps. Then we address the Brinciari fixed-point theorem, which is a generalization of this theorem. We will then see more powerful and somewhat deeper theorems. We can thus study successively the fixed-point theorem of Brouwer (valid in a finite number of dimensions) and then the fixed-point theorem of Schauder (which is the generalization to an infinite number dimensions). Unlike Banach's theorem, the proofs of the latter two results are not constructive, which explains why they require somewhat more sophisticated tools. Many different proofs of these results exist and one may be interested in one or more of them.

10.1 The Banach fixed-point theorem

Banach's fixed-point theorem (also known as the contracting-application theorem) is a simple-to-prove theorem and applies to complete spaces; it has many applications. These applications include theorems on the existence of solutions for differential equations or integral equations and the study of the convergence of certain methods, like Newton's in solving nonlinear equations.

Suppose that $\mathbf{X}$ is a metric space with a metric d. Let also $\mathbf{A} \subseteq \mathbf{X}$.

Definition 10.1. We say that a map $\mathbb{T} : \mathbf{A} \longmapsto \mathbf{X}$ is a contraction map if there exists $k \in (0,1)$ such that

$$d(\mathbb{T}x_1, \mathbb{T}x_2) \leq kd(x_1, x_2)$$

for any $x_1, x_2 \in \mathbf{X}$.

Exercise 10.1. Let $\mathbb{T} : \mathbf{A} \longmapsto \mathbf{X}$ be a contraction map. Prove that it is continuous on $\mathbf{A}$.

Definition 10.2. We say that a map $\mathbb{T} : \mathbf{A} \longmapsto \mathbf{X}$ has a fixed point $x_0 \in \mathbf{A}$ if

$$\mathbb{T}x_0 = x_0.$$

Theorem 10.1. *Let* $\mathbf{X}$ *be a complete metric space and*

$$\mathbb{T} : X \longmapsto X$$

be a contraction map,

$$d(\mathbb{T}x_1, \mathbb{T}x_2) \leq kd(x_1, x_2)$$

for any $x_1, x_2 \in \mathbf{X}$ *and for some constant* $k \in (0,1)$. *Then* $\mathbb{T}$ *has unique fixed point in* $\mathbf{X}$.

https://doi.org/10.1515/9783110657722-010

Proof.

1. Firstly, we will prove the uniqueness of the fixed point of the map $\mathbb{T}$. Let x_1, x_2 be two fixed points of $\mathbb{T}$. Then $\mathbb{T}(x_1) = x_1$, $\mathbb{T}(x_2) = x_2$ and

$$d(\mathbb{T}x_1, \mathbb{T}x_2) \leq kd(x_1, x_2),$$

whereupon

$$d(x_1, x_2) \leq kd(x_1, x_2).$$

Hence, $d(x_1, x_2) = 0$. Therefore $x_1 = x_2$.

2. Now we will prove the existence of a fixed point of the map $\mathbb{T}$. Let $x_0 \in X$ be arbitrarily chosen. Define the sequence $\{x_n\}_{n\in\mathbf{N}}$ as follows:

$$x_{n+1} = \mathbb{T}x_n, \quad n \in \mathbf{N}_0,$$

where $\mathbf{N}_0 = \mathbf{N} \cup \{0\}$. We will prove that it is a Cauchy sequence in $\mathbf{X}$. Let $p, q \in \mathbf{N}$ be such that $q > p$. Then

$$d(x_p, x_q) \leq d(x_p, x_{p+1}) + \cdots + d(x_{q-1}, x_q).$$

Thus,

$$\begin{aligned} d(x_p, x_{p+1}) &= d(\mathbb{T}x_{p-1}, \mathbb{T}x_p) \\ &\leq kd(x_{p-1}, x_p) \\ &\leq k^2 d(x_{p-2}, x_{p-1}) \\ &\leq \dots \\ &\leq k^p d(x_0, x_1). \end{aligned}$$

Then

$$\begin{aligned} d(x_p, x_q) &\leq (k^p + k^{p+1} + \cdots + k^{q-1}) d(x_0, x_1) \\ &= k^p d(x_0, x_1) \sum_{l=0}^{q-1} k^l \\ &< \frac{k^p}{1-k} d(x_0, x_1). \end{aligned}$$

If $d(x_0, x_1) = 0$, then $x_0 = x_1 = \mathbb{T}x_0$ and x_0 is a fixed point of $\mathbb{T}$. If $d(x_0, x_1) \neq 0$, then for any $\varepsilon > 0$ there exists $n_0 \in \mathbf{N}$ such that

$$\frac{k^p}{1-k} d(x_0, x_1) < \varepsilon$$

for any $p > n_0$. Therefore

$$d(x_p, x_q) < \epsilon d(x_0, x_1)$$

for any $p, q \in \mathbf{N}$, $q > p > n_0$. Therefore $\{x_n\}_{n\in\mathbf{N}}$ is a Cauchy sequence in $\mathbf{X}$. Because $\mathbf{X}$ is a complete metric space, we conclude that the sequence $\{x_n\}_{n\in\mathbf{N}}$ is convergent to $x \in \mathbf{X}$. Therefore

$$x_n \to x, \quad \text{as} \quad n \to \infty,$$

and

$$\mathbb{T}x_n \to \mathbb{T}x, \quad \text{as} \quad n \to \infty.$$

Hence,

$$\mathbb{T}x = x.$$

This completes the proof. □

10.2 The Brinciari fixed-point theorem

In this section we present Brinciari's fixed-point theorem which is a generalization of the Banach fixed-point theorem. This is done by introducing a full-size contraction. In 2002 Branciari demonstrated the following theorem.

Theorem 10.2. *Let* $\mathbf{X}$ *be a complete metric space and let the map*

$$\mathbb{T} : \mathbf{X} \longmapsto \mathbf{X}$$

satisfy

$$\int_0^{d(\mathbb{T}x,\mathbb{T}y)} \phi(t)dt \le c \int_0^{d(x,y)} \phi(t)dt \tag{10.1}$$

for $x, y \in \mathbf{X}$, $0 \le c < 1$, *where*

$$\phi : \mathbf{R}_+ \longmapsto \mathbf{R}_+$$

is an integrable function in the Lebesgue sense such that

$$\int_0^b \phi(t)dt > 0 \tag{10.2}$$

for any $b > 0$. *Then* $\mathbb{T}$ *admits a unique fixed point* z *in* X. *On the other hand, for every* $x_0 \in X$, *we have*

$$\lim_{n\to\infty} \mathbb{T}^n x_0 = z.$$

Proof. Let $x \in \mathbf{X}$ be arbitrarily chosen and fixed.
Step 1. We will prove that

$$\int_0^{d(\mathbb{T}^n x, \mathbb{T}^{n+1} x)} \phi(t)dt \leq c^n \int_0^{d(x,\mathbb{T}x)} \phi(t)dt$$

for any $n \in \mathbf{N}_0$. Note that

$$\begin{aligned} \int_0^{d(\mathbb{T}^n x, \mathbb{T}^{n+1} x)} \phi(t)dt &\leq c \int_0^{d(\mathbb{T}^{n-1} x, \mathbb{T}^n x)} \phi(t)dt \\ &\leq \dots \\ &\leq c^n \int_0^{d(x,\mathbb{T}x)} \phi(t)dt. \end{aligned}$$

Therefore

$$\lim_{n\to\infty} \int_0^{d(\mathbb{T}^n x, \mathbb{T}^{n+1} x)} \phi(t)dt = 0. \tag{10.3}$$

Step 2. Suppose that

$$\lim_{n\to\infty} \sup d(\mathbb{T}^n x, \mathbb{T}^{n+1} x) = a,$$

for some $a > 0$. Let $\epsilon > 0$ be arbitrarily chosen. Then there exist $k_\varepsilon \in \mathbf{N}$ and a sequence $(\mathbb{T}^{n_k} x)_{k \geq k_\varepsilon}$ such that, for all $k \geq k_\varepsilon$, we have

$$\left| d(\mathbb{T}^{n_k} x, \mathbb{T}^{n_k+1} x) - a \right| \leq \varepsilon$$

and

$$d(\mathbb{T}^{n_k} x, \mathbb{T}^{n_k+1} x) > \frac{a}{2}.$$

Since ϕ is positive, using (10.2) and (10.3), we get

$$\begin{aligned} 0 &= \lim_{k\to\infty} \int_0^{d(\mathbb{T}^{n_k} x, \mathbb{T}^{n_{k+1}} x)} \phi(t)dt \\ &\geq \int_0^{a/2} \phi(t)dt. \end{aligned}$$

Therefore,

$$\lim_{n\to\infty} d(\mathbb{T}^n x, \mathbb{T}^{n+1} x) = 0.$$

Step 3. Now we will show that the sequence $\{\mathbb{T}^n x\}_{n\in\mathbf{N}}$ is a Cauchy sequence in $\mathbf{X}$. Assume the contrary. Then there exist $\varepsilon > 0$, $k_\varepsilon > 0$, $h \in \mathbf{N}$ and two subsequences $\{m_k\}_{k\in\mathbf{N}}$ and $\{n_k\}_{k\in\mathbf{N}}$ with $m_k > n_k > k_\varepsilon$ such that

$$d(\mathbb{T}^{m_k}x, \mathbb{T}^{n_k}x) \geq \varepsilon, \quad d(\mathbb{T}^{h}x, \mathbb{T}^{n_k}x) < \varepsilon, \quad \text{for} \quad \text{all} \quad h \in n_k + 1, \dots, m_k - 1.$$

Note that

$$\begin{aligned}\varepsilon &\leq d(\mathbb{T}^{m_k}x, \mathbb{T}^{n_k}x)\\ &\leq d(\mathbb{T}^{m_k}x, \mathbb{T}^{m_k-1}x) + d(\mathbb{T}^{m_k-1}x, \mathbb{T}^{n_k}x)\\ &< d(\mathbb{T}^{m_k}x, \mathbb{T}^{m_k-1}x) + \varepsilon.\end{aligned}$$

This gives

$$d(\mathbb{T}^{m_k}x, \mathbb{T}^{n_k}x) \to \varepsilon, \quad \text{as} \quad k \to \infty. \tag{10.4}$$

Moreover, there exists $u \in \mathbf{N}$ such that, for any $k \in \mathbf{N}$, $k > u$, we have

$$d(\mathbb{T}^{m_k+1}x, \mathbb{T}^{n_k+1}x) < \varepsilon.$$

Indeed, if there exists a subsequence $\{m_{k_l}\}_{l\in\mathbf{N}}$ such that

$$d(\mathbb{T}^{m_{k_l}+1}x, \mathbb{T}^{n_{k_l}+1}x) \geq \varepsilon,$$

then we have

$$\begin{aligned}\varepsilon &\leq d(\mathbb{T}^{m_{k_l}+1}x, \mathbb{T}^{n_{k_l}+1}x)\\ &\leq d(\mathbb{T}^{m_{k_l}+1}x, \mathbb{T}^{n_{k_l}}x) + d(\mathbb{T}^{m_{k_l}}x, \mathbb{T}^{n_{k_l}}x) + d(\mathbb{T}^{m_{k_l}}x, \mathbb{T}^{n_{k_l}+1}x).\end{aligned}$$

Using (10.3) and (10.4), we find

$$d(\mathbb{T}^{m_{k_l}+1}x, \mathbb{T}^{n_{k_l}+1}x) \to \varepsilon, \quad \text{as} \quad l \to \infty. \tag{10.5}$$

Applying (10.1), we obtain

$$\int_0^{d(\mathbb{T}^{m_k+1}x, \mathbb{T}^{n_k+1}x)} \phi(t)dt \leq c \int_0^{d(\mathbb{T}^{m_k}x, \mathbb{T}^{n_k}x)} \phi(t)dt. \tag{10.6}$$

Using (10.4), (10.5) and (10.6), one finds

$$\int_0^{\varepsilon} \phi(t)dt \leq c \int_0^{\varepsilon} \phi(t)dt, \tag{10.7}$$

which is a contradiction. Therefore, for some $u \in N$, we have

$$d(\mathbb{T}^{m_k+1}x, \mathbb{T}^{n_k+1}x) < \varepsilon$$

for all $k > u$. Now we will prove that there exist a positive number $\delta_\varepsilon \in (0, \varepsilon)$ and $k_\varepsilon \in \mathbf{N}$ such that, for every $k > k_\varepsilon$, we have

$$d(\mathbb{T}^{m_k+1}x, \mathbb{T}^{n_k+1}x) < \varepsilon - \delta_\varepsilon.$$

Assuming the existence of a sequence $\{k_l\}_{l\in\mathbf{N}} \subset \mathbf{N}$ such that

$$d(\mathbb{T}^{m_{k_l}+1}x, \mathbb{T}^{n_{k_l}+1}x) < \varepsilon$$

and using (10.2), we have

$$\int\limits_0^{d(\mathbb{T}^{m_{k_l}+1}x, \mathbb{T}^{n_{k_l}+1}x)} \phi(t)dt \leq c \int\limits_0^{d(\mathbb{T}^{m_{k_l}}x, \mathbb{T}^{n_{k_l}}x)} \phi(t)dt.$$

Let l tend to infinity. Then

$$\int\limits_0^{\varepsilon} \phi(t)dt \leq c \int\limits_0^{\varepsilon} \phi(t)dt.$$

In conclusion of this step, we will prove that the sequence $\{\mathbb{T}^n x\}_{n\in\mathbf{N}}$ is a Cauchy sequence in $\mathbf{X}$. Indeed, for any $k \in \mathbf{N}$, we have

$$\varepsilon \leq d(\mathbb{T}^{m_k}x, \mathbb{T}^{n_k}x) \leq d(\mathbb{T}^{m_k}x, \mathbb{T}^{m_k+1}x) + d(\mathbb{T}^{m_{k_l}+1}x, \mathbb{T}^{n_k+1}x) + d(\mathbb{T}^{n_k+1}x, \mathbb{T}^{n_k}x).$$

Then

$$\varepsilon < d(\mathbb{T}^{m_k}x, \mathbb{T}^{m_k+1}x) + (\varepsilon - \delta_\varepsilon) + d(\mathbb{T}^{n_k+1}x, \mathbb{T}^{n_k}x).$$

Passing to the limit when k tends to infinity, we find

$$\varepsilon < \varepsilon - \delta_\varepsilon,$$

which is a contradiction. Hence, the sequence $\{\mathbb{T}^n x\}_{n\in\mathbf{N}}$ is a Cauchy sequence in X.

Step 4. In this step we will prove the existence and uniqueness of the fixed point of the map $\mathbb{T}$. Since $\mathbf{X}$ is a complete metric space, then there exists a point $z \in \mathbf{X}$ such that

$$z = \lim_{n\to\infty} \mathbb{T}^n x.$$

Moreover, z is a fixed point of the map $\mathbb{T}$. Indeed, suppose that

$$d(z, \mathbb{T}z) > 0.$$

Then

$$0 < d(z, \mathbb{T}z) \leq d(z, \mathbb{T}^{n+1}z) + d(\mathbb{T}^{n+1}z, \mathbb{T}z).$$

Note that

$$\lim_{n\to\infty} d(z, \mathbb{T}^{n+1}z) = 0.$$

We will prove that

$$\lim_{n\to\infty} d(\mathbb{T}^{n+1}z, \mathbb{T}z) = 0.$$

Observe that

$$\int_0^{d(\mathbb{T}^{n+1}z,\mathbb{T}z)} \phi(t)dt \leq c \int_0^{d(\mathbb{T}^n x,z)} \phi(t)dt \to 0, \quad \text{as} \quad n \to \infty.$$

If we suppose that

$$d(\mathbb{T}^{n+1}x, \mathbb{T}z)$$

does not converge to zero when $n \to \infty$, then there is a subsequence $\{\mathbb{T}^{n_k+1}x\}_{k\in\mathbf{N}}$ such that $d(\mathbb{T}^{n_k+1}x, \mathbb{T}z) \geq \varepsilon$ for some $\varepsilon > 0$. Hence, we have

$$\begin{aligned} 0 &< \int_0^{\varepsilon} \phi(t)dt \\ &\leq \int_0^{d(\mathbb{T}^{n_k+1}x,\mathbb{T}z)} \phi(t)dt \\ &\leq c \int_0^{d(\mathbb{T}^{n_k}x,\mathbb{T}z)} \phi(t)dt \\ &\to 0, \quad \text{as} \quad k \to \infty, \end{aligned}$$

which is a contradiction.

$$\int_0^{\varepsilon} \phi(t)dt > 0.$$

Therefore $z = Tz$. The uniqueness of z follows from the condition (10.1). This completes the proof. □

Remark 10.1. If one takes $\phi(t) = 1$ in the Brinciari theorem, then we get the Banach fixed-point theorem.

Example 10.1. Let

$$\mathbf{X} = \left\{\frac{1}{n} : n \in \mathbf{N}\right\} \cup \{0\}.$$

We provide $\mathbf{X}$ with the metric

$$d(x,y) = |x-y|, \quad x,y \in \mathbf{X}.$$

Since $\mathbf{X}$ is a closed subset of $\mathbf{R}$, $\mathbf{X}$ is a complete metric space. Consider the map

$$\mathbb{T} : \mathbf{X} \longmapsto \mathbf{X}$$

defined as follows:

$$\mathbb{T}x = \begin{cases} \frac{1}{n+1} & \text{if } x = \frac{1}{n}, \quad n \in \mathbf{N}, \\ 0 & \text{if } x = 0. \end{cases}$$

Also, define the function

$$\phi : \mathbf{R}_+ \longmapsto \mathbf{R}_+$$

by

$$\phi(t) = \begin{cases} t^{1/t-2}(1-\ln t) & \text{if } t \in (0,e), \\ 0 & \text{if } t \in \{0\} \cup [e,\infty]. \end{cases}$$

Note that

$$\int_0^e \phi(t)dt = e^{1/e}, \quad t > 0.$$

1. If $x = y$, we have

$$d(x,y) = 0,$$
$$d(\mathbb{T}x, \mathbb{T}y) = 0,$$

and

$$\int_0^{d(\mathbb{T}x,\mathbb{T}y)} \phi(t)dt \le c \int_0^{d(x,y)} \phi(t)dt$$

for all $c \in (0,1)$.

2. If $x = 0$, $y = \frac{1}{n}$, we get

$$\begin{aligned}\int_0^{d(Tx,Ty)} \phi(t)dt &= \left(\frac{1}{n+1}\right)^{n+1} \\ &= \frac{1}{n+1}\left(\frac{1}{n+1}\right)^n \\ &\leq \frac{1}{2}\left(\frac{1}{n}\right)^n \\ &= \frac{1}{2}\int_0^{d(x,y)} \phi(t)dt.\end{aligned}$$

3. If $x = \frac{1}{n}$, $y = \frac{1}{m}$, $n, m \in \mathbf{N}$, we obtain

$$\begin{aligned}\int_0^{d(Tx,Ty)} \phi(t)dt &= \left|\frac{1}{n+1} - \frac{1}{m+1}\right|^{1/|\frac{1}{n+1}-\frac{1}{m+1}|} \\ &= \left(\frac{|n-m|}{(n+1)(m+1)}\right)^{(n+1)(m+1)/|n-m|} \\ &= \left(\frac{|n-m|}{(n+1)(m+1)}\right)^{(n+m+1)/|n-m|}\left(\frac{nm}{(n+1)(m+1)}\right)^{nm/|n-m|}\left(\frac{|n-m|}{nm}\right)^{nm/|n-m|} \\ &\leq \frac{1}{2}\left(\frac{|n-m|}{nm}\right)^{nm/|n-m|} \\ &= \frac{1}{2}\int_0^{d(x,y)} \phi(t)dt.\end{aligned}$$

Hence, by the Brinciari theorem, we conclude that 0 is the unique fixed point of $\mathbb{T}$. Assume that

$$d(\mathbb{T}x, \mathbb{T}y) \leq cd(x,y), \quad c \in (0,1), \quad x, y \in \mathbf{X}.$$

Then, for $x = \frac{1}{n}$ and $y = \frac{1}{n+1}$, we have

$$\begin{aligned}d(\mathbb{T}x, \mathbb{T}y) &= \frac{1}{(n+1)(n+2)}, \\ d(x,y) &= \frac{1}{n(n+1)},\end{aligned}$$

and

$$\sup_{x,y\in\mathbf{X},x\neq y} \frac{d(\mathbb{T}x, \mathbb{T}y)}{d(x,y)} = 1,$$

which is a contradiction.

10.3 The Brouwer fixed-point theorem

A remark to consider: (three dimensions) the mathematician Luitzen Egbertus Jan Brouwer remarked, by mixing his coffee with milk, that the central point of the surface of the liquid, in the midst of the whirlwind created by the rotatory movement of the spoon, remained motionless. He examined the problem in this way: At any moment there is a point on the surface which is not changed.

We will examine the problem in n dimensions following Brouwer. Let

$$\overline{\mathbf{B}}_m = \{x \in \mathbf{R}^n : \|x\| \leq 1\} \tag{10.8}$$

and

$$\mathbf{S}^{m-1} = \partial \mathbf{B}_m.$$

We will use the maximum norm on $\mathbf{B}_m$ defined by

$$\|f\| = \max\{|f(x)| : x \in \overline{B}_m\}, \tag{10.9}$$

Theorem 10.3. *Any continuous map* $f : \overline{\mathbf{B}}_m \longmapsto \overline{\mathbf{B}}_m$ *admits at least one fixed point in* $\overline{\mathbf{B}}_m$.

Proof. We will use the maximum norm on $\overline{B}_m$ defined by

$$\|f\| = \max\{|f(x)| : x \in \overline{B}_m\}. \tag{10.10}$$

Note that for every $\varepsilon > 0$ there exists a polynomial P such that

$$\|f - P\| < \varepsilon.$$

We have

$$\|P\| \leq 1 + \varepsilon.$$

Let

$$Q(x) = \frac{P(x)}{1+\varepsilon}, \quad x \in \overline{\mathbf{B}}_m.$$

We have $Q : \overline{\mathbf{B}}_m \longmapsto \overline{\mathbf{B}}_m$ and

$$\|f - Q\| \leq 2\varepsilon.$$

Suppose now that $x \in \overline{\mathbf{B}}_m$ is a fixed point of Q. Then

$$|x - f(x)| = |Q(x) - f(x)| < 2\varepsilon. \tag{10.11}$$

Thus f admits a fixed point $x \in \overline{\mathbf{B}}_m$. Assume that $f \in \mathbf{C}^1(\overline{\mathbf{B}}_m)$. Let

$$P(\lambda) = a\lambda^2 + 2b\lambda + c, \quad a > 0,$$

be a polynomial satisfying the conditions $P(0) \leq 1$ and $P(1) \leq 1$. Since P is convex we have exactly two values λ_1 and λ_2 such that $P(\lambda_1) = P(\lambda_2) = 1$. More precisely, we have $\lambda_1 \leq 0 < 1 \leq \lambda_2$ and $P(\lambda) < 1$ for $\lambda_1 < \lambda < \lambda_2$. So, $\lambda_1 = A - \sqrt{C}$, $\lambda_2 = A + \sqrt{C}$ with $A = \frac{-b}{a}$, $C = (\frac{b}{a})^2 + \frac{1-c}{a} \geq \frac{1}{4}$ because $\lambda_2 - \lambda_1 \geq 1$.

Now we suppose that f does not admit a fixed point. Because $\overline{\mathbf{B}}_m$ is compact, there exists a $\gamma > 0$ such that

$$|f(x) - x| \geq \gamma$$

in $\overline{\mathbf{B}}_m$. For all $x \in \overline{\mathbf{B}}_m$, the quadratic polynomial

$$P(\lambda) = |x + \lambda(f(x) - x)|^2$$

satisfies

$$P(0) = c = |x|^2 \leq 1, \quad P(1) = |f(x)|^2 \leq 1, \quad a = |f(x) - x|^2 \geq \gamma^2,$$

and

$$b = x(f(x) - x).$$

The function $\lambda_1 = \lambda_2(x)$ is negative and belongs to $C^1(\overline{\mathbf{B}}_m)$ since $\lambda_1 = A - \sqrt{C}$. We define the function $g \in \mathbf{C}^1(\overline{\mathbf{B}}_m)$ as follows:

$$g(x) = \lambda_1(x)(f(x) - x)$$

and

$$h(t, x) = x + tg(x)$$

for $0 \leq t \leq 1$ and consider the integral

$$V(t) = \int_{\overline{B}_m} \det \frac{\partial h(t,x)}{\partial x} dx = \int_{\overline{B}_m} \det\left(Id + t\frac{\partial g(x)}{\partial x} \right) dx \tag{10.12}$$

here $\frac{\partial h}{\partial x}$ and $\frac{\partial g}{\partial x}$ are the Jacobians of order $n \times n$ of h and g, respectively.

We will show that

$$V(t) = |\overline{\mathbf{B}}_m|, \quad t \in [0, 1],$$

where $|\overline{\mathbf{B}}_m|$ is the volume of $\overline{\mathbf{B}}_m$. By the definition, $V(0) = |\overline{\mathbf{B}}_m|$. Note that we have

$$|h(1, x)|^2 = |x + \lambda_1(x)(f(x) - x)|^2 = P(\lambda_1) = 1.$$

Moreover, $h(1,\cdot) = \overline{\mathbf{B}}_m$ for $x \in S^{m-1} = \partial\overline{\mathbf{B}}_m$. So, for $x \in \overline{\mathbf{B}}_m$, the matrix $\frac{\partial h(1,x)}{\partial x}$ is singular. Otherwise $h(1,0)$ becomes a bijection and it associates any neighborhood of $x \in \overline{\mathbf{B}}_m$ with a neighborhood $h(1,x)$. Note that the function g of class C^1 satisfies the condition of Lipschitz,

$$|g(x) - g(x')| \le L|x - x'| \quad \text{in} \quad \overline{\mathbf{B}}_m.$$

Moreover, $g(x) = 0$ for $x \in \partial\overline{\mathbf{B}}_m$, since in this case $P(0) = |x|^2 = 1$ and hence $\lambda_1(x) = 0$. Let Q be the projection on the unit ball

$$Qx = x, \quad \text{for} \quad |x| \le 1,$$
$$Qx = \frac{x}{|x|}, \quad \text{for} \quad |x| > 1.$$

We have

$$|Qx - Qx'| \le |x - x'|,$$

Then the function $\overline{g}(x) = g(Qx)$ satisfies the Lipschitz condition in $\mathbf{R}^n$ with the same constant L ($\overline{g}$ is simply an extension of g to $\mathbf{R}^n$ by 0 outside of $\overline{\mathbf{B}}_m$). Now we will prove that for $0 \le t \le \frac{1}{L}$, the mapping $h(t,\cdot)$ is a bijection of $\overline{\mathbf{B}}_m$ into $\overline{\mathbf{B}}_m$. To show that, write

$$\overline{h}(1,t) = x + t\overline{g}(x),$$

and let $a \in \mathbf{R}^n$. The equation $\overline{h}(t,x) = a$ is equivalent to $x = a - t\overline{g}(x)$. Since the right hand side is contraction with the Lipschitz constant < 1, there exists a single fixed point $x = x_a$ with $\overline{h}(t,a) = a$, so $h(t,\cdot)$ is a bijection of $\mathbf{R}^n$ into itself. However, $h(t,x)$ is the identity of $\mathbf{R}^n \setminus \overline{\mathbf{B}}_m$ and equal to $h(t,\cdot)$ in $\overline{\mathbf{B}}_m$. Then $h(t,\cdot)$ is a bijection of $\overline{\mathbf{B}}_m$ into $\overline{\mathbf{B}}_m$. The substitution rule of the n-dimensional integral means that $V(t) = C$, since $h(t,\cdot)$ is a bijection of $\overline{\mathbf{B}}_m$ into $\overline{\mathbf{B}}_m$ and $\det \frac{\partial h(t,x)}{\partial x} > 0$. So, there is an interval

$$0 \le t \le \varepsilon < \frac{1}{L},$$

where V is constant, and since V is a polynomial with respect to t of degree N, $V(t) = |\overline{B}_m|$, $0 \le t \le 1$. This completes the proof. □

Remark 10.2. James Dugundji showed in 1951 that Brouwer's theorem characterized normed spaces of finite dimension by proving that every map of the unit ball of a normed space $\mathbf{X}$ in itself has a fixed point if and only if $\mathbf{X}$ is of finite dimension.

10.4 The Schauder fixed-point theorem

The Schauder fixed-point theorem and its multiple variants or generalizations are used daily to study the existence and multiplicity of solutions of nonlinear equations of all natures, for example the Navier–Stokes equations in hydrodynamics.

Theorem 10.4. *Let $\mathbf{E}$ be a Banach space, $\mathbf{D} \subset \mathbf{E}$ be a closed convex set. Then any continuous and compact map $\mathbb{T} : \mathbf{D} \longmapsto \mathbf{D}$ admits at least one fixed point.*

Proof. It is sufficient to find a point $x \in \mathbf{D}$ such that for any $\varepsilon > 0$ we have

$$\|x - Tx\| < \varepsilon.$$

Let $\varepsilon > 0$ be arbitrarily chosen. The set $\mathbf{B} = \overline{\mathbb{T}(\mathbf{D})}$ is compact. Then there exists a finite covering of balls $\{\mathbf{S}_\varepsilon(b_i)\}_{i=1}^p$. Let

$$\mathbf{F} = \{b_i, \dots, b_p\} \subset \mathbf{B}$$

and $\mathbf{C} = \operatorname{span} \mathbf{F}$. Note that $\mathbf{C}$ is compact and convex in $\mathbf{D}$. Define the continuous map

$$\Phi : \mathbf{B} \longmapsto \mathbf{C}$$

as follows:

$$\Phi(x) = \begin{cases} 0, & \text{if } \|x - b_i\| \ge \varepsilon, \\ \varepsilon - \|x - b_i\|, & \text{if } \|x - b_i\| < \varepsilon, \end{cases} \qquad x \in \mathbf{B}.$$

We can represent the map Φ in the form

$$\Phi(x) = \sum_{i=1}^p \lambda_i(x) b_i \quad \text{with} \quad \lambda_i(x) = \frac{U_i(x)}{U(x)}, \qquad x \in \mathbf{B},$$

and

$$U(x) = \sum_{i=1}^p U_i(x), \qquad x \in \mathbf{B}.$$

Note that for every $x \in \mathbf{B}$ there exists a $b_K \in \{b_1, \dots, b_p\}$ with

$$\|x - b_K\| < \varepsilon.$$

We have $U(x) > 0$ for $x \in \mathbf{B}$ and Φ is continuous on $\mathbf{B}$. Also, $\lambda_i(x) \ge 0$, $i \in \{1, \dots, p\}$, $x \in \mathbf{B}$, and

$$\sum_{i=1}^p \lambda_i(x) = 1, \qquad x \in \mathbf{B}.$$

Therefore $\Phi(\mathbf{B}) \subset \mathbf{C}$. Moreover, using

$$x = \sum_{i=1}^P \lambda_i(x) x, \qquad x \in \mathbf{B},$$

we get

$$\begin{aligned} \|\Phi(x) - x\| &= \left\| \sum_{i=1}^{p} \lambda_i(x)(b_i - x) \right\| \\ &\leq \sum_{i=1}^{p} \lambda_i(x)\|b_i - x\| < \varepsilon, \quad \text{for} \quad x \in \mathbf{B}. \end{aligned} \tag{10.13}$$

This shows that

$$\|b_i - x\| < \varepsilon$$

otherwise, if $\|b_i - x\| \geq \varepsilon$, then $\lambda_i = 0$. Then $S = \Phi \circ \mathbb{T} : \mathbf{D} \longmapsto \mathbf{C}$. Its restriction on $\mathbf{C}$ is a continuous map from $\mathbf{C}$ into $\mathbf{C}$. Since it is convex and compact, there exists $x_0 = S(x_0) = \Phi(\mathbb{T}x_0) \in \mathbf{C}$. From equation (10.13), we obtain

$$\|x_0 - \mathbb{T}x_0\| = \|\Phi(\mathbb{T}x_0) - \mathbb{T}x_0\| < \varepsilon,$$

i. e., x_0 is a fixed point of the map $\mathbb{T}$. This completes the proof. □

10.5 Non-compact Type Krasnosel'skii fixed-point theorems

In [3], Ball studied the propagation of elastic waves generated by an earthquake in the earth crust modeled by a channel separated from the atmosphere and the mantel by two horizontal interfaces. Geographical studies have shown the validity of radiative transfer in this frequency regime to describe the phase space energy density of seismic waves. For long times and large distances, the radiative transfer in weakly absorbing media was approximated by a diffusion equation. It is shown in [3] that this diffusion is valid in the following sense: the radiative transfer solution factors asymptotically in the limit of vanishing mean free paths as the product of a two-dimensional diffusion term in the horizontal directions an a one-dimensional transport term in the vertical direction. A boundary value problem of the obtained model was investigated by Latrach [19] for the existence of solutions for isotropic scattering kernels on $\mathbf{L}^p$-spaces for $p \in (0, \infty)$. The general problem for the existence of solutions on $\mathbf{L}^1$-spaces was not fully resolved. Some efforts have been made in [2, 20] in some special cases as regards the isotropic scattering kernel. Here, our aim is to represent an existence result on $\mathbf{L}^1$-space in more general situation for the isotropic kernel.

The second source of our consideration stems from a class of Darboux problems. This type of problems involves a mixed partial derivative. The typical methods to treat the existence of the solutions for Darboux problems involve functional differential equations, and the method of upper and lower solutions based on Schauder's fixed point theorem; see [14, 25–27] and the references therein. In this section, we propose new approaches to dealing with the local and global existence for a class of Darboux problems.

Many varieties of difference models (with delay) and Volterra integral models describing physiological processes, production of blood cells, respiration, cardiac arrhythmias etc. were investigated as regards the existence of positive (periodic) solutions using global bifurcation techniques and Krasnosel'skii fixed-point theorems on cones; see for instance [21, 28] and the references therein. In this section, we will consider a class of nonlinear difference equations and a class of Volterra integral equations using a new approach for investigating the existence of positive (periodic) solutions and we will give new ranges for the parameters, participating in this class difference equations, which ensure the existence of positive periodic solutions.

We observe these aforementioned existence problems, arising from integral and transport equation, and a class of difference equations, can be transformed abstractly into fixed problems for sum of two operators $\mathbb{T} + \mathbb{S}$. A prototype tool to address such fixed-point problems is the well-known Krasnosel'skii fixed point theorem, which is a continuation of the Banach contraction mapping principle and the Schauder fixed-point theorem.

Theorem A (Krasnosel'skii [18]). *Let* $\mathbf{K}$ *be a nonempty closed, convex, and bounded subset of a Banach space* $\mathbf{E}$*. Suppose that* $\mathbb{T}$ *and* $\mathbb{S}$ *map* $\mathbf{K}$ *into* $\mathbf{E}$ *such that*

(i) $\mathbb{T}$ *is a contraction with constant* $\alpha < 1$*;*

(ii) $\mathbb{S}$ *is compact; and*

(iii) *any* $x, y \in \mathbf{K}$ *imply* $\mathbb{T}x + \mathbb{S}y \in \mathbf{K}$.

Then there exists $x^* \in \mathbf{K}$ *with* $\mathbb{S}x^* + \mathbb{T}x^* = x^*$.

This overarching result has initiated numerous studies and has been extended in different directions by modifying assumption (i), (ii), (iii) or even the underlying space. See [9, 11, 30]. It was mentioned in [9] that the condition (iii) is too stringent and can be replaced by a mild one, in which Burton proposed the following improvement for (iii): if $x = \mathbb{T}x + \mathbb{S}y$ with $y \in \mathbf{K}$, then $x \in \mathbf{K}$. Subsequently, in [11], Dhage replaced (i) by the following requirement: $\mathbb{T}$ is a bounded linear operator on $\mathbf{E}$, and $\mathbb{T}^p$ is a nonlinear contraction for some $p \in N$. More recently, in [30], the authors firstly replaced the contraction map by an expansion and then replaced the compactness of the operator $\mathbb{S}$ by a k-set contractive one, and they obtained some new fixed-point results.

For the sum of two operators, many kinds of generalizations and variants of Krasnoselskii's fixed-point theorem have been obtained; see for example [2, 5, 9, 11, 19, 23, 24, 30] and the references therein. It is well known that, in some previous related work, the compactness of $\mathbb{S}$ plays a crucial role in their arguments. The reason is that their discussions are based on the Schauder fixed-point theorem.

We note that, although there are so many theoretical generations, those with practical applications are infrequent [2, 19, 22, 23]. From application point of view,

Sadovskii's fixed-point theorem associated with measure of non-compact and some of the above-mentioned work, we first explore several further extensions of the Krasnoselskii Theorem in the direction of [30]. It is shown that the Krasnoselskii fixed-point theorem can be expressed as quite general forms; see Theorems 10.7, 10.8 and 10.9 and their corollaries. Furthermore, these abstract results not only lead us to define further generalized contractions and expansions, but they also facilitate the application of dissipative operators; see Theorem 10.10.

Next, more importantly, it is shown that the generalized formulations of these Theorems are applicable to a large class of problems. To exhibit the power of them, we study the existence and uniqueness of solution for some kind of perturbed Volterra-type integral equation in Section 10.4, the existence of solutions for a class transport equations, for a class of difference equations and for a Darboux problem in Sections 10.5, 10.6, and 10.7, respectively.

The central point of this section is to address some practical problems arising from integral and transport equation, and a class of difference equations. To this end, we appeal to extended fixed-point theorems of Krasnosel'skii type. These extensions encompass a number of previously known generalizations or modifications of the Krasnosel'skii fixed-point theorem or the Sadovskii fixed-point theorem. Lots of practical applications are provided to illustrate the theories. Of course, the abstract techniques and results of this section can be applied to various kinds of other problems which are not investigated here.

Let $\mathbf{E}$ be a Banach space and Ω_E the collection of bounded subsets of $\mathbf{E}$.

Definition 10.3. The Kuratowskii measure of non-compactness is the map $\alpha_{\mathbf{E}} : \Omega_{\mathbf{E}} \to [0, \infty)$ (or simply α) defined by

$$\alpha_{\mathbf{E}}(\mathbf{A}) = \inf\Big\{\delta > 0 \;\Big|\; \text{there is a finite number of subsets } \mathbf{A}_i \subset \mathbf{A} \\ \text{such that} \quad \mathbf{A} \subset \bigcup_i \mathbf{A}_i \text{ and } \operatorname{diam}(\mathbf{A}_i) \le \delta\Big\},$$

where $\mathbf{A} \in \Omega_{\mathbf{E}}$, $\operatorname{diam}(\mathbf{A}_i)$ denotes the diameter of the set $\mathbf{A}_i$.

For convenience, we list some properties of α which we will tacitly use in the sequel. Let $\mathbf{A}, \mathbf{B} \in \Omega_{\mathbf{E}}$. Then

(1) $\alpha(\mathbf{A}) = 0$ if and only if $\mathbf{A}$ is relatively compact;
(3) If $\mathbf{A} \subset \mathbf{B}$, then $\alpha(\mathbf{A}) \le \alpha(\mathbf{B})$;
(4) $\alpha(\mathbf{A} \cup \mathbf{B}) = \max\{\alpha(\mathbf{A}), \alpha(\mathbf{B})\}$;
(5) $\alpha(\lambda\mathbf{A}) = |\lambda|\alpha(\mathbf{A})$ for $\lambda \in \mathbf{R}$, where $\lambda\mathbf{A} = \{\lambda x : x \in \mathbf{A}\}$;
(6) $\alpha(\mathbf{A} + \mathbf{B}) \le \alpha(\mathbf{A}) + \alpha(\mathbf{B})$, where $\mathbf{A} + \mathbf{B} = \{x + y : x \in \mathbf{A}, y \in \mathbf{B}\}$;

Let $\mathbf{X}$, $\mathbf{Y}$ be two Banach spaces and Ω be a subset of $\mathbf{X}$.

Definition 10.4. A continuous and bounded map $\mathbb{N} : \Omega \longmapsto \mathbf{Y}$ is k-set contractive if for any bounded set $\mathbf{A} \subset \Omega$ we have $\alpha_{\mathbf{Y}}(N(\mathbf{A})) \leq k\alpha_{\mathbf{X}}(\mathbf{A})$. $\mathbb{N}$ is strictly k-set contractive if N is k-set contractive and $\alpha_{\mathbf{Y}}(\mathbb{N}(\mathbf{A})) < k\alpha_{\mathbf{X}}(\mathbf{A})$ for all bounded sets $\mathbf{A} \subset \Omega$ with $\alpha_{\mathbf{X}}(\mathbf{A}) \neq 0$. $\mathbb{N}$ is a condensing map if $\mathbb{N}$ is strictly 1-set contractive.

Notice that $\mathbb{N}$ is a compact map if and only if $\mathbb{N}$ is a 0-set contractive one.

Remark 10.3. In the literature, a continuous and bounded map $\mathbb{N} : \Omega \longmapsto \mathbf{Y}$ is called strict-set contraction if $\mathbb{N}$ is k-set contractive with $k < 1$. Obviously, a strict-set contraction is a condensing map. The concept of (strictly) k-set contractive map with $k < 1$ or not is useful, see Proposition 10.1.

Theorem B (Sadovskii [1, 4, 12]). *Let $\mathbf{K}$ be a closed, bounded and convex subset of a Banach space $\mathbf{E}$ and $\mathbb{N} : \mathbf{K} \longmapsto \mathbf{K}$ be a condensing map. Then $\mathbb{N}$ has a fixed point in $\mathbf{K}$.*

10.6 Fixed-point results for the sum $\mathbb{T} + \mathbb{S}$

Throughout this section, we always denote by $\mathbf{E}$ a Banach space. This section is devoted to the study of the fixed point problem of the sum operators $\mathbb{T} + \mathbb{S}$ or the existence of solution of the abstract operator equation $\mathbb{T}x + \mathbb{S}x = x$ in some subset of $\mathbf{E}$ required in the sequel. Let us begin with some preliminary definitions and lemmas.

Definition 10.5. Let $(\mathbf{X}, d)$ be a metric space and $\mathbf{M}$ be a subset of $\mathbf{X}$ and $\mathbb{T} : \mathbf{M} \longmapsto \mathbf{X}$ a map. Assume that there exists a constant $\beta \geq 0$ such that

$$d(\mathbb{T}x, \mathbb{T}y) \geq \beta d(x, y), \quad \forall x, y \in \mathbf{M}.$$

Then we say that $\mathbb{T}$ is weakly expansive. In particular, we call $\mathbb{T}$ expansive if $\beta > 1$.

Remark 10.4. We note that a (weakly) expansive map $\mathbb{T} : \mathbf{M} \longmapsto \mathbf{X}$ may not be continuous. If $\mathbb{T} : \mathbf{M} \subset \mathbf{X} \longmapsto \mathbf{X}$ is a weakly expansive map, we will denote

$$\operatorname{lip}(\mathbb{T}) = \max\{\beta \geq 0 : d(\mathbb{T}x, \mathbb{T}y) \geq \beta d(x, y), x, y \in \mathbf{M}\}.$$

As usual, $\operatorname{Lip}(\mathbb{T})$ is the Lipschitzian constant for $\mathbb{T}$ if $\mathbb{T}$ is a Lipschitzian map.

In what follows we shall employ Lemmas 10.1 and 10.2, which have been established in [30].

Lemma 10.1. *Let $\mathbf{X}$ be a complete metric space and $\mathbf{M}$ a closed subset of $\mathbf{X}$. Assume that the mapping $\mathbb{T} : \mathbf{M} \longmapsto \mathbf{X}$ is expansive and $\mathbb{T}(\mathbf{M}) \supset \mathbf{M}$. Then there exists a unique point $x^* \in \mathbf{M}$ such that $\mathbb{T}x^* = x^*$.*

Lemma 10.2. *Let $(\mathbf{X}, \|\cdot\|)$ be a linear normed space, $\mathbf{M} \subset \mathbf{X}$. Assume that the map $\mathbb{T} : \mathbf{M} \longmapsto \mathbf{X}$ is expansive with constant $h > 1$. Then the inverse of $\mathbb{F} := \mathbb{I} - \mathbb{T} : \mathbf{M} \longmapsto$*

$(\mathbb{I} - \mathbb{T})(\mathbf{M})$ *exists and*

$$\|\mathbb{F}^{-1}x - \mathbb{F}^{-1}y\| \le \frac{1}{h-1}\|x - y\|, \quad x, y \in \mathbb{F}(\mathbf{M}).$$

Lemma 10.3. *Let $\mathbb{T} : \mathbf{E} \longmapsto \mathbf{E}$ be Lipschitz with constant $\beta > 0$. Assume that for each $y \in \mathbf{E}$ the map $\mathbb{T}_y : \mathbf{E} \longmapsto \mathbf{E}$ defined by $\mathbb{T}_y x = \mathbb{T}x + y$ satisfies that $\mathbb{T}_y^p$ is expansive for some $p \in \mathbf{N}$ and onto. Then $(\mathbb{I} - \mathbb{T})$ maps $\mathbf{E}$ onto $\mathbf{E}$, the inverse of $\mathbb{F} := \mathbb{I} - \mathbb{T} : \mathbf{E} \longmapsto \mathbf{E}$ exists and*

$$\|\mathbb{F}^{-1}x - \mathbb{F}^{-1}y\| \le \gamma_p\|x - y\|, \quad x, y \in \mathbf{E}, \tag{10.14}$$

where

$$\gamma_p = \frac{\beta^p - 1}{(\beta - 1)[\operatorname{lip}(\mathbb{T}^p) - 1]}.$$

Proof. Let $y \in \mathbf{E}$ be an arbitrary point. Since $\mathbb{T}_y^p$ is expansive it follows

$$\|\mathbb{T}_y^p x - \mathbb{T}_y^p z\| \ge \operatorname{lip}(\mathbb{T}_y^p)\|x - z\|, \quad \forall x, z \in \mathbf{E}.$$

We now claim that both $(\mathbb{I} - \mathbb{T})$ and $(\mathbb{I} - \mathbb{T}^p)$ map $\mathbf{E}$ onto $\mathbf{E}$. Indeed, notice that $\mathbb{T}_y^p$ is onto, thus Lemma 10.1 ensures there is a unique $x^* \in \mathbf{E}$ such that $\mathbb{T}_y^p x^* = x^*$. It then follows readily that $\mathbb{T}_y x^*$ is also a fixed point of $\mathbb{T}_y^p$. In view of uniqueness, we obtain $\mathbb{T}_y x^* = x^*$ and x^* is the unique fixed point of $\mathbb{T}_y$. Hence, we have

$$(\mathbb{I} - \mathbb{T})x^* = y,$$

which implies that $\mathbb{I} - \mathbb{T} : \mathbf{E} \longmapsto \mathbf{E}$ is onto. The assumption implies that $\mathbb{T}^p$ is expansive and onto. Then an application of Lemma 10.1 to $\tilde{\mathbb{T}}_y x = \mathbb{T}^p x + y$ shows there is a unique x^* so that $\tilde{\mathbb{T}}_y x^* = x^*$, implying $\mathbb{I} - \mathbb{T}^p : \mathbf{E} \longmapsto \mathbf{E}$ is onto. So the claim is proved. Next, for each $x, y \in \mathbf{E}$, by the expansiveness of $\mathbb{T}^p$, one easily obtains

$$\|(\mathbb{I} - \mathbb{T}^p)x - (\mathbb{I} - \mathbb{T}^p)y\| \ge [\operatorname{lip}(\mathbb{T}^p) - 1]\|x - y\| > 0,$$

which shows that $(\mathbb{I} - \mathbb{T}^p)$ is one-to-one. Summing the above arguments, we derive that $(\mathbb{I} - \mathbb{T}^p)^{-1}$ exists on $\mathbf{E}$. Therefore, we infer that $(\mathbb{I} - \mathbb{T})^{-1}$ exists on $\mathbf{E}$ due to the fact that

$$(\mathbb{I} - \mathbb{T})^{-1} = (\mathbb{I} - \mathbb{T}^p)^{-1}\sum_{k=0}^{p-1}\mathbb{T}^k. \tag{10.15}$$

From Lemma 10.2, it follows

$$\|(\mathbb{I} - \mathbb{T}^p)^{-1}x - (\mathbb{I} - \mathbb{T}^p)^{-1}y\| \le \frac{1}{\operatorname{lip}(\mathbb{T}^p) - 1}\|x - y\|, \quad \forall x, y \in (\mathbb{I} - \mathbb{T}^p)(\mathbf{E}),$$

that is,

$$\operatorname{Lip}\big((\mathbb{I}-\mathbb{T}^p)^{-1}\big) \le \frac{1}{\operatorname{lip}(\mathbb{T}^p)-1}. \tag{10.16}$$

A series of induction calculations show that

$$\|\mathbb{T}^k x - \mathbb{T}^k y\| \le \beta^k \|x-y\|, \quad \forall x,y \in \mathbf{E} \text{ and } k \in \mathbf{N}, \tag{10.17}$$

and

$$\operatorname{lip}(\mathbb{T}^p)\|x-y\| \le \|\mathbb{T}^p x - \mathbb{T}^p y\| \le \beta^p \|x-y\|, \quad \forall x,y \in \mathbf{E}.$$

Recalling $\operatorname{lip}(\mathbb{T}^p) > 1$, we get $\beta > 1$. So we conclude from (10.15), (10.16) and (10.17) that

$$\begin{aligned}\operatorname{Lip}\big((\mathbb{I}-\mathbb{T})^{-1}\big) &\le \operatorname{Lip}\big((\mathbb{I}-\mathbb{T}^p)^{-1}\big)\sum_{k=0}^{p-1}\operatorname{Lip}(\mathbb{T}^k)\\ &\le \frac{1}{\operatorname{lip}(\mathbb{T}^p)-1}\sum_{k=0}^{p-1}\beta^k = \frac{\beta^p-1}{(\beta-1)(\operatorname{lip}(\mathbb{T}^p)-1)}.\end{aligned}$$

This proves the lemma. □

Corollary 10.1. *Let $\mathbb{T}:\mathbf{E}\longmapsto\mathbf{E}$ be a bounded linear operator. Assume that $\mathbb{T}^p$ is expansive for some $p\in\mathbf{N}$ and onto. Then the conclusion of Lemma* 10.3 *holds. In such a case,* $\operatorname{Lip}(\mathbb{T}) = \|\mathbb{T}\|$.

Proof. Let $y\in\mathbf{E}$ be fixed. Notice that $\mathbb{T}$ is linear, therefore $\operatorname{Lip}(\mathbb{T}) = \|\mathbb{T}\|$. By induction, one easily deduces that

$$\mathbb{T}_y^k x = \mathbb{T}^k x + \mathbb{T}^{k-1} y + \cdots + \mathbb{T}y + y, \quad \text{for all } k \in \mathbf{N}.$$

This shows

$$\|\mathbb{T}_y^k x - \mathbb{T}_y^k z\| = \|\mathbb{T}^k x - \mathbb{T}^k z\|, \quad \text{for all } k\in\mathbf{N} \text{ and } x,z\in\mathbf{E}.$$

Consequently, $\mathbb{T}_y^p$ is expansive and onto, so Lemma 10.3 holds. This completes the proof. □

A standard argument yields the following result.

Lemma 10.4. *Let $\mathbf{M}$ be a subset of $\mathbf{E}$. Assume that $\mathbb{T}:\mathbf{M}\longmapsto\mathbf{E}$ is k-Lipschitzian map, i. e.,*

$$\|\mathbb{T}x - \mathbb{T}y\| \le k\|x-y\|, \quad x,y\in\mathbf{M}. \tag{10.18}$$

Then, for each bounded subset Ω of $\mathbf{M}$, we have $\alpha(\mathbb{T}(\Omega)) \le k\alpha(\Omega)$.

Now we are ready to state and prove the first main abstract result of this section.

Theorem 10.5. *Let* $\mathbf{K} \subset \mathbf{E}$ *be a nonempty, bounded, closed convex subset. Suppose that* $\mathbb{T} : \mathbf{E} \longmapsto \mathbf{E}$ *and* $\mathbb{S} : \mathbf{K} \longmapsto \mathbf{E}$ *such that:*
(i) $\mathbb{T}$ *fulfills the conditions of Lemma* 10.3*;*
(ii) $\mathbb{S}$ *is a strictly* γ_p^{-1}*-set contractive map (or a* γ*-set contractive map with* $\gamma < \gamma_p^{-1}$*);*
(iii) $[x = \mathbb{T}x + \mathbb{S}y, y \in \mathbf{K}] \Longrightarrow x \in \mathbf{K}$.

Then there exists a point $x^* \in \mathbf{K}$ *with* $\mathbb{S}x^* + \mathbb{T}x^* = x^*$.

Proof. Since $\mathbb{T} : \mathbf{E} \longmapsto \mathbf{E}$ satisfies all conditions of Lemma 10.3, $\mathbb{I} - \mathbb{T}$ maps $\mathbf{E}$ onto $\mathbf{E}$. Because $\mathbf{K} \subset \mathbf{E}$ and $\mathbb{S} : \mathbf{K} \longmapsto \mathbf{E}$, it follows that for every $x \in \mathbf{K}$ there exists $y \in \mathbf{E}$ such that

$$y - \mathbb{T}y = \mathbb{S}x \quad \Longleftrightarrow (\mathbb{I} - \mathbb{T})y = \mathbb{S}x.$$

By Lemma 10.3 again, there exists $(\mathbb{I} - \mathbb{T})^{-1}$, and thus from (iii) and the above equality, we get $y = (\mathbb{I} - \mathbb{T})^{-1}\mathbb{S}x \in \mathbf{K}$.

Now, let $\mathbf{A}$ be a subset of $\mathbf{K}$. From (10.14) and (10.18), one can easily infer that

$$\alpha(((\mathbb{I} - \mathbb{T})^{-1}\mathbb{S})(\mathbf{A})) \leq \gamma_p \alpha(\mathbb{S}(\mathbf{A})),$$

which, together with (ii), implies that $(\mathbb{I} - \mathbb{T})^{-1}\mathbb{S} : \mathbf{K} \longmapsto \mathbf{K}$ is a condensing map. Applying Sadovskii's Theorem B, we see that there exists $x^* \in \mathbf{K}$ such that $(\mathbb{I} - \mathbb{T})^{-1}\mathbb{S}x^* = x^*$. This is the same as $\mathbb{S}x^* + \mathbb{T}x^* = x^*$. The proof of the theorem is thus complete. □

An easy consequence of Corollary 10.1 and Theorem 10.5 is the following.

Corollary 10.2. *In Theorem* 10.5*, if only* (i) *is replaced by that* (i′)*:* $\mathbb{T} : \mathbf{E} \longmapsto \mathbf{E}$ *is a linear and bounded operator, and* $\mathbb{T}^p$ *is expansive for some* $p \in \mathbf{N}$ *and onto. Then there exists a point* $x^* \in \mathbf{K}$ *with* $\mathbb{S}x^* + \mathbb{T}x^* = x^*$.

We will naturally consider the case when $\mathbb{T}^p$ (for some $p \in \mathbf{N}$) is a contractive map. For this purpose, the following well-known result, which is analogous to Lemma 10.2, is a basic tool.

Lemma 10.5. *Let* $(\mathbf{X}, \|\cdot\|)$ *be a normed vector space,* $\mathbf{M} \subset \mathbf{X}$*. Assume that the map* $\mathbb{T} : \mathbf{M} \longmapsto \mathbf{X}$ *is contractive with constant* $\gamma < 1$*, then the inverse of* $\mathbb{F} := \mathbb{I} - \mathbb{T} : \mathbf{M} \longmapsto (\mathbb{I} - \mathbb{T})(\mathbf{M})$ *exists and*

$$\left\|\mathbb{F}^{-1}x - \mathbb{F}^{-1}y\right\| \leq \frac{1}{1-\gamma}\|x - y\|, \quad x, y \in \mathbb{F}(\mathbf{M}). \tag{10.19}$$

The following notion of nonlinear contraction will also be used in the sequel.

Definition 10.6 (Boyd and Wong [8]). Let $\mathbf{M}$ be a subset of $\mathbf{E}$. The map $\mathbb{T} : \mathbf{M} \longmapsto \mathbf{E}$ is called a nonlinear contraction, if there exists a continuous and nondecreasing function $\phi : \mathbf{R}_+ \longmapsto \mathbf{R}_+$ satisfying $\phi(r) < r$ for $r > 0$, such that

$$\|\mathbb{T}x - \mathbb{T}y\| \leq \phi(\|x - y\|), \quad \forall x, y \in \mathbf{M}.$$

Of course, every contraction is a nonlinear contraction, the converse is not true. Nonlinear contractions play essential role in various generalizations of the Banach fixed-point theorem. Based on the fact that every nonlinear contraction $\mathbb{T} : \mathbf{E} \longmapsto \mathbf{E}$ has a unique fixed point [8], we give the following lemma and present all the details for convenience.

Lemma 10.6. *Let $\mathbb{T} : \mathbf{E} \longmapsto \mathbf{E}$ be Lipschitz with constant $\beta \geq 0$.*

(a) *Assume that for each $y \in \mathbf{E}$ the map $\mathbb{T}_y : \mathbf{E} \longmapsto \mathbf{E}$ defined by $\mathbb{T}_y x = \mathbb{T}x + y$ satisfies the requirement that $\mathbb{T}_y^p$ is contractive for some $p \in \mathbf{N}$. Then $(\mathbb{I} - \mathbb{T})$ maps $\mathbf{E}$ onto $\mathbf{E}$, the inverse of $\mathbb{F} := \mathbb{I} - \mathbb{T} : \mathbf{E} \longmapsto \mathbf{E}$ exists and*

$$\|\mathbb{F}^{-1}x - \mathbb{F}^{-1}y\| \leq \rho_p \|x - y\|, \quad x, y \in \mathbf{E}, \tag{10.20}$$

where

$$\rho_p = \begin{cases} \frac{p}{1-\mathrm{Lip}(\mathbb{T}^p)}, & \text{if } \beta = 1, \\ \frac{1}{1-\beta}, & \text{if } \beta < 1, \\ \frac{\beta^p - 1}{(\beta-1)[1-\mathrm{Lip}(\mathbb{T}^p)]}, & \text{if } \beta > 1. \end{cases}$$

(b) *In particular, if $\mathbb{T} : \mathbf{E} \longmapsto \mathbf{E}$ is linear, bounded and $\mathbb{T}^p$ is a nonlinear contraction for some $p \in \mathbf{N}$. Then $\mathbb{F}^{-1} : \mathbf{E} \longmapsto \mathbf{E}$ is continuous and bounded. Moreover, the spectral radius of $\mathbb{T}$, $r(\mathbb{T})$, is smaller than $[\phi(1)]^p$.*

Remark 10.5. In the case of (b), we are unable to obtain an estimate similar to (10.20).

Proof. (a) Based on the Banach contraction mapping principle, using similar arguments to Lemma 10.3, one can easily deduce that both $(\mathbb{I} - \mathbb{T})$ and $(\mathbb{I} - \mathbb{T}^p)$ map $\mathbf{E}$ onto $\mathbf{E}$. Now, for any $x, y \in \mathbf{E}$, since $\mathbb{T}^p$ is contractive it follows from the triangle inequality that

$$\|(\mathbb{I} - \mathbb{T}^p)x - (\mathbb{I} - \mathbb{T}^p)y\| \geq [1 - \mathrm{Lip}(\mathbb{T}^p)]\|x - y\|,$$

which illustrates that $(\mathbb{I} - \mathbb{T}^p)$ is one-to-one. Hence $(\mathbb{I} - \mathbb{T}^p)^{-1}$ exists on $\mathbf{E}$, and consequently, $(\mathbb{I} - \mathbb{T})^{-1}$ exists on $\mathbf{E}$ due to (10.15). From (10.15), (10.17) and (10.19), one concludes that

$$\mathrm{Lip}((\mathbb{I} - \mathbb{T})^{-1}) \leq \begin{cases} \frac{p}{1-\mathrm{Lip}(\mathbb{T}^p)}, & \text{if } \beta = 1, \\ \frac{1}{1-\beta}, & \text{if } \beta < 1, \\ \frac{\beta^p - 1}{(\beta-1)[1-\mathrm{Lip}(\mathbb{T}^p)]}, & \text{if } \beta > 1. \end{cases}$$

This proves (10.20).

(b) By invoking a fixed-point result of Boyd and Wong [8], together with the arguments just presented, one derives that $\mathbb{F}^{-1} : \mathbf{E} \longmapsto \mathbf{E}$ is continuous and bounded. Now, since $\mathbb{T}^p$ is nonlinearly contractive, it follows that $\|\mathbb{T}^p x\| \leq \phi(\|x\|)$ and so

$$\|\mathbb{T}^p\| = \sup_{\|x\|\leq 1} \|\mathbb{T}^p x\| \leq \sup_{\|x\|\leq 1} \phi(\|x\|) \leq \phi(1).$$

Observe that $\lim_{n\longmapsto\infty} \|\mathbb{T}^n\|^{1/n}$ exists and is equal to $r(\mathbb{T})$. Thus, we obtain

$$r(\mathbb{T}) = \lim_{k\to\infty} \|\mathbb{T}^{kp}\|^{\frac{1}{kp}} \leq \|\mathbb{T}^p\|^{\frac{1}{p}} \leq [\phi(1)]^{\frac{1}{p}}.$$

The lemma is completely proved. □

Corollary 10.3. *Let $\mathbb{T} : \mathbf{E} \longmapsto \mathbf{E}$ be a linear and bounded operator. Assume that $\mathbb{T}^p$ is contractive for some $p \in \mathbf{N}$. Then the conclusions of* (a) *in Lemma* 10.6 *hold.*

Making use of Lemma 10.4 and Corollary 10.3, we shall see that a k-set contractive map with $k \geq 1$ defined on $\mathbf{E}$ may have a fixed point. Such an interesting phenomenon is exhibited in the following proposition (For an exact and concrete example, we refer to Section 10.4).

Proposition 10.1. *Let $\mathbb{T}$ be as in Corollary* 10.3. *Then $\mathbb{T}$ has a unique fixed point in $\mathbf{E}$ and $\mathbb{T}$ is a $\|\mathbb{T}\|$-set contractive map. Obviously, the number $\|\mathbb{T}\|$ may not be small than* 1.

Together with Lemmas 10.6, 10.4 and the ideas to prove Theorem 10.5, one can easily derive the following result.

Theorem 10.6. *Let $\mathbf{K} \subset \mathbf{E}$ be a nonempty, bounded, closed convex subset. Suppose that $\mathbb{T} : \mathbf{E} \longmapsto \mathbf{E}$ and $\mathbb{S} : \mathbf{K} \longmapsto \mathbf{E}$ such that:*
(i) *$\mathbb{T}$ satisfies the conditions* (a) *of Lemma* 10.6;
(ii) *$\mathbb{S}$ is a strictly ρ_p^{-1}-set contractive map (or a ρ-set contractive map with $\rho < \rho_p^{-1}$);*
(iii) $[x = \mathbb{T}x + \mathbb{S}y, y \in \mathbf{K}] \Longrightarrow x \in \mathbf{K}$.

Then the sum $\mathbb{S} + \mathbb{T}$ possesses at least one fixed point in $\mathbf{K}$.

Remark 10.6. Theorems 10.5 and 10.6, in a certain sense, develop the corresponding theorems 2.7 and 2.12 in [30], respectively.

Corollary 10.4. *In Theorem* 10.6, *if only* (i) *is replaced by* (b) *of Lemma* 10.4, *then $\mathbb{S} + \mathbb{T}$ has at least one fixed point in $\mathbf{K}$.*

Remark 10.7. Corollary 10.4 extends a variant of Theorem A in Nashed and Wong [22].

Corollary 10.5 (Dhage [11]). *Let $\mathbf{K}$, $\mathbb{T}$, $\mathbb{S}$ and* (iii) *be the same as Theorem* 10.6. *In addition, assume that* (i′), *$\mathbb{T} : \mathbf{E} \longmapsto \mathbf{E}$ satisfies the conditions* (2) *of Lemma* 10.6, *and that* (ii′), *$\mathbb{S} : \mathbf{K} \longmapsto \mathbf{E}$ is compact. Then $\mathbb{S} + \mathbb{T}$ has at least one fixed point in $\mathbf{K}$.*

Inspired by the proofs of Theorems 10.5 and 10.6, we now can formulate an abstract existence theorem, which summarizes Theorems 10.5 and 10.6.

Theorem 10.7. *Let* $\mathbf{K} \subset \mathbf{E}$ *be a nonempty, bounded, closed convex subset. Suppose that* $\mathbb{T} : \mathbf{E} \longmapsto \mathbf{E}$ *and* $\mathbb{S} : \mathbf{K} \longmapsto \mathbf{E}$ *such that*
(i) $(\mathbb{I} - \mathbb{T})$ *is Lipschitz invertible with constant* $\gamma > 0$;
(ii) $\mathbb{S}$ *is a strictly* γ^{-1}*-set contractive map (or a* ρ*-set contractive map with* $\rho < \gamma^{-1}$*);*
(iii) $\mathbb{S}(\mathbf{K}) \subset (\mathbb{I} - \mathbb{T})(\mathbf{E})$ *and* $[x = \mathbb{T}x + \mathbb{S}y, y \in \mathbf{K}] \Longrightarrow x \in \mathbf{K}$.

Then the equation $\mathbb{S}x + \mathbb{T}x = x$ *has at least one solution in* $\mathbf{K}$.

Remark 10.8. Clearly, one of the advantages of Theorem 10.7 is that the compactness of $\mathbb{S}$ is not necessarily required. Moreover, the number γ^{-1} may not be small than 1. Therefore, it extends essentially a number of previously known generalizations of Theorem A, such as those due to Burton [9], Nashed and Wong [22], Dhage [11, Theorem 1.5], and some results in [30].

Observe that, in Krasnoselskii's theorem, the operator $\mathbb{T}$ is contractive and hence uniformly continuous. We dedicate our work in the sequel to relaxing such a restriction. In the case when $\mathbb{I} - \mathbb{T}$ is one-to-one, these generalizations complement and refine non-compact-type Krasnoselskii fixed-point theorems in [31]. Thus, they encompass and extend a lot of existing Krsnoselskii-type fixed-point theorems in the strong topology setup. The proofs are based on the technique associated with measures of non-compact type. For convenience and completeness, we will provide all the details here. To achieve this, the following notation will be necessary.

Let $\mathbf{M}$, $\mathbf{K}$ be two subsets of $\mathbf{E}$, $\mathbb{T} : \mathbf{M} \longmapsto \mathbf{E}$ and $\mathbb{S} : \mathbf{K} \longmapsto \mathbf{E}$ two maps. We shall denote by $\mathcal{F} = \mathcal{F}(\mathbf{M}, \mathbf{K}; \mathbb{T}, \mathbb{S})$ the following set:

$$\mathcal{F} = \{x \in \mathbf{M} : x = \mathbb{T}x + \mathbb{S}y \text{ for some } y \in \mathbf{K}\}.$$

Theorem 10.8. *Let* $\mathbf{K}$ *be a nonempty, bounded, closed convex subset of* $\mathbf{E}$ *with* $\mathbf{K} \subset \mathbf{D}(\mathbb{T}) \subset \mathbf{E}$, *and* $\mathbb{T} : \mathbf{D}(\mathbb{T}) \longmapsto \mathbf{E}$ *a map. Suppose that* $\mathbb{S} : \mathbf{K} \longmapsto \mathbf{E}$ *is continuous such that*
(i) $(\mathbb{I} - \mathbb{T})$ *is one-to-one;*
(ii) $\alpha(\mathbb{T}(\mathbf{A}) + \mathbb{S}(\mathbf{A})) < \alpha(\mathbf{A})$ *for all* $\mathbf{A} \subset \mathbf{K}$ *with* $\alpha(\mathbf{A}) > 0$;
(iii) *if* $\{x_n\} \subset \mathcal{F}(\mathbf{D}(\mathbb{T}), \mathbf{K}; \mathbb{T}, \mathbb{S})$ *with* $x_n \to x$ *and* $\mathbb{T}x_n \to y$, *then* $x \in \mathbf{D}(\mathbb{T})$ *and* $y = \mathbb{T}x$;
(iv) $\mathbb{S}(\mathbf{K}) \subset (\mathbb{I} - \mathbb{T})(\mathbf{D}(\mathbb{T}))$ *and* $[x = \mathbb{T}x + \mathbb{S}y, y \in \mathbf{K}] \Longrightarrow x \in \mathbf{K}$.

Then the sum $\mathbb{S} + \mathbb{T}$ *has at least one fixed point in* $\mathbf{K}$.

Proof. Since $(\mathbb{I} - \mathbb{T}) : \mathbf{D}(\mathbb{T}) \longmapsto \mathbf{E}$ is one-to-one, the inverse of $(\mathbb{I} - \mathbb{T})^{-1}$ exists on $(\mathbb{I} - \mathbb{T})(\mathbf{D}(\mathbb{T}))$. From $\mathbb{S} : \mathbf{K} \longmapsto \mathbf{E}$ and $\mathbb{S}(\mathbf{K}) \subset (\mathbb{I} - \mathbb{T})(\mathbf{D}(\mathbb{T}))$ we conclude that the operator $\mathbb{N} = (\mathbb{I} - \mathbb{T})^{-1}\mathbb{S} : \mathbf{K} \longmapsto \mathbf{D}(\mathbb{T})$ is well defined and that $\mathcal{F}$ is nonempty.

For each $x \in \mathcal{F}$, by the definition of $\mathcal{F}$, there exists $y \in \mathbf{K}$ such that $x = \mathbb{T}x + \mathbb{S}y$, i. e., $x = \mathbb{N}y$. This shows $\mathcal{F} \subset \mathbb{N}(\mathbf{K})$.

On the other hand, if $x \in \mathbb{N}(\mathbf{K})$ then there exists $y \in \mathbf{K}$ so that $\mathbb{N}y = x$ or $x = (\mathbb{I} - \mathbb{T})^{-1}\mathbb{S}y$ or $(\mathbb{I} - \mathbb{T})x = \mathbb{S}y$. Consequently $x \in \mathcal{F}$, from which $\mathbb{N}(K\mathbf{K}) \subset \mathcal{F}$ and then $\mathcal{F} = \mathbb{N}(\mathbf{K})$.

Let $x \in \mathcal{F}$. Then there exists $y \in \mathbf{K}$ such that $x = \mathbb{T}x + \mathbb{S}y$. The second part of (iv) then gives $x \in \mathbf{K}$. Therefore, $\mathcal{F} \subset \mathbf{K}$ and thus $\mathbb{N}$ maps $\mathbf{K}$ into itself.

Let now $x_0 \in \mathbf{K}$ and

$$\mathcal{A} = \{\mathbf{A} : x_0 \in \mathbf{A} \subset \mathbf{K}, \mathbf{A} \text{ is a closed convex set and } \mathbb{N}(\mathbf{A}) \subset \mathbf{A}\}.$$

Since $x_0 \in \mathbf{K}$, $\mathbf{K} \subset \mathbf{K}$, $\mathbf{K}$ is a closed convex set and $\mathcal{F} = \mathbb{N}(\mathbf{K}) \subset \mathbf{K}$, we obtain $\mathbf{K} \in \mathcal{A}$, i. e., $\mathcal{A} \neq \leftarrow$.

Moreover, for any $\mathbf{A} \in \mathcal{A}$ we have

$$(\mathbb{I} - \mathbb{T})^{-1}\mathbb{S}(\mathbf{A}) = (\mathbb{I} - \mathbb{T} + \mathbb{T})(\mathbb{I} - \mathbb{T})^{-1}\mathbb{S}(\mathbf{A}) = \mathbb{S}(\mathbf{A}) + \mathbb{T}(\mathbb{I} - \mathbb{T})^{-1}\mathbb{S}(\mathbf{A}).$$

The definition of $\mathcal{A}$ gives $(\mathbb{I} - \mathbb{T})^{-1}\mathbb{S}(\mathbf{A}) = \mathbb{N}(\mathbf{A}) \subset \mathbf{A}$, and so we get from the above equality

$$(\mathbb{I} - \mathbb{T})^{-1}\mathbb{S}(\mathbf{A}) \subset \mathbb{T}(\mathbb{I} - \mathbb{T})^{-1}\mathbb{S}(\mathbf{A}) + \mathbb{S}(\mathbf{A}) \subset \mathbb{T}(\mathbf{A}) + \mathbb{S}(\mathbf{A}).$$

This fact, together with (ii), yields

$$\alpha(\mathbb{N}(\mathbf{A})) \leq \alpha(\mathbb{T}(\mathbf{A}) + \mathbb{S}(\mathbf{A})) < \alpha(\mathbf{A}) \text{ for all } \mathbf{A} \in \mathcal{A} \text{ with } \alpha(\mathbf{A}) > 0. \tag{10.21}$$

Put $\mathbf{A}_0 = \bigcap_{\mathbf{A}\in\mathcal{A}} \mathbf{A}$. Then $x_0 \in \mathbf{A}_0 \subset \mathbf{K}$, $\mathbf{A}_0$ is a closed convex set and $\mathbb{N}(\mathbf{A}_0) \subset \mathbf{A}_0$, and therefore $\mathbf{A}_0 \in \mathcal{A}$. Notice that $\overline{\operatorname{span}}\{\mathbb{N}(\mathbf{A}_0), x_0\} \subset \mathbf{A}_0$. Hence, we have

$$\mathbb{N}(\overline{\operatorname{span}}\{\mathbb{N}(\mathbf{A}_0), x_0\}) \subset \mathbb{N}(\mathbf{A}_0) \subset \overline{\operatorname{span}}\{\mathbb{N}(\mathbf{A}_0), x_0\},$$

which implies that $\overline{\operatorname{span}}\{\mathbb{N}(\mathbf{A}_0), x_0\} \in \mathcal{A}$. It then follows from the definition of $\mathbf{A}_0$ that $\overline{\operatorname{span}}\{\mathbb{N}(\mathbf{A}_0), x_0\} = \mathbf{A}_0$. Thus, by the properties of α, we obtain

$$\alpha(\mathbf{A}_0) = \alpha(\overline{\operatorname{span}}\{\mathbb{N}(\mathbf{A}_0), x_0\}) = \alpha(\{\mathbb{N}(\mathbf{A}_0), x_0\}) = \alpha(\mathbb{N}(\mathbf{A}_0)). \tag{10.22}$$

Recalling that $\mathbf{A}_0 \in \mathcal{A}$, we the deduce from (10.21) and (10.22) that $\alpha(\mathbf{A}_0) = 0$. Consequently, $\mathbf{A}_0$ is a nonempty compact convex subset of $\mathbf{K}$ and $\mathbb{N}(\mathbf{A}_0) \subset \mathbf{A}_0$.

We next examine that $\mathbb{N} : \mathbf{A}_0 \longmapsto \mathbf{A}_0$ is continuous. Indeed, let $\{x_n\}$ be a sequence in $\mathbf{A}_0$ with $x_n \to x$. Set $y_n = (\mathbb{I} - \mathbb{T})^{-1}\mathbb{S}x_n$ and $y = (\mathbb{I} - \mathbb{T})^{-1}\mathbb{S}x$ (this is well defined since $x \in \mathbf{A}_0 \subset \mathbf{K}$). Then $(\mathbb{I} - \mathbb{T})y_n = \mathbb{S}x_n$ and $(\mathbb{I} - \mathbb{T})y = \mathbb{S}x$. Hence $y_n, y \in \mathbf{A}_0 \cap \mathcal{F}$, and so $\{y_n\}$ has a subsequence $\{y_{n_k}\}$ converging to some $y_0 \in \mathbf{A}_0$. Evidently, by the continuity of $\mathbb{S}$,

$$\mathbb{T}y_{n_k} = y_{n_k} - (\mathbb{I} - \mathbb{T})y_{n_k} \to y_0 - \mathbb{S}x = y_0 - (\mathbb{I} - \mathbb{T})y. \tag{10.23}$$

It follows from (10.23) and (iii) that $y_0 - (\mathbb{I} - \mathbb{T})y = \mathbb{T}y_0$, and thus $y_0 = y = (\mathbb{I} - \mathbb{T})^{-1}Sx$ since $\mathbb{I} - \mathbb{T}$ is injective. Summing up the above arguments, we have derived

$$(\mathbb{I} - \mathbb{T})^{-1}\mathbb{S}x_{n_k} \to (\mathbb{I} - \mathbb{T})^{-1}\mathbb{S}x.$$

We next claim that

$$(\mathbb{I} - \mathbb{T})^{-1}\mathbb{S}x_n \longrightarrow (\mathbb{I} - \mathbb{T})^{-1}\mathbb{S}x.$$

Assume the contrary case; then there exists a neighborhood $\mathbf{U}$ of $(\mathbb{I} - \mathbb{T})^{-1}\mathbb{S}x$ and a subsequence $\{x_{n_j}\}$ of $\{x_n\}$ such that $(\mathbb{I} - \mathbb{T})^{-1}\mathbb{S}x_{n_j} \notin \mathbf{U}$ for all $j \geq 1$. Naturally, $\{x_{n_j}\}$ converges to x; then reasoning as before we may extract a subsequence $\{x_{n_{j_k}}\}$ of $\{x_{n_j}\}$ so that $(\mathbb{I}-\mathbb{T})^{-1}\mathbb{S}x_{n_{j_k}} \to (\mathbb{I}-\mathbb{T})^{-1}\mathbb{S}x$. But this is a contradiction, since $(\mathbb{I}-\mathbb{T})^{-1}\mathbb{S}x_{n_j} \notin \mathbf{U}$ for all $j \geq 1$. The claim is hence confirmed, and finally $(\mathbb{I}-\mathbb{T})^{-1}\mathbb{S} : \mathbf{A}_0 \longmapsto \mathbf{A}_0$ is continuous.

The Schauder fixed-point theorem guarantees that $\mathbb{N} = (\mathbb{I} - \mathbb{T})^{-1}\mathbb{S}$ has at least one fixed point in $\mathbf{A}_0$. This ends the proof. □

Remark 10.9. It is easy to see that various kinds of generalized contractions verify conditions (i) and (iii); if $\mathbb{T}$ is a contraction, then Theorem 10.8 extends Theorem 10.7. Especially, if $\mathbb{T}$ is the zero operator on $\mathbf{E}$, then Theorem 10.8 is the well-known Sadovskii fixed-point theorem.

The (closedness) condition (iii) is much weaker than the condition that $\mathbb{T}$ is continuous. Clearly, if $\mathbb{T}$ is continuous then it is closed. Conversely, this may not be true, as can be seen from the fact that a closed linear operator is not necessarily continuous. If $(\mathbb{I} - \mathbb{T})^{-1}$ exists and it is continuous, then the condition (i) is fulfilled, and more importantly the condition (iii) is totally redundant. This shows that Theorem 10.8 holds irrespective of the continuity of $\mathbb{T}$ in such a case. Furthermore, instead of the requirements of (ii) and $\mathbf{K} \subset \mathbf{D}(\mathbb{T})$, we can impose conditions on $(\mathbb{I} - \mathbb{T})^{-1}$ and $\mathbb{S}$ so that $\alpha((\mathbb{I} - \mathbb{T})^{-1}\mathbb{S}(\mathbf{A})) < \alpha(\mathbf{A})$ for all $\mathbf{A} \subset \mathbf{K}$ with $\alpha(\mathbf{A}) > 0$ as was done previously in the proof of the theorem. These observations lead to the following consequence of Theorem 10.8.

Corollary 10.6. *Let K be a nonempty, bounded, closed convex subset of $\mathbf{E}$ and $\mathbb{T} : \mathbf{D}(\mathbb{T}) \subset \mathbf{E} \longmapsto \mathbf{E}$ a map. Suppose that $\mathbb{S} : \mathbf{K} \longmapsto \mathbf{E}$ is continuous such that*

(i) *$(\mathbb{I} - \mathbb{T})^{-1}$ exists and it is continuous;*
(ii) *$\alpha((\mathbb{I} - \mathbb{T})^{-1}\mathbb{S}(\mathbf{A})) < \alpha(\mathbf{A})$ for all $\mathbf{A} \subset \mathbf{K}$ with $\alpha(\mathbf{A}) > 0$;*
(iii) *$\mathbb{S}(\mathbf{K}) \subset (\mathbb{I} - \mathbb{T})(\mathbf{D}(\mathbb{T}))$ and $[x = \mathbb{T}x + \mathbb{S}y, y \in \mathbf{K}] \Longrightarrow x \in \mathbf{K}$.*

Then the sum $\mathbb{S} + \mathbb{T}$ admits one fixed point in $\mathbf{K}$.

Although Theorem 10.8 includes the case that $\mathbb{S}$ is compact, we will revisit it for such a particular case. It turns out that the boundedness of $\mathbf{K}$ and that the requirement of $\mathbf{K} \subset \mathbf{D}(\mathbb{T})$ are not needed, if we impose a compactness condition on $\mathcal{F}(\mathbf{D}(\mathbb{T}), \mathbf{K}; \mathbb{T}, \mathbb{S})$.

Theorem 10.9. *Let $\mathbf{K} \subset \mathbf{E}$ be a nonempty, closed convex subset and $\mathbb{T} : \mathbf{D}(\mathbb{T}) \subset \mathbf{E} \longmapsto \mathbf{E}$ a mapping. Suppose that $\mathbb{S} : \mathbf{K} \longmapsto \mathbf{E}$ is continuous such that*

(i) *$(\mathbb{I} - \mathbb{T})$ is one-to-one;*
(ii) *the set $\mathcal{F}(\mathbf{D}(\mathbb{T}), \mathbf{K}; \mathbb{T}, \mathbb{S})$ is relatively compact;*

(iii) *if* $\{x_n\} \subset \mathcal{F}$ *for which* $x_n \to x$ *and* $\mathbb{T}x_n \to y$, *then* $x \in \mathbf{D}(\mathbb{T})$ *and* $y = \mathbb{T}x$;
(iv) $\mathbb{S}(\mathbf{K}) \subset (\mathbb{I} - \mathbb{T})(\mathbf{D}(\mathbb{T}))$ *and* $[x = \mathbb{T}x + \mathbb{S}y, y \in \mathbf{K}] \Longrightarrow x \in \mathbf{K}$.

Then the sum $\mathbb{S} + \mathbb{T}$ *has one fixed point in* $\mathbf{K}$.

Proof. It is sufficient to show that the operator $(\mathbb{I} - \mathbb{T})^{-1}\mathbb{S} : \mathbf{K} \longmapsto \mathbf{K}$ is compact and continuous. Thanks to the fact $\mathcal{F} = (\mathbb{I} - \mathbb{T})^{-1}\mathbb{S}(\mathbf{K})$ and (ii), we see that $(\mathbb{I} - \mathbb{T})^{-1}\mathbb{S} : \mathbf{K} \longmapsto \mathbf{K}$ is compact. For the continuity, let $y_n, y \in \mathbf{K}$ with $y_n \to y$, and let $x_n = (\mathbb{I} - \mathbb{T})^{-1}\mathbb{S}y_n$ and $x = (\mathbb{I} - \mathbb{T})^{-1}\mathbb{S}y$. The definition of $\mathcal{F}$ implies that $x_n \in \mathcal{F}$ and $(\mathbb{I} - \mathbb{T})x_n \to \mathbb{S}y$ by the continuity of $\mathbb{S}$. In view of $x_n \in \mathcal{F}$ and $\mathcal{F}$ is pre-compact, $\{x_n\}$ has a subsequence $\{x_{n_k}\}$ converging to some x_0. Accordingly $\mathbb{T}x_{n_k} \to x_0 - \mathbb{S}y$. The closedness of $\mathbb{T}$ in $\mathcal{F}$ (cf. item (iii)) therefore tells us that $x_0 - \mathbb{S}y = \mathbb{T}x_0$, i. e., $x_0 = (\mathbb{I} - \mathbb{T})^{-1}\mathbb{S}y$. Since $\mathbb{I} - \mathbb{T}$ is injective it follows $x_0 = x$.

The same argument as at the end of Theorem 10.8 shows $x_n \to x$, and consequently $(\mathbb{I} - \mathbb{T})^{-1}\mathbb{S} : \mathbf{K} \longmapsto \mathbf{K}$ is continuous. □

Corollary 10.7. *The conclusion of Theorem* 10.9 *continues to be valid, if only the condition* (ii) *is replaced by the following assumptions.*
(ii′) $\mathbb{S}(\mathbf{K})$ *resides in a compact subset of* $\mathbf{E}$;
(ii″) *if* $\{x_n\}$ *is a sequence in* $\mathcal{F}(\mathbf{D}(\mathbb{T}), \mathbf{K}; \mathbb{T}, \mathbb{S})$ *and* $(\mathbb{I} - \mathbb{T})x_n \to y$, *then* $\{x_n\}$ *possesses a convergent subsequence* $\{x_{n_k}\}$;

If $(\mathbb{I} - \mathbb{T})^{-1}$ exists and it is continuous, then the conditions (i) and (ii″) are fulfilled, and as noted before, the condition (iii) is unnecessary. Thus, Theorems 10.8 and 10.9 and their corollaries facilitate the application of another kind of operator, namely, dissipative operator. Here, we provide an application of Corollary 10.6, which complements the result in [29].

For $x \in \mathbf{E}$, define the duality set of x, a subset of the dual space $\mathbf{E}^*$ of $\mathbf{E}$, by

$$J(x) = \{x^* \in \mathbf{E}^* : \|x^*\|_{\mathbf{E}^*}^2 = \|x\|_{\mathbf{E}}^2 = \langle x^*, x\rangle\}.$$

Let $\mathbb{T} : \mathbf{D}(\mathbb{T}) \subset \mathbf{E} \longmapsto \mathbf{E}$ be a (possibly) nonlinear operator. Then $\mathbb{T}$ is said to be dissipative if for each $x, y \in \mathbf{D}(\mathbb{T})$ there exists $x^* \in J(x - y)$ such that $\mathrm{Re}\langle x^*, \mathbb{T}x - \mathbb{T}y\rangle \le 0$. This notion is a nonlinear version of linear dissipative operators, introduced in [7] and [16] independently. For a Hilbert space H this is equivalent to $\mathrm{Re}(x - y, \mathbb{T}x - \mathbb{T}y) \le 0$ for all $x, y \in \mathbf{D}(\mathbb{T})$. Using this equivalent characterization, the Laplacian operator, Δ, defined on the dense subspace of compactly supported smooth functions on the domain $\Omega \subset \mathbf{R}^n$, is a dissipative operator.

Proposition 10.2. *Assume that* $\mathbb{T} : \mathbf{D}(\mathbb{T}) \subset \mathbf{E} \longmapsto \mathbf{E}$ *is a dissipative operator. Then* $(\mathbb{I} - \mathbb{T})$ *is invertible on* $(\mathbb{I} - \mathbb{T})(\mathbf{D}(\mathbb{T}))$ *and its inverse is non-expansive, and the condition* (ii″) *is satisfied. Additionally, if* $\mathbb{S} : \mathbf{B}(0, \rho) \longmapsto \mathbf{B}(-\mathbb{T}0, \rho)$ *is a mapping for some* $\rho > 0$, *where* $\mathbf{B}(x_0, \rho) = \{x \in \mathbf{E} : \|x - x_0\| \le \rho\}$, *then*

$$[x = \mathbb{T}x + \mathbb{S}y, y \in \mathbf{B}(0, \rho)] \Longrightarrow x \in \mathbf{B}(0, \rho). \tag{10.24}$$

Proof. Since $\mathbb{T} : \mathbf{D}(\mathbb{T}) \subset \mathbf{E} \longmapsto \mathbf{E}$ is a dissipative operator, we obtain (cf. [16, Lemma 1.1])

$$\|x-y\| \leq \|x-y-\lambda(\mathbb{T}x-\mathbb{T}y)\| \quad \text{for all } \lambda > 0 \quad \text{and } x,y \in \mathbf{D}(\mathbb{T}). \tag{10.25}$$

Setting $\lambda = 1$ in (10.25), we see that $(\mathbb{I}-\mathbb{T})$ is injective and that

$$\|(\mathbb{I}-\mathbb{T})^{-1}w - (\mathbb{I}-\mathbb{T})^{-1}z\| \leq \|w-z\|, \quad \forall w,z \in (\mathbb{I}-\mathbb{T})(\mathbf{D}(\mathbb{T})), \tag{10.26}$$

which says exactly that $(\mathbb{I}-\mathbb{T})^{-1}$ is non-expansive on $(\mathbb{I}-\mathbb{T})(\mathbf{D}(\mathbb{T}))$.

Now, suppose $x = \mathbb{T}x + \mathbb{S}y$ with $y \in \mathbf{B}(0,\rho)$. It then follows from (10.25) and the assumption that $\mathbb{S}(\mathbf{B}_\rho) \subset \mathbf{B}(-\mathbb{T}0,\rho)$ that $\|x\| \leq \|x-(\mathbb{T}x-\mathbb{T}0)\| = \|\mathbb{S}y-(-\mathbb{T}0)\| \leq \rho$; that is, $x \in \mathbf{B}_\rho$ and (10.24) is verified.

Finally, let $x_n \in \mathcal{F}(\mathbf{D}(\mathbb{T}),\mathbf{K};\mathbb{T},\mathbb{S})$ with $(\mathbb{I}-\mathbb{T})x_n \to y$. Putting $y_n = (\mathbb{I}-\mathbb{T})x_n$, we deduce from (10.25) that

$$\|x_n - x_m\| \leq \|x_n - x_m - (\mathbb{T}x_n - \mathbb{T}x_m)\| = \|y_n - y_m\|,$$

which illuminates that $\{x_n\}$ is a Cauchy sequence, because $\{y_n\}$ is a convergent sequence in E. Therefore, $\{x_n\}$ converges in $\mathbf{E}$. The condition (ii$''$) is thus proved. □

Theorem 10.10. *Let $\mathbb{T} : \mathbf{D}(\mathbb{T}) \subset \mathbf{E} \longmapsto \mathbf{E}$ be dissipative and let $\mathbb{S} : \mathbf{B}(0,\rho) \longmapsto \mathbf{B}(-\mathbb{T}0,\rho)$ be condensing for some $\rho > 0$. If $\mathbb{S}(\mathbf{B}(0,\rho)) \subset (\mathbb{I}-\mathbb{T})(\mathbf{D}(\mathbb{T}))$, then the sum $\mathbb{S}+\mathbb{T}$ has at least one fixed point in $\mathbf{B}(0,\rho)$.*

Proof. Because $\mathbb{T}$ is dissipative, $(\mathbb{I}-\mathbb{T})^{-1}$ exists and it is continuous. Because $\mathbb{S}$ maps $\mathbf{B}(0,\rho)$ into $\mathbf{B}(-\mathbb{T}0,\rho)$, equation (10.24) implies $[x = \mathbb{T}x+\mathbb{S}y, y \in \mathbf{B}(0,\rho)] \Longrightarrow x \in \mathbf{B}(0,\rho)$.

Next, we will show that $\alpha((\mathbb{I}-\mathbb{T})^{-1}\mathbb{S}(\mathbf{A})) < \alpha(\mathbf{A})$ for all $\mathbf{A} \subset \mathbf{B}_\rho$ with $\alpha(\mathbf{A}) > 0$. To see this, take any $\mathbf{A} \subset \mathbf{B}_\rho$ with $\alpha(\mathbf{A}) > 0$, it follows from (10.26), Lemma 10.4 and the assumption $\mathbb{S}$ is condensing that

$$\alpha((\mathbb{I}-\mathbb{T})^{-1}\mathbb{S}(\mathbf{A})) \leq \alpha(\mathbb{S}(\mathbf{A})) < \alpha(\mathbf{A}).$$

By the use of Corollary 10.6 one achieves the proof. □

Remark 10.10. The closed ball $\mathbf{B}(0,\rho)$ can be replaced by a nonempty, closed, convex and unbounded subset of $\mathbf{E}$. In this case, a Leray–Schauder type of condition should be satisfied; see [29]. Here, we note there is a typo in the condition (ii) of [29, Theorem 2.2], it suffices that "$\mathbb{T}$ is nonlinear and $\mathbb{S}(\mathbf{K}) \subset \mathbf{R}(\mathbb{I}-\mathbb{T}) = (\mathbb{I}-\mathbb{T})(\mathbf{K})$".

In what follows, we consider the case when $\mathbb{T} \in \mathcal{L}(\mathbf{E})$ and $\|\mathbb{T}^p\| = \mathrm{Lip}(\mathbb{T}^p) \leq 1$ for some $p \in \mathbf{N}$. Clearly, the above arguments cannot be applied in such case. Thus, in order to study such cases some additional assumptions should be imposed. We first investigate the case when $\mathbb{T}^p$ is a non-expansive mapping on $\mathbf{E}$, i. e., it satisfies

$$\|\mathbb{T}^p x - \mathbb{T}^p y\| \leq \|\mathbb{T}^p\|\|x-y\| \quad \text{for all } x,y \in \mathbf{E}. \tag{10.27}$$

Theorem 10.11. *Let* $\mathbf{K} \subset \mathbf{E}$ *be a nonempty, compact, convex subset. Suppose that* $\mathbb{T} \in \mathcal{L}(\mathbf{E})$ *and satisfies* (10.27) *with constant* $\|\mathbb{T}^p\| \le 1$ *and* $\mathbb{S} : \mathbf{K} \longmapsto \mathbf{E}$ *is continuous. In addition, assume also the following condition holds:*

$$\text{there is a sequence } \lambda_n \in (0,1) \text{ with } \lambda_n \to 1 \text{ such that } [x = \lambda_n \mathbb{T}x + \mathbb{S}y, y \in \mathbf{K}] \Longrightarrow x \in \mathbf{K}.$$

Then $\mathbb{T} + \mathbb{S}$ *has a fixed point in* $\mathbf{K}$.

Proof. Let $\mathbb{T}_n = \lambda_n \mathbb{T} : \mathbf{E} \longmapsto \mathbf{E}$. Then we have

$$\|\mathbb{T}_n^p x - \mathbb{T}_n^p y\| = \lambda_n^p \|\mathbb{T}^p x - \mathbb{T}^p y\| \le \lambda_n^p \|\mathbb{T}^p\| \|x - y\|, \quad \forall x, y \in \mathbf{E}.$$

Therefore, $\mathbb{T}_n^p : \mathbf{E} \longmapsto \mathbf{E}$ is contractive since $\lambda_n^p \|\mathbb{T}^p\| \le \lambda_n^p < 1$ for all $n \in \mathbf{N}$. By Corollary 10.3, we see that $(\mathbb{I} - \mathbb{T}_n)$ maps $\mathbf{E}$ onto $\mathbf{E}$, the inverse of $(\mathbb{I} - \mathbb{T}_n)$ exists on $\mathbf{E}$, and

$$\|(\mathbb{I} - \mathbb{T}_n)^{-1} x - (\mathbb{I} - \mathbb{T}_n)^{-1} y\| \le \rho(p, n) \|x - y\|, \quad x, y \in \mathbf{E}, \tag{10.28}$$

where

$$\rho(p, n) = \begin{cases} \frac{p}{1 - \lambda_n^p \|\mathbb{T}^p\|}, & \text{if } \lambda_n \|\mathbb{T}\| = 1, \\ \frac{1}{1 - \lambda_n \|\mathbb{T}\|}, & \text{if } \lambda_n \|\mathbb{T}\| < 1, \\ \frac{\lambda_n^p \|\mathbb{T}\|^p - 1}{(\lambda_n \|\mathbb{T}\| - 1)[1 - \lambda_n^p \|\mathbb{T}^p\|]}, & \text{if } \lambda_n \|\mathbb{T}\| > 1. \end{cases}$$

It follows from (10.28) that $(\mathbb{I} - \mathbb{T}_n)$ is Lipschitz invertible with constant $\rho(p, n) > 0$. Since K is compact, $\mathbb{S} : \mathbf{K} \longmapsto \mathbf{E}$ is compact] or 0-set contractive. Now, applying Theorem 10.7 to $\lambda_n \mathbb{T}$ and $\mathbb{S}$ for each $n \ge 1$, one sees that there is $x_n^* \in \mathbf{K}$ such that

$$\mathbb{S}x_n^* + \lambda_n \mathbb{T}x_n^* = x_n^*. \tag{10.29}$$

By the compactness of $\mathbf{K}$, up to a subsequence we may assume that $x_n^* \to x^*$ in $\mathbf{K}$. Passing to the limit as $n \to \infty$ in (10.29) we complete the proof. □

We next consider the particular case when $\mathbb{T} \in \mathcal{L}(E)$ and $\mathbb{T}^p$ is non-contractive, i. e.,

$$\|\mathbb{T}^p x - \mathbb{T}^p y\| \ge \|x - y\| \quad \text{for all } x, y \in \mathbf{E}. \tag{10.30}$$

Theorem 10.12. *Let* $\mathbf{K} \subset \mathbf{E}$ *be a nonempty, compact convex subset. Suppose that* $\mathbb{T} \in \mathcal{L}(\mathbf{E})$, $\mathbb{T}^p$ *maps* $\mathbf{E}$ *onto* $\mathbf{E}$ *and satisfies* (10.30) *and that* $\mathbb{S} : \mathbf{K} \longmapsto \mathbf{E}$ *is continuous. In additional, assume also the following condition holds.*

$$\text{there is a sequence } \lambda_n > 1 \text{ with } \lambda_n \to 1 \text{ such that } [x = \lambda_n \mathbb{T}x + \mathbb{S}y, y \in \mathbf{K}] \Longrightarrow x \in \mathbf{K}. \tag{10.31}$$

Then $\mathbb{T} + \mathbb{S}$ *has a fixed point in* $\mathbf{K}$.

Proof. Notice that $\lambda_n \mathbb{T} : \mathbf{E} \longmapsto \mathbf{E}$ fulfills all the requirements of Corollary 10.1. Arguing as in the proof of Theorem 10.11 by using (10.31) and then applying Theorem 10.5 or 10.7, one can easily derive the desired result. □

At the end of this section, we shall see that Theorems 10.8 and 10.9 and their corollaries will motivate us to define a large class of contractions and expansions. Let Φ denote the class of all functions $\phi : [0, \infty) \longmapsto [0, \infty)$ fulfilling:

(i) ϕ is continuous, and $\phi(r) < r$ for all $r > 0$; or
(ii) ϕ is nondecreasing, and $\lim_{n\to\infty} \phi^n(r) = 0$ for each $r > 0$.

Let $\phi \in \Phi$. Then it is an easy matter to show that $\phi(0) = 0$ and $\phi(r) < r$ for every $r > 0$.

Let $\mathbf{X}$ be a complete metric space, $\mathbf{M} \subset \mathbf{X}$, and $\mathbb{T} : \mathbf{M} \longmapsto \mathbf{X}$ a mapping. $\mathbb{T}$ is called a p–Φ-contraction if there are an integer p and $\phi \in \Phi$ such that $d(\mathbb{T}^p x, \mathbb{T}^p y) \leq \phi(d(x, y))$ for all $x, y \in \mathbf{M}$; $\mathbb{T}$ is called a p–Φ-expansion if there are an integer p and $\phi \in \Phi$ such that $\phi(d(\mathbb{T}^p x, \mathbb{T}^p y)) \geq d(x, y)$ for all $x, y \in \mathbf{M}$.

These definitions differ from and extend those of Garcia-Falset [13]. Fixed-point results for such generalized contractions and expansions are collected in the following proposition. Hence, the above established Krasonselskii fixed-point theorems may be adapted to them.

Proposition 10.3. *Let $\mathbf{M}$ be a closed subset of a complete metric space $\mathbf{X}$, and let $\mathbb{T} : \mathbf{M} \longmapsto \mathbf{X}$ be a mapping. Then $\mathbb{T}$ has a unique fixed point in $\mathbf{M}$ if either one of the following conditions is satisfied.*

(i) *$\mathbb{T}$ is a p-Φ-contraction and $\mathbb{T}^p(\mathbf{M}) \subset \mathbf{M}$.*
(ii) *$\mathbb{T}$ is a p-Φ-expansion and $\mathbb{T}^p(\mathbf{M}) \supset \mathbf{M}$.*

Moreover, in each case, one sees that $(\mathbb{I} - \mathbb{T})^{-1}$ exists and it is continuous.

Proof. (i) Since $\mathbb{T}$ is a p-Φ-contraction and $\mathbb{T}^p(\mathbf{M}) \subset \mathbf{M}$, by the monograph [17], we know that $\mathbb{T}^p$ has a unique fixed point $x^* \in \mathbf{M}$. Thus, $\mathbb{T}^p(\mathbb{T}x^*) = \mathbb{T}x^*$, and so $\mathbb{T}x^* = x^*$ by the uniqueness.

If there is $z^* \in \mathbf{M}$ such that $\mathbb{T}z^* = z^*$, then

$$\mathbb{T}^p z^* = \mathbb{T}^{p-1}(\mathbb{T}z^*) = \mathbb{T}^{p-1} z^* = \cdots = \mathbb{T}z^* = z^*,$$

which shows that z^* is also a fixed point of $\mathbb{T}$ in $\mathbf{M}$. Because the fixed point in $\mathbf{M}$ is unique, it follows $x^* = z^*$. Consequently x^* is the unique fixed point of $\mathbb{T}$. Using essentially the same reasoning as Lemma 10.6, one can readily infer that $(\mathbb{I} - \mathbb{T})^{-1}$ is exists and it is continuous.

(ii) Since $\mathbb{T}$ is a p-Φ-expansion and $\mathbb{T}^p(\mathbf{M}) \supset \mathbf{M}$, we get $\phi(d(\mathbb{T}^p x, \mathbb{T}^p y)) \geq d(x, y)$ for all $x, y \in \mathbf{M}$. This shows that $\mathbb{T}^p : \mathbf{M} \longmapsto \mathbb{T}^p(\mathbf{M}) \supset \mathbf{M}$ is one-to-one. Therefore, $(\mathbb{T}^p)^{-1} : \mathbf{M} \longmapsto \mathbf{M}$ exists, and

$$d((\mathbb{T}^p)^{-1} x, (\mathbb{T}^p)^{-1} y) \leq \phi(d(\mathbb{T}^p((\mathbb{T}^p)^{-1} x), \mathbb{T}^p((\mathbb{T}^p)^{-1} y))) = \phi(d(x, y))$$

for all $x, y \in \mathbb{T}^p(\mathbf{M})$ and so for all $x, y \in \mathbf{M}$. The previous paragraph says there exists a unique $y^* \in \mathbf{M}$ such that $(\mathbb{T}^p)^{-1}y^* = y^*$, i. e., $\mathbb{T}^p y^* = y^*$ and y^* is the unique fixed point of $\mathbb{T}$. A similar proof to that of Lemma 10.3 shows that $(\mathbb{I} - \mathbb{T})^{-1}$ exists and that it is continuous. □

10.7 Fixed-point results to one parameter operator equations and eigenvalues problems

As applications to some of the main results, the purpose of this section is to present some existence results for the following nonlinear abstract operator equation in Banach space:

$$\lambda \mathbb{T}x + \mathbb{S}x = x, \tag{10.32}$$

where $\mathbb{T}, \mathbb{S} : \mathbf{E} \longmapsto \mathbf{E}$ and $\lambda \geq 0$ is a parameter. In order to do this, we first establish some local version of the above results. Then we consider the eigenvalue problems of Krasnosel'skii type in the critical case, that is, we investigate the mapping $\mathbb{T} : \mathbf{M} \subset \mathbf{E} \longmapsto \mathbf{E}$ is non-expansive. The first result concerning equation (10.32) is as follows.

Theorem 10.13. *Let $\mathbf{K} \subset \mathbf{E}$ be a nonempty, bounded, closed convex subset, $\mathbb{S} : \mathbf{K} \longmapsto \mathbf{E}$ a map and $\mathbb{T} : \mathbf{E} \longmapsto \mathbf{E}$ a Lipschitz with constant $l \geq 0$. Suppose there is $\lambda_0 \geq 0$ such that*
(i) *$\mathbb{S}$ is a μ-set contractive map with $\mu < 1$ (a condensing map);*
(ii) *$[x = \lambda \mathbb{T}x + \mathbb{S}y, y \in \mathbf{K}] \Longrightarrow x \in \mathbf{K}$ for all $\lambda \leq \lambda_0$.*

Then there exists $\lambda_1 \geq 0$ such that equation (10.32) *is solvable for all $\lambda \in [0, \lambda_1]$ ($\lambda = 0$).*

Proof. Choose $\lambda_1 \geq 0$ so that $\lambda_1 l < 1$ and $\mu < 1 - \lambda l$ for all $\lambda \leq \lambda_1$. Now, $\lambda\mathbb{T} : \mathbf{E} \longmapsto \mathbf{E}$ is a contraction with constant $\lambda l < 1$ for $\lambda \leq \lambda_1$. It is straightforward to see that all the conditions of Theorem 10.7 or 10.11 are satisfied $\lambda \leq \lambda_1$. □

Next we shall modify some assumptions to study equation (10.32).

Theorem 10.14. *Let $\mathbf{K} \subset \mathbf{E}$ be a nonempty, bounded, closed convex subset, $\mathbb{S} : \mathbf{K} \longmapsto \mathbf{E}$ a map and $\mathbb{T} : \mathbf{E} \longmapsto \mathbf{E}$ a weakly expansive with constant $\beta > 0$. Suppose there is $\lambda_0 \geq 0$ such that*
(i) *$\mathbb{S}$ is a k-set contractive map;*
(ii) *$\mathbb{S}(\mathbf{K}) \subset (\mathbb{I} - \lambda\mathbb{T})(\mathbf{E})$ and $[x = \lambda \mathbb{T}x + \mathbb{S}y, y \in \mathbf{K}] \Longrightarrow x \in \mathbf{K}$ for all $\lambda \geq \lambda_0$.*

Then there exists $\lambda_1 \geq 0$ such that equation (10.32) *is solvable for all $\lambda \geq \lambda_1$.*

Proof. Choose $\lambda_1 \geq 0$ so that $\lambda_1\beta > 1$ and $k < \lambda\beta - 1$ for all $\lambda \geq \lambda_1$. Now, $\lambda\mathbb{T} : \mathbf{E} \longmapsto \mathbf{E}$ is an expansion with constant $\lambda\beta > 1$ for $\lambda \geq \lambda_1$. This says the condition (i) of Theorem 10.7 holds. From $k < \lambda\beta - 1$ for all $\lambda \geq \lambda_1$ it follows that the condition (ii) of Theorem 10.7 holds. The result then follows from Theorem 10.7. □

Let us now begin to consider the eigenvalue problems of Krasnosel'skii type. We obtain the following results.

Theorem 10.15. *Let $\mathbf{K} \subset \mathbf{E}$ be a nonempty, bounded, closed convex subset, $\mathbb{S} : \mathbf{K} \longmapsto \mathbf{E}$ a map and $\mathbb{T} : \mathbf{E} \longmapsto \mathbf{E}$ a non-expansion. Suppose that there exists $\lambda > 1$ such that*
(i) *$\mathbb{S}$ is a strictly $(\lambda - 1)$-set contractive map (or a k-set contractive map with $k < \lambda - 1$);*
(ii) *$[\lambda x = \mathbb{T}x + \mathbb{S}y, y \in \mathbf{K}] \Longrightarrow x \in \mathbf{K}$.*

Then there exists $x^* \in \mathbf{K}$ with $\mathbb{S}x^* + \mathbb{T}x^* = \lambda x^*$.

Proof. Let $\mu = 1/\lambda$. Then $\mu\mathbb{T} : \mathbf{E} \longmapsto \mathbf{E}$ is a contraction with constant $\mu \in (0, 1)$ and $\mu\mathbb{S} : \mathbf{K} \longmapsto \mathbf{E}$ is a strictly $\mu(\lambda - 1)$-set contractive map. One can easily verify that all the assumptions of Theorem 10.7 or 10.8 are satisfied for $\mu\mathbb{T}$ and $\mu\mathbb{S}$. Hence the result follows. □

In the end of this section, we investigate the case when $\mathbb{T}$ is a non-contractive mapping on $\mathbf{M} \subset \mathbf{E}$, i. e., an operator which satisfies $\|\mathbb{T}x - \mathbb{T}y\| \geq \|x - y\|$ for all $x, y \in \mathbf{M}$.

Theorem 10.16. *Let $\mathbf{K} \subset \mathbf{E}$ be a nonempty, bounded, closed convex subset, $\mathbb{S} : \mathbf{K} \longmapsto \mathbf{E}$ a map and $\mathbb{T} : \mathbf{E} \longmapsto \mathbf{E}$ a non-contractive one. Suppose that there exists $\lambda \in (0, 1)$ such that:*
(i) *$\mathbb{S}$ is a strictly $(1 - \lambda)$-set contractive map (or a μ-set contractive map with $\mu < 1 - \lambda$);*
(ii) *$\mathbb{S}(\mathbf{K}) \subset (\lambda\mathbb{I} - \mathbb{T})(\mathbf{E})$ and $[\lambda x = \mathbb{T}x + \mathbb{S}y, y \in \mathbf{K}] \Longrightarrow x \in \mathbf{K}$.*

Then there exists $x^* \in \mathbf{K}$ with $\mathbb{S}x^* + \mathbb{T}x^* = \lambda x^*$.

Proof. This is a direct consequence of Theorem 10.5 or 10.7 or 10.8. □

10.8 Application to perturbed Volterra integral equation

Let $\mathbf{E} = \mathbf{C}([a, b])$ with the usual supremum norm $\|x\| = \max_{t\in[a,b]} |x(t)|$. We also denote by $\mathbf{B}_R$ the set $\{x \in \mathbf{E} : \|x\| \leq R\}$. In the present section, our main objective is to prove some existence and unique (in a special case) results for the following perturbed Volterra integral equation of the form:

$$u(t) = \int_a^t k(t, s)u(s)ds + f(t, u(t)), \quad t \in [a, b], u \in E, \tag{10.33}$$

where the kernel k defined on $\Delta = \{(t, s) : a \leq t \leq b, a \leq s \leq t\}$ is essentially bounded and measurable and $f : [a, b] \times \mathbf{R} \longmapsto \mathbf{R}$ is continuous. When $f(t, u) \equiv g(t)$, equation (10.33) is the classical linear Volterra integral equation of the second kind. It is well known that the theory of equations of such a case is very developed both theoretically and numerically. For a comprehensive theory of linear Volterra integral equation, we refer to the monograph [15]. Nevertheless, for the purpose of illustrating the power of our abstract results established in Section 10.2, we would like to address the solvability and uniqueness (in a special case) of equation (10.33) in a generalized form. To perform such a task, we shall introduce the definition of the special measure of non-compactness in **E** which was introduced and studied in [4]. To do this let us fix a subset $\mathbf{X} \in \Omega_{\mathbf{E}}$. For $\epsilon > 0$ and $x \in \mathbf{X}$ denote by $w(x, \epsilon)$ the modulus of continuity of x, i. e.,

$$w(x, \epsilon) = \sup\{|x(t) - x(s)| : t, s \in [a, b], \quad |t - s| < \epsilon\}.$$

Further, put

$$w(\mathbf{X}, \epsilon) = \sup\{w(x, \epsilon) : x \in X\},$$
$$w_0(X) = \lim_{\epsilon \to 0} w(X, \epsilon).$$

It may be shown [4] that $w_0(\mathbf{X})$ is a measure of the non-compactness in the space **E**.

Let us now introduce the operators $\mathbb{T}, \mathbb{S} : \mathbf{E} \longmapsto \mathbf{E}$ as follows:

$$(\mathbb{T}x)(t) = \int_a^t k(t, s)x(s)ds, \tag{10.34}$$

$$(\mathbb{S}y)(t) = f(t, y(t)). \tag{10.35}$$

Then one can easily show that $\mathbb{S} : \mathbf{E} \longmapsto \mathbf{E}$ is continuous and bounded since f is continuous.

For each $x, y \in \mathbf{E}$, one readily derives from (10.34) that

$$(\mathbb{T}x)(t) - (\mathbb{T}y)(t) \leq \int_a^t |k(t, s)| \|x - y\| ds \leq c(t - a)\|x - y\|, \tag{10.36}$$

where $c = \operatorname{ess\,sup}_{(t,s)\in\Delta} |k(t, s)| < \infty$. By induction, one can deduce from (10.34) and (10.36) that

$$|(\mathbb{T}^n x)(t) - (\mathbb{T}^n y)(t)| \leq \frac{[c(t - a)]^n}{n!} \|x - y\|.$$

Hence

$$\|\mathbb{T}^n x - \mathbb{T}^n y\| \leq \frac{[c(b - a)]^n}{n!} \|x - y\|. \tag{10.37}$$

Notice that

$$\lim_{n \to \infty} \frac{[c(b - a)]^n}{n!} = 0.$$

It follows from (10.37) that there exists $p \in \mathbf{N}$ such that $\mathbb{T}^p$ is a contraction. On the other hand, one can also obtain from (10.34)

$$\|\mathbb{T}x - \mathbb{T}y\| \leq M\|x - y\|, \tag{10.38}$$

where

$$M = \max_{a \le t \le b} \int_a^t |k(t,s)| ds.$$

Together with (10.37), (10.38) and Corollary 10.3, we see that $(\mathbb{I} - \mathbb{T})$ maps $\mathbf{E}$ onto $\mathbf{E}$, the inverse of $\mathbb{I} - \mathbb{T} : \mathbf{E} \longmapsto \mathbf{E}$ exists and

$$\left\| (\mathbb{I} - \mathbb{T})^{-1} x - (\mathbb{I} - \mathbb{T})^{-1} y \right\| \le \rho_p \|x - y\|, \quad \forall x, y \in \mathbf{E}, \tag{10.39}$$

where

$$\rho_p = \begin{cases} \frac{p}{1-\mathrm{Lip}(\mathbb{T}^p)}, & \text{if } M = 1, \\ \frac{1}{1-M}, & \text{if } M < 1, \\ \frac{M^p - 1}{(M-1)[1-\mathrm{Lip}(\mathbb{T}^p)]}, & \text{if } M > 1. \end{cases} \tag{10.40}$$

In this section, we shall study equation (10.33) by considering three cases: $M < 1$, $M = 1$ and $M > 1$. Our strategy is to apply Theorem 10.5 or 10.7 or 10.8 to derive the fixed point of the sum $\mathbb{T} + \mathbb{S}$.

Case of $M < 1$. We obtain the existence of one and only one positive solution of equation (10.33) in this case. In order to do so, assume that the functions involved in equation (10.33) fulfill the following conditions:

(H1) k is nonnegative on Δ;

(H2) there are two constants $B > A \ge 0$ such that

$$(1 - M')A \le f(t,x) \le (1 - M)B, \quad \forall (t,x) \in [a,b] \times [A,B],$$

where $M' = \min_{a \le t \le b} \int_a^t k(t,s) ds$;

(H3) for each fixed $t \in [a,b]$, $x, y \in [A,B]$ with $x \ne y$, we have

$$|f(t,x) - f(t,y)| \le \phi(|x - y|),$$

where $\phi : \mathbf{R}_+ \longmapsto \mathbf{R}_+$ is a nondecreasing continuous function satisfying $\phi(r) < (1 - M)r$ for all $r > 0$.

Theorem 10.17. *Suppose that the conditions* (H1)–(H3) *hold. Then equation* (10.33) *has one and only one positive solution* $u \in \mathbf{C}([a,b])$ *satisfying* $A \le u(t) \le B$ *for all* $t \in [a,b]$.

Proof. Define first the set

$$\mathbf{K} = \{x \in \mathbf{E} : A \le x(t) \le B, \quad t \in [a,b]\}.$$

Then $\mathbf{K}$ is a closed, convex and bounded subset of $\mathbf{E}$. Let $x, y \in K$. We have from (10.34), (10.35) and (H3)

$$(\mathbb{T}x)(t) + (\mathbb{S}y)(t) = f(t, y(t)) + \int_0^t k(t,s)x(s) ds. \tag{10.41}$$

On the other hand,

$$(\mathbb{T}x)(t) + (\mathbb{S}y)(t) = \int_a^t k(t,s)x(s)ds + f(t, y(t)) \geq AM' + A(1 - M') = A. \tag{10.42}$$

It follows from (10.41) and (10.42) that $\mathbb{T}x + \mathbb{S}y \in \mathbf{K}$ for all $x, y \in \mathbf{K}$. Hence, the condition (iii) of Theorem 10.7 is satisfied.

To prove that $\mathbb{T}$ satisfies the hypothesis (i) of Theorem 10.7. It follows from (10.38) that $\mathbb{T} : \mathbf{E} \longmapsto \mathbf{E}$ is a contraction with constant $M < 1$. We see from (10.39) that $(\mathbb{I} - \mathbb{T})$ is Lipschitz invertible with constant $(1 - M)^{-1}$, i. e., the assumption (i) of Theorem 10.7 is fulfilled.

Next, we show that $\mathbb{S}$ is a strictly $(1 - M)$-set contractive map. To this end, let $\mathbf{X}$ be a subset of $\mathbf{K}$ and $x \in \mathbf{X}$. Then, for a given $\epsilon > 0$ and $t, s \in [a, b]$ such that $|t - s| < \epsilon$, without loss of generality, assume that $x(t) \neq x(s)$. Therefore, one derives that

$$\begin{aligned} |(\mathbb{S}x)(t) - (\mathbb{S}x)(s)| &= |f(t, x(t)) - f(s, x(s))| \\ &\leq |f(t, x(t)) - f(t, x(s))| + |f(t, x(s)) - f(s, x(s))| \\ &\leq \phi(|x(t) - x(s)|) + w_f(\epsilon, \cdot), \end{aligned} \tag{10.43}$$

where

$$w_f(\epsilon, \cdot) = \sup\{|f(t, r) - f(s, r)| : t, s \in [a, b], |t - s| < \epsilon \text{ and } r \in [A, B]\}.$$

Notice that ϕ is continuous and nondecreasing. Thus, it follows from (10.43) that

$$w(\mathbb{S}x, \epsilon) \leq \phi(w(x, \epsilon)) + w_f(\epsilon, \cdot). \tag{10.44}$$

Taking into account that the function $f(t, x)$ is uniformly continuous on $[a, b] \times [A, B]$, we conclude that $w_f(\epsilon, \cdot) \to 0$ as $\epsilon \to 0$. Consequently, one deduces from (10.44) that

$$w_0(\mathbb{S}\mathbf{X}) \leq \phi(w_0(\mathbf{X})),$$

which illustrates that $\mathbb{S}$ is a strictly $(1 - M)$-set contractive map. Now, invoking Theorem 10.7 we see that equation (10.33) has at least one solution in $\mathbf{K}$.

Finally, let $u, v \in \mathbf{K}$ be any two solutions of equation (10.33). Then it follows from (10.38) that

$$\begin{aligned} |u(t) - v(t)| &\leq \left| \int_a^t k(t,s)[u(s) - v(s)]ds \right| + |f(t, u(t)) - f(t, v(t))| \\ &\leq M\|u - v\| + |f(t, u(t)) - f(t, v(t))|. \end{aligned} \tag{10.45}$$

Suppose now that there exists $t_0 \in [a, b]$ such that $u(t_0) \neq v(t_0)$. One infers from (10.45) and (H3) that

$$\|u - v\| \leq M\|u - v\| + \phi(\|u - v\|),$$

which is a contraction. This accomplishes the proof. □

Corollary 10.8. *In Theorem* 10.17, *if only* (H3) *is replaced by a generalized assumption* (H3$'$)*: S is a strictly* (1–*M*)*-set contractive map or a γ-set contractive map with* $\gamma < (1-M)$. *Then equation* (10.33) *has at least one solution* $u \in \mathbf{C}([a,b])$ *satisfying* $A \le u(t) \le B$ *for all* $t \in [a,b]$.

Corollary 10.9. *Suppose the condition* (H3) *holds, in addition, if f is bounded on* $[a,b] \times \mathbf{R}$, *then equation* (10.33) *has a unique solution in* $\mathbf{C}([a,b])$.

Let us now investigate the case when $M \ge 1$. To this end, we set

$$p = \min\left\{n \in N : \frac{[c(b-a)]^n}{n!} < 1\right\}.$$

Then one has from (10.40)

$$\rho_p = \begin{cases} \frac{pp!}{p! - [c(b-a)]^p}, & \text{if } M = 1, \\ \frac{(M^p - 1)p!}{(M-1)\{p! - [c(b-a)]^p\}}, & \text{if } M > 1. \end{cases} \tag{10.46}$$

We now assume that the functions concerning equation (10.33) satisfy the following hypotheses:

(H4) there exists $R > 0$ such that $\rho_p f_R \le R$, where $f_R = \sup\{|f(t,y)| : (t,y) \in [a,b] \times [-R,R]\}$;

(H5) for each fixed $t \in [a,b]$ we have

$$|f(t,x) - f(t,y)| \le \phi_p(|x-y|), \quad \forall x, y \in [-R,R],$$

where $\phi_p : \mathbf{R}_+ \longmapsto \mathbf{R}_+$ is a nondecreasing continuous function satisfying $\phi_p(r) < \rho_p^{-1} r$ for all $r > 0$ and ρ_p is defined in (10.46).

By invoking Theorem 10.7, we derive the following result.

Theorem 10.18. *Suppose that the conditions* (H4) *and* (H5) *hold. Then equation* (10.33) *has at least one solution in* $\mathbf{C}([a,b])$.

Proof. For each $y \in \mathbf{E}$, one can see from the above arguments that the equation

$$x = \mathbb{T}x + y$$

has a unique solution in **E**. Now, if $x = \mathbb{T}x + \mathbb{S}y$ with $y \in \mathbf{B}_R$, then one has

$$x(t) = (\mathbb{I} - \mathbb{T})^{-1} f(t, y(t)). \tag{10.47}$$

From (10.39), (10.46), (H4) and (10.47), one can easily deduce that $\|x\| \le \rho_p f_R \le R$, i. e., $x \in \mathbf{B}_R$. The remained arguments are similar to that of Theorem 10.17 and are therefore omitted. □

Corollary 10.10. *If only the condition* (H5) *is interchanged by a generalized assumption* (H5′)*: S is a strictly ρ_p^{-1}-set contractive map or a γ_p-set contractive map with $\gamma_p < \rho_p^{-1}$, then the conclusion of Theorem* 10.18 *is also valid.*

Corollary 10.11. *The conclusion of Theorem* 10.18 *also holds true if instead of* (H4) *we see that f is bounded on $[a,b] \times \mathbf{R}$.*

Remark 10.11. Theorem 10.17 or Corollary 10.9 implies that the Volterra integral equation

$$u(t) = \int_a^t k(t,s)u(s)ds + g(t) \tag{10.48}$$

has a unique solution in $\mathbf{E}$. Thus, under the conditions of Theorem 10.17 or Corollary 10.9, equation (10.33) is a "harmless perturbation" of equation (10.48). However, in other cases, it is not known by the authors whether or not equation (10.33) is still a "harmless perturbation" of equation (10.48). It should be mentioned that the conclusion of Theorem 10.18 might not be obtained by many previously known results because of the condition $M \geq 1$. It might also be noticed that the operator S defined above, generally, is not compact.

Having arrived at the end of this section, it is worthwhile to point out that the abstract techniques and results of the previous sections can be applied to various kinds of other problems which are not investigated here. In particular, by employing Theorem 10.5 or 10.7 or 10.8 or 10.9 or 10.10, the nonlinear integral equation

$$u(t) = \int_a^t k(t,s)g(s,u(s))ds + f(t,u(t)), \quad t \in [a,b],$$

where u takes values in a Banach space $\mathbf{E}$, can be studied totally analogously to Theorems 10.17 and 10.18.

10.9 Application to transport equations

The main aim of this section is to propose an existence result for the radiation transfer equations in channel on $\mathbf{L}^1$ spaces

$$v_3 \frac{\partial \psi}{\partial x}(x,v) + \sigma(x,v)\psi(x,v) - \lambda\psi(x,v) = \int_K r(x,v,v',\psi(x,v'))dv', \quad \text{in } \mathbf{D}, \tag{10.49}$$

where $\mathbf{D} = (0,1)\times\mathbf{K}$, $\mathbf{K}$ is the unit sphere of $\mathbf{R}^3$, $x \in (0,1)$, $v = (v_1, v_2, v_3)$. Equation (10.49) describes the asymptotic behavior of the energy distribution inside the channel in the variables x and v. The unknown function ψ represents the energy density.

The boundary condition

$$\psi^0 := \psi(0,v)_{|_K} = H^1(\psi(1,v)_{|_K}), \psi^1 := \psi(1,v)_{|_K} = H^2(\psi(0,v)_{|_K}), \tag{10.50}$$

describes how the incident energy at the boundary is reflected back inside the domain.

Our assumptions are as follows:

(H1) $H^i : \mathbf{C}(\mathbf{K}) \longrightarrow \mathbf{C}(\mathbf{K}), H^i(0) = 0, |H^i(\phi) - H^i(\psi)| \le q|\phi - \psi|$ on $\mathbf{K}$ for every $\phi, \psi \in \mathbf{C}(\mathbf{K})$, $i = 1, 2$, q is a fixed positive constant,

(H2) $r \in \mathbf{C}((0,1)\times\mathbf{K}\times\mathbf{K}\times\mathbf{C})$, $|r(x,v,v',\psi) - r(x,v,v',\phi)| \le a(x,v')|\phi - \psi|$, $|r(x,v,v',\psi)| \le qa(x,v')$ for every $x \in (0,1)$, $v, v' \in \mathbf{K}$, $\phi, \psi \in \mathbf{C}$, $a, \sigma \in \mathbf{C}(\mathbf{D})$, $\sigma(x,v) = 0$ for every $(x,v) \in \mathbf{D} : v_3 \le \frac{1}{2}$, $r(y,v,v',\psi) = 0$ for every $(y,v,v',\psi) \in (0,1)\times\mathbf{K}\times\mathbf{K}\times\mathbf{C}$ such that $v_3 \le \frac{1}{2}$,

(H3) $q + \sup_{v\in\mathbf{K}} \int_0^1 |\sigma(y,v)|dy + |\lambda| + \int_0^1 \int_{\mathbf{K}} a(y,v)dvdy < \frac{1}{2}$,

$\lambda \in \mathbf{C}$ and ψ is a complex valued unknown function.

The problem for existence of $\mathbf{L}^1$ solutions of (10.49) and (10.50) was open. For a first time there was found an answer to it in [2] in the particular case

$$r(x,v,v',\psi(x,v')) = \xi(x,v,v')f(x,v,\psi(x,v')),$$

where $f : [0,1]\times\mathbf{K}\times\mathbf{C} \longmapsto \mathbf{C}$, $\xi : [0,1]\times\mathbf{K}\times\mathbf{K} \longmapsto \mathbf{R}$ are measurable functions. In [2] are given conditions for ξ and f so that the problem (10.49) and (10.50) has an $\mathbf{L}^1$-solution.

Here we propose a solution of this problem in a more general situation than in [2]. Our result can be considered as an improvement of the result in [2].

To find the answer of the considered problem we will consider the problem

$$v_3\frac{\partial\psi}{\partial x}(\alpha,x,v) + \sigma(x,v)\psi(\alpha,x,v) - \lambda\psi(\alpha,x,v) = \int_K r(x,v,v',\psi(\alpha,x,v'))dv' \tag{10.51}$$

$$\text{in} \quad [0,1]\times\mathbf{D},$$

$\psi(\alpha,x,v)$ is an unknown complex function,

$$\psi(\alpha,\alpha,v) = \psi^\alpha(v) \quad \forall\alpha \in [0,1], v \in \mathbf{K}, \tag{10.52}$$

where

$$\begin{aligned}\psi^\alpha(v) &= (1-\alpha)\psi(0,0,v)_{|_K} + \alpha\psi(1,1,v)_{|_K}\\ &= (1-\alpha)H^1(\psi(1,1,v)_{|_K}) + \alpha H^2(\psi(0,0,v)_{|_K}).\end{aligned}$$

Since every solution of equations (10.51) and (10.52) is a solution of equations (10.49) and (10.50), we will work on (10.51) and (10.52) instead of (10.49) and (10.50).

Theorem 10.19. *We suppose* (H1), (H2) *and* (H3) *hold. Then the problem* (10.49) *and* (10.50) *has a solution* $\psi \in \mathbf{C}(\mathbf{D})$.

Proof. Let $q_1 > q$ be arbitrarily chosen and fixed. Let also $b \in (0,1)$ be arbitrarily chosen and fixed.

Let $\mathbf{E} = \{\psi \in \mathbf{C}([0,1] \times \mathbf{D}) : \psi(\alpha, x, v) = 0 \text{ for } v_3 \leq \frac{1}{2}\}$ be endowed with supremum norm, and let $\mathbf{K}_1 = \{u \in \mathbf{E} : |u| \leq q_1\}$.

For $\psi \in \mathbf{E}$ we define the operators

$$\begin{aligned}
\mathbb{T}\psi(\alpha, x, v) &= (1 + bv_3)\psi(\alpha, x, v),\\
\mathbb{S}\psi(\alpha, x, v) &= -bv_3\psi^{\alpha} + b\int_{\alpha}^{x} \sigma(y, v)\psi(\alpha, y, v)dy - \lambda b \int_{\alpha}^{x} \psi(\alpha, y, v)dy\\
&\quad - b\int_{\alpha}^{x}\int_{K} r(y, v, v', \psi(\alpha, y, v'))dv'dy.
\end{aligned}$$

If $\psi \in \mathbf{K}_1$ is a fixed point of $\mathbb{T} + \mathbb{S}$, putting $x = \alpha$ in $\psi = \mathbb{T}\psi + \mathbb{S}\psi$ we see that ψ satisfies (10.52); and differentiating with respect in x the equality $\psi = \mathbb{T}\psi + \mathbb{S}\psi$ we conclude that the function ψ is a solution of equation (10.51). In particular, it is a solution of the problem (10.49) and (10.50) because $\psi(0, 0, v) = \psi^0$ and $\psi(1, 1, v) = \psi^1$ for $v \in \mathbf{K}$. For $\psi, \phi \in E$ we have

$$\begin{aligned}
|\mathbb{T}\psi(\alpha, x, v) - \mathbb{T}\phi(\alpha, x, v)| &= |(1 + bv_3)(\psi(\alpha, x, v) - \phi(\alpha, x, v))|\\
&\leq (1 + b)|\psi(\alpha, x, v) - \phi(\alpha, x, v)|\\
&\leq (1 + b) \sup_{[0,1]\times D} |\psi(\alpha, x, v) - \phi(\alpha, x, v)|\\
&= (1 + b)\|\psi - \phi\|,
\end{aligned}$$

from which

$$\|\mathbb{T}\psi - \mathbb{T}\phi\| \leq (1 + b)\|\psi - \phi\|.$$

Therefore $\mathbb{T} : \mathbf{E} \longrightarrow \mathbf{E}$ is a Lipschitz operator with a constant $1 + b$. Also, for every $\varphi \in \mathbf{E}$ for the operator $\mathbb{T}_{\varphi} = \mathbb{T} + \varphi$ we have

$$\begin{aligned}
|\mathbb{T}_{\varphi}\psi - \mathbb{T}_{\varphi}\phi| &= |(1 + bv_3)(\psi(\alpha, x, v) - \phi(\alpha, x, v))|\\
&\geq \left(1 + \frac{b}{2}\right)|\psi(\alpha, x, v) - \phi(\alpha, x, v)|,
\end{aligned}$$

for $\phi, \psi \in \mathbf{E}$, and

$$\|\mathbb{T}_{\varphi}\psi - \mathbb{T}_{\varphi}\phi\| \geq \left(1 + \frac{b}{2}\right)\|\psi - \phi\|.$$

Consequently $\mathbb{T}_{\varphi} : \mathbf{E} \longrightarrow \mathbf{E}$ is an expansive operator with a constant $1 + \frac{b}{2}$. For any given $v \in \mathbf{E}$, if we put $\psi = \frac{v - \varphi}{1 + bv_3} \in \mathbf{E}$, then $\mathbb{T}_{\varphi}\psi = v$. This shows that $\mathbb{T}_{\varphi} : \mathbf{E} \longrightarrow \mathbf{E}$ is onto.

For the operator $\mathbb{S}$, we have $\mathbb{S} : \mathbf{K}_1 \longrightarrow \mathbf{E}$, and for $\phi, \psi \in \mathbf{K}_1$, using (H2) we deduce that

$$\begin{aligned}
&|\mathbb{S}\psi(\alpha,x,v) - \mathbb{S}\phi(\alpha,x,v)| \\
&= \Bigg| bv_3(\psi^\alpha - \phi^\alpha) \\
&\quad - b\int_\alpha^x \sigma(y,v)(\psi(\alpha,y,v) - \phi(\alpha,y,v))dy + \lambda b\int_\alpha^x (\psi(\alpha,y,v) - \phi(\alpha,y,v))dy \\
&\quad + b\int_\alpha^x \int_{\mathbf{K}} (r(y,v,v',\psi(\alpha,y,v')) - r(y,v,v',\phi(\alpha,y,v')))dv'dy \Bigg| \\
&\le b\Bigg(q\|\psi - \phi\| + \int_0^1 |\sigma(y,v)||\psi(\alpha,y,v) - \phi(\alpha,y,v)|dy + |\lambda|\int_0^1 |\psi(\alpha,y,v) - \phi(\alpha,y,v)|dy \\
&\quad + \int_0^1\int_K |r(y,v,v',\psi(\alpha,y,v')) - r(y,v,v',\phi(\alpha,y,v'))|dv'dy \Bigg) \\
&\le b\Bigg(q\|\phi - \psi\| + \int_0^1 |\sigma(y,v)||\psi(\alpha,y,v) - \phi(\alpha,y,v)|dy + |\lambda|\int_0^1 |\psi(\alpha,y,v) - \phi(\alpha,y,v)|dy \\
&\quad + \int_0^1\int_{\mathbf{K}} a(y,v')|\psi(\alpha,y,v') - \phi(\alpha,y,v')|dv'dy \Bigg) \\
&\le b\Bigg(q + \int_0^1 |\sigma(y,v)|dy + |\lambda| + \int_0^1\int_{\mathbf{K}} a(y,v')dv'dy \Bigg)\|\psi - \phi\|
\end{aligned}$$

from which it follows

$$\|\mathbb{S}\phi - \mathbb{S}\psi\| \le b\Bigg(q + \sup_{v\in\mathbf{K}}\int_0^1 |\sigma(y,v)|dy + |\lambda| + \int_0^1\int_{\mathbf{K}} a(y,v')dv'dy \Bigg)\|\phi - \psi\|.$$

This, combined with (H3), asserts that $\mathbb{S} : \mathbf{K}_1 \longrightarrow \mathbf{E}$ is a strictly $\frac{b}{2}$- set contractive operator.

Let $\phi \in \mathbf{K}_1$ be fixed. We will show that the equation $\psi = \mathbb{T}\psi + \mathbb{S}\phi$ has a solution $\psi_1 \in \mathbf{K}_1$. Indeed, let $\psi_1(\alpha,x,v) = 0$ for $v_3 \le \frac{1}{2}$ and for $v_3 \ge \frac{1}{2}$

$$\begin{aligned}
\psi_1(\alpha,x,v) = \phi^\alpha - \frac{1}{v_3}\Bigg(&\int_\alpha^x \sigma(y,v)\phi(\alpha,y,v)dy - \lambda\int_\alpha^x \phi(\alpha,y,v)dy \\
&- \int_\alpha^x\int_{\mathbf{K}} r(y,v,v',\phi(\alpha,y,v'))dv'dy \Bigg).
\end{aligned}$$

It follows readily from (H1), (H2) and (H3) follows that $\psi_1 \in \mathbf{E}$. We will show actually that $\psi_1 \in \mathbf{K}_1$. By definition $\psi_1(\alpha, x, v) = 0$ for $v_3 \le \frac{1}{2}$; for $v_3 \ge \frac{1}{2}$, using (H1)–(H3) we infer that

$$\begin{aligned}
|\psi_1(\alpha, x, v)| &= \left|\phi^\alpha - \frac{1}{v_3}\left(\int_\alpha^x \sigma(y, v)\phi(\alpha, y, v)dy + \lambda\int_\alpha^x \phi(\alpha, y, v)dy \right.\right.\\
&\quad \left.\left. + \int_\alpha^x\int_{\mathbf{K}} r(y, v, v', \phi(\alpha, y, v'))dv'dy\right)\right| \\
&\le \frac{1}{v_3}\left(|\phi^\alpha| + \int_0^1 |\sigma(y, v)||\phi(\alpha, y, v)|dy + |\lambda|\int_0^1 |\phi(\alpha, y, v)|dy \right.\\
&\quad \left. + \int_0^1\int_{\mathbf{K}} |r(y, v, v', \phi(\alpha, y, v'))|dv'dy\right) \\
&\le 2\left(qq_1 + q_1\int_0^1 |\sigma(y, v)|dy + |\lambda|q_1 + q_1\int_0^1\int_{\mathbf{K}} a(y, v')dv'dy\right) \\
&< q_1.
\end{aligned}$$

An application of Theorem 2.1 shows the existence of a $\psi \in \mathbf{K}_1$ such that $\psi = \mathbb{T}\psi + \mathbb{S}\psi$. Consequently, the problem (10.51) and (10.52) and hence (10.49) and (10.50) has a solution $\psi \in \mathbf{K}_1$. □

Corollary 10.12. *Let* (H1), (H2) *and* (H3) *hold. Then the problem* (10.49), (10.50) *has a solution* $\psi \in \mathbf{L}^1(\mathbf{D})$.

Proof. The previous theorem ensures that the problem (10.49) and (10.50) has a solution $\psi \in \mathbf{C}(\mathbf{D})$. Since $|\psi| \le q_1$ in $\mathbf{D}$ it follows trivially that $\psi \in \mathbf{L}^1(\mathbf{D})$. □

10.10 Application to a class of difference equations

Here we will consider the difference equation

$$\Delta u(n) = a(n)u(n) - \lambda b(n)f(u(n - \tau(n))) + g(n), \quad n \in \mathbf{Z}, \tag{10.53}$$

where Δ is the difference operator defined by $\Delta u(n) = u(n+1) - u(n)$,
(H1) $a : \mathbf{Z} \longmapsto [0, \infty)$ and $b : \mathbf{Z} \longmapsto (0, \infty)$ are ω-periodic functions for some $\omega > 0$,
(H2) $f : \mathbf{R} \longmapsto \mathbf{R}$ is onto and, there exist $0 < d_1 \le d_2$ such that

$$d_1|u - v| \le |f(u) - f(v)| \le d_2|u - v|, \quad \forall u, v \in \mathbf{R},$$

(H3) $\tau : \mathbf{Z} \longmapsto \mathbf{Z}$ is ω-periodic function, $\mathbb{I} - \tau : \mathbf{Z} \longmapsto \mathbf{Z}$ is onto and $(\mathbb{I} - \tau)^{-1}$ exists.

Here and below we will suppose that the period $\omega > 0$ is arbitrarily chosen and fixed.

We will prove that equation (10.53) has an ω-periodic solution. Our main result is as follows.

Theorem 10.20. *Suppose* (H1)–(H3) *hold. Then equation* (10.53) *has a unique ω-periodic solution whenever*

$$|\lambda| > \frac{2(1+\sup_{n\in\mathbf{Z}} a(n))}{d_1 \inf_{n\in\mathbf{Z}} b(n)}.$$

In addition, if $-\lambda b(n)f(0)+g(n)$ is not identically equal to zero, then equation (10.53) *has an ω-periodic solution which is not identically equal to zero.*

Remark 10.12. Equation (10.53) is investigated in [28] (and the references therein) in the case when $g \equiv 0$ and it is proved that if $f \in \mathbf{C}([0,\infty),[0,\infty))$, $f(s) > 0$ for $s > 0$, $\sum_{n=1}^{\omega} a(n) > 0$, $\sum_{n=1}^{\omega} b(n) > 0$ and there exist $f_0 = \lim_{|s|\longrightarrow 0} \frac{f(s)}{s}$, $f_\infty = \lim_{|s|\longrightarrow\infty} \frac{f(s)}{s}$ and $\frac{1}{\sigma B f_\infty} < \lambda < \frac{1}{A f_0}$ or $\frac{1}{\sigma A f_0} < \lambda < \frac{1}{B f_\infty}$, $A = \max_{n\in\mathbf{Z}} \sum_{s=0}^{\omega-1} G(n,s)b(s)$, $B = \min_{n\in\mathbf{Z}} \sum_{s=0}^{\omega-1} G(n,s)b(s)$, $\sigma = \prod_{i=1}^{\omega}(1+a(i))^{-1}$, equation (10.53) has a positive periodic solution. Here $G(n,s)$ is the corresponding Green function of equation (10.53).

Here we propose new conditions, new range of λ and new approach for investigating of this problem.

Proof. Equation (10.53) can be rewritten in the form

$$u(n) = \frac{1}{1+a(n)}u(n+1) + \frac{\lambda b(n)}{1+a(n)} f(u(n-\tau(n))) - \frac{g(n)}{1+a(n)}.$$

We will work on the periodic function space $\mathbf{E} = \{u : \mathbf{Z} \longmapsto \mathbf{R}, u(n+\omega) = u(n)\}$, endowed with supremum norm. For $u \in \mathbf{E}$, we define the operator

$$\mathbb{T}u(n) = \frac{1}{1+a(n)}u(n+1) + \frac{\lambda b(n)}{1+a(n)} f(u(n-\tau(n))) - \frac{g(n)}{1+a(n)}.$$

Then $\mathbb{T} : \mathbf{E} \longmapsto \mathbf{E}$, and for $u, v \in \mathbf{E}$

$$\begin{aligned}
&\big|\mathbb{T}u(n) - \mathbb{T}v(n)\big| \\
&\leq \frac{1}{1+a(n)}\big|u(n+1) - v(n+1)\big| + \frac{|\lambda| b(n)}{1+a(n)}\big|f(u(n-\tau(n))) - f(v(n-\tau(n)))\big| \\
&\leq \sup_{n\in\mathbf{Z}}\big|u(n) - v(n)\big| + d_2|\lambda| \sup_{n\in\mathbf{Z}} b(n)\big|u(n-\tau(n)) - v(n-\tau(n))\big| \\
&\leq \|u - v\| + d_2|\lambda| \sup_{n\in\mathbf{Z}} b(n) \sup_{n\in\mathbf{Z}}\big|u(n) - v(n)\big| \\
&= \Big(1 + d_2|\lambda| \sup_{n\in\mathbf{Z}} b(n)\Big)\|u - v\|,
\end{aligned}$$

and so

$$\|\mathbb{T}u - \mathbb{T}v\| \leq \Big(1 + d_2|\lambda| \sup_{n\in\mathbf{Z}} b(n)\Big)\|u - v\|.$$

Now, let $y \in \mathbf{E}$ be fixed. For $u \in \mathbf{E}$, we define the operator

$$\mathbb{T}_y u(n) = \frac{1}{1+a(n)}u(n+1) + \frac{\lambda b(n)}{1+a(n)}f(u(n-\tau(n))) - \frac{g(n)}{1+a(n)} + y(n).$$

Then $\mathbb{T}_y : \mathbf{E} \longmapsto \mathbf{E}$ and for $u, v \in \mathbf{E}$

$$\left|\mathbb{T}_y u(n) - \mathbb{T}_y v(n)\right| \geq \frac{d_1|\lambda| \inf_{n\in\mathbf{Z}} b(n)}{1+\sup_{n\in\mathbf{Z}} a(n)}\left|u(n-\tau(n)) - v(n-\tau(n))\right| - \|u-v\|,$$

i. e.,

$$\left\|\mathbb{T}_y u(n) - \mathbb{T}_y v(n)\right\| \geq \frac{d_1|\lambda| \inf_{n\in\mathbf{Z}} b(n)}{1+\sup_{n\in\mathbf{Z}} a(n)} \sup_{n\in\mathbf{Z}}\left|u(n-\tau(n)) - v(n-\tau(n))\right| - \|u-v\|.$$

Recalling that $\mathbb{I} - \tau : \mathbf{Z} \longmapsto \mathbf{Z}$ is onto, we obtain

$$\frac{\|\mathbb{T}_y u - \mathbb{T}_y v\| \geq (d_1|\lambda| \inf_{n\in\mathbf{Z}} b(n)}{1+\sup_{n\in\mathbf{Z}} a(n) - 1)\|u-v\|.} \tag{10.54}$$

Because $\frac{d_1|\lambda| \inf_{n\in\mathbf{Z}} b(n)}{1+\sup_{n\in\mathbf{Z}} a(n)} - 1 > 1$, we conclude that $\mathbb{T}_y : \mathbf{E} \longmapsto \mathbf{E}$ is expansive.

Let $y_1 \in \mathbf{E}$ be fixed. We consider the equation

$$\frac{1}{1+a(n)}u(n+1) + \frac{\lambda b(n)}{1+a(n)}f(u(n-\tau(n))) - \frac{g(n)}{1+a(n)} + y(n) = y_1(n)$$

or

$$\begin{aligned} u(n) = & -\lambda b(n) f(u(n-1-\tau(n-1))) + g(n-1) \\ & - (1+a(n-1))y(n-1) + (1+a(n-1))y_1(n-1). \end{aligned}$$

For $u \in \mathbf{E}$, we define the operator

$$\begin{aligned} \mathbb{T}_y^1 u(n) = & -\lambda b(n-1) f(u(n-1-\tau(n-1))) + g(n-1) \\ & - (1+a(n-1))y(n-1) + (1+a(n-1))y_1(n-1). \end{aligned}$$

Then for $u, v \in \mathbf{E}$ we have

$$\left\|\mathbb{T}_y^1 u - \mathbb{T}_y^1 v\right\| \leq |\lambda| d_2 \sup_{n\in\mathbf{Z}} b(n)\|u-v\|.$$

Consequently $\mathbb{T}_y^1 : \mathbf{E} \longmapsto \mathbf{E}$ is $|\lambda| d_2 \sup_{n\in Z} b(n)$-Lipschitz operator.

Let $y_2 \in \mathbf{E}$ be fixed. For $u \in \mathbf{E}$ we define the operator

$$\begin{aligned} \mathbb{T}_{yy_2}^1 u = & -\lambda b(n-1) f(u(n-1-\tau(n-1))) + g(n-1) \\ & - (1+a(n-1))y(n-1) + (1+a(n-1))y_1(n-1) + y_2(n). \end{aligned}$$

Then as above one has

$$\|\mathbb{T}^1_{yy_2}u - \mathbb{T}^1_{yy_2}v\| \geq d_1|\lambda| \inf_{n\in\mathbf{Z}} b(n)\|u - v\|,$$

which implies that $\mathbb{T}^1_{yy_2} : \mathbf{E} \longmapsto \mathbf{E}$ is expansive since $d_1|\lambda| \inf_{n\in\mathbf{Z}} b(n) > 1$.

We now claim $\mathbb{T}^1_{yy_2} : \mathbf{E} \longmapsto \mathbf{E}$ is onto. To see this, for any $y_3 \in \mathbf{E}$, we consider the equation

$$\begin{aligned}&-\lambda b(n-1)f(u(n-1-\tau(n-1))) + g(n-1)\\ &\quad - (1+a(n-1))y(n-1) + (1+a(n-1))y_1(n-1) + y_2(n) = y_3(n),\end{aligned}$$

or

$$\begin{aligned}f(u(n-1-\tau(n-1))) = {}&\frac{1}{\lambda b(n-1)}(g(n-1)\\ &- (1+a(n-1))(y(n-1) - y_1(n-1)) + y_2(n) - y_3(n)).\end{aligned}$$

By the assumptions (H2) and (H3), this equation is solvable and its unique solution is given explicitly by

$$\begin{aligned}u(n) = f^{-1}\Big(&\frac{1}{\lambda b((I-\tau)^{-1}(n))}(g((I-\tau)^{-1}(n)) - (1 + a((I-\tau)^{-1}(n))(y((I-\tau)^{-1}(n))\\ &- y_1((I-\tau)^{-1}(n))) + y_2((I-\tau)^{-1}n+1) - y_3((I-\tau)^{-1}n+1)))\Big) \in \mathbf{E},\end{aligned}$$

proving the claim. From this and Lemma 10.3 it follows that $\mathbb{I} - \mathbb{T}^1_y : \mathbf{E} \longmapsto \mathbf{E}$ is onto. In particular, $\mathbb{T}^1_y$ has a fixed point u_2 in $\mathbf{E}$; the definition of $\mathbb{T}^1_y$ then gives u_2 is a solution of $\mathbb{T}_y u = y_1$. This implies that $\mathbb{T}_y : \mathbf{E} \longmapsto \mathbf{E}$ is onto. Now, Lemma 10.3 again shows that the operator $\mathbb{I} - \mathbb{T} : \mathbf{E} \longmapsto \mathbf{E}$ is onto. Therefore, there exists $u_3 \in \mathbf{E}$ such that

$$(\mathbb{I} - \mathbb{T})u_3 = 0,$$

and thus equation (10.53) has a solution $u_3 \in \mathbf{E}$. The uniqueness of solution (10.53) follows from (10.54). Finally, assume that $-\lambda b(n)f(0) + g(n)$ is not identically equal to zero. If we suppose that $u_3 \equiv 0$ then

$$-\lambda b(n)f(0) + g(n) = 0 \quad \text{for} \quad \forall n \in \mathbf{Z},$$

which is a contradiction. □

Now we will consider the case when $g \equiv 0$, more precisely we will consider the problem

$$\begin{aligned}\Delta u(n) &= a(n)u(n) - \lambda b(n)f(u(n-\tau(n))), \quad n \in \mathbf{Z},\\ u(n+\omega) &= u(n), \quad \forall n \in \mathbf{Z},\end{aligned} \tag{10.55}$$

where a and b satisfy (H1), τ satisfies (H3) and f satisfies

(H4) $f : \mathbf{E} \longmapsto \mathbf{E}, f : \mathbf{K} \longmapsto \mathbf{K}_1$ is onto and

$$|f(u) - f(v)| \geq d_3|u - v| \quad \forall u, v \in \mathbf{E},$$

for some positive constant d_3.

Here

$$\mathbf{K} = \{u \in \mathbf{E} : 0 < q_1 \leq u(n) \leq q_2 \quad \forall n \in \mathbf{Z}\},$$
$$\mathbf{K}_1 = \left\{u \in \mathbf{E} : -\frac{q_2}{\inf_{n\in\mathbf{Z}} b(n)} \leq u(n) \leq \frac{2q_2}{\inf_{n\in\mathbf{Z}} b(n)} \quad \forall n \in \mathbf{Z}\right\},$$

for some positive constants q_1 and q_2, $q_1 < q_2$.

Theorem 10.21. *Suppose* (H1), (H3) *and* (H4) *hold. Then there exists* $\lambda_1 \geq 0$ *such that the problem* (10.55) *is solvable for all* $\lambda \geq \lambda_1$.

Remark 10.13. We note that here in our result we have not condition for f to be continuous function as in [28].

Proof. For $u \in \mathbf{E}$ we define the operators

$$Su(n) = \frac{1}{1 + a(n)} u(n + 1),$$
$$Tu(n) = \frac{b(n)}{1 + a(n)} f(u(n - \tau(n))).$$

Firstly we will note that if $u \in \mathbf{E}$ is a fixed point of the operator $\mathbb{S} + \lambda\mathbb{T}$ then u is a solution of the problem (10.55). Indeed,

$$\begin{aligned}
u(n) &= \mathbb{S}u(n) + \lambda\mathbb{T}u(n) \iff \\
u(n) &= \frac{u(n+1)}{1 + a(n)} + \lambda\frac{b(n)}{1 + a(n)} f(u(n - \tau(n))) \iff \\
(1 + a(n))u(n) &= u(n + 1) + \lambda b(n) f(u(n - \tau(n))) \iff \\
a(n)u(n) &= u(n + 1) - u(n) + \lambda b(n) f(u(n - \tau(n)) \iff \\
\Delta u(n) &= a(n)u(n) - \lambda b(n) f(u(n - \tau(n))).
\end{aligned}$$

Also, $\mathbb{S} : \mathbf{K} \longmapsto \mathbf{E}$, $\mathbb{T} : \mathbf{E} \longmapsto \mathbf{E}$ and for $u_1, u_2 \in \mathbf{E}$ we have

$$\begin{aligned}
|\mathbb{T}u_1(n) - \mathbb{T}u_2(n)| &= \left|\frac{b(n)}{1 + a(n)} f(u_1(n - \tau(n))) - \frac{b(n)}{1 + a(n)} f(u_2(n - \tau(n)))\right| \\
&= \left|\frac{b(n)}{1 + a(n)} (f(u_1(n - \tau(n))) - f(u_2(n - \tau(n))))\right| \\
&= \frac{b(n)}{1 + a(n)} |f(u_1(n - \tau(n))) - f(u_2(n - \tau(n)))| \\
&\geq d_3 \frac{\inf_{n\in\mathbf{Z}} b(n)}{1 + \sup_{n\in\mathbf{Z}} a(n)} |u_1(n - \tau(n)) - u_2(n - \tau(n))|, \quad \forall n \in \mathbf{Z},
\end{aligned}$$

i. e.,

$$\left|\mathbb{T}u_1(n) - \mathbb{T}u_2(n)\right| \geq d_3 \frac{\inf_{n\in\mathbf{Z}} b(n)}{1 + \sup_{n\in\mathbf{Z}} a(n)} \left|u_1(n - \tau(n)) - u_2(n - \tau(n))\right|, \quad \forall n \in \mathbf{Z},$$

from this

$$\sup_{n\in\mathbf{Z}}\left|\mathbb{T}u_1(n) - \mathbb{T}u_2(n)\right| \geq d_3 \frac{\inf_{n\in\mathbf{Z}} b(n)}{1 + \sup_{n\in\mathbf{Z}} a(n)} \sup_{n\in\mathbf{Z}}\left|u_1(n - \tau(n)) - u_2(n - \tau(n))\right|,$$

and since τ satisfies (H3) from the previous inequality we get

$$\|\mathbb{T}u_1 - \mathbb{T}u_2\| \geq d_3 \frac{\inf_{n\in\mathbf{Z}} b(n)}{1 + \sup_{n\in\mathbf{Z}} a(n)} \|u_1 - u_2\|. \tag{10.56}$$

Consequently $\mathbb{T} : \mathbf{E} \longmapsto \mathbf{E}$ is a weakly expansive operator with constant

$$d_3 = \frac{\inf_{n\in\mathbf{Z}} b(n)}{1 + \sup_{n\in\mathbf{Z}} a(n)}.$$

For $u_1, u_2 \in \mathbf{K}$ we have

$$\begin{aligned}\left|\mathbb{S}u_1(n) - \mathbb{S}u_2(n)\right| &= \left|\frac{1}{1 + a(n)} u_1(n+1) - \frac{1}{1 + a(n)} u_2(n+1)\right| \\ &= \frac{1}{1 + a(n)} \left|u_1(n+1) - u_2(n+1)\right| \\ &\leq \left|u_1(n+1) - u_2(n+1)\right| \quad \forall n \in \mathbf{Z},\end{aligned}$$

whereupon

$$\sup_{n\in\mathbf{Z}}\left|\mathbb{S}u_1(n) - \mathbb{S}u_2(n)\right| \leq \sup_{n\in\mathbf{Z}}\left|u_1(n+1) - u_2(n+1)\right|$$

or

$$\|\mathbb{S}u_1 - \mathbb{S}u_2\| \leq \|u_1 - u_2\|,$$

therefore $\mathbb{S} : \mathbf{K} \longmapsto \mathbf{E}$ is a 1-set contractive operator.

Let now $\lambda_0 > 1$ be fixed so that

$$\lambda_0 \frac{\inf_{n\in\mathbf{Z}} b(n)}{1 + \sup_{n\in\mathbf{Z}} a(n)} d_3 > 1$$

and $\lambda \geq \lambda_0$ be arbitrarily chosen and fixed.

We fix $v \in \mathbf{K}$ and we consider on $\mathbf{K}$ the equation

$$u(n) = \lambda \mathbb{T}u(n) + \mathbb{S}v(n).$$

Let

$$\mathbb{T}_1 u(n) = \lambda \mathbb{T} u(n) + \mathbb{S} v(n), \quad u \in \mathbf{K}.$$

We have $\mathbb{T}_1 : \mathbf{K} \longmapsto \mathbf{E}$ and for $u_1, u_2 \in \mathbf{K}$, using (10.56),

$$\begin{aligned}\|\mathbb{T}_1 u_1 - \mathbb{T}_1 u_2\| &= \|\lambda \mathbb{T} u_1 - \lambda \mathbb{T} u_2\| \\ &= \lambda \|\mathbb{T} u_1 - \mathbb{T} u_2\| \\ &\geq \lambda_0 d_3 \frac{\inf_{n \in \mathbf{Z}} b(n)}{1 + \sup_{n \in \mathbf{Z}}} \|u_1 - u_2\|,\end{aligned}$$

from this and our choice of λ_0 we conclude that $\mathbb{T}_1 : \mathbf{K} \longmapsto \mathbf{E}$ is an expansive operator, also we will note that $\mathbf{K}$ is a closed subset of $\mathbf{E}$.

Let $v_1 \in \mathbf{K}$. For $u \in \mathbf{K}$ we will consider the equation

$$\begin{aligned} v_1(n) &= \mathbb{T}_1 u(n) \quad \Longleftrightarrow \\ v_1(n) &= \lambda b(n) f(u(n - \tau(n))) + \mathbb{S} v(n) \quad \Longleftrightarrow \\ v_1(n) - \mathbb{S} v(n) &= \lambda b(n) f(u(n - \tau(n))) \quad \Longleftrightarrow \\ f(u(n - \tau(n))) &= \frac{v_1(n) - \mathbb{S} v(n)}{\lambda b(n)} \quad \Longleftrightarrow \\ f(u(n)) &= \frac{v_1((\mathbb{I} - \tau)^{-1} n) - \mathbb{S} v((\mathbb{I} - \tau)^{-1} n)}{\lambda b((I - \tau)^{-1} n)}. \end{aligned} \tag{10.57}$$

For

$$\frac{v_1((\mathbb{I} - \tau)^{-1} n) - \mathbb{S} v((\mathbb{I} - \tau)^{-1} n)}{\lambda b((\mathbb{I} - \tau)^{-1} n)}$$

we have the following estimates

$$\begin{aligned} \frac{v_1((\mathbb{I} - \tau)^{-1} n) - \mathbb{S} v((\mathbb{I} - \tau)^{-1} n)}{\lambda b((\mathbb{I} - \tau)^{-1} n)} &\leq \frac{q_2 + \frac{v((\mathbb{I} - \tau)^{-1}(n) + 1)}{1 + a((\mathbb{I} - \tau)^{-1}(n))}}{\lambda_0 \inf_{n \in \mathbf{Z}} b(n)} \\ &\leq \frac{2 q_2}{\lambda_0 \inf_{n \in Z} b(n)} \leq \frac{2 q_2}{\inf_{n \in \mathbf{Z}} b(n)}, \\ \frac{v_1((\mathbb{I} - \tau)^{-1} n) - \mathbb{S} v((\mathbb{I} - \tau)^{-1} n)}{\lambda b((\mathbb{I} - \tau)^{-1} n)} &\geq - \frac{\mathbb{S} v((\mathbb{I} - \tau)^{-1} n)}{\lambda b((\mathbb{I} - \tau)^{-1} n)} \\ &\geq - \frac{q_2}{\lambda_0 \inf_{n \in \mathbf{Z}} b(n)} \\ &\geq - \frac{q_2}{\inf_{n \in \mathbf{Z}} b(n)}, \end{aligned}$$

therefore

$$\frac{v_1((\mathbb{I} - \tau)^{-1} n) - \mathbb{S} v((\mathbb{I} - \tau)^{-1} n)}{\lambda b((\mathbb{I} - \tau)^{-1} n)} \in \mathbf{K}_1$$

and since $f : \mathbf{K} \longmapsto \mathbf{K}_1$ is onto, there exists $u \in \mathbf{K}$ so that (10.57) holds. Consequently there exists $u \in \mathbf{K}$ such that $v_1(n) = \mathbb{T}_1 u(n) \in \mathbb{T}_1(\mathbf{K})$ for every $n \in \mathbf{Z}$. Since v_1 was arbitrarily chosen we conclude that $\mathbf{K} \subset \mathbb{T}_1(\mathbf{K})$. From this and Lemma 10.1 follows that $\mathbb{T}_1$ has unique fixed point $u_3 \in \mathbf{K}$, $u_3 = \mathbb{T}_1 u_3$ or

$$u_3(n) = \lambda \mathbb{T} u_3(n) + \mathbb{S} v(n) \quad \forall n \in \mathbf{Z}.$$

Because $\lambda \geq \lambda_0$ was arbitrarily chosen, then from Theorem 3.2 we conclude that there exists $\lambda_1 \geq 0$ such that the problem (10.55) is solvable for every $\lambda \geq \lambda_1$. □

Now we will consider the problem

$$\begin{cases} \Delta u(n) = a(n)u(n) - \lambda b(n) f(u(n - \tau(n))) + \mu u(n), & n \in \mathbf{Z}, \\ u(n + \omega) = u(n) & \forall n \in \mathbf{Z}, \end{cases} \tag{10.58}$$

where a and b satisfy (H1), τ satisfies (H3) and f satisfy

(H5) $f : \mathbf{E} \longmapsto \mathbf{E}$, $|f(u)| \leq Q$ for every $u \in \mathbf{E}$,

$$|f(u) - f(v)| \leq d_4 |u - v| \quad \text{for} \quad \forall u, v \in \mathbf{E},$$

for some positive constants Q and d_4,

(H6) $\lambda > 0$ is a parameter for which

$$\lambda d_4 \sup_{n \in \mathbf{Z}} b(n) \leq 1,$$

μ is a positive parameter.

Theorem 10.22. *Suppose* (H1), (H3), (H5) *and* (H6) *hold. Then there exists* $\mu_1 > 1$ *such that for every* $\mu \geq \mu_1$ *the problem* (10.58) *is solvable for every* $\mu \geq \mu_1$.

Proof. We define the set

$$\mathbf{K}_2 = \{u \in \mathbf{E} : |u(n)| \leq Q \quad \text{for} \quad \forall n \in \mathbf{Z}\}.$$

For $u \in \mathbf{E}$ we define the operators

$$\mathbb{S}u(n) = u(n+1) - (1 + a(n))u(n),$$
$$\mathbb{T}u(n) = \lambda b(n) f(u(n - \tau(n))).$$

We will note that $\mathbb{S} : \mathbf{K}_2 \longmapsto \mathbf{E}$, $\mathbb{T} : \mathbf{E} \longmapsto \mathbf{E}$ and for $u_1, u_2 \in \mathbf{E}$ we have

$$\begin{aligned} |\mathbb{T}u_1(n) - \mathbb{T}u_2(n)| &= |\lambda b(n) f(u_1(n - \tau(n))) - \lambda b(n) f(u_2(n - \tau(n)))| \\ &= |\lambda b(n)(f(u_1(n - \tau(n))) - f(u_2(n - \tau(n))))| \\ &= \lambda b(n) |f(u_1(n - \tau(n))) - f(u_2(n - \tau(n)))| \\ &\leq \lambda d_4 b(n) |u_1(n - \tau(n)) - u_2(n - \tau(n))| \quad \text{for} \quad \forall n \in \mathbf{Z}, \end{aligned}$$

from this

$$\sup_{n\in\mathbf{Z}} \left|\mathbb{T}u(n) - \mathbb{T}u_2(n)\right| \le \lambda d_4 \sup_{n\in Z} b(n) \sup_{n\in\mathbf{Z}} \left|u_1(n-\tau(n)) - u_2(n-\tau(n))\right|$$

or

$$\|\mathbb{T}u_1 - \mathbb{T}u_2\| \le \lambda d_4 \sup_{n\in\mathbf{Z}} b(n)\|u_1 - u_2\| \le \|u_1 - u_2\|.$$

Therefore $\mathbb{T} : \mathbf{E} \longmapsto \mathbf{E}$ is a non-expansive operator.

Let $\mu_1 > 1$ be chosen so that

$$\mu_1 > 3 + \sup_{n\in\mathbf{Z}} a(n), \quad \mu_1 > \left(2 + \sup_{n\in\mathbf{Z}} a(n) + \lambda \sup_{n\in\mathbf{Z}} b(n)\right).$$

Let also $\mu \ge \mu_1$ and $v \in \mathbf{K}_2$ be fixed.

We consider the equation

$$\mu u(n) = \mathbb{T}u(n) + \mathbb{S}v(n)$$

for $u \in \mathbf{K}_2$ or

$$u(n) = \frac{1}{\mu}\lambda b(n) f(u(n-\tau(n))) + \frac{1}{\mu}\mathbb{S}v(n)$$

for $u \in \mathbf{K}_2$.

Let

$$\mathbb{T}_1 u(n) = \frac{1}{\mu}(\lambda b(n) f(u(n-\tau(n))) + Sv(n))$$

for $u \in \mathbf{K}_2$. Then

$$\begin{aligned}
\left|\mathbb{T}_1 u(n)\right| &\le \frac{1}{\mu_1}(\lambda \sup_{n\in\mathbf{Z}} b(n)\left|f(u(n-\tau(n))\right| + \left|v(n+1)\right| + (1+a(n))\left|v(n)\right|) \\
&\le \frac{1}{\mu_1}\left(\lambda \sup_{n\in\mathbf{Z}} b(n) Q + \left(2 + \sup_{n\in\mathbf{Z}} a(n)\right) Q\right) \\
&= \frac{1}{\mu_1}\left(\lambda \sup_{n\in\mathbf{Z}} b(n) + \left(2 + \sup_{n\in\mathbf{Z}} a(n)\right)\right) Q \\
&\le Q,
\end{aligned}$$

therefore $\mathbb{T}_1 : \mathbf{K}_2 \longmapsto \mathbf{K}_2$.

Also, for $u_1, u_2 \in \mathbf{K}_2$

$$\begin{aligned}
\left|\mathbb{T}_1 u_1(n) - \mathbb{T}_1 u_2(n)\right| &= \frac{1}{\mu}\lambda b(n)\left|f(u_1(n-\tau(n))) - f(u_2(n-\tau(n)))\right| \\
&\le \frac{1}{\mu}\lambda b(n) d_4 \left|u_1(n-\tau(n)) - u_2(n-\tau(n))\right| \quad \text{for} \quad \forall n \in Z,
\end{aligned}$$

consequently

$$\sup_{n\in\mathbf{Z}}\left|\mathbb{T}_1u_1(n) - \mathbb{T}_1u_2(n)\right| \le d_4\frac{\lambda}{\mu_1}\sup_{n\in\mathbf{Z}} b(n)\left|u_1(n-\tau(n)) - u_2(n-\tau(n))\right|$$

or

$$\begin{aligned}\|\mathbb{T}_1u_1 - \mathbb{T}_1u_2\| &\le \frac{\lambda}{\mu_1}d_4\sup_{n\in Z} b(n)\|u_1-u_2\| \\ &\le \frac{1}{\mu_1}\|u_1-u_2\|.\end{aligned}$$

In other words the operator $\mathbb{T}_1 : \mathbf{K}_2 \longmapsto \mathbf{K}_2$ is a contractive operator. Therefore there exists a unique $u \in \mathbf{K}_2$ such that $u = \mathbb{T}_1u$ or there exists a unique $u \in \mathbf{K}_2$ so that $\mu u(n) = \mathbb{T}u(n) + \mathbb{S}v(n)$.

Let $u_1, u_2 \in \mathbf{K}_2$. Then

$$\begin{aligned}\left|\mathbb{S}u_1(n) - \mathbb{S}u_2(n)\right| &= \left|u_1(n+1) - (1+a(n))u_1(n) - u_2(n+1) + (1+a(n))u_2(n)\right| \\ &= \left|(u_1(n+1) - u_2(n+1)) - (1+a(n))(u_1(n) - u_2(n))\right| \\ &\le \left|u_1(n+1) - u_2(n+1)\right| + (1+a(n))\left|u_1(n) - u_2(n)\right| \\ &\le \left(2 + \sup_{n\in\mathbf{Z}} a(n)\right)\|u_1 - u_2\| \quad \forall n \in \mathbf{Z},\end{aligned}$$

from which

$$\|\mathbb{S}u_1 - \mathbb{S}u_2\| \le \left(2 + \sup_{n\in\mathbf{Z}} a(n)\right)\|u_1 - u_2\|$$

and since $2 + \sup_{n\in\mathbf{Z}} a(n) < \mu_1 - 1$ then the operator $\mathbb{S} : \mathbf{K}_2 \longmapsto \mathbf{E}$ is $k = 2 + \sup_{n\in\mathbf{Z}} a(n) < \mu_1 - 1$-set contractive operator.

From this and Theorem 3.3 follows that for every $\mu \ge \mu_1$ the problem (10.58) is solvable. □

Now we will consider the problem (10.58) in the case when

(H7) $a, b : \mathbf{Z} \longmapsto (0, \infty)$, $a(n+\omega) = a(n)$, $b(n+\omega) = b(n)$ for every $n \in \mathbf{Z}$, $\inf_{n\in\mathbf{Z}} a(n) > 1$,

(H8) $f : \mathbf{E} \longmapsto \mathbf{E}$, $0 \le f(u) \le Q_1$ for every $u \in \mathbf{E}$,

$$\left|f(u) - f(v)\right| \le d_5|u - v| \quad \text{for} \quad \forall u, v \in \mathbf{E},$$

for some positive constants Q_1 and d_5,

(H9) $\lambda > 0$ is a parameter for which

$$\lambda\sup_{n\in\mathbf{Z}} b(n)d_5 < 1, \quad \lambda\sup_{n\in\mathbf{Z}} b(n)Q_1 < q_3\inf_{n\in\mathbf{Z}} a(n),$$

for some positive constant q_3,

(H10) $\tau : \mathbf{Z} \longmapsto \mathbf{Z}$, $\tau(n+\omega) = \tau(n)$ for every $n \in \mathbf{Z}$.

Theorem 10.23. *Let* (H7), (H8), (H9) *and* (H10) *hold. Then there exists* $\mu \in (0,1)$ *for which the problem* (10.58) *is solvable.*

Proof. Let

$$\mathbf{K}_3 = \{u \in \mathbf{E} : 0 \le u(n) \le q_3 \quad \text{for} \quad \forall n \in \mathbf{Z}\}.$$

For $u \in \mathbf{E}$ we define the operators

$$\begin{aligned} \mathbb{S}u(n) &= \lambda b(n) f(u(n-\tau(n))), \\ \mathbb{T}u(n) &= u(n+1) - (1+a(n))u(n). \end{aligned}$$

We note that $\mathbb{S} : \mathbf{K}_3 \longmapsto \mathbf{E}$, $\mathbb{T} : \mathbf{E} \longmapsto \mathbf{E}$.

For $u_1, u_2 \in \mathbf{E}$ we have

$$\begin{aligned} |\mathbb{T}u_1(n) - \mathbb{T}u_2(n)| &= |u_1(n+1) - (1+a(n))u_1(n) - u_2(n+1) + (1+a(n))u_2(n)| \\ &= |-(1+a(n))(u_1(n) - u_2(n)) + (u_1(n+1) - u_2(n+1))| \\ &\ge (1+a(n))|u_1(n) - u_2(n)| - |u_1(n+1) - u_2(n+1)| \\ &\ge \Big(1 + \inf_{n\in\mathbf{Z}} a(n)\Big)|u_1(n) - u_2(n)| - \|u_1 - u_2\| \quad \text{for} \quad \forall n \in \mathbf{Z}, \end{aligned}$$

i. e.,

$$|\mathbb{T}u_1(n) - \mathbb{T}u_2(n)| \ge \Big(1 + \inf_{n\in\mathbf{Z}} a(n)\Big)|u_1(n) - u_2(n)| - \|u_1 - u_2\| \quad \text{for} \quad \forall n \in \mathbf{Z},$$

from this

$$\sup_{n\in\mathbf{Z}}|\mathbb{T}u_1(n) - \mathbb{T}u_2(n)| \ge \Big(1 + \inf_{n\in\mathbf{Z}} a(n)\Big)|u_1(n) - u_2(n)| - \|u_1 - u_2\| \quad \text{for} \quad \forall n \in \mathbf{Z},$$

or

$$\|\mathbb{T}u_1 - \mathbb{T}u_2\| \ge \Big(1 + \inf_{n\in\mathbf{Z}} a(n)\Big)|u_1(n) - u_2(n)| - \|u_1 - u_2\| \quad \text{for} \quad \forall n \in \mathbf{Z}.$$

Therefore

$$\|\mathbb{T}u_1 - \mathbb{T}u_2\| \ge \Big(1 + \inf_{n\in\mathbf{Z}} a(n)\Big) \sup_{n\in\mathbf{Z}}|u_1(n) - u_2(n)| - \|u_1 - u_2\|,$$

or

$$\|\mathbb{T}u_1 - \mathbb{T}u_2\| \ge \Big(1 + \inf_{n\in\mathbf{Z}} a(n)\Big)\|u_1 - u_2\| - \|u_1 - u_2\| = \inf_{n\in\mathbf{Z}} \|u_1 - u_2\|$$

and since $\inf_{n\in\mathbf{Z}} a(n) > 1$, from the previous inequality we conclude that the operator $\mathbb{T} : \mathbf{E} \longmapsto \mathbf{E}$ is a non-contractive one.

Let $\mu \in (0,1)$ be fixed so that

$$\mu < 1 - \lambda \sup_{n\in\mathbf{Z}} b(n) d_5.$$

Also, for $u_1, u_2 \in \mathbf{K}_3$ we have

$$\begin{aligned}|\mathbb{S}u_1(n) - \mathbb{S}u_2(n)| &= \lambda b(n)|f(u_1(n-\tau(n))) - f(u_2(n-\tau(n)))| \\ &\le \lambda \sup_{n\in\mathbf{Z}} b(n) d_5 |u_1(n-\tau(n)) - u_2(n-\tau(n))| \\ &\le \lambda \sup_{n\in\mathbf{Z}} b(n) d_5 \|u_1 - u_2\| \quad \text{for} \quad \forall n \in \mathbf{Z},\end{aligned}$$

from which

$$\sup_{n\in\mathbf{Z}} |\mathbb{S}u_1(n) - \mathbb{S}u_2(n)| \le \lambda \sup_{n\in\mathbf{Z}} b(n) d_5 \|u_1 - u_2\|$$

or

$$\|\mathbb{S}u_1 - \mathbb{S}u_2\| \le \lambda \sup_{n\in\mathbf{Z}} b(n) d_5 \|u_1 - u_2\|.$$

Therefore $\mathbb{S} : \mathbf{K}_3 \longmapsto \mathbf{E}$ is a $\mu_1 = \lambda \sup_{n\in\mathbf{Z}} b(n) d_5 < 1 - \mu$-set contractive map.

Let $v \in \mathbf{K}_3$ be fixed. We consider the equation

$$\begin{aligned}\mu u(n) &= \mathbb{T}u(n) + \mathbb{S}v(n) \quad \text{for} \quad u \in \mathbf{K}_3 \quad \text{or} \\ \mu u(n) &= u(n+1) - (1 + a(n))u(n) + \lambda b(n) f(v(n-\tau(n))) \quad \text{for} \quad u \in \mathbf{K}_3 \quad \text{or} \\ (1 + \mu + a(n))u(n) &= u(n+1) + \lambda b(n) f(v(n-\tau(n))) \quad \text{for} \quad u \in \mathbf{K}_3 \quad \text{or} \\ u(n) &= \frac{1}{1+\mu+a(n)} (u(n+1) + \lambda b(n) f(v(n-\tau(n)))) \quad \text{for} \quad u \in \mathbf{K}_3.\end{aligned}$$

For $u \in \mathbf{K}_3$ we define the operator

$$\mathbb{T}_2 u(n) = \frac{1}{1+\mu+a(n)} (u(n+1) + \lambda b(n) f(v(n-\tau(n)))).$$

For $u \in \mathbf{K}_3$ we have

$$\mathbb{T}_2 u(n) \ge 0 \quad \text{for} \quad \forall n \in \mathbf{Z}$$

and

$$\mathbb{T}_2 u(n) \le \frac{q_3 + \lambda \sup_{n\in\mathbf{Z}} b(n) Q_1}{1 + \mu + \inf_{n\in\mathbf{Z}} a(n)} \quad \text{for} \quad \forall n \in \mathbf{Z}. \tag{10.59}$$

We have

$$\begin{cases} \frac{q_3 + \lambda \sup_{n\in\mathbf{Z}} b(n) Q_1}{1+\mu+\inf_{n\in\mathbf{Z}} a(n)} \le q_3 \quad \Longleftrightarrow \\ q_3 + \lambda \sup_{n\in\mathbf{Z}} b(n) Q_1 \le q_3 + q_3\mu + q_3 \inf_{n\in\mathbf{Z}} a(n) \quad \Longleftrightarrow \\ \lambda \sup_{n\in\mathbf{Z}} b(n) Q_1 \le q_3\mu + q_3 \inf_{n\in\mathbf{Z}} a(n), \end{cases} \tag{10.60}$$

which is true because (H9). From this and (10.59) follows that

$$\mathbb{T}_2 u(n) \le q_3 \quad \text{for} \quad \forall n \in \mathbf{Z}.$$

Therefore $\mathbb{T}_2 : \mathbf{K}_3 \longmapsto \mathbf{K}_3$.

For $u_1, u_2 \in \mathbf{K}_3$ we have

$$\begin{aligned}|\mathbb{T}_2 u_1(n) - \mathbb{T}_2 u_2(n)| &= \frac{1}{1+\mu+a(n)} |u_1(n+1) - u_2(n+1)| \\ &\le \frac{1}{1+\mu+\inf_{n\in Z} a(n)} \|u_1 - u_2\| \quad \text{for} \quad \forall n \in \mathbf{Z},\end{aligned}$$

from which

$$\sup_{n\in\mathbf{Z}} |\mathbb{T}_2 u_1(n) - \mathbb{T}_2 u_2(n)| \le \frac{1}{1+\mu+\inf_{n\in\mathbf{Z}} a(n)} \|u_1 - u_2\|$$

or

$$\|\mathbb{T}_2 u_1 - \mathbb{T}_2 u_2\| \le \frac{1}{1+\mu+\inf_{n\in\mathbf{Z}} a(n)} \|u_1 - u_2\|.$$

Consequently $\mathbb{T}_2 : \mathbf{K}_3 \longmapsto \mathbf{K}_3$ is a contractive operator and therefore there exists a unique $u_4 \in \mathbf{K}_3$ such that $\mathbb{T}_2 u_4 = u_4$ or there exists a unique $u_4 \in \mathbf{K}_3$ such that

$$\mu u_4(n) = u_4(n+1) - (1+a(n))u_4(n) + \lambda b(n) f(v(n-\tau(n))).$$

From this and Theorem 3.4 it follows that there exists $\mu \in (0,1)$ for which the problem (10.58) is solvable. □

10.11 Application to a Darboux problem

Here we consider the following Darboux problem:

$$u_{xy}(x,y) = \lambda u(x,y) + \mu g(x,y,u(x,y)), \quad x \ge 0, y \ge 0, \tag{10.61}$$

$$u(x,0) = \phi(x), \quad u(0,y) = \psi(y), \quad x \ge 0, y \ge 0, \tag{10.62}$$

where

$$\phi, \psi \in \mathbf{C}^1([0,\infty)), \quad \phi(0) = \psi(0), \tag{10.63}$$

$g \in \mathbf{C}([0,\infty) \times [0,\infty) \times \mathbf{R})$ and there exist $M > 0$ and $R_0 > 0$ such that

$$\text{for any } R > R_0, |g(x,y,z)| \le MR, \text{uniformly for bounded}, x, y \text{ and} \quad |z| \le R. \tag{10.64}$$

From the assumption (10.64), we infer there exist $\lambda_1 \in (0,1)$ and $\mu_1 > 0$ such that, for any given $A \ge 0$ there exists $q = q(A) > R_0$ so that

$$A + \mu_1 |g(x,y,z)| \le (1-\lambda_1) q, \text{ uniformly for bounded}, x, y \text{ and} \quad |z| \le q. \tag{10.65}$$

Indeed, for any given $\lambda_1 \in (0,1)$, pick an $\epsilon_1 > 0$ so that $1 - \lambda_1 - \epsilon_1 > 0$. Then (10.65) will be satisfied for $\mu_1 = (1 - \lambda_1 - \epsilon_1)/M$ and $q \ge \max(A/\epsilon_1, R_0)$ by (10.63).

Remark 10.14. In particular, if $g(x, y, 0) = 0$ and $g(x, y, u)$ is Lipschitz in u uniformly for bounded x and y, then it satisfies (10.64).

Remark 10.15. Here we propose a new approach for investigating of the problem (10.61) and (10.62), different from the approach which is used in [6, 10, 25, 27], where the corresponding local problem is investigated. Our approach is more universal than the well-known present approach.

Theorem 10.24. *Let the functions g, ϕ and ψ satisfy the conditions* (10.63) *and* (10.64). *Then the problem* (10.61) *and* (10.62) *has a solution $u \in \mathbf{C}^1([0,\infty) \times [0,\infty))$, u_{xy} exists and $u_{xy} \in \mathbf{C}([0,\infty) \times [0,\infty))$, for every $\lambda \in [0, \lambda_1]$ and for every $\mu \in [0, \mu_1]$, where λ_1 and μ_1 are determined by* (10.65).

In the case when $\lambda = 0$, $\mu = 1$, $g = g(u(x, y))$, $\phi \equiv \psi \equiv 0$ in [26] a local existence result is proved of $\mathbf{C}^1$-solutions. Evidently our result is connected with the more general case and our result ensures global and local existence.

The proof of our result is broken into a series of lemmas and propositions. The main idea of the proof is as follows: Firstly, we prove the existence of a solution on $[0,1] \times [0,1]$, say u^{11}; secondly, the existence of a solution on $[0,1] \times [1,2]$, say u^{12}, then to build a solution on $[0,1] \times [2,3]$-u^{13} etc.; in this way, the existence of a solution on $[0,1] \times [0,\infty)$-$\tilde{u}^1$; then the existence of a solution on $[1,2] \times [0,1]$, say u^{21}, after which a solution on $[1,2] \times [1,2]$ is obtained, say u^{22}, in this way we build a solution on $[1,2] \times [0,\infty)$, say $\tilde{u}^2$, etc., and inductively the solution on $[2,3] \times [0,\infty)$, the solution on $[3,4] \times [0,\infty)$ etc. Initial data of every next part of the solution will depend on the previous constructed part. As the schematic figure shows:

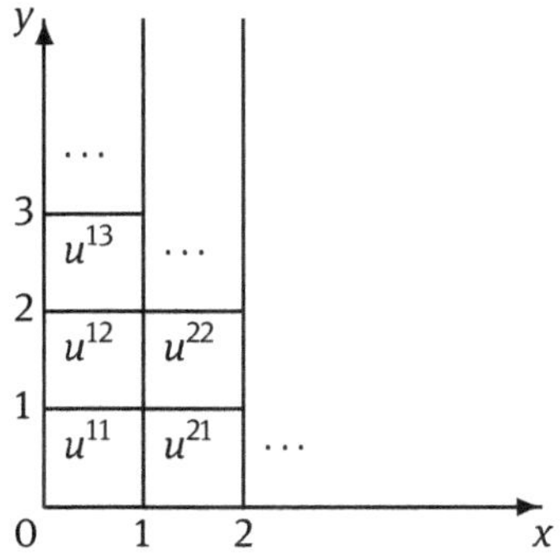

Firstly, we will show that the Darboux problem

$$u_{xy}(x,y) = \lambda u(x,y) + \mu g(x,y,u(x,y)), \quad x \in [0,1], y \in [0,1], \tag{10.66}$$

$$u(x,0) = \phi(x), \quad u(0,y) = \psi(y), \quad x \in [0,1], y \in [0,1], \tag{10.67}$$

has a solution $u \in \mathbf{C}^1([0,1] \times [0,1])$, u_{xy} exists and $u_{xy} \in \mathbf{C}([0,1] \times [0,1])$, for every $\lambda \in [0, \lambda_1]$ and every $\mu \in [0, \mu_1]$.

To this end, for $A^{11} = \max_{[0,1]\times[0,1]} |\psi(y) + \phi(x) - \phi(0)| \geq 0$, according to (10.65), we can find a $q_{11} > R_0$ so that

$$A^{11} + \mu_1 \big| g([0,1],[0,1],[-q_{11}, q_{11}]) \big| \leq (1-\lambda_1) q_{11}, \tag{10.68}$$

where we will use the following notation:

$$\big| g([a,b],[c,d],[-r,r]) \big| = \max_{(x,y,z)\in[a,b]\times[c,d]\times[-r,r]} \big| g(x,y,z) \big|.$$

Then we let $\mathbf{E}^{11} = \mathbf{C}([0,1]\times[0,1])$, $\mathbf{K}^{11} = \{u \in \mathbf{E}^{11} : |u| \leq q_{11} \text{ in } [0,1]\times[0,1]\}$, endowed with the maximum norm. We note that $\mathbf{E}^{11}$ is a completely normed space, $\mathbf{K}^{11}$ is a closed, bounded, convex subset of $\mathbf{E}^{11}$.

Given an $\epsilon_1 \in (0,1)$. For $u \in \mathbf{K}^{11}$ we define the operators

$$\mathbb{S}^{11} u = (1-\epsilon_1) u + \epsilon_1 \left(\psi(y) + \phi(x) - \phi(0) + \mu \int_0^x \int_0^y g(z,s,u(z,s))\,ds\,dz \right),$$

$$\mathbb{T}^{11} u = \epsilon_1 \int_0^x \int_0^y u(z,s)\,ds\,dz.$$

Lemma 10.7. *If $u \in \mathbf{K}^{11}$ is a fixed point of the operator $\mathbb{S}^{11} + \lambda\mathbb{T}^{11}$ then u is a solution of the problem* (10.66) *and* (10.67).

Proof. We have

$$\begin{aligned} u &= \mathbb{S}^{11} u + \lambda \mathbb{T}^{11} u \quad \Longleftrightarrow \\ \epsilon_1 u &= \epsilon_1 \left(\psi(y) + \phi(x) - \phi(0) + \mu \int_0^x \int_0^y g(z,s,u(z,s))\,ds\,dz \right) + \lambda\epsilon_1 \int_0^x \int_0^y u(z,s)\,ds\,dz \quad \Longrightarrow \\ u &= \left(\psi(y) + \phi(x) - \phi(0) + \mu \int_0^x \int_0^y g(z,s,u(z,s))\,ds\,dz \right) + \lambda \int_0^x \int_0^y u(z,s)\,ds\,dz. \end{aligned}$$

Differentiating the previous equality in x and y we get

$$u_{xy} = \lambda u(x,y) + \mu g(x,y,u(x,y)),$$

i. e., $u(x,y)$ satisfies equation (10.66).

After we put $x = 0$ in the equality

$$\begin{aligned} u(x,y) = {} & \left(\psi(y) + \phi(x) - \phi(0) + \mu \int_0^x \int_0^y g(z,s,u(z,s))\,ds\,dz \right) \\ & + \lambda \int_0^x \int_0^y u(z,s)\,ds\,dz, \end{aligned} \tag{10.69}$$

we obtain $u(0,y) = \psi(y)$ for every $y \in [0,1]$; and we put $y = 0$ in (10.69) and use $\phi(0) = \psi(0)$ to get $u(x,0) = \phi(x)$ for every $x \in [0,1]$. Consequently, u satisfies the boundary conditions (10.66). □

Lemma 10.8. *The operator* $\mathbb{S}^{11} : \mathbf{K}^{11} \longmapsto \mathbf{E}^{11}$ *is a* $1 - \epsilon_1 < 1$*-set contractive.*

Proof. For $u \in \mathbf{K}^{11}$ we clearly have $\mathbb{S}^{11}u \in \mathbf{E}^{11}$. Therefore $\mathbb{S}^{11} : \mathbf{K}^{11} \longmapsto \mathbf{E}^{11}$. Also, for $u \in \mathbf{K}$ we decompose $\mathbb{S}^{11}$ as

$$\mathbb{S}_1 u = (1-\epsilon_1)u + \epsilon_1(\psi(y) + \phi(x) - \phi(0)), \quad \mathbb{S}_2 u = \epsilon_1 \mu \int_0^x \int_0^y g(z,s,u(z,s))dsdz.$$

Then $\mathbb{S}^{11} = \mathbb{S}_1 + \mathbb{S}_2$ and $\|\mathbb{S}_1 u - \mathbb{S}_1 v\| = (1-\epsilon_1)\|u-v\|$, and so $\mathbb{S}_1$ is $1-\epsilon_1$-set contractive. To achieve the proof of the lemma, we just need to show that $\mathbb{S}_2 : \mathbf{K}^{11} \longmapsto \mathbf{E}^{11}$ is compact. To see this, for any bounded set $\mathbf{B} \subset \mathbf{E}^{11}$ with $\|u\| \le M$ for all $u \in \mathbf{B}$, we have

$$|\mathbb{S}_2 u| \le \epsilon_1 \mu |g([0,1],[0,1],[-M,M])|,$$

which shows that $\mathbb{S}_2(\mathbf{B})$ is uniformly bounded; also

$$|(\mathbb{S}_2 u)_x| \le \epsilon_1 \mu \left(|\phi'(x)| + \left| \int_0^y g(x,s,u(x,s))ds \right| \right) \le \epsilon_1 \mu (\|\phi'\| + |g([0,1],[0,1],[-M,M])|)$$

and

$$|(\mathbb{S}_2 u)_y| \le \epsilon_1 \mu \left(|\psi'(y)| + \left| \int_0^x g(z,y,u(z,y))dz \right| \right) \le \epsilon_1 \mu (\|\psi'\| + |g([0,1],[0,1],[-M,M])|).$$

These two inequalities say that $\mathbb{S}(\mathbf{B})$ is equicontinuous in $\mathbf{E}^{11}$. Now, a standard application of the Ascoli–Arzela theorem shows that $\mathbb{S}_2 : \mathbf{K}^{11} \longmapsto \mathbf{E}^{11}$ is compact. □

Lemma 10.9. *The operator* $\mathbb{T}^{11} : \mathbf{E}^{11} \longmapsto \mathbf{E}^{11}$ *is an* ϵ_1*-Lipschitz operator.*

Proof. Let $u, v \in \mathbf{E}$. Then

$$|\mathbb{T}^{11}u - \mathbb{T}^{11}v| \le \epsilon_1 \int_0^x \int_0^y |u(z,s) - v(z,s)| dsdz; \le \epsilon_1 \|u - v\|;$$

from this it follows readily that

$$\|\mathbb{T}^{11}u - \mathbb{T}^{11}v\| \le \epsilon_1 \|u - v\|.$$

□

Lemma 10.10. *Let* λ_1, μ_1 *and* q_{11} *satisfy* (10.68) *and let* $v \in \mathbf{K}^{11}$ *be fixed. Then for every* $\lambda \in [0,\lambda_1]$ *and* $\mu \in [0,\mu_1]$ *we see that the equation*

$$u = \lambda \mathbb{T}^{11} u + \mathbb{S}^{11} v$$

has a unique solution $u \in \mathbf{K}^{11}$.

Proof. For $u \in \mathbf{K}^{11}$, define

$$\mathbb{T}_1 u = \lambda \mathbb{T}^{11} u + \mathbb{S}^{11} v.$$

Then it follows from (10.68) that

$$\begin{aligned}|\mathbb{T}_1 u| &\le \lambda \epsilon_1 \int_0^x \int_0^y |u(z,s)| ds dz + (1-\epsilon_1)|v| \\ &\quad + \epsilon_1 \left(|\psi(y) + \phi(x) - \phi(0)| + \mu \int_0^x \int_0^y |g(z,s,v(z,s))| ds dz \right) \\ &\le \lambda \epsilon_1 q_{11} + (1-\epsilon_1) q_{11} + \epsilon_1 (A^{11} + \mu |g([0,1],[0,1],[-q_{11}, q_{11}])|) \\ &\le \lambda_1 \epsilon_1 q_{11} + (1-\epsilon_1) q_{11} + \epsilon_1 (A^{11} + \mu_1 |g([0,1],[0,1],[-q_{11}, q_{11}])|) \le q_{11}.\end{aligned}$$

Therefore $\mathbb{T}_1 : \mathbf{K}^{11} \longmapsto \mathbf{K}^{11}$. Also, for $u_1, u_2 \in \mathbf{K}^{11}$ we have

$$|\mathbb{T}_1 u_1 - \mathbb{T}_1 u_2| \le \lambda_1 \epsilon_1 \int_0^x \int_0^y |u_1(z,s) - u_2(z,s)| ds dz \le \lambda_1 \epsilon_1 \|u_1 - u_2\|$$

and then

$$\|\mathbb{T}_1 u_1 - \mathbb{T}_1 u_2\| \le \lambda_1 \epsilon_1 \|u_1 - u_2\|.$$

From our choice of ϵ_1 and λ_1 it follows that $\mathbb{T}_1 : \mathbf{K}^{11} \longmapsto \mathbf{K}^{11}$ is a contractive map. Therefore, there exists a unique $u \in \mathbf{K}^{11}$ so that $\mathbb{T}_1 u = u$. □

Proposition 10.4. *The Darboux problem* (10.66) *and* (10.67) *has a solution* $u^{11} \in \mathbf{C}^1([0,1] \times [0,1])$, u^{11}_{xy} *exists and* $u^{11}_{xy} \in \mathbf{C}([0,1] \times [0,1])$, *for every* $\lambda \in [0, \lambda_1]$ *and every* $\mu \in [0, \mu_1]$, *where* μ_1 *are determined in* (10.68) *or* (10.65) *above.*

Proof. From Lemmas 10.7–10.10 and Theorem 3.1, it follows that for every $\lambda \in [0, \lambda_1]$ and $\mu \in [0, \mu_1]$ the problem (10.65), (10.66) has a solution $u^{11} \in K^{11}$. Then u^{11} satisfies (10.67), therefore $u^{11} \in \mathbf{C}^1([0,1] \times [0,1])$. In the light of (10.69), there exists u^{11}_{xy} and $u^{11}_{xy} \in \mathbf{C}([0,1] \times [0,1])$. □

Now, we will prove that the problem

$$u_{xy}(x,y) = \lambda u(x,y) + \mu g(x,y,u(x,y)), \quad x \in [0,1], y \in [1,2], \tag{10.70}$$

$$u(x,1) = u^{11}(x,1), \quad u(0,y) = \psi(y), \quad x \in [0,1], y \in [1,2], \tag{10.71}$$

has a solution $u \in \mathbf{C}^1([0,1] \times [1,2])$, u_{xy} exists and $u_{xy} \in \mathcal{C}([0,1] \times [1,2])$, for every $\lambda \in [0, \lambda_1]$ and every $\mu \in [0, \mu_1]$. λ_1 and μ_1 are as in (10.68) above.

Evidently we have

$$u(0,1) = u^{11}(0,1) = \psi(1).$$

For $A^{12} = \max_{[0,1]\times[1,2]} |\psi(y) + u^{11}(x,1) - u^{11}(0,1)| \geq 0$, according to (10.65), we can find a $q_{12} > R_0$ so that

$$A^{12} + \mu_1 |g([0,1],[1,2],[-q_{12},q_{12}])| \leq (1-\lambda_1)q_{12}, \tag{10.72}$$

Then we let $\mathbf{E}^{12} = \mathbf{C}([0,1]\times[1,2])$, $\mathbf{K}^{12} = \{u \in \mathbf{E}^{12} : |u| \leq q_{12} \text{ in } [0,1]\times[1,2]\}$, endowed with the maximum norm. We note that $\mathbf{E}^{12}$ is a completely normed space, $\mathbf{K}^{12}$ is a closed, bounded, convex subset of $\mathbf{E}^{12}$.

For $u \in \mathbf{K}^{12}$ we define the operators

$$\mathbb{S}^{12}u = (1-\epsilon_1)u + \epsilon_1\left(\psi(y) + u^{11}(x,1) - u^{11}(0,1) + \mu\int_0^x\int_1^y g(z,s,u(z,s))dsdz\right),$$

$$\mathbb{T}^{12}u = \epsilon_1\int_0^x\int_1^y u(z,s)dsdz.$$

Lemma 10.11. *If $u \in \mathbf{K}^{12}$ is a fixed point of the operator $\mathbb{S}^{12} + \lambda\mathbb{T}^{12}$ then u is a solution of the problem* (10.70) *and* (10.71).

Proof. We have

$$u = \mathbb{S}^{12}u + \lambda\mathbb{T}^{12}u \iff$$

$$u = \left(\psi(y) + u^{11}(x,1) - u^{11}(0,1) + \mu\int_0^x\int_1^y g(z,s,u(z,s))dsdz\right) + \lambda\int_0^x\int_1^y u(z,s)dsdz.$$

Differentiating the above equality in x and y we obtain

$$u_{xy} = \lambda u(x,y) + \mu g(x,y,u(x,y)),$$

i. e., $u(x,y)$ satisfies equation (10.70).

After we put $x = 0$ in the equality

$$\begin{aligned} u(x,y) = {} & \left(\psi(y) + u^{11}(x,1) - u^{11}(0,1) + \mu\int_0^x\int_1^y g(z,s,u(z,s))dsdz\right) \\ & + \lambda\int_0^x\int_1^y u(z,s)dsdz, \end{aligned} \tag{10.73}$$

we obtain $u(0,y) = \psi(y)$ for every $y \in [1,2]$, and we put $y = 1$ in (10.73) and use $\psi(1) = u^{11}(0,1)$ to obtain $u(x,1) = u^{11}(x,1)$ for every $x \in [0,1]$. Hence, u satisfies (10.71). □

Lemma 10.12. *The operator $\mathbb{S}^{12} : \mathbf{K}^{12} \longmapsto \mathbf{E}^{12}$ is a $1-\epsilon_1 < 1$-set contractive.*

Lemma 10.13. *The operator $\mathbb{T}^{12} : \mathbf{E}^{12} \longmapsto \mathbf{E}^{12}$ is an ϵ_1-Lipschitz operator.*

Lemma 10.14. *Let λ_1, μ_1 and q_{12} satisfy* (10.72) *and let $v \in \mathbf{K}^{12}$ be fixed. Then for every $\lambda \in [0, \lambda_1]$ and $\mu \in [0, \mu_1]$ we see that the equation*

$$u = \lambda \mathbb{T}^{12} u + \mathbb{S}^{12} v$$

has a solution $u \in \mathbf{K}^{12}$.

Proof. For $u \in \mathbf{K}$, let

$$\mathbb{T}_2 u = \lambda \mathbb{T}^{12} u + \mathbb{S}^{12} v.$$

Then by (10.72) it follows

$$\begin{aligned}
|\mathbb{T}_2 u| &\le \lambda |\mathbb{T}^{12} u| + |\mathbb{S}^{12} v| \\
&\le \lambda \epsilon_1 \int_0^x \int_1^y |u(z,s)| ds dz + (1-\epsilon_1)|v| \\
&\quad + \epsilon_1 \left(|\psi(y) + u^{11}(x,1) - u^{11}(0,1)| + \mu \int_0^x \int_1^y |g(z,s,u(z,s))| ds dz \right) \\
&\le \lambda \epsilon_1 q_{12} + (1-\epsilon_1) q_{12} + \epsilon_1 (A^{12} + \mu |g([0,1],[1,2],[-q_{12},q_{12}]|) \\
&\le \lambda_1 \epsilon_1 q_{12} + (1-\epsilon_1) q_{12} + \epsilon_1 (A^{12} + \mu_1 |g([0,1],[1,2],[-q_{12},q_{12}]|) \le q_{12}.
\end{aligned}$$

Therefore $\mathbb{T}_2 : \mathbf{K}^{12} \longmapsto \mathbf{K}^{12}$. Also, for $u_1, u_2 \in \mathbf{K}^{12}$ we have

$$|\mathbb{T}_2 u_1 - \mathbb{T}_2 u_2| = |\mathbb{T}^{12} u_1 - \mathbb{T}^{12} u_2| \le \epsilon_1 \|u_1 - u_2\|$$

and thus

$$\|\mathbb{T}_2 u_1 - \mathbb{T}_2 u_2\| \le \lambda_1 \epsilon_1 A \|u_1 - u_2\|.$$

The choices of ϵ_1 and λ_1 imply that $\mathbb{T}_2 : \mathbf{K}^{12} \longmapsto \mathbf{K}^{12}$ is a contractive map. Consequently there exists a unique $u \in \mathbf{K}^{12}$ so that $\mathbb{T}_1^{12} u = u$. □

Proposition 10.5. *The Darboux problem* (10.70) *and* (10.71) *has a solution $u^{12} \in \mathbf{C}^1([0,1] \times [0,1])$, u_{xy}^{12} exists and $u_{xy}^{12} \in \mathbf{C}([0,1] \times [0,1])$, for every $\lambda \in [0, \lambda_1]$ and every $\mu \in [0, \mu_1]$, where μ_1 are determined in* (10.72) *or* (10.65) *above. Furthermore,*

$$\tilde{u}^{11}(x,y) = \begin{cases} u^{11}(x,y), & x, y \in [0,1], \\ u^{12}(x,y), & x \in [0,1], y \in [1,2], \end{cases} \tag{10.74}$$

is a solution of the Darboux problem

$$u_{xy} = \lambda u(x,y) + \mu g(x,y,u(x,y)), \quad x \in [0,1], y \in [0,2], \tag{10.75}$$

$$u(0,y) = \psi(y), u(x,0) = \phi(x), \quad x \in [0,1], y \in [0,2], \tag{10.76}$$

for which $\tilde{u}^{11} \in \mathbf{C}^1([0,1] \times [0,2])$, $\tilde{u}_{xy}^{11}$ exists and $\tilde{u}_{xy}^{11} \in \mathbf{C}([0,1] \times [0,2])$.

Proof. In view of Lemmas 10.11–10.14 and Theorem 3.1, for every $\lambda \in [0, \lambda_1]$ and $\mu \in [0, \mu_1]$ the problem (10.70), (10.71) has a solution $u^{12} \in \mathbf{K}^{11}$. Then u^{12} satisfies (10.73), and therefore $u^{12} \in \mathbf{C}^1([0,1] \times [1,2])$, u^{12}_{xy} exists and $u^{12}_{xy} \in \mathbf{C}([0,1] \times [1,2])$.

From (10.71) it follows that

$$u^{11}(x,1) = u^{12}(x,1), u^{11}_x(x,1) = u^{12}_x(x,1) \quad \forall x \in [0,1].$$

This, coupled with (10.70), gives

$$u^{11}_{xy}(x,1) = u^{12}_{xy}(x,1) \quad \forall x \in [0,1].$$

Thus, one can easily see that $\tilde{u}^{11}$ defined in (10.74) solves the Darboux problem (10.75) and (10.76). □

Now, we will prove that the problem

$$u_{xy}(x,y) = \lambda g(x,y,u(x,y)) + \mu u(x,y), \quad x \in [0,1], y \in [2,3], \tag{10.77}$$

$$u(x,2) = u^{12}(x,2), \quad u(0,y) = \psi(y), \quad x \in [0,1], y \in [2,3], \tag{10.78}$$

has a solution $u \in \mathbf{C}^1([0,1] \times [2,3])$, u_{xy} exists and $u_{xy} \in \mathbf{C}([0,1] \times [2,3])$, for every $\lambda \in [0, \lambda_1]$ and every $\mu \in [0, \mu_1]$. λ_1 and μ_1 are as above.

Evidently we have

$$u(0,2) = u^{12}(0,2) = \psi(2).$$

For $A^{13} = \max_{[0,1]\times[2,3]} |\psi(y) + u^{12}(x,2) - u^{12}(0,2)| \geq 0$, thanks to (10.65), we can choose a $q_{13} > R_0$ so that

$$A^{13} + \mu_1 |g([0,1],[2,3],[-q_{13}, q_{13}])| \leq (1 - \lambda_1) q_{13},$$

Then we let $\mathbf{E}^{13} = \mathbf{C}([0,1]\times[2,3])$, $\mathbf{K}^{13} = \{u \in \mathbf{E}^{13} : |u| \leq q_{13} \quad \text{in} \quad [0,1]\times[2,3]\}$, endowed with the maximum norm. Performing a similar analysis to above, one obtains a fixed point u^{13} of $\mathbb{S}^{13} + \lambda \mathbb{T}^{13}$ in $\mathbf{K}^{13}$ for every $\lambda \in [0, \lambda_1]$ and every $\mu \in [0, \mu_1]$, which is a solution of (10.77) and (10.78). Here

$$\mathbb{S}^{13} u = (1 - \epsilon_1) u + \epsilon_1 \left(\psi(y) + u^{12}(x,2) - u^{12}(0,2) + \mu \int_0^x \int_2^y g(z,s,u(z,s)) ds dz \right),$$

$$\mathbb{T}^{13} u = \epsilon_1 \int_0^x \int_2^y u(z,s) ds dz.$$

By (10.78), this solution u^{13} fulfills

$$u^{13}(x,2) = u^{12}(x,2), u^{13}_x(x,2) = u^{12}_x(x,2) \quad \forall x \in [0,1],$$

and so from (10.77) it follows

$$u^{13}_{xy}(x,2) = u^{12}_{xy}(x,2) \quad \forall x \in [0,1].$$

Therefore

$$\tilde{u}^{12}(x,y) = \begin{cases} u^{11}(x,y), & x,y \in [0,1], \\ u^{12}(x,y), & x \in [0,1], y \in [1,2], \\ u^{13}(x,y), & x \in [0,1], y \in [2,3], \end{cases}$$

is a solution of the problem

$$\begin{aligned} u_{xy} &= \lambda u(x,y) + \mu g(x,y,u(x,y)), \quad x \in [0,1], y \in [0,3], \\ u(0,y) &= \psi(y), u(x,0) = \phi(x), \quad x \in [0,1], y \in [0,3], \end{aligned}$$

for which $\tilde{u}^{12} \in \mathbf{C}^1([0,1] \times [0,3])$, $\tilde{u}^{12}_{xy}$ exists and $\tilde{u}^{12}_{xy} \in \mathbf{C}([0,1] \times [0,3])$.

Continuing the above process, we obtain the following proposition.

Proposition 10.6. *The solution*

$$\tilde{u}^{1}(x,y) = \begin{cases} u^{11}(x,y), & x,y \in [0,1], \\ u^{12}(x,y), & x \in [0,1], y \in [1,2], \\ u^{13}(x,y), & x \in [0,1], y \in [2,3], \\ \dots \end{cases}$$

is a solution of the problem

$$\begin{aligned} u_{xy} &= \lambda u(x,y) + \mu g(x,y,u(x,y)), \quad x \in [0,1], y \in [0,\infty), \\ u(0,y) &= \psi(y), u(x,0) = \phi(x), \quad x \in [0,1], y \in [0,\infty), \end{aligned}$$

for which $\tilde{u}^1 \in \mathbf{C}^1([0,1]\times[0,\infty))$, $\tilde{u}^1_{xy}$ *exists and* $\tilde{u}^1_{xy} \in \mathbf{C}([0,1]\times[0,\infty))$ *for every* $\lambda \in [0,\lambda_1]$ *and* $\mu \in [0,\mu_1]$, *where* μ_1 *are determined in* (10.68) *or* (10.65) *above.*

Now we will prove that the problem

$$u_{xy}(x,y) = \lambda u(x,y) + \mu g(x,y,u(x,y)), \quad x \in [1,2], y \in [0,1], \tag{10.79}$$

$$u(x,0) = \phi(x), \quad u(1,y) = u^{11}(1,y), \quad x \in [1,2], y \in [0,1], \tag{10.80}$$

has a solution $u \in \mathbf{C}^1([1,2] \times [0,1])$, u_{xy} exists and $u_{xy} \in \mathbf{C}([1,2] \times [0,1])$, for every $\lambda \in [0,\lambda_1]$ and every $\mu \in [0,\mu_1]$, where λ_1 and μ_1 are as above.

To see this, for $A^{21} = \max_{[1,2]\times[0,1]} |u^{11}(1,y) + \phi(x) - \phi(1)| \geq 0$, thanks to (10.65), one can select a $q_{21} > R_0$ so that

$$A^{21} + \mu_1 |g([1,2],[0,1],[-q_{21},q_{21}])| \leq (1-\lambda_1)q_{21}.$$

We let $\mathbf{E}^{21} = \mathbf{C}([1,2] \times [0,1])$, $\mathbf{K}^{21} = \{u \in \mathbf{E}^{21} : |u| \leq q_{21} \quad \text{in} \quad [1,2] \times [0,1]\}$, endowed with the maximum norm.

Since $u^{11}(1,0) = \phi(1)$, the problem (10.79) and (10.80) is addressed by seeking a fixed point of the sum $\mathbb{S}^{21} + \lambda\mathbb{T}^{21}$ in $\mathbf{K}^{21}$, where

$$\mathbb{S}^{21}u = (1-\epsilon_1)u + \epsilon_1\left(u^{11}(1,y) + \phi(x) - \phi(1) + \mu\int_1^x\int_0^y g(z,s,u(z,s))dsdz\right),$$

$$\mathbb{T}^{21}u = \epsilon_1\int_1^x\int_0^y u(z,s)dsdz.$$

Lemma 10.15. *If $u \in \mathbf{K}^{21}$ is a fixed point of the operator $\mathbb{S}^{21} + \lambda\mathbb{T}^{21}$ then u is a solution of the problem* (10.79) *and* (10.80).

Proof. One has

$$u = \mathbb{S}^{21}u + \lambda\mathbb{T}^{21}u \quad \Longleftrightarrow$$

$$u = \left(u^{11}(1,y) + \phi(x) - \phi(1) + \mu\int_1^x\int_0^y g(z,s,u(z,s))dsdz\right) + \lambda\int_1^x\int_0^y u(z,s)dsdz.$$

Differentiation of the previous equality in x and y shows

$$u_{xy} = \lambda u(x,y) + \mu g(x,y,u(x,y)),$$

i. e., $u(x,y)$ satisfies equation (10.79).

After we put $x = 1$ in the equality

$$\begin{aligned} u(x,y) = {} & \left(u^{11}(1,y) + \phi(x) - \phi(1) + \mu\int_1^x\int_0^y g(z,s,u(z,s))dsdz\right) \\ & + \lambda\int_1^x\int_0^y u(z,s)dsdz, \end{aligned} \tag{10.81}$$

we obtain $u(1,y) = u^{11}(1,y)$ for every $y \in [0,1]$, and we put $y = 0$ in (10.81) and use $u^{11}(1,0) = \phi(1)$ to see that $u(x,0) = \phi(x)$ for every $x \in [1,2]$. Hence u satisfies (10.80). □

Lemma 10.16. *The operator $\mathbb{S}^{21} : \mathbf{K}^{21} \longmapsto \mathbf{E}^{21}$ is a $1-\epsilon_1 < 1$-set contractive.*

Lemma 10.17. *The operator $\mathbb{T}^{21} : \mathbf{E}^{21} \longmapsto \mathbf{E}^{21}$ is ϵ_1-Lipschitz operator.*

Lemma 10.18. *Let λ_1, μ_1 and q_{21} satisfy* (10.81) *and let $v \in \mathbf{K}^{12}$ be fixed. Then for every $\lambda \in [0,\lambda_1]$ and $\mu \in [0,\mu_1]$ we see that the equation*

$$u = \lambda\mathbb{T}^{21}u + \mathbb{S}^{21}v$$

has a solution $u \in \mathbf{K}^{21}$.

Proof. Let

$$\mathbb{T}_1^{21}u = \lambda\mathbb{T}^{21}u + \mathbb{S}^{21}v$$

for $u \in \mathbf{K}$. From (10.81) we have

$$\begin{aligned}|\mathbb{T}_1^{21}u| &\le \lambda\epsilon_1 \int_1^x\int_0^y |u(z,s)|dsdz + (1-\epsilon_1)|v| \\ &\quad + \epsilon_1\left(|u^{11}(1,y) + \phi(x) - \phi(1)| + \mu\int_1^x\int_0^y |g(z,s,u(z,s))|dsdz\right) \\ &\le \lambda\epsilon_1 q_{21} + (1-\epsilon_1)q_{21} + \epsilon_1(A^{21} + \mu|g([1,2],[0,1],[-q_{21},q_{21}])|) \\ &\le \lambda_1\epsilon_1 q_{21} + (1-\epsilon_1)q_{21} + \epsilon_1(A^{21} + \mu_1|g([1,2],[0,1],[-q_{21},q_{21}])|) \le q_{21}.\end{aligned}$$

This shows $\mathbb{T}_1^{21} : \mathbf{K}^{21} \longmapsto \mathbf{K}^{21}$. Also, for $u_1, u_2 \in \mathbf{K}^{21}$ we have

$$\|\mathbb{T}_1^{21}u_1 - \mathbb{T}_1^{21}u_2\| \le \lambda_1\epsilon_1\|u_1 - u_2\|$$

and thus $\mathbb{T}_1^{21} : \mathbf{K}^{21} \longmapsto \mathbf{K}^{21}$ is a contractive map. Consequently there exists a unique $u \in \mathbf{K}^{21}$ so that $\mathbb{T}_1^{21}u = u$. □

Proposition 10.7. *The Darboux problem* (10.79) *and* (10.80) *has a solution* $u^{21} \in \mathbf{C}^1([1,2]\times[0,1])$, u_{xy}^{21} *exists and* $u_{xy}^{21} \in \mathbf{C}([1,2]\times[0,1])$ *for every* $\lambda \in [0,\lambda_1]$ *and every* $\mu \in [0,\mu_1]$.

Proof. Applying Lemmas 10.14–10.18 and Theorem 3.1, we know that $\lambda T + S$ has a fixed point $u^{21} \in \mathbf{K}^{21}$ for every $\lambda \in [0,\lambda_1]$ and $\mu \in [0,\mu_1]$, which is a solution of the problem (10.79) and (10.80). Since u^{21} satisfies (10.81), $u^{21} \in \mathbf{C}^1([1,2]\times[0,1])$, and there exists u_{xy}^{21} and $u_{xy}^{21} \in \mathbf{C}([1,2]\times[0,1])$. □

From (10.79) it follows that

$$u^{11}(1,y) = u^{21}(1,y), u_y^{11}(1,y) = u_y^{21}(1,y) \quad \forall y \in [0,1],$$

from this and from (10.79) it follows that

$$u_{xy}^{11}(1,y) = u_{xy}^{21}(1,y) \quad \forall y \in [0,1].$$

Now we will prove that the problem

$$u_{xy}(x,y) = \lambda u(x,y) + \mu g(x,y,u(x,y)), \quad x \in [1,2], y \in [1,2], \tag{10.82}$$

$$u(x,1) = u^{21}(x,1), \quad u(1,y) = u^{12}(1,y), \quad x \in [1,2], y \in [1,2], \tag{10.83}$$

has a solution $u^{22} \in C^1([1,2]\times[1,2])$, u_{xy}^{22} exists and $u_{xy}^{22} \in \mathbf{C}([1,2]\times[1,2])$, for every $\lambda \in [0,\lambda_1]$ and every $\mu \in [0,\mu_1]$. λ_1 and μ_1 are as above.

To this end, for $A^{22} = \max_{[1,2]\times[1,2]} |u^{12}(1,y) + u^{21}(x,1) - u^{21}(1,1)| \geq 0$, due to (10.65), we choose a $q_{22} > R_0$ so that

$$A^{22} + \mu_1 |g([1,2],[1,2],[-q_{22}, q_{22}])| \leq (1-\lambda_1) q_{22}. \tag{10.84}$$

Then we let $\mathbf{E}^{22} = \mathbf{C}([1,2]\times[1,2])$, $\mathbf{K}^{22} = \{u \in \mathbf{E}^{22} : |u| \leq q_{22} \quad \text{in} \quad [1,2]\times[1,2]\}$, endowed with the maximum norm.

Since $u^{21}(1,1) = u^{12}(1,1) = u^{11}(1,1)$, as above, the problem (10.82) and (10.83) is addressed by looking for a fixed point of $\lambda\mathbb{T}^{22} + \mathbb{S}^{22}$ in $\mathbf{K}^{22}$, where

$$\mathbb{S}^{22}u = (1-\epsilon_1)u + \epsilon_1\left(u^{12}(1,y) + u^{21}(x,1) - u^{21}(1,1) + \mu\int_1^x\int_1^y g(z,s,u(z,s))dsdz\right),$$

$$\mathbb{T}^{22}u = \epsilon_1\int_1^x\int_1^y u(z,s)dsdz.$$

By (10.83) this solution u^{22} satisfies

$$u^{12}(1,y) = u^{22}(1,y), \quad u_y^{12}(1,y) = u_y^{22}(1,y), \quad \forall y \in [1,2],$$

from which with (10.84) it follows that

$$u_{xy}^{12}(1,y) = u_{xy}^{22}(1,y), \quad \forall y \in [1,2],$$

and

$$u^{22}(x,1) = u^{21}(x,1), \quad u_x^{22}(x,1) = u_x^{21}(x,1), \quad \forall x \in [1,2].$$

Then by equation (10.83)

$$u_{xy}^{22}(x,1) = u_{xy}^{21}(x,1) \quad \forall x \in [1,2].$$

In this way, we construct a solution

$$\tilde{u}^{22} = \begin{cases} u^{21}(x,y), & x \in [1,2], y \in [0,1], \\ u^{22}(x,y), & x,y \in [1,2], \end{cases}$$

of the problem

$$\begin{aligned} u_{xy}(x,y) &= \lambda u(x,y) + \mu g(x,y,u(x,y)), \quad x \in [1,2], y \in [0,2], \\ u(x,0) &= \phi(x), \quad u(1,y) = \tilde{u}^1(1,y), \quad x \in [1,2], y \in [0,2]. \end{aligned}$$

Moreover, $\tilde{u}^{22} \in \mathbf{C}^1([1,2]\times[0,2])$, $\tilde{u}_{xy}^{22}$ exists and $\tilde{u}_{xy}^{22} \in \mathbf{C}([1,2]\times[0,2])$.

Inductively, the function

$$\tilde{u}^2 = \begin{cases} u^{21}(x,y), & x \in [1,2], y \in [0,1], \\ u^{22}(x,y), & x,y \in [1,2], \\ u^{23}(x,y), & x \in [1,2], y \in [2,3], \\ \dots \end{cases}$$

is a solution of the problem

$$\begin{aligned} u_{xy}(x,y) &= \lambda u(x,y) + \mu g(x,y,u(x,y)), \quad x \in [1,2], y \geq 0, \\ u(x,0) &= \phi(x), \quad u(1,y) = \tilde{u}^1(1,y), \quad x \in [1,2], y \geq 0, \end{aligned}$$

for which $\tilde{u}^2 \in \mathbf{C}^1([1,2] \times [0,\infty))$, there exists $\tilde{u}^2_{xy}$ and $\tilde{u}^2_{xy} \in \mathbf{C}([1,2] \times [0,\infty))$.

Repeating the process again and again, we obtain a solution to our original problem (10.61) and (10.62).

Proposition 10.8. *The function*

$$\tilde{u}(x,y) = \begin{cases} \tilde{u}^1(x,y), & x \in [0,1], y \geq 0, \\ \tilde{u}^2(x,y), & x \in [1,2], y \geq 0, \\ \dots \end{cases}$$

is a solution of the problem (10.61) *and* (10.62) *for which* $\tilde{u} \in \mathbf{C}^1([0,\infty) \times [0,\infty))$, *there exists* $\tilde{u}_{xy}$ *and* $\tilde{u}_{xy} \in \mathbf{C}([0,\infty) \times [0,\infty))$.

Remark 10.16. For the solution $u^{m+1,n+1}$, $m \geq 1$, $n \geq 1$, we see that it solves the problem

$$\begin{aligned} u_{xy}(x,y) &= \lambda u(x,y) + \mu g(x,y,u(x,y)), \quad x \in [m,m+1], y \in [n,n+1], \\ u(m,y) &= u^{m,n+1}(m,y), \quad u(x,n) = u^{m+1,n}(x,n), \quad x \in [m,m+1], y \in [n,n+1], \end{aligned}$$

$\mathbf{E}^{m+1,n+1} = \mathbf{C}([m,m+1] \times [n,n+1])$, $\mathbf{K}^{m+1,n+1} = \{u \in \mathbf{E}^{m+1,n+1} : |u| \leq q_{m+1,n+1}$ in $\in$ $[m,m+1] \times [n,n+1]\}$, endowed with the maximum norm. We note that $\mathbf{E}^{m+1,n+1}$ is a completely normed space, $\mathbf{K}^{m+1,n+1}$ is a closed, bounded, convex subset of $\mathbf{E}^{m+1,n+1}$, $q_{m+1,n+1}$ is determined by

$$A^{m+1,n+1} + \mu_1 |g([m,m+1],[n,n+1],[-q_{m+1,n+1}, q_{m+1,n+1}])| \leq (1-\lambda_1) q_{m+1,n+1}$$

where

$$A^{m+1,n+1} = \max_{(x,y)\in[m,m+1]\times[n,n+1]} |u^{m,n+1}(m,y) + u^{m+1,n}(x,n) - u^{m+1,n}(m,n)|.$$

The solution is a fixed point of $\lambda\mathbb{T}^{m+1,n+1}+\mathbb{S}^{m+1,n+1}$ in $\mathbf{K}^{m+1,n+1}$ for every $\lambda\in[0,\lambda_1]$ and every $\mu\in[0,\mu_1]$, where

$$\begin{aligned}\mathbb{S}^{m+1,n+1}u&=(1-\epsilon_1)u\\&\quad+\epsilon_1\left(u^{m,n+1}(m,y)+u^{m+1,n}(x,n)-u^{m+1,n}(m,n)+\mu\int_m^x\int_n^y g(z,s,u(z,s))dsdz\right),\\\mathbb{T}^{m+1,n+1}u&=\epsilon_1\int_m^x\int_n^y u(z,s)dsdz.\end{aligned}$$

A Sets and mappings

A.1 Union and intersection of sets

A set is a collection of distinct objects, considered as an object in its own right. The objects that make up a set (also known as the set's elements or members) can be anything: numbers, people, letters of the alphabet, other sets, and so on. Sets are conventionally denoted by capital letters. For a set $\mathbf{A}$, the membership of the element x in $\mathbf{A}$ is denoted by $x \in \mathbf{A}$, and the nonmembership of x in $\mathbf{A}$ is denoted by $x \notin \mathbf{A}$. We will call a member of the set $\mathbf{A}$ a point of $\mathbf{A}$. Two sets are the same if they have the same memberships. Let $\mathbf{A}$ and $\mathbf{B}$ be two sets. We say that $\mathbf{A}$ is a subset of the set $\mathbf{B}$ if each member of $\mathbf{A}$ is a member of the set $\mathbf{B}$. We denote this by $\mathbf{A} \subseteq \mathbf{B}$. Also, we say that the set $\mathbf{A}$ is contained in the set $\mathbf{B}$ or the set $\mathbf{B}$ contains the set $\mathbf{A}$. A subset $\mathbf{A}$ of the set $\mathbf{B}$ is called a proper subset of the set $\mathbf{B}$ if $\mathbf{A} \neq \mathbf{B}$. We will write $\mathbf{A} \subset \mathbf{B}$. The union of the sets $\mathbf{A}$ and $\mathbf{B}$ is the set

$$\mathbf{A} \cup \mathbf{B} = \{x : x \in \mathbf{A} \quad \text{or} \quad x \in \mathbf{B}\}.$$

The intersection of the sets $\mathbf{A}$ and $\mathbf{B}$ is the set

$$\mathbf{A} \cap \mathbf{B} = \{x : x \in \mathbf{A} \quad \text{and} \quad x \in \mathbf{B}\}.$$

The complement of the set $\mathbf{A}$ in the set $\mathbf{B}$ is the set

$$\mathbf{B} \setminus \mathbf{A} = \{x : x \in \mathbf{B} \quad \text{and} \quad x \notin \mathbf{A}\}.$$

The set that has no elements is said to be the empty set and it will be denoted by $\emptyset$. A set that is not equal to the empty set is called nonempty. A set that has a single element is called a singleton set. For a given set $\mathbf{A}$, the set of all subsets of $\mathbf{A}$ is denoted by $\mathcal{P}(\mathbf{A})$ or $2^{\mathbf{A}}$ and it is called the power set of $\mathbf{A}$. We will use often the words "collection" and "family" as synonymous to the word "set". Let $\mathcal{F}$ be a collection of sets. We define the union of $\mathcal{F}$, denoted by $\bigcup_{\mathbf{A} \in \mathcal{F}} \mathbf{A}$, to be the set of points that belong to at least one of the sets in $\mathcal{F}$. We define the intersection of $\mathcal{F}$, denoted by $\bigcap_{\mathbf{A} \in \mathcal{F}} \mathbf{A}$, to be the set of points that belong to every set in $\mathcal{F}$. The collection of sets $\mathcal{F}$ is said to be disjoint provided the intersection of any two sets in $\mathcal{F}$ is empty.

Theorem A.1 (De Morgan's identities). *Let* $\mathbf{X}$ *be a set and* $\mathcal{F}$ *be a family of sets. Then*

$$\mathbf{X} \setminus \Big(\bigcup_{\mathbf{A} \in \mathcal{F}} \mathbf{A} \Big) = \bigcap_{\mathbf{A} \in \mathcal{F}} (\mathbf{X} \setminus \mathbf{A}) \tag{A.1}$$

and

$$\mathbf{X} \setminus \Big(\bigcap_{\mathbf{A} \in \mathcal{F}} \mathbf{A} \Big) = \bigcup_{\mathbf{A} \in \mathcal{F}} (\mathbf{X} \setminus \mathbf{A}). \tag{A.2}$$

https://doi.org/10.1515/9783110657722-011

Proof. Let $x \in \mathbf{X} \setminus (\bigcup_{\mathbf{A}\in\mathcal{F}} \mathbf{A})$ be arbitrarily chosen. Then $x \in \mathbf{X}$ and $x \notin \bigcup_{\mathbf{A}\in\mathcal{F}} \mathbf{A}$. Hence, $x \in \mathbf{X}$ and $x \notin \mathbf{A}$ for any $\mathbf{A} \in \mathcal{F}$. Therefore $x \in \mathbf{X} \setminus \mathbf{A}$ for any $\mathbf{A} \in \mathcal{F}$ and $x \in \bigcap_{\mathbf{A}\in\mathcal{F}}(\mathbf{X} \setminus \mathbf{A})$. Because $x \in \mathbf{X} \setminus (\bigcup_{\mathbf{A}\in\mathcal{F}} \mathbf{A})$ was arbitrarily chosen and we see that it is an element of $\bigcap_{\mathbf{A}\in\mathcal{F}}(\mathbf{X} \setminus \mathbf{A})$, we conclude that

$$\mathbf{X} \setminus \Big(\bigcup_{\mathbf{A}\in\mathcal{F}} \mathbf{A} \Big) \subseteq \bigcap_{\mathbf{A}\in\mathcal{F}} (\mathbf{X} \setminus \mathbf{A}). \tag{A.3}$$

Let $x \in \bigcap_{\mathbf{A}\in\mathcal{F}}(\mathbf{X} \setminus \mathbf{A})$ be arbitrarily chosen. Then $x \in \mathbf{X} \setminus \mathbf{A}$ for any $\mathbf{A} \in \mathcal{F}$. Hence, $x \in \mathbf{X}$ and $x \notin \mathbf{A}$ for any $\mathbf{A} \in \mathcal{F}$. Therefore $x \notin \bigcup_{\mathbf{A}\in\mathcal{F}} \mathbf{A}$ and $x \in \mathbf{X} \setminus (\bigcup_{\mathbf{A}\in\mathcal{F}} \mathbf{A})$. Because $x \in \bigcap_{\mathbf{A}\in\mathcal{F}}(\mathbf{X}\setminus\mathbf{A})$ was arbitrarily chosen and we see that it is an element of $\mathbf{X}\setminus(\bigcup_{\mathbf{A}\in\mathcal{F}} \mathbf{A})$, we conclude that

$$\bigcap_{\mathbf{A}\in\mathcal{F}} (\mathbf{X} \setminus \mathbf{A}) \subseteq \mathbf{X} \setminus \Big(\bigcup_{\mathbf{A}\in\mathcal{F}} \mathbf{A} \Big).$$

From the previous relation and from (A.3), we get equation (A.1). Now we will prove equation (A.2). Let $x \in \mathbf{X} \setminus (\bigcap_{\mathbf{A}\in\mathcal{F}} \mathbf{A})$ be arbitrarily chosen. Then $x \in \mathbf{X}$ and $x \notin \bigcap_{\mathbf{A}\in\mathcal{F}} \mathbf{A}$. Hence, there is $\mathbf{A} \in \mathcal{F}$ such that $x \notin \mathbf{A}$. From this, $x \in \mathbf{X} \setminus \mathbf{A}$ and $x \in \bigcup_{\mathbf{A}\in\mathcal{F}}(\mathbf{X} \setminus \mathbf{A})$. Because $x \in \mathbf{X} \setminus (\bigcap_{\mathbf{A}\in\mathcal{F}} \mathbf{A})$ was arbitrarily chosen and we see that it is an element of $\bigcup_{\mathbf{A}\in\mathcal{F}}(\mathbf{X} \setminus \mathbf{A})$, we obtain the relation

$$\mathbf{X} \setminus \Big(\bigcap_{\mathbf{A}\in\mathcal{F}} \mathbf{A} \Big) \subseteq \bigcup_{\mathbf{A}\in\mathcal{F}} (\mathbf{X} \setminus \mathbf{A}). \tag{A.4}$$

Let $x \in \bigcup_{\mathbf{A}\in\mathcal{F}}(\mathbf{X} \setminus \mathbf{A})$ be arbitrarily chosen. Then, there is $\mathbf{A} \in \mathcal{F}$ such that $x \in \mathbf{X} \setminus \mathbf{A}$. Hence, $x \in \mathbf{X}$ and $x \notin \mathbf{A}$. Therefore $x \in \mathbf{X}$ and $x \notin \bigcap_{\mathbf{A}\in\mathcal{F}} \mathbf{A}$. From this, $x \in \mathbf{X} \setminus (\bigcap_{\mathbf{A}\in\mathcal{F}} \mathbf{A})$. Since $x \in \bigcup_{\mathbf{A}\in\mathcal{F}}(\mathbf{X} \setminus \mathbf{A})$ was arbitrarily chosen and we see that it is an element of $\mathbf{X} \setminus (\bigcap_{\mathbf{A}\in\mathcal{F}} \mathbf{A})$, we conclude that

$$\bigcup_{\mathbf{A}\in\mathcal{F}} (\mathbf{X} \setminus \mathbf{A}) \subseteq \mathbf{X} \setminus \Big(\bigcap_{\mathbf{A}\in\mathcal{F}} \mathbf{A} \Big).$$

From the previous relation and from (A.4), we obtain equation (A.2). This completes the proof. □

For a set Λ, assume that for each $\lambda \in \Lambda$, a set $\mathbf{E}_\lambda$ is defined. Let $\mathcal{F}$ be the collection of sets $\mathbf{E}_\lambda, \lambda \in \Lambda$. We write $\mathcal{F} = \{\mathbf{E}_\lambda\}_{\lambda\in\Lambda}$ and we will say that this is an indexing or parametrization of $\mathcal{F}$ by the index set or the parameter set Λ.

A.2 Mappings between sets

Let **A** and **B** be two sets. A mapping or a function from **A** into **B** is a correspondence that assigns to each member of **A** a member of **B**. If **B** is a set of real numbers, we will use the word "function". We denote a mapping by $f : \mathbf{A} \longmapsto \mathbf{B}$ and, for each $x \in \mathbf{A}$, we denote by $f(x)$ the member of $\mathbf{B}$ to which x is assigned. The set $f(\mathbf{A}) = \{f(a) : a \in \mathbf{A}\}$ is called the image of $\mathbf{A}$ under f. The set $\mathbf{A}$ is called the domain of f and the set $f(\mathbf{A})$ is called the image or range of f. If $f(\mathbf{A}) = \mathbf{B}$, then the mapping f is said to be onto. If for each member b of $f(\mathbf{A})$ there is exactly one member a of $\mathbf{A}$ for which $b = f(a)$, the mapping f is said to be one-to-one. A mapping $f : \mathbf{A} \longmapsto \mathbf{B}$ that is one-to-one and onto is said to be invertible. Let $f : \mathbf{A} \longmapsto \mathbf{B}$ be invertible. Then for each $b \in \mathbf{B}$ there is exactly one member $a \in \mathbf{A}$ for which $f(a) = b$ and it is denoted by $f^{-1}(b)$. This assignment defines the mapping $f^{-1} : \mathbf{B} \longmapsto \mathbf{A}$ and it is called the inverse of f. Two sets $\mathbf{A}$ and $\mathbf{B}$ are said to be equipotent provided there is an invertible mapping from $\mathbf{A}$ and $\mathbf{B}$. Consider the sets $\mathbf{A}$, $\mathbf{B}$, $\mathbf{C}$ and $\mathbf{D}$, and the mappings $f : \mathbf{A} \longmapsto \mathbf{B}$ and $g : \mathbf{C} \longmapsto \mathbf{D}$ such that $f(\mathbf{A}) \subseteq \mathbf{C}$. Then the composition $g \circ f : \mathbf{A} \longmapsto \mathbf{D}$ is defined by $(g \circ f)(x) = g(f(x))$ for each $x \in \mathbf{A}$. For a set $\mathbf{A}$, we define the identity mapping $\mathbf{Id} : \mathbf{A} \longmapsto \mathbf{A}$ as follows: $\mathbf{Id}(x) = x$ for each $x \in \mathbf{A}$. Sometimes we will denote it by $\mathbf{Id}_{\mathbf{A}}$.

Theorem A.2. *A mapping $f : \mathbf{A} \longmapsto \mathbf{B}$ is invertible if and only if there is a mapping $g : \mathbf{B} \longmapsto \mathbf{A}$ for which*

$$g \circ f = \mathbf{Id}_{\mathbf{A}} \quad \textit{and} \quad f \circ g = \mathbf{Id}_{\mathbf{B}}. \tag{A.5}$$

Proof.

1. Let $f : \mathbf{A} \longmapsto \mathbf{B}$ be invertible. Then f is one-to-one and onto. Define the mapping $g : \mathbf{B} \longmapsto \mathbf{A}$ by

   $$g(b) = a \quad \text{if} \quad f(a) = b.$$

   Then

   $$f \circ g(b) = f(g(b)) = f(a) = b,$$

   i. e.,

   $$f \circ g = \mathbf{Id}_{\mathbf{B}}.$$

   Also,

   $$g \circ f(a) = g(f(a)) = g(b) = a,$$

   i. e.,

   $$g \circ f = \mathbf{Id}_{\mathbf{A}}.$$

2. Let there be a mapping $g : \mathbf{B} \longmapsto \mathbf{A}$ such that (A.5) holds. Let $b \in \mathbf{B}$ be arbitrarily chosen. Then $g(b) \in \mathbf{A}$ and $f(g(b)) = b$. Since $b \in \mathbf{B}$ was arbitrarily chosen, we

conclude that $f : \mathbf{A} \longmapsto \mathbf{B}$ is onto. Let now $b \in f(\mathbf{A})$ be arbitrarily chosen. Assume that there are $a_1, a_2 \in \mathbf{A}$ such that

$$b = f(a_1) \quad \text{and} \quad b = f(a_2).$$

Then

$$\begin{aligned} g(b) &= g(f(a_1)) = a_1, \\ g(b) &= g(f(a_2)) = a_2. \end{aligned}$$

Hence, $a_1 = a_2$. Consequently $f : \mathbf{A} \longmapsto \mathbf{B}$ is one-to-one. This completes the proof. □

Let $f : \mathbf{A} \longmapsto \mathbf{B}$. For a set $\mathbf{E}$ we define the set

$$f^{-1}(\mathbf{E}) = \{a \in \mathbf{A} : f(a) \in \mathbf{E}\}.$$

Theorem A.3. *Let $f : \mathbf{A} \longmapsto \mathbf{B}$. Then for any sets $\mathbf{E}_1$ and $\mathbf{E}_2$ we have*

$$f^{-1}(\mathbf{E}_1 \cup \mathbf{E}_2) = f^{-1}(\mathbf{E}_1) \cup f^{-1}(\mathbf{E}_2), \tag{A.6}$$

$$f^{-1}(\mathbf{E}_1 \cap \mathbf{E}_2) = f^{-1}(\mathbf{E}_1) \cap f^{-1}(\mathbf{E}_2), \tag{A.7}$$

$$f^{-1}(\mathbf{E}_1 \setminus \mathbf{E}_2) = f^{-1}(\mathbf{E}_1) \setminus f^{-1}(\mathbf{E}_2). \tag{A.8}$$

Proof.

1. We will prove (A.6). Let $x \in f^{-1}(\mathbf{E}_1 \cup \mathbf{E}_2)$ be arbitrarily chosen. Then $f(x) \in \mathbf{E}_1 \cup \mathbf{E}_2$. Hence, $f(x) \in \mathbf{E}_1$ or $f(x) \in \mathbf{E}_2$. Therefore $x \in f^{-1}(\mathbf{E}_1)$ or $x \in f^{-1}(\mathbf{E}_2)$. From this, $x \in f^{-1}(\mathbf{E}_1) \cup f^{-1}(\mathbf{E}_2)$. Because $x \in f^{-1}(\mathbf{E}_1 \cup \mathbf{E}_2)$ was arbitrarily chosen and we see that it is an element of $f^{-1}(\mathbf{E}_1) \cup f^{-1}(\mathbf{E}_2)$, we get the relation

   $$f^{-1}(\mathbf{E}_1 \cup \mathbf{E}_2) \subseteq f^{-1}(\mathbf{E}_1) \cup f^{-1}(\mathbf{E}_2). \tag{A.9}$$

   Let $x \in f^{-1}(\mathbf{E}_1) \cup f^{-1}(\mathbf{E}_2)$ be arbitrarily chosen. Then $x \in f^{-1}(\mathbf{E}_1)$ or $x \in f^{-1}(\mathbf{E}_2)$. Hence, $f(x) \in \mathbf{E}_1$ or $f(x) \in \mathbf{E}_2$. Therefore $f(x) \in \mathbf{E}_1 \cup \mathbf{E}_2$ and from this $x \in f^{-1}(\mathbf{E}_1 \cup \mathbf{E}_2)$. Since $x \in f^{-1}(\mathbf{E}_1) \cup f^{-1}(\mathbf{E}_2)$ was arbitrarily chosen and we see that it is an element of $f^{-1}(\mathbf{E}_1 \cup \mathbf{E}_2)$, we obtain the relation

   $$f^{-1}(\mathbf{E}_1) \cup f^{-1}(\mathbf{E}_2) \subseteq f^{-1}(\mathbf{E}_1 \cup \mathbf{E}_2).$$

   From the previous relation and from (A.9), we get equation (A.6).
2. Now we will prove (A.7). Let $x \in f^{-1}(\mathbf{E}_1 \cap \mathbf{E}_2)$ be arbitrarily chosen. Then $f(x) \in \mathbf{E}_1 \cap \mathbf{E}_2$. Hence, $f(x) \in \mathbf{E}_1$ and $f(x) \in \mathbf{E}_2$. From this, $x \in f^{-1}(\mathbf{E}_1)$ and $x \in f^{-1}(\mathbf{E}_2)$. Consequently $x \in f^{-1}(\mathbf{E}_1) \cap f^{-1}(\mathbf{E}_2)$. Since $x \in f^{-1}(\mathbf{E}_1 \cap \mathbf{E}_2)$ was arbitrarily chosen and we see that it is an element of the set

$f^{-1}(\mathbf{E}_1) \cap f^{-1}(\mathbf{E}_2)$, we obtain the relation

$$f^{-1}(\mathbf{E}_1 \cap \mathbf{E}_2) \subseteq f^{-1}(\mathbf{E}_1) \cap f^{-1}(\mathbf{E}_2). \tag{A.10}$$

Let $x \in f^{-1}(\mathbf{E}_1) \cap f^{-1}(\mathbf{E}_2)$ be arbitrarily chosen. Then $x \in f^{-1}(\mathbf{E}_1)$ and $x \in f^{-1}(\mathbf{E}_2)$. Hence, $f(x) \in \mathbf{E}_1$ and $f(x) \in \mathbf{E}_2$. Therefore $f(x) \in \mathbf{E}_1 \cap \mathbf{E}_2$. Consequently $x \in f^{-1}(\mathbf{E}_1 \cap \mathbf{E}_2)$. Because $x \in f^{-1}(\mathbf{E}_1) \cap f^{-1}(\mathbf{E}_2)$ was arbitrarily chosen and we see that it is an element of the set $f^{-1}(\mathbf{E}_1 \cap \mathbf{E}_2)$, we get the relation

$$f^{-1}(\mathbf{E}_1) \cap f^{-1}(\mathbf{E}_2) \subseteq f^{-1}(\mathbf{E}_1 \cap \mathbf{E}_2).$$

From the previous relation and from (A.10), we obtain equation (A.7).

3. Now we will prove equation (A.8). Let $x \in f^{-1}(\mathbf{E}_1 \setminus \mathbf{E}_2)$ is arbitrarily chosen. Then $f(x) \in \mathbf{E}_1 \setminus \mathbf{E}_2$. Hence, $f(x) \in \mathbf{E}_1$ and $f(x) \notin \mathbf{E}_2$. Therefore $x \in f^{-1}(\mathbf{E}_1)$ and $x \notin f^{-1}(\mathbf{E}_2)$. Consequently $x \in f^{-1}(\mathbf{E}_1) \setminus f^{-1}(\mathbf{E}_2)$. Because $x \in f^{-1}(\mathbf{E}_1 \setminus \mathbf{E}_2)$ was arbitrarily chosen and we see that it is an element of the set $f^{-1}(\mathbf{E}_1) \setminus f^{-1}(\mathbf{E}_2)$, we get the relation

$$f^{-1}(\mathbf{E}_1 \setminus \mathbf{E}_2) \subseteq f^{-1}(\mathbf{E}_1) \setminus f^{-1}(\mathbf{E}_2). \tag{A.11}$$

Let $x \in f^{-1}(\mathbf{E}_1) \setminus f^{-1}(\mathbf{E}_2)$ is arbitrarily chosen. Then $x \in f^{-1}(\mathbf{E}_1)$ and $x \notin f^{-1}(\mathbf{E}_2)$. Hence, $f(x) \in \mathbf{E}_1$ and $f(x) \notin \mathbf{E}_2$. From this, $f(x) \in \mathbf{E}_1 \setminus \mathbf{E}_2$ and $x \in f^{-1}(\mathbf{E}_1 \setminus \mathbf{E}_2)$. Since $x \in f^{-1}(\mathbf{E}_1) \setminus f^{-1}(\mathbf{E}_2)$ was arbitrarily chosen and we see that it is an element of $f^{-1}(\mathbf{E}_1 \setminus \mathbf{E}_2)$, we obtain the relation

$$f^{-1}(\mathbf{E}_1) \setminus f^{-1}(\mathbf{E}_2) \subseteq f^{-1}(\mathbf{E}_1 \setminus \mathbf{E}_2).$$

From the previous relation and from (A.11), we obtain equation (A.8). This completes the proof. □

For a mapping $f : \mathbf{A} \longmapsto \mathbf{B}$ and $\mathbf{A}_1 \subseteq \mathbf{A}$, the restriction of f to $\mathbf{A}_1$, denoted by $f|_{\mathbf{A}_1}$, is the mapping from $\mathbf{A}_1$ to $\mathbf{B}$ which assigns $f(x)$ to each $x \in \mathbf{A}_1$.

The sets **A** and **B** are said to be equipotent provided there is an invertible mapping from **A** to **B**. Sets which are equipotent are, from the set-theoretic point of view, indistinguishable.

A.3 Countable and uncountable sets

A set **A** is said to be finite provided either it is empty or there is a natural number n for which **A** is equipotent to the set $\{1, \ldots, n\}$. We say that **A** is countably infinite provided **A** is equipotent to the set **N** of natural numbers. A set that is finite or countably infinite is said to be countable. A set that is not countable is called uncountable.

Theorem A.4. *A subset of a countable set is countable.*

Proof. Let **B** be a countable set and **A** be a nonempty subset of **B**.

1. Suppose that **B** is finite. Assume that there is a natural number n such that **B** and $\{1,\dots,n\}$ are equipotent. Let f be a one-to-one correspondence between **B** and $\{1,\dots,n\}$. Let $g(1)$ be the first natural number l, $l \in \{1,\dots,n\}$, for which $f(l) \in \mathbf{A}$. If $\mathbf{A} = \{f(g(1))\}$, then $f \circ g$ is a one-to-one correspondence between $\{1\}$ and **A**. Otherwise, we define $g(2)$ to be the first natural number l, $l \in \{1,\dots,n\}$, such that $f(l) \in \mathbf{A} \setminus \{f(g(1))\}$. This inductive selective process terminates after at most N selections, $N \le n$. Therefore $f \circ g$ is a one-to-one correspondence between $\{1,\dots,N\}$ and **A**. Therefore **A** is finite.
2. Let **B** be countably infinite. Assume that f is a one-to-one correspondence between **N** and **B**. Define $g(1)$ to be the first natural number l for which $f(l) \in \mathbf{A}$. Arguing as in the first case, if this selection terminates, then **A** is finite. Otherwise, this selection process does not terminate and g is properly defined on all of **N**. Note that $f \circ g$ is a one-to-one correspondence between **N** and a subset of **A**. Observe that $g(l) \ge l$ for all $l \in \mathbf{N}$. For each $x \in \mathbf{A}$, there is some natural number k such that $x = f(k)$. Hence, $x \in \{f(g(1)),\dots,f(g(k))\}$. Thus the image of $f \circ g$ is **A** and **A** is countably infinite. This completes the proof. □

Exercise A.1. Prove that the set **Q** of the rational numbers is countably infinite.

Exercise A.2. Prove that the union of a countable collection of countable sets is countable.

A set **O** of real numbers is called open provided for each $x \in \mathbf{O}$, there is an $r > 0$ for which $(x - r, x + r)$ is contained in **O**.

Theorem A.5. *The intersection of any finite collection of open sets is an open set.*

Proof. Let $\mathbf{O}_1, \dots, \mathbf{O}_k$ be open sets and

$$\mathbf{O} = \bigcap_{l=1}^{k} \mathbf{O}_l.$$

We take $x \in \mathbf{O}$ arbitrarily. Then $x \in \mathbf{O}_l$ for any $l \in \{1,\dots,k\}$. Because $\mathbf{O}_l$, $l \in \{1,\dots,k\}$, are open sets, there are $r_l > 0$, $l \in \{1,\dots,k\}$, such that

$$(x - r_l, x + r_l) \subset \mathbf{O}_l.$$

Let

$$r = \min_{l \in \{1,\dots,k\}} r_l.$$

Then

$$(x - r, x + r) \subset \mathbf{O}.$$

This completes the proof. □

Exercise A.3. Prove that the union of any collection of open sets is an open set.

Theorem A.6. *Every nonempty open set is the disjoint union of a countable collection of open intervals.*

Proof. Let $\mathbf{O}$ be a nonempty open subset of $\mathbf{R}$ and $x \in \mathbf{O}$. There are y and z such that $y > x$ and $z < x$, and

$$(x, y) \subset \mathbf{O}, \quad (z, x) \subset \mathbf{O}.$$

Define

$$a_x = \inf\{z : (z, x) \subset \mathbf{O}\}, \quad b_x = \sup\{y : (x, y) \subset \mathbf{O}\}.$$

We set

$$\mathbf{I}_x = (a_x, b_x).$$

Then $\mathbf{I}_x$ is an open set that contains x. We will prove that

$$\mathbf{I}_x \subset \mathbf{O}, \quad a_x \notin \mathbf{O}, \quad b_x \notin \mathbf{O}. \tag{A.12}$$

Let $w \in \mathbf{I}_x$, say $x < w < b_x$. By the definition of b_x, it follows that there is a number y such that $w \in (x, y)$ and $(x, y) \subset \mathbf{O}$. Then $w \in \mathbf{O}$. Since $w \in \mathbf{I}_x$ was arbitrarily chosen and we see that it is an element of $\mathbf{O}$, we conclude that $\mathbf{I}_x \subset \mathbf{O}$. Assume that $b_x \in \mathbf{O}$. Then there is an $r > 0$ such that $(b_x - r, b_x + r) \subset \mathbf{O}$. Thus $(x, b_x + r) \subset \mathbf{O}$, which is a contradiction. Therefore $b_x \notin \mathbf{O}$. As above, we see that $a_x \notin \mathbf{O}$. Consider the collection of open intervals $\{\mathbf{I}_x\}_{x \in \mathbf{O}}$. Since $x \in \mathbf{O}$ is a member of $\mathbf{I}_x$ and each $\mathbf{I}_x$ is contained in $\mathbf{O}$, we conclude that

$$\mathbf{O} = \bigcup_{x \in \mathbf{O}} \mathbf{I}_x.$$

By (A.12), it follows that $\{\mathbf{I}_x\}_{x \in \mathbf{O}}$ is disjoint. Because each of the intervals $\mathbf{I}_x$ contains a rational number, there is a one-to-one correspondence between $\{\mathbf{I}_x\}_{x \in \mathbf{O}}$ and a subset of the set of the rational numbers. Therefore $\{\mathbf{I}_x\}_{x \in \mathbf{O}}$ is a countable disjoint collection of open intervals. This completes the proof. □

A set of real numbers is called closed if its complement in $\mathbf{R}$ is open.

Exercise A.4. Prove that the union of any finite collection of closed sets is closed.

Exercise A.5. Prove that the intersection of any collection of closed sets is closed.

A collection $\{\mathbf{E}_\lambda\}_{\lambda \in \Lambda}$ is said to be a cover of the set $\mathbf{E}$ if $\mathbf{E} \subseteq \bigcup_{\lambda \in \Lambda} \mathbf{E}_\lambda$. If each of the sets $\mathbf{E}_\lambda$, $\lambda \in \Lambda$, is open, we say that $\{\mathbf{E}_\lambda\}_{\lambda \in \Lambda}$ is an open cover of the set $\mathbf{E}$. If each of the sets $\mathbf{E}_\lambda$, $\lambda \in \Lambda$, is closed, we say that $\{\mathbf{E}_\lambda\}_{\lambda \in \Lambda}$ is a closed cover of the set $\mathbf{E}$. A nonempty set of real numbers is said to be bounded above if there is a real number b such that $x \leq b$ for any $x \in \mathbf{E}$. The real number b is called upper bound of $\mathbf{E}$. A nonempty set of real numbers $\mathbf{E}$ is said to be bounded below if there is a real number a such that $x \geq a$ for any $x \in \mathbf{E}$. The real number a is called the lower bound of the set $\mathbf{E}$. A nonempty set $\mathbf{E}$ of real numbers is said to be bounded if it is bounded below and it is bounded above.

Theorem A.7 (The Heine–Borel theorem). *Let* $\mathbf{B}$ *be a closed and bounded set of real numbers. Then every open cover of* $\mathbf{B}$ *has a finite subcover.*

Proof.
1. Let $\mathbf{B} = [a,b]$ and $\mathcal{F}$ is an open cover of $\mathbf{B}$. Define the set $\mathbf{E}$ to be the set of numbers $x \in [a,b]$ such that $[a,x]$ can be covered by a finite number of the sets of $\mathcal{F}$. Since $a \in \mathbf{E}$, we see that the set $\mathbf{E}$ is nonempty. Also, $\mathbf{E}$ is a bounded above set by b. Hence, $\mathbf{E}$ has a supremum. Let $c = \sup \mathbf{E}$. Because $c \in [a,b]$, there is an $\mathbf{O} \in \mathcal{F}$ such that $c \in \mathbf{O}$. Since $\mathbf{O}$ is open, there is an $\epsilon > 0$ so that $(c-\epsilon, c+\epsilon) \subset \mathbf{O}$. Note that $c - \epsilon$ is not a supremum for $\mathbf{E}$. Then there is an $x \in \mathbf{E}$ such that $x > c - \epsilon$. Since $x \in \mathbf{E}$, there is a finite collection $\{\mathbf{O}_1, \dots, \mathbf{O}_k\}$ of sets in $\mathcal{F}$ that covers $[a,x]$. Consequently $\{\mathbf{O}_1, \dots, \mathbf{O}_k, \mathbf{O}\}$ covers the interval $[a, c+\epsilon)$. If $c < b$, then c is not an upper bound for $\mathbf{E}$. Therefore $c = b$. Thus $[a,b]$ can be covered by a finite number of sets from $\mathcal{F}$.
2. Let $\mathbf{B}$ be any closed and bounded set and $\mathcal{F}$ be an open cover of $\mathbf{B}$. Since $\mathbf{B}$ is bounded, it is contained in some interval $[a,b]$. Note that $\mathbf{O} = \mathbf{R} \setminus \mathbf{B}$ is an open set. Let $\mathcal{F}^*$ be the collection of open sets obtained by adding $\mathbf{O}$ to $\mathcal{F}$. Since $\mathcal{F}$ covers $\mathbf{B}$, we see that $\mathcal{F}^*$ covers $[a,b]$. By the previous case, it follows that there is a finite subcollection of $\mathcal{F}^*$ that covers $[a,b]$ and hence $\mathbf{B}$. By removing $\mathbf{O}$ from the finite subcover of $\mathbf{B}$, if $\mathbf{O}$ belongs to the finite subcover, we have a finite collection of sets in $\mathcal{F}$ that covers $\mathbf{B}$. This completes the proof. □

Theorem A.8 (The nested set theorem). *Let* $\{\mathbf{F}_n\}_{n\in\mathbf{N}}$ *be a descending countable collection of nonempty closed sets of real numbers for which* $\mathbf{F}_1$ *is bounded. Then*

$$\bigcap_{n=1}^{\infty} \mathbf{F}_n \neq \emptyset.$$

Proof. Assume the contrary. Then for each real number x there is a natural number n for which $x \notin \mathbf{F}_n$. Then $x \in \mathbf{O}_n = \mathbf{R} \setminus \mathbf{F}_n$. Therefore $\mathbf{R} = \bigcup_{n=1}^{\infty} \mathbf{O}_n$, i. e., $\{\mathbf{O}_n\}_{n\in\mathbf{N}}$ is an open cover of $\mathbf{R}$ and hence also of $\mathbf{F}_1$. By the Heine–Borel theorem, it follows that there is a natural number N for which $\mathbf{F}_1 \subseteq \bigcup_{n=1}^{N} \mathbf{O}_n$. Because $\{\mathbf{F}_n\}_{n\in\mathbf{N}}$ is descending, the collection $\{\mathbf{O}_n\}_{n\in\mathbf{N}}$ is ascending. Therefore

$$\bigcup_{n=1}^{N} \mathbf{O}_n = \mathbf{O}_N.$$

Hence, $\mathbf{F}_1 \subseteq \mathbf{R} \setminus \mathbf{F}_N$. This is a contradiction because $\mathbf{F}_N \subset \mathbf{F}_1$. This completes the proof. □

A.4 Continuous real-valued functions on a real variable

Let f be a real-valued function defined on a set $\mathbf{E}$ of real numbers. We say that f is continuous at $x \in \mathbf{E}$ if for each $\epsilon > 0$ there is a $\delta > 0$ for which

$$\text{if} \quad x' \in \mathbf{E} \quad \text{and} \quad |x' - x| < \delta, \quad \text{then} \quad |f(x') - f(x)| < \epsilon.$$

The function f is said to be continuous on $\mathbf{E}$ if it is continuous at each point in its domain $\mathbf{E}$.

Theorem A.9. *Let f be a real-valued function defined on a set $\mathbf{E}$ of real numbers. Then f is continuous on $\mathbf{E}$ if and only if for any open set $\mathbf{O}$ there is an open set $\mathbf{U}$ such that $f^{-1}(\mathbf{O}) = \mathbf{E} \cap \mathbf{U}$.*

Proof.

1. Let there for any open set $\mathbf{O}$ exist an open set $\mathbf{U}$ such that $f^{-1}(\mathbf{O}) = \mathbf{E} \cap \mathbf{U}$. We take $x \in \mathbf{E}$ and $\epsilon > 0$ arbitrarily. Because the interval

$$\mathbf{I} = \big(f(x) - \epsilon, f(x) + \epsilon\big)$$

is an open set, there exists an open set $\mathbf{U}$ such that

$$f^{-1}(\mathbf{I}) = \{x_1 \in \mathbf{E} : f(x) - \epsilon < f(x_1) < f(x) + \epsilon\} = \mathbf{E} \cap \mathbf{U}.$$

Hence, $x \in \mathbf{E} \cap \mathbf{U}$. Since $\mathbf{U}$ is an open set, there is a $\delta > 0$ such that $(x - \delta, x + \delta) \subseteq \mathbf{U}$. Thus, if $x_1 \in \mathbf{E}_1$ and $|x_1 - x| < \delta$, then $|f(x) - f(x_1)| < \epsilon$. Therefore f is continuous at x. Because $x \in \mathbf{E}$ was arbitrarily chosen, we conclude that f is continuous on $\mathbf{E}$.

2. Let f be continuous on $\mathbf{E}$. We take an open set $\mathbf{O}$ arbitrarily and $x \in f^{-1}(\mathbf{O})$. Then $f(x) \in \mathbf{O}$. Because $\mathbf{O}$ is open, there is an $\epsilon > 0$ such that

$$\big(f(x) - \epsilon, f(x) + \epsilon\big) \subseteq \mathbf{O}.$$

Because f is continuous at x, there is a $\delta > 0$ such that if $|x - x_1| < \delta$, then $f(x) - \epsilon < f(x_1) < f(x) + \epsilon$. Define $\mathbf{I}_x = (x - \delta, x + \delta)$. Then $f(\mathbf{E} \cap \mathbf{I}_x) \subseteq \mathbf{O}$. Define

$$\mathbf{U} = \bigcup_{x \in f^{-1}(\mathbf{O})} \mathbf{I}_x.$$

Because $\mathbf{I}_x$ are open sets for any $x \in f^{-1}(\mathbf{O})$ and the union of open sets is open, we see that $\mathbf{U}$ is an open set. Now we will prove that

$$\mathbf{E} \cap \mathbf{U} = f^{-1}(\mathbf{O}). \tag{A.13}$$

Let $y \in \mathbf{E} \cap \mathbf{U}$ be arbitrarily chosen. Then $y \in \mathbf{E}$ and $y \in \mathbf{U}$. Hence, there is an $x \in f^{-1}(\mathbf{O})$ such that $y \in \mathbf{I}_x$. Therefore $y \in \mathbf{E} \cap \mathbf{I}_x$. From this, $f(y) \in f(\mathbf{E} \cap \mathbf{I}_x) \subseteq \mathbf{O}$. Then $y \in \mathbf{E}$ and $f(y) \in \mathbf{O}$. Consequently $y \in f^{-1}(\mathbf{O})$. Because $y \in \mathbf{E} \cap \mathbf{U}$ was arbitrarily chosen and we see that it is an element of $f^{-1}(\mathbf{O})$, we obtain the relation

$$\mathbf{E} \cap \mathbf{U} \subseteq f^{-1}(\mathbf{O}). \tag{A.14}$$

Let now $y \in f^{-1}(\mathbf{O})$ be arbitrarily chosen. Then $y \in \mathbf{E}$ and $f(y) \in \mathbf{O}$. Since $y \in f^{-1}(\mathbf{O})$, it follows that $y \in \mathbf{I}_y$ and from this $y \in \mathbf{U}$. Therefore $y \in \mathbf{E} \cap \mathbf{U}$. Because $y \in f^{-1}(\mathbf{O})$ was arbitrarily chosen and we see that it is an element of the set $\mathbf{E} \cap \mathbf{U}$, we find the relation

$$f^{-1}(\mathbf{O}) \subseteq \mathbf{E} \cap \mathbf{U}.$$

From the previous relation and from equation (A.14), we obtain equation (A.13). This completes the proof. □

B Functions of bounded variation

Definition B.1. Let f be a real-valued function on the closed bounded interval $[a,b]$ and $\mathbf{P} = \{x_0, \dots, x_k\}$ be a partition of $[a,b]$, $a = x_0 < x_1 < \dots < x_k = b$. Define the variation of f with respect to $\mathbf{P}$ by

$$\bigvee_a^b(f,\mathbf{P}) = \sum_{l=1}^{k} |f(x_l) - f(x_{l-1})|$$

and the total variation of f on $[a,b]$ by

$$\bigvee_a^b(f) = \sup\left\{\bigvee_a^b(f,\mathbf{P}) : \mathbf{P} \quad \text{is} \quad \text{a} \quad \text{partition} \quad \text{of} \quad [a,b]\right\}.$$

Definition B.2. A real-valued function f on the closed bounded interval $[a,b]$ is said to be of bounded variation on $[a,b]$ provided

$$\bigvee_a^b(f) < \infty.$$

Theorem B.1. *Let f be a Lipschitz function on the closed bounded interval $[a,b]$, i. e., there exists a constant $L > 0$ such that*

$$|f(x) - f(y)| \le L|x - y|$$

for any $x, y \in [a,b]$. Then f is of bounded variation on $[a.b]$.

Proof. For an arbitrary partition $\mathbf{P} = \{x_0, \dots, x_k\}$ of the interval $[a,b]$, we have

$$\begin{aligned}\bigvee_a^b(f,\mathbf{P}) &= \sum_{l=1}^{k} |f(x_l) - f(x_{l-1})| \\ &\le L\sum_{l=1}^{k}(x_l - x_{l-1}) \\ &= L(b-a).\end{aligned}$$

Because $\mathbf{P}$ was arbitrarily chosen partition of the interval $[a,b]$, we conclude that $\bigvee_a^b(f) < \infty$. This completes the proof. □

Theorem B.2. *Let f be a monotonic function on the closed bounded interval $[a,b]$. Then f is of bounded variation on $[a,b]$.*

Proof. Let f be an increasing function on $[a,b]$. The case when f is a decreasing function on $[a,b]$ we leave to the reader as an exercise. Take a partition $\mathbf{P} = \{x_0, \dots, x_k\}$ of the interval $[a,b]$. Then

$$\bigvee_a^b(f,\mathbf{P}) = \sum_{l=1}^{k} |f(x_l) - f(x_{l-1})| = \sum_{l=1}^{k}(f(x_l) - f(x_{l-1})) = f(b) - f(a).$$

https://doi.org/10.1515/9783110657722-012

Because **P** was an arbitrarily chosen partition of $[a,b]$, we conclude that $\bigvee_a^b(f) < \infty$. This completes the proof. □

Theorem B.3. *Let $[a,b]$ be a closed bounded interval and f be of bounded variation on $[a,b]$. Then*

$$\bigvee_a^c(f) \le \bigvee_a^b(f)$$

for any $c \in (a,b]$.

Proof. For any partition **P** of $[a,c]$ we have

$$\bigvee_a^c(f,\mathbf{P}) \le \bigvee_a^b(f).$$

Hence,

$$\bigvee_a^c(f) \le \bigvee_a^b(f).$$

This completes the proof. □

Theorem B.4. *Let f be a function of bounded variation on $[a,b]$. Then, for any $c \in [a,b]$, we have*

$$\bigvee_a^b(f) = \bigvee_a^c(f) + \bigvee_c^b(f). \tag{B.1}$$

Proof. Let $\mathbf{P}_1$ and $\mathbf{P}_2$ be arbitrary partitions of $[a,c]$ and $[c,b]$, respectively. Then $\mathbf{P}_1 \cup \mathbf{P}_2$ is a partition of $[a,b]$ and

$$\bigvee_a^b(f) \ge \bigvee_a^b(f,\mathbf{P}_1 \cup \mathbf{P}_2) = \bigvee_a^c(f,\mathbf{P}_1) + \bigvee_c^b(f,\mathbf{P}_2).$$

Hence,

$$\bigvee_a^b(f) \ge \bigvee_a^c(f) + \bigvee_c^b(f). \tag{B.2}$$

Let **P** be a partition of $[a,b]$ and $\mathbf{P}'$ be a refinement of **P** obtained by adjoining c to **P**. Then

$$\bigvee_a^b(f,\mathbf{P}) \le \bigvee_a^b(f,\mathbf{P}') \le \bigvee_a^c(f) + \bigvee_c^b(f).$$

Hence,

$$\bigvee_a^b(f) \le \bigvee_a^c(f) + \bigvee_c^b(f).$$

From this and from (B.2), we get (B.1). □

Theorem B.5 (Jordan's theorem). *A function f is of bounded variation on the closed bounded interval $[a,b]$ if and only if it is difference of two increasing functions.*

Proof.

1. Let f is of bounded variation on $[a,b]$. Using Theorem B.3, we see that $\bigvee_a^x(f)$ is an increasing function on $[a,b]$. Let

$$g(x) = f(x) + \bigvee_a^x(f), \quad x \in [a,b].$$

Take $x_1, x_2 \in [a,b]$ and $x_1 \geq x_2$. Hence, using Theorem B.4, we get

$$\begin{aligned} f(x_1) - f(x_2) &\leq |f(x_1) - f(x_2)| \\ &\leq \bigvee_{x_2}^{x_1}(f) = \bigvee_a^{x_1}(f) - \bigvee_a^{x_2}(f), \end{aligned}$$

whereupon

$$f(x_2) + \bigvee_a^{x_2}(f) \leq f(x_1) + \bigvee_a^{x_1}(f),$$

i. e.,

$$g(x_2) \leq g(x_1).$$

Consequently the function g is an increasing function on $[a,b]$. Hence,

$$f(x) = g(x) - \bigvee_a^x(f), \quad x \in [a,b].$$

2. Let

$$f(x) = f_1(x) - f_2(x), \quad x \in [a,b],$$

where f_1 and f_2 are increasing functions on $[a,b]$. Then, for any partition $\mathbf{P} = \{x_0, \ldots, x_k\}$ of $[a,b]$, we have

$$\begin{aligned} \bigvee_a^b(f, \mathbf{P}) &= \sum_{l=1}^k |f(x_l) - f(x_{l-1})| \\ &= \sum_{l=1}^k |f_1(x_l) - f_1(x_{l-1}) - (f_2(x_l) - f_2(x_{l-1}))| \\ &\leq \sum_{l=1}^k |f_1(x_l) - f_1(x_{l-1})| + \sum_{l=1}^k |f_2(x_l) - f_2(x_{l-1})| \\ &= f_1(b) - f_1(a) + f_2(b) - f_2(a) < \infty. \end{aligned}$$

Consequently $\bigvee_a^b(f) < \infty$. This completes the proof. □

Bibliography

[1] R. R. Akhmerov, M. I. Kamenskii, A. S. Potapov, A. E. Rodkina and B. N. Sadovskii, Measures of noncompactness and condensing operators, Birkhäuser, Basel, 1992.
[2] A. Amar, A. Jeribi and M. Mnif, On a generalization of the Schauder and Krasnosel'skii fixed points theorems on Dunford–Pettis spaces and applications, Math. Methods Appl. Sci. 28 (2005), 1737–1756.
[3] G. Ball, Diffusion approximation of radiative transfer equations in a channel, Transp. Theory Stat. Phys. 30 (2–3) (2001), 269–293.
[4] J. Banas and K. Goebel, Measures of noncompactness in Banach spaces, Marcel Dekker, New York, 1980.
[5] C. S. Barroso and E. V. Teixeira, A topological and geometric approach to fixed points results for sum of operators and applications, Nonlinear Anal. 60 (2005), 625–650.
[6] P. Binding, The differential equation $x' = f \circ x$, J. Differ. Equ. 31 (1979), 183–199.
[7] F. E. Bowder, Nonlinear accretive operators in Banach spaces, Bull. Am. Math. Soc. 73 (1967) 470–476.
[8] D. W. Boyd and J. S. W. Wong, On nonlinear contractions, Proc. Am. Math. Soc. 20 (1969), 458–464.
[9] T. Burton, A fixed-point theorem of Krasnolel'skii, Appl. Math. Lett. 11 (1), (1998), 85–88.
[10] K. Deimling, A Caratheodory theory for systems of integral equations, Ann. Mat. Pura Appl. 86 (4) (1970), 217–260.
[11] B. C. Dhage, Remarks on two fixed-point theorems involving the sum and the product of two operators, Comput. Math. Appl. 46 (2003), 1779–1785.
[12] J. Dugundji and A. Granas, Fixed point theory, Monografie Matematyczne, PWN, Warsaw, 1982.
[13] J. Garcia-Falset, Existence of fixed points and measures of weak noncompactness, Nonlinear Anal. 71 (2009), 2625–2633.
[14] A. Granas and J. Dugundji, Fixed point theory, Springer-Verlag, New York, 2003.
[15] G. Gripenberg, S. O. Londen and O. Staffans, Volterra integral and functional equations, Cambridge Univ. Press, 1990.
[16] T. Kato, Nonlinear semigroups and evolution equations. J. Math. Soc. Jpn. 19 (1967), 508–520.
[17] W. A. Kirk, Contraction mappings and extensions, Handbook of metric fixed point theory, pp. 1–34, Kluwer Acad. Publ., Dordrecht, 2001.
[18] M. A. Krasnosel'skii, Two remarks on the method of successive approximations, Usp. Mat. Nauk 10 (1955), 123–127.
[19] K. Latrach, On a nonlinear stationary problem arising in transport theory, J. Math. Phys. 37 (1996), 1336–1348.
[20] K. Latrach, M. Taoudi and A. Zeghal, Some fixed point theorems of the Schauder and the Krasnosel'skii type and applications to nonlinear transport equation, J. Differ. Equ. 221 (2006), 256–271.
[21] R. Ma, Global behavior of the components of nodal solutions of assymptotically linear eigenvalue problems, Appl. Math. Lett. 21 (2008), 754–760.
[22] M. Z. Nashed and J. S. W. Wong, Some variants of a fixed point theorem of Krasnoselskii and applications to nonlinear integral equations, J. Math. Mech. 18 (1969), 767–777.
[23] D. O'regan, Fixed-point theory for the sum of two operators, Appl. Math. Lett. 9 (1) (1996), 1–8.
[24] S. Park, Generalizations of the Krasnoselskii fixed point theorem, Nonlinear Anal. 67 (2007), 3401–3410.
[25] P. Pikuta, J. Appl. Anal. 12 (1) (2006), 147–152.

https://doi.org/10.1515/9783110657722-013

[26] P. Pikuta, Local solutions to Darboux problem with a discontinuous right-hand side, Condens. Matter Phys. 11, No. 4 (56) (2008), 755–760.
[27] P. Pikuta, W. Rzymowski, J. Math. Anal. Appl. 277 (2003), 122–129.
[28] Y. N. Raffoul, Positive periodic solutions of nonlinear functional difference equations, Electron. J. Differ. Equ. 55 (2002), 1–8.
[29] T. Xiang, Krasnosel'skii fixed point theorem for dissipative operators, Electron. J. Differ. Equ. (2011), No. 147, 5 pp.
[30] T. Xiang and R. Yuan, A class of expansive-type Krasnoselskii fixed point theorem, Nonlinear Anal. 71 (2009), 3229–3239.
[31] T. Xiang and R. Yuan, Critical type of Krasnosel'skii fixed point theorem. Proc. Am. Math. Soc. 139 (2011), 1033–1044.

Index

https://doi.org/10.1515/9783110657722-014

www.ingramcontent.com/pod-product-compliance
Lightning Source LLC
Chambersburg PA
CBHW080659220326
41598CB00033B/5260
9783110657692